BIOLOGY FOR LIVING

BIOLOGY FOR LIVING

Bruce Wallace and George M. Simmons, Jr.

Department of Biology, Virginia Polytechnic Institute and State University, Blacksburg, Virginia

THE JOHNS HOPKINS UNIVERSITY PRESS
Baltimore and London

The Johns Hopkins University Press, 701 West 40th Street, Baltimore, Maryland 21211
The Johns Hopkins Press Ltd., London

∞
The paper used in this publication meets the minimum requirements of American
National Standard for Information Sciences–Permanence of Paper for Printed Library
Materials, ANSI Z39.48-1984.

LIBRARY OF CONGRESS CATALOGING-IN-PUBLICATION DATA

Wallace, Bruce, 1920–
   Biology for living.

   Bibliography: p.
   Includes index.
   1. Biology.   I. Simmons, George M.   II. Title.
[DNLM: 1. Biology.   QH 308.2 w1873b]
QH308.2.W33   1987   574   86-10292
ISBN 0-8018-3221-7 (alk. paper)

$ 32.95
BK

This text is dedicated to those students and their instructors who see freshmen biology not as a Herculean task designed to test the student's mental prowess, but as an opportunity to master the knowledge and reasoning skills essential for one's major role in life—that of an informed and concerned citizen. That there may be many such students and instructors in no way dilutes this dedication, which expands in proportion to demand.

CONTENTS

PREFACE

The head of an adult fruit fly, *Drosophila melanogaster,* represents six embryonic segments; the thorax represents three. The hibernating shrew resorbs its intervertebral discs, thus shortening its skeleton. As a result, the shrew's skin, now too large for its shrunken body, becomes rumpled and serves as an improved insulation against the cold. The hummingbird, which "hibernates" every chilly night, ruffles its feathers for better insulation. A "fairy ring"—that nearly circular array of toadstools that seems to have been planted by elves—consists of fruiting bodies growing from the rim of a large, circular, underground fungal mat. Genes with related functions are often located at 90° or 180° from one another on the circular chromosome of the common colon bacillus (*Escherichia coli*), thus suggesting that the chromosome of this organism has arisen by two doublings of a much smaller chromosome of a more primitive, ancestral bacterium. Sites on DNA at which RNA transcription begins are often palindromic (*palindrome:* an anagram that reads the same in either direction, such as "Able was I ere I saw Elba") in structure.

The above paragraph consists entirely of biological facts. The world of biology abounds with facts. A correspondent has written to the journal *Bioscience* boasting that in thirty years of teaching biology, he has taught only facts and how these facts are obtained by scientists. The question arises, however, "What facts did he teach?" This question has its sequel: "And, for what purpose?"

The facts of biology are so numerous that no one, not even a professional biologist, knows them all. A molecular biologist working on the transfer of genes from one organism to another may never have heard of a fairy ring—let alone have seen one, or know how one is formed.

The botanist who owns a colored slide of a fairy ring growing in a beautiful forest glade (a perfect home for elves and other wee folk) may know nothing about the heat-saving behaviors of hummingbirds and hibernating shrews. Neither the botanist nor the field ecologist is likely to know about the repeated sequences of nucleotide bases in DNA (some in reverse order—palindromic—others in direct order), about the enzymes that recognize such sequences, nor about the possible solutions such sequences and enzymes may provide for problems confronting the developing embryos of higher organisms.

If the facts of biology are too numerous for even professional biologists to memorize, what facts are to be taught to college students? Especially to those students who have chosen careers not immediately dependent upon the science of biology. Excluded from these nonprofessional students are premedical and preveterinary students, nursing and other health-service students, agricultural majors, future foresters, and various environmental specialists. Students whose careers depend upon biology, and upon an awareness of modern developments in the biological sciences, must keep abreast of biological research (of the "state of the art"). They must be able to read their professional literature and to converse with their colleagues. They must have a multitude of facts—organized facts—on tap, ready for recall at any moment. They must also know the vocabulary that permits them to read, write, or talk about their science.

Students whose futures lie in business, industry, or service in some level of government must also be familiar with biology. The need arises because they are living organisms, as are their spouses and children. The need arises because their own health and their families' health depends upon the control of disease and the availability of food and shelter. Disease-causing organisms are a part of the biological world, as are the plants and animals that sustain and shelter us. Ironically, these students—more than most professional biologists—will make decisions having the greatest impact on the living world. Nonmajors who go into commerce, industry, or politics will decide on the care with which strip-mining is done, offshore oil is drilled, or pipelines and other conveyer systems are built. Politicians, interacting with attorneys and businessmen, determine the extent to which such activities are regulated by federal, state, and local governments.

Human beings are a part of the natural world. Their roots are in nature; their material sustenance is derived from nature. The energy that powers each human body, like that which powers the cheetah or the eagle, has been obtained from the sun through the miracle of photosynthesis. Miracle? Yes, in the sense that even the best equipped laboratories have yet to combine water and carbon dioxide and show a net gain of energy.

Green plants provide the energy required for all life on earth: small diatoms and other one-celled plants utilize the energy of sunlight as it falls on the oceans and sustain the immense variety of marine life; trees, shrubs, grasses, and other green terrestrial plants sustain grazing animals (herbivores), which in turn provide food for predators. At river deltas and marshy coastlines, marine and land systems blend and become dependent upon one another.

Although they are living organisms, and, as such, represent one facet of that complex relationship of nature known as the "web of life," human beings are often overwhelmed by the immensity of the societies in which they live, and by the seemingly insoluble economic and political problems that beset these societies. Human social structure is a form of organization that rests on a foundation of individual and ecological biology. We should not forget that scarcely 60 minutes elapsed between the sinking of the *Titanic* and the loss of 1,517 lives; human beings are not marine mammals. Similarly, 2,000 or more persons died in their sleep on December 3, 1984, in Bhopal, India when a nearby factory spewed methyl isocyanate (a chemical used in making pesticides) into the atmosphere; human lungs, like those of the cattle and other animals that died that night, are easily damaged and rendered useless.

This book is written for the nonbiologist. It is written in the belief that among the multitude of facts known to biologists, some are more important to nonmajors than are others. More important than facts, however, are the underlying concepts and the logical relationships that connect disparate facts. Every biology teacher and every biology textbook author faces a dilemma: to appear up-to-date by emphasizing the modern scientific advances that occur daily and that are dutifully reported by newspapers and television reporters, or to appear old-fashioned by stressing concepts and the logic of science at the expense of well-publicized and exciting "break-throughs." Emphasis can be placed on the *timely* or the *timeless*—that is, on the current events of science or on its structure, which extends from its foundation to the I-beams only now being fitted into place. We shall opt largely for the timeless. The very speed with which today's science advances defeats any logical reason for emphasizing the timely: when a textbook has arrived in the classroom today, the author's originally timely vignettes are passé; when today's freshmen and sophomore students have graduated from college, those exciting and up-to-date wonders of science that they memorized with great effort will have largely slipped into oblivion. If, on the other hand, a student understands how science progressed from *there* to *here*, he may understand how it will have arrived at any given point in the future. Whereas the "timely" ages before our very eyes, the "timeless" provides a means for understanding the current events of any age.

A document recently received from an admired colleague suggests that six professors are now required to teach a beginning course in biology. No one biologist, it seems, now possesses all the appropriate knowledge. Isn't that ridiculous for a course in which ninety percent or more of all students have no intention of becoming biologists? The authors of this text are an evolutionary geneticist and a limnologist. We know only incidentally, for example, of the body's natural killer cells and their role in the destruction of potentially metastasizing cancer cells; immunology is not our area of expertise. Nevertheless, if a marketing expert or a computer analyst of the coming decade, because of our efforts, knows of these cells only what we know today (and has retained that knowledge because we made these cells seem important), that would be wonderful. Without those efforts, that person might have retained little or no knowledge about the immune system or about biology generally; if his only fault tomorrow is that he repeats our errors of today, let's wish him Godspeed.

The biology that follows, then, represents a selected sample of biological facts, a sample selected for seeming relevance to the non–biology major. These facts will bear on the biology of the individual and of the ecological web of life that constitutes the biosphere, and on those aspects of biology that bear on society at large. Public health, racial attitudes, and the measurement of intelligence are but three examples of the latter category. Throughout, however, every effort is made to avoid piling detail on needless detail; concepts, procedures, and parallels are emphasized in every chapter. Vocabulary receives minimal attention in the pages that follow; proper terminology is used if it facilitates understanding—it is used neither for creating quiz questions nor for adhering to editorial handbooks. Words may be needed to identify structures, functions, and processes, but they should not detract from reality.

The purpose of this book is to alter the behaviors of students; that is, to provide students with information that is internalized and kept available for recall throughout their lives. Professor Garrett Hardin, a perceptive West Coast biologist, claims that five years elapse between the time that a person learns something and the time that he behaves as if he has learned it. Five years! Most students of standard biology will have forgotten nearly everything they memorized for mid-term and final examinations within months, long before the five years have gone by.

The internalization of knowledge follows from personal involvement. To appreciate a textbook photograph of a lung made useless by emphysema, the student must visualize not a photograph, but a real lung, his or her lung. The textbook illustration of the effects of drought and famine showing a vacant-eyed mother and her bloated children must become the student's vacant stare and the student's swollen stomach. Following the merest fluctuations in average meteorological conditions, Africa's famine of 1985 could be anybody's famine anywhere in the world.

The text refers repeatedly to the student's journal. Ideally, the student's journal would consist of a permanently bound ledger in which daily entries are made. Unfortunately, a bound ledger makes no provision for periodic review by the instructor or his teaching assistants; hence, a loose-leaf notebook must suffice. Nevertheless, the pages might still be numbered consecutively in advance so as to provide continuity throughout the course. Much of the material entered into the journal will be derived from and based upon each student, individually. Blood pressure? It will be the student's. Pulse, before and after exercise? It will be the student's. Blood cell counts? Again, the student's own blood. Sickling or non-sickling hemoglobin? Again— you guessed it. The personal factual items can form the basis of many classroom or laboratory discussions: Why does any one student exhibit variation through time? Why is there variation from student to student? How should one react to individual differences?

An involved student is one most likely to internalize and subsequently use the facts that are learned in any course; this is doubly true if the facts bear on the student himself or herself, on friends and relatives, or on the surrounding community. This, then, is a text based on personal involvement. Involvement (and the journal that bears witness to this involvement) is much more important than rote learning.

In addition to the three segments of the text already alluded to (the individual, the web of life, and the society in which the student lives), the text contains two additional segments, or parts. The fifth and last is an analysis of the course human beings are likely to follow in the near future—a somewhat light-hearted but rational attempt at fortune-telling, so to speak. The first part places science, biological science in particular, in some sort of intellectual framework. This segment, Part I, is not to be glossed over as if it were a routine introduction intended for the eyes of neither instructor nor student. The actual biology can wait a moment. Spend some time on the topics covered in Part I. The high-school biology curriculum was modernized and made truly scientific during the 1960s. Most persons less than forty years old have been exposed to this modern curriculum. Nevertheless, as the continuing debate over "equal time" for Biblical creationism in high-school biology texts and other educational materials has revealed, the citizens of this nation are still woefully ignorant about the nature of science, and of the relation of science to ethics or value judgments. Many would place boundaries on scientific inquiry and would declare certain areas of potential knowledge "out of bounds" or "off limits." Part I helps the student probe, and perhaps understand, the boundaries of science. If this understanding can be achieved only by sacrificing a lecture on the comparative structures of the mammalian and the molluscan eye, so be it.

Reversing the coin for a moment, one can ask of this text: What has been omitted? How were the decisions to omit these items arrived at?

The answers to these questions can be understood if one recalls that for the nonmajor, the so-called "introductory" biology course is, in fact, the *terminal* biology course; the introduction was consummated in high school. If biologists have any take-home message to give these college students, freshman biology provides the *last* opportunity. Imagine, then, the matters that pertain to self, the local social group (from extended family through the neighborhood), and society (town, city, state, national, and international) that the student will eventually encounter. (Recall, in passing, that a college student beginning his or her sophomore year is a voting member of society!) If these matters pose problems whose solutions require information that is now taught only in advanced biology courses, that information must be incorporated into the nonmajors' "introductory-terminal" course. To make this information comprehensible, however, requires the inclusion of basic material that normally would be covered in intermediate or other advanced courses. Every explanation must be complete.

What has been omitted? Those items of the standard texts whose main purpose is merely to acquaint the student with the world of biology. The exclusion of this material is not a tremendous loss: Modern high-school biology courses do an excellent job at this level. Numerous television programs and beautifully illustrated magazine articles cover this material far better than most classroom teachers possibly can.

Not entirely omitted from this text, but nearly so, is "creation science." Comments about creationism have been made where it seems appropriate. Darwin's responses to the then-prevailing Biblical views are also reported. Still, it is not the purpose of this text to destroy the faith that any person feels he or she must accept in order to cope with life and with living. The text refers to creationism only as a reaction to those modern creationists who attempt to give their faith credibility as a science.

Still another point may be made concerning this text: It assumes that each freshman matures. At the end of his or her first year, twenty-five percent of the student's college maturation will have occurred. Returning sophomores, as all observant professors know, are noticeably more self-assured than are incoming freshmen, who are essentially recent high-school graduates. When does this maturation occur? Not during summer holidays! The maturation of freshmen occurs throughout the first year of college. That fact is recognized in this text; the material becomes more sophisticated and more abstract as one proceeds through the book. Neither the teacher nor the incoming freshman student, having peeked at an advanced chapter, should react with alarm or dismay. Throughout the course, the student and the text mature together; an understanding and sympathetic instructor can help each maintain rapport with the other.

Finally, a few words can be said regarding two features of this text that will be apparent to anyone who even casually riffles its pages. The first has already been alluded to in conjunction with the journal: passages that bear the heading *"Now, it's your turn"* pose questions or describe situations suitable for discussion in the student's journal. Not every question nor every issue need be addressed by each student for grading purposes, of course. For the benefit of those students who do ponder the often mysterious connection between classroom biology and "real" life, a great many issues that might otherwise be overlooked have been raised.

One or more *Informational Essays* follow the text of nearly every chapter in this book. They are more than the verbal "asides" that appear in the narrow columns on many pages. Some, like the first one, provide substantive material that bears on matters mentioned in the text; such material may be reprinted from other sources. Other Informational Essays present the background information (often quantitative) that is needed to better understand material discussed in the text; the presentation required for providing a technical understanding, in these cases, would have interrupted the flow of the narrative had it been inserted within the text itself. Recalling what was said earlier about the *timely* and the *timeless*, and imagining a future revision of this book, the text should need fewer revisions (= timeless) than the Informational Essays (= timely).

ACKNOWLEDGMENTS

If a book can be said to possess a conception, a gestation period, and a birth, this book—now born—owes its conception to Dennis Curtin. With understandable pride over his role in publishing Jacob Bronowski's *Ascent of Man,* Dennis suggested that biology might be taught to nonmajors in an improved, unconventional format; *Biology for Living* is the outcome of that suggestion.

Many persons have been involved in this text's prenatal development. Chief among these are Ingrith Olsen and John A. Moore. Despite considerable rewriting and reorganization, Ingrith will still recognize the intellectual contributions she once made to Part II of this text.

Three presidents of the Biological Sciences Curriculum Study (BSCS) must be given special thanks for their support during the formative days of this text and for actions that have made its publication possible: William V. Mayer, Jack Carter, and Joseph McInerney. True to their belief that providing students with quality educational material is the main task of BSCS, these persons have by words and by deeds urged (and aided in) the completion of this text. Jane Larsen, former head of the BSCS art department, will recognize some of her own handiwork in the book. Thanks, Jane.

The final typescript was completed through the combined efforts of Nathan Simmons, Connie Richardson, and Sue Rasmussen. Original charts and artwork have been provided by Sharon Chiang, Ruth Steinberger, and Helen Graeff. Many professional colleagues, friends and strangers alike, have generously provided photographs and other illustrative material for our use; we are grateful for these kindnesses. We also appreciate the courtesies extended to us by the many individuals who represent the federal and international agencies, commercial firms, and companies that are acknowledged in the figure legends. John Riina, a friend of long standing, and Anders Richter, editorial director of the Johns Hopkins University Press, were in their various ways instrumental in bringing this effort to fruition.

It is fitting, perhaps, to clarify here the roles played by the two authors: Most of the writing has fallen to Bruce Wallace, who hereby assumes responsibility for its shortcomings and for the errors it might contain; George Simmons, using a preliminary manuscript as a textbook, showed that *Biology for Living* not only was suitable for classroom use but also was enjoyed immensely by students. Both of us are extremely grateful to the seventy to eighty students of that experimental class, especially so to those who thanked us personally for taking the time to write *Biology for Living*.

BIOLOGY FOR LIVING

INTRODUCTION

Robert Benchley, a great American humorist of the twenties and thirties, once wrote of the difficulty of beginning anything. Building a suspension bridge, he claimed, would not faze him; the fatal decision—the question that would kill the entire project—would be, "Where do I turn that first spadeful of dirt?" Textbook authors face the same question: "How do I start?" Here are some solutions some authors have adopted:

A rabbit sat under a raspberry bush. [That *must* be a classic.]

To be alive is to face the problems of living.

The subject of this book is biological science—the study of life. But what is life?

A knowledge of life requires a knowledge of chemistry because living things are physical and chemical machines organized to display those characteristics we take to be life: growth, reproduction, and sensitivity. [Now, that introduction is a far cry from the rabbit and its favorite hiding place!]

Very few men have ever left our planet earth; and even fewer have seriously had to wonder whether they could get back safely. [We would gamble several dollars that *all* men who have left our planet and hundreds, even thousands, more have wondered whether they could get back safely.]

You will live your life in one of the most interesting, dramatic, and dangerous periods ever faced by human beings.

No formal scientific training is needed to help us distinguish mosquitos from mushrooms, magpies from men.

Wherever Benchley's suspension bridge may lead, there is certainly no one point at which the first spadeful of dirt *must* be turned. Poking fun, especially in good humor, is easy; nevertheless, *Biology for Living* still needs an introduction. Our logic is straightforward: biology is one of the natural sciences. Let us begin, then, by making it clear what science is, and what it is not. At any one moment, what is or is not science can be outlined clearly. Nonetheless, the following claim by Karl Pearson, a nineteenth-century mathematician, is still to be taken seriously: "The scope of science is to ascertain truth in every possible branch of knowledge. *There is no sphere of inquiry which lies outside the legitimate field of science.*" (Italics added for emphasis.)

You will read that science can tell you how tall, how heavy, and how many, but that it cannot tell you how beautiful. Even as these lines are written, however, the evening newscaster tells his listeners why lovers crave chocolate. Is there really a chemical that causes one to see beauty—even if no beauty is there? And is that chemical to be found in chocolate, even as it is found in the brain?

1

SCIENCE: AN HONORABLE ENTERPRISE

Personal Involvement

Suppose your brother-in-law telephones one evening to say that your sister (his wife) has been rushed to the hospital for an emergency operation. Your first reaction is to ask what's going on: "Look," you say, "is it her heart? Her stomach? Her uterus? What is the surgery for?"

With each question you realize your sister's husband simply cannot provide an adequate answer. He never learned how the human body—or any other body—is built, or how it operates. He does not know when things are right, or when they are wrong. Nevertheless, in the hospital he signed a statement saying that the doctor in charge could do anything necessary in treating the patient. Before signing, he may have been handed a card specifying the patient's rights. But of what use was this card to him when he knew nothing at all of elementary biology? His role with respect to the surgery being performed on his wife is no different from that of an ancient Egyptian who called in members of the priesthood when a child or relative was ill, or of any parent in a primitive society who calls for the local witch doctor. Your brother-in-law's consent represents an act of abject faith, not one of informed consent.

In a lengthy interview with a reporter from the *New York Times*, former Secretary of State Kissinger once acknowledged that an official report prepared for his use in conducting foreign affairs claimed that there would *always* be a surplus of energy. Four years after the report was written, an Arab oil embargo closed gasoline stations throughout the United States. At the same time, a shortage of natural gas was also acknowledged. Until 1972, Kissinger continued, he and other government officials believed that world food supplies were inexhaustible. Since then, official figures on malnutrition and starvation among the

A Typical Authorization for Surgical Treatment

DATE_____TIME_____ AM / PM

I, the undersigned, a patient in the above named hospital, hereby authorize Dr. _____
(and whomever he may designate as his assistants) to administer such treatment as is necessary, and to perform the following operation: _____
(name of operation and/or procedures) and such additional operations or procedures as are considered therapeutically necessary on the basis of findings during the course of said operation. I also consent to the administration of such anesthetics as are necessary with the exception of _____
(none, spinal anesthesia, or other). Any tissues or parts surgically removed may be disposed of by the hospital in accordance with accustomed practice.

I hereby certify that I have read, or have had read to me, and fully understand the above AUTHORIZATION FOR SURGICAL TREATMENT, the reasons why the above-named surgery is considered necessary, its advantages and the possible complications, if any, as well as possible alternative modes of treatment, which were explained to me by Dr._____.
I also certify that no guarantee or assurance has been made as to the results that may be obtained.

Witness_____ Signed_____
                       (Patient or nearest relative)

Witness_____ Signed_____
                       (Relationship)

Authorization must be signed by the patient, or by the nearest relative in the case of a minor or when patient is physically or mentally incompetent.

3

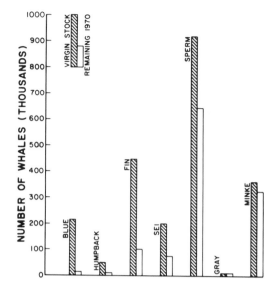

Figure 1.1. The decline in world whale populations, illustrated by comparing the numbers of each type available before the advent of modern hunting techniques and the numbers remaining in 1970. Blue whales, for example, which originally numbered over 200,000 in all, have been reduced to an estimated 10,000; only five percent of the original population still remains in the earth's oceans. Only ten percent of the humpback and twenty percent of the fin whales remain. The number of sperm whales has decreased by one-third. The larger percentages of sperm, gray, and minke still remaining reflect the preferences of whalers; they began by harvesting blue whales, but as blues declined, they switched to humpbacks, fins, and seis.

peoples of the world have continued to climb. Oil supplies have fluctuated, so that at times we in the United States have had an excess (when prices drop) or have experienced shortages (when prices rise), but the *amount* of oil (and of coal and natural gas) available continues to decline. Those who add milk to their coffee directly from the two-quart carton know that the uncontrollable flood and the unseemly slosh come from the nearly empty carton, not from the full one.

The purpose in describing these two vignettes is to preface a plea: a plea for well-informed students, a plea for their continuous involvement in transpiring events. Biology, among other things, is a science of involvement. Individual life is tenuous. The life of a species is also tenuous, especially that of a species of organisms that for one reason or another has come to the attention of human beings. For example, fishing fleets have depleted one species of fish after another; even the oceans can no longer be looked upon as an inexhaustible source of food. Do not be one of those who someday exclaim, "Goodness, where did all the fish go? We used to have so many!"

To understand the treatment—surgical or otherwise—that clinicians propose for a patient, the patient (or the responsible persons) must understand something of human biology; in an emergency, the patient's next of kin are often those who need to know.

Knowing the human body and how it operates leads to an understanding of the physiological needs of the nearly six billion persons who inhabit the earth. The step from (1) knowing how much water one person needs, how many calories one person needs, or how much urine one person excretes to (2) knowing the demands for drinking water, calories, and drainage pipes worldwide involves only multiplication. The next step, from knowing worldwide demand to calculating how long available resources will last, depends upon the nature of the resources. For fixed-quantity resources, the total amount divided by the amount consumed per year yields the number of years the supply will last. Circulating resources, like gold or diamonds, will "last" a long time indeed. Material that is consumed—burned, as coal is—cannot be saved, nor can it be renewed. If, as in the case of timber or lake and marine fish, the resource lives and reproduces, rates of harvest exist that allow the resource to replenish itself after each generation. Still other, higher rates of harvest may appear to be profitable, but they may mean the ultimate destruction of a "renewable" resource.

The plea to the student is twofold: be knowledgeable and be alert. Biology is the science of life (*your* life!). We urge you to take the material of this course seriously. Although a foolish skepticism merely impedes the discovery of honest answers to perplexing problems, intelligent skepticism can lead otherwise uninformed or greedy persons to abandon expedient or short-sighted practices. However, informed skepticism requires that you remain alert, that you make an effort to stay abreast of events occurring around you. An understanding of "current events," once the bane of high school civics courses, is necessary for informed citizenship. Everyone should make a habit of reading newspapers and news magazines, and everyone should read each item with an informed skepticism. Awareness counts, however, only if the person becomes involved.

Former Secretary of State Henry Kissinger, whom we mentioned earlier, is not a biologist. No president of the United States has ever been a biologist. No cabinet member (with the possible exception of Henry Wallace, once Secretary of Agriculture) has ever been a biologist. The presidents and board members of industrial firms, of insurance companies, and of banks are not biologists, and yet these persons make decisions that influence the course of the nation and of the world. From the red-lining of city maps (a largely outlawed procedure by which bankers, merely by controlling the flow of mortgage loans, influenced the ultimate fate of urban neighborhoods) to the promotion of synthetic infant formulas in developing nations, from the construction of resort hotels on saltwater marshes to the dumping of in-

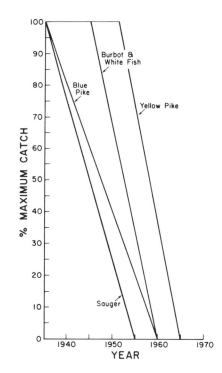

Figure 1.2. The decline in the annual catch of five species of freshwater fish in Lake Erie over a period of three decades. Because poundages per species differ, the maximum yield for each species has been set at 100%. The total of the five species caught in 1935–40 was about 18,000,000 pounds; in 1965–70 the catch had dropped to 400,000 pounds—about two percent of its previous size. The collapse of the Lake Erie fishing industry was caused by several factors: overfishing, predation by the newly introduced lamprey (a jawless predatory fish), eutrophication, and pollution of the lake by industrial, agricultural, and household substances, many of which were toxic. Following a strenuous effort undertaken in the seventies to clean up Lake Erie, several species of commercially valuable fish have reappeared. Carp, catfish, and yellow perch were relatively unaffected by the conditions that eliminated the other species.

Figure 1.3. A montage of headlines dealing with scientific subjects covered in four issues (science section, to be sure) of a daily newspaper. Today's citizen is expected to be knowledgeable about a variety of science-related topics. On the other hand, newspaper articles, excellent as many are, should not be automatically accepted as providing the *whole* story; many important facets are frequently overlooked.

Figure 1.4. An estuary. Many persons regard the estuary as an unsightly area needing "development." Others would use it as a dumping area for garbage, toxic wastes, or landfill. In truth, an estuary is one of the most nutritionally productive areas of the world, supporting the young of many marine animals and feeding countless waterfowl and shore birds. The productivity of saltwater estuaries is prodigious: whereas most agricultural land produces 3–10 grams of dry matter per square meter per day, estuaries and salt marshes produce 10–25 grams per day. To destroy such a productive area is to destroy a natural (and national) resource. *(Photo courtesy of the Dow Chemical Company)*

dustrial wastes in abandoned anthracite mines in Pennsylvania, paths are chosen by nonbiologists that direct all our futures.

As a student using this book, you probably consider yourself a nonbiologist. You may, it is true, never occupy a position of power in your state or in the nation, but some of your classmates may rise to such positions. While you are in class together, before knowing who will and who will not eventually influence public decisions the most, urge one another to learn, to think, and to remain alert.

Figure 1.5. Estuaries and saltwater marshes are destroyed during the "improvement" of seashore property. Shown here is downtown Tampa, Florida; not shown are the thousands of waterside residences with personal marinas, each of which was developed at the expense of marine life. The ledger within which economic development is entered against food production (and the estuary is by far one of the most productive areas) has not yet been balanced: the Tampa skyline looks impressive, but at what cost? *(Courtesy of the Tampa News Bureau)*

NOW, IT'S YOUR TURN (1-1)[1]

1. Have you ever taken part in the public discussion of topics like consumerism, women's rights, gay rights, solar or wind energy, nuclear energy, or organic farming? If so, note in your journal the topic, how you became involved in it, and (if you wish) the views you propounded. If appropriate, enter your reasons for not taking part in a particular discussion.

2. Several current public issues impinge on biology: the right-to-life movement, creationism, and the save-the-whales and save-the-

[1]To call the student's attention to his (or her) journal, and to remind the student (1) that the life of neither a teacher nor a textbook writer is easy and (2) that the efforts of neither accomplish anything without the student's cooperative response, exercises concerning the student's journal are introduced with the words, "Now, it's your turn."

redwoods campaigns are but a few. Take any one of these issues (or another one that has not been listed) and record your views as you see (or understand) them today.

What Is Science?

A biology course intended for effective, socially alert, and active citizens enters unfamiliar and, as like as not, hostile territory. The generally accepted purpose of education is to preserve the community's status quo. Should anyone disagree with this last statement, let him extol Communism in a local high school in America. Or, vice versa, let a Soviet teacher extol capitalism in a Moscow or Leningrad public school.

A science course that asks students to think about public issues trespasses into private lives and questions of personal beliefs. How are value judgments to be handled? Or religious beliefs? What are the bounds of science and scientific arguments? These questions must be faced or any course advocating involved citizenship will fail.

What *is* science? At its core, science is an activity that is based on an *assumed* relationship between human beings and the universe within which they live. It is the belief (as well as the activities that follow from that belief) that the world can be understood by posing questions, whose answers are often sought by means of experiments. It assumes that the answers that are obtained are objective. That is, Nature does not lie. (However, if questions are badly posed and experiments are badly designed, the experimental data obtained may be misunderstood at best or untrue at worst. The fault in this case lies with the *scientist*.) Supposedly, Nature holds no malice nor does Nature deceive. Therefore, through trial and error, and through the efforts of countless men and women, information has been, and continues to be, gathered that tells us how different parts of the world, including living organisms like ourselves, function. This is scientific information. We have referred to the gathering of this information as an honorable enterprise.

Scientific facts tell us *how* the world functions. They do not tell us *why*. Science, that is, does not provide meaning. Scientific experiments tell us how large, how many, how fast, how far, and in what order. These experiments cannot tell us why birds, flowers, and butterflies are beautiful. They cannot measure beauty in sight, taste, or sound. They cannot tell us why a silent Spring would be a sad Spring. Matters such as these are resolved by the use of our emotions.

Within the world of science, the matter of accepting facts is intensely personal. Badly designed experiments yield "facts" that can easily mislead. As an example, consider the scientist studying grasshoppers. When he shouted "JUMP!" at normal grasshoppers, they jumped.

Some Mistaken Observations

Two examples, one modern and one ancient, illustrate the misinterpretation of observational data in science.

Early in the study of bacteriophage (viruses that attack the colon bacterium *E. coli*), it was found that individual phage particles are "tadpole" shaped. Photographs suggested that these particles were attracted to bacterial cells even when they were far removed: "attraction at a distance." Phage were thought to "swim," like sperm, head-first toward bacteria, and to enter the bacterium head-first.

Eventually, it was learned that phage particles attach to bacterial cells not by their "heads," but by their "tails." This finding was made possible by the use of liquid CO_2 as a mounting fluid. Adjustments of pressure and temperature can turn liquid CO_2 into gaseous CO_2 with no liquid–gas interface.

The apparent "attraction at a distance" is now ascribed to the systematic toppling of upright phage particles by the surface tension of the original mounting fluid, which, as the liquid evaporated, toppled the upright particles so that their heads pointed toward nearby bacterial cells.

The second erroneous "observation" was that which led to the conclusion that sperm heads contain miniature organisms resembling adult individuals. Each human sperm head was said to contain a small *homunculus,* a tiny human being. Logically, it followed that each male homunculus contained sperm that in turn contained miniature humans, and that these even tinier homunculi also contained sperm cells that contained even smaller homunculi, and so on indefinitely. The person who "saw" and drew the homunculus in a sperm head made what seemed to be a positive observation. In this case, the persons who did *not* see this small being prevailed, even though theirs was negative evidence. On their side was the absurdity of the logical consequences that followed *if* homunculi really existed.

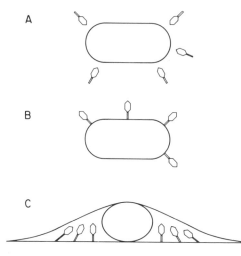

Figure 1.6 (*above*). The "attraction at a distance" that was seemingly exhibited by bacteriophage (*A*) implied that they swam spermlike toward a nearby bacterium. Improved mounting techniques that removed surface tension as a distorting force revealed that phage attach to bacteria by their tails, not by their heads (*B*). The "attraction" was then seen (*C*) to be an artifact caused by the surface tension of the evaporating mounting liquid; phage particles, which were standing tail-down, were toppled head-first toward any nearby bacterial cell.

Figure 1.7 (*right*). The human sperm according to Hartsoeker, a seventeenth-century microscopist. Supposedly, within the sperm head was a small infant with a disproportionately large head (notice the nose, which is tucked between the knees). Only with the aid of its mother could the small infant within the sperm escape its prison. Note that although this observation proved to be wrong, it was—at that time—positive evidence. One might have claimed that there *had* to be a small person in each sperm, or from where else could babies come? *(Courtesy of the Cornell University Library)*

Figure 1.8. A scanning electron micrograph of human (*left*) and bovine (*right*) sperm. The theory that led to the claim that each human sperm contained a small person was extended to all species. Needless to say, no small organism—human, bovine, or otherwise—exists in sperm heads. Only the plans or information needed to make a new individual is contained in the sperm's DNA. The bent sperm tail illustrated here is characteristic of one particular bull; it is a marker that permits animal breeders to know which of two possible bulls has actually inseminated a cow. *(Photo of human sperm, Cambridge University Press; photo of bull sperm courtesy of R. G. Saacke, Virginia Tech)*

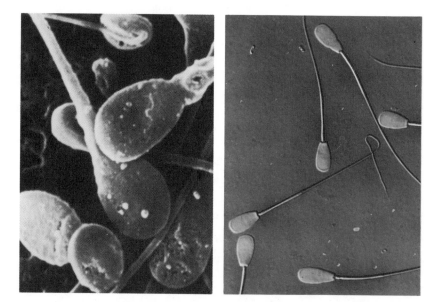

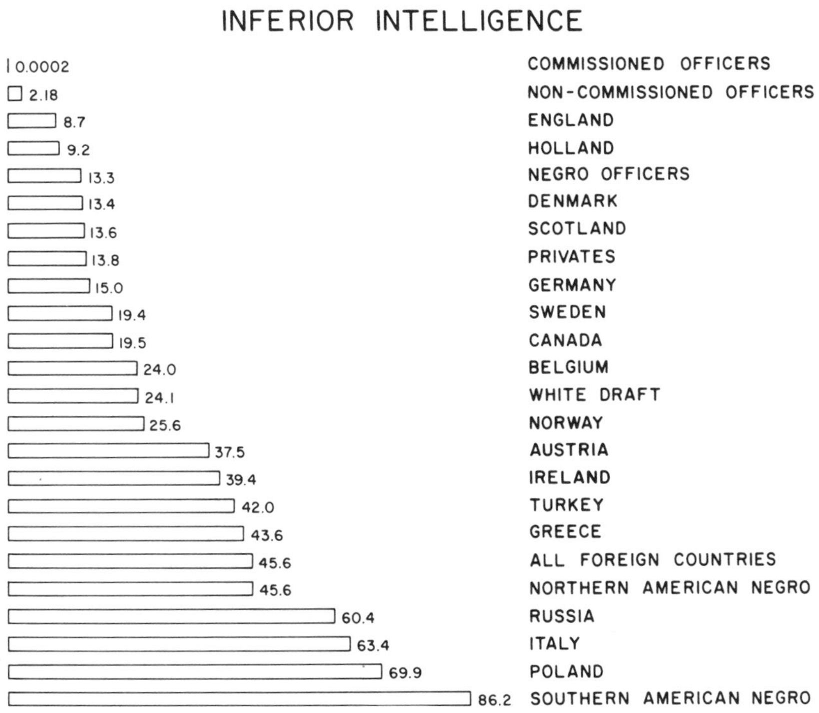

INFERIOR INTELLIGENCE

0.0002	COMMISSIONED OFFICERS
2.18	NON-COMMISSIONED OFFICERS
8.7	ENGLAND
9.2	HOLLAND
13.3	NEGRO OFFICERS
13.4	DENMARK
13.6	SCOTLAND
13.8	PRIVATES
15.0	GERMANY
19.4	SWEDEN
19.5	CANADA
24.0	BELGIUM
24.1	WHITE DRAFT
25.6	NORWAY
37.5	AUSTRIA
39.4	IRELAND
42.0	TURKEY
43.6	GREECE
45.6	ALL FOREIGN COUNTRIES
45.6	NORTHERN AMERICAN NEGRO
60.4	RUSSIA
63.4	ITALY
69.9	POLAND
86.2	SOUTHERN AMERICAN NEGRO

Figure 1.9. A chart that was published in the United States *Congressional Record* of March 8, 1924 showing the distribution of inferior intelligence among those persons who had taken the Army Mental Tests. Both the validity and the usefulness of such data have been questioned for fifty years or more; nevertheless, claims similar to those made in this chart are still promoted by some persons.

When he removed the insects' legs and again shouted "JUMP!" at them, they didn't move. Therefore, he concluded, grasshoppers hear with their legs.

The flaw in the scientist's claim about grasshoppers is obvious; that is why the story is a popular child's joke. What, though, about a claim that crowding is unrelated to urban crime? The "facts" are that while the populations of inner cities have declined, crime has increased. Is there an obvious flaw in that argument?

What about the long-assumed inferiority of nonwhite races? This belief has persisted for centuries among Caucasians and has been defended by generations of outstanding (white) scientists. Is the average intelligence of blacks inferior to that of whites (and that of Japanese higher than both)? This question is likely to be controversial even as you read this sentence. What can a person believe?

Could you believe that your red blood cells, if placed side by side, would form a chain stretching two-thirds of the distance to the moon? Why not make the necessary calculations and test this assertion for yourself? Claims based on "facts" may in the future prove to be wrong because the observations on which they are based were irrelevant to the conclusion(s) drawn from them.

For example, a person can safely reject claims that invoke perpetual motion or other forms of magic. Physicists have developed laws that have long withstood the tests of time. Two of these involve the conser-

Figure 1.10. Theories generally held by scientists, such as the theory of evolution, may at times conflict with ideas held by other persons. The source of the conflicting view may be religious faith or even intuition (" . . . it stands to reason"). When the majority of persons living in a community refuse to accept a scientific theory, local control of education raises serious problems for professional educators, as one can surmise from the montage of headlines presented here. *(Courtesy of the Biological Sciences Curriculum Study)*

vation of matter and the conservation of energy. These laws have been combined into the first law of thermodynamics, which is that energy can be neither created nor destroyed. An arsenal of atomic and nuclear weapons reminds us, however, that matter can be transformed into energy according to Einstein's famous equation $E = mc^2$, where E is energy, m is mass, and c is the speed of light.

The second law of thermodynamics states that *useful* energy is always lost, never gained. It vanishes largely unseen as heat, ultimately as infrared radiation. When gasoline is burned thoroughly (as in an efficient stove) carbon dioxide and water are formed. These two substances cannot be recombined in a factory in order to regain gasoline. More energy would be expended in the factory than would be recovered when the newly made gasoline was burned once more. That is why used gasoline will never be recycled. Green plants, however, do combine carbon dioxide with water to make sugar, but in doing so they use only a fraction of the total energy they receive from the sun in order to carry out the essential chemical process; a great deal of the energy impinging upon plants from the sun is lost as heat. Gasoline, of course, is a decomposition product of fossil plants.

Between easily verifiable statements and obviously false ones lie other, often conflicting, statements and claims involving complex sets of interrelated facts. In assessing these, each person must use his or her best judgment. The biological facts assembled in this textbook are intended to help in a search for the truth. Do not be over-confident, however. Despite the many facts you encounter here and elsewhere during your life, you will often reach erroneous conclusions. Your errors may eventually be corrected through the use of better information. Irrational conclusions, when proven wrong, are much too often simply replaced by other irrational ones.

This text will direct you to outside sources of information and will encourage you to consult and to use these sources. For example, sup-

Figure 1.11. Many early travelers, when describing the exotic beings they encountered on their voyages, did not hesitate to represent any congenital abnormality or deformity resulting from a postnatal accident as the standard appearance of certain adult human beings. As a result, those who did not travel were subjected to a good deal of nonsense like that shown here. Among today's counterparts of those who stayed at home are (1) those who still believe that the moon landing of 1969 was an elaborate hoax, and (2) Asiatic and African school children who study biology from textbooks prepared for use in North American and European schools. *(Courtesy of Leslie Fiedler)*

pose you encounter a claim that long-horned cattle will solve the beef-supply problem in the United States. The reason, according to those making this claim, is that long-horned cattle thrive on wild range plants and do not require corn or other grains. To evaluate this claim, one must first know that *the information required for its evaluation is available*. The necessary information (such as total acres of range land, productivity of range land, and the number of animals that can be sustained per acre) can be found in many of the annual books of facts that are sold at magazine stands. Consequently, you can easily check this claim for yourself.

Figure 1.12. These are the longhorned cattle of the Western round-up and of old-time cattle drives. Today's beef cattle are the heavy, thick-set, shorthorned Herefords. The longhorns' ability to thrive on untended range land has led some persons to suggest that they might solve the food problem created by a rapidly expanding world population. With the aid of a few readily available reference books, any informed person can check the reliability of such a suggestion. *(Frederic Remington, ca. 1875)*

NOW, IT'S YOUR TURN (1-2)

1. Here are some facts about red blood cells that will allow you to verify the statement that, if aligned side by side, the red blood cells of a person would stretch two-thirds of the way to the moon: (1) the diameter of red blood cells equals 0.0072 mm; (2) the number of red blood cells per mm^3 equals 6,000,000; and (3) the volume of blood in an adult equals six liters (1 liter = 1,000 cm^3). You might want to convince yourself that the smaller each red blood cell might be (assuming that the total mass of all red blood cells is constant), the farther they would extend if aligned side by side. To do so, consider the length of wire that can be manufactured from a cubic centimeter of copper: the finer the wire, the longer it will be.

2. The syndicated feature "Believe It or Not!" generally includes factual material. In what sense can facts be considered "unbelievable"?

3. Three persons raise peas, and observe differences between varieties; one variety, for example, has smooth seeds, while another has wrinkled ones. The first person notes that crosses between plants grown from smooth seeds sometimes produce only smooth seeds, and sometimes both smooth and wrinkled seeds. The second person makes the same observation but adds that the wrinkled seeds in such cases are always less numerous than the smooth ones. The third person makes accurate counts of the two kinds of seed, notes proportions, and not only describes the results but also *accounts* for them by a theoretical *model* that permits him to predict the types of peas produced by a variety of crosses. Are all three of these persons scientists? If so, would you say that one is a better scientist than the other two? Is the evaluation "better" a scientific evaluation? Compose an evaluation of these persons based on the information provided here, and defend your evaluation if you do indeed rank them by their achievements. (Your defense will require a definition of the term "better.")

To many persons, the advance of science has been related to the emergence of *the* scientific method. This formalized method consists of observation and an accumulation of data, followed by the building of a general hypothesis. The validity of this hypothesis is tested by making predictions, which are then subjected to experimental test. If these tests uphold the hypothesis, it becomes a theory or law. If any test fails, the hypothesis is either discarded or modified and is replaced by one that encompasses the discordant data.

As we learn in the text, individual scientists depart from this formal procedure. Many proceed without writing or verbalizing intermediate steps. Some collect data but never attempt to generalize. Many enter the formal process on the basis of intuitive feelings (or mathematical reasoning) and omit the early observations. The "scientific method" is an idealized procedure that is followed by some, but by no means all, scientists.

Scientific papers, by tradition, also follow a standard format: an *Introduction*, in which the historical background to a problem is discussed together with a hypothesis that seems to follow from earlier observations and that is to be tested; the section dealing with the *Materials and Methods* by which the test was made; the experimental *Results* that were obtained (presented largely in tabular forms or as diagrams); and a *Discussion* of the experimental results that lead to certain *Conclusions*. Most papers either begin or end with a brief *Summary* for the readers' convenience. In truth, patterns of experimentation scarcely if ever follow the orderly presentation of journal publications; the latter provide, in most instances, a mythical account of the progress of scientific research.

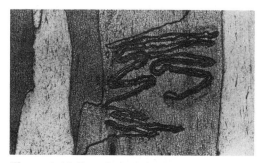

Figure 1.13. Tunnels made by certain insect larvae as they bore beneath the bark of the Australian "scribble" tree (a form of eucalyptus). As the outer layer of bark is shed each year, last year's scribbles are exposed to view. Although differing in details, the patterns made by these insects show a similarity, especially in the obvious way each tracing folds back on itself. Presumably, no one has studied the cause of the overall similarity in these patterns. Can you make a suggestion? Can you phrase a testable hypothesis? How might you test your hypothesis experimentally?

The Scientific Method

For generations, students have been told about *the* scientific method; that is, about a method used by scientists in their efforts to unlock nature's secrets. Like much that young students learn, the supposed scientific method is an oversimplification that makes the teaching of a subject absurdly simple while making the subject simply absurd. There is not *one* scientific method; on the contrary, there are many.

Scientists go about their work, and think about their observations, in diverse ways. Some, in effect, mix reagents and say "Let's see what happens." The things they mix, of course, are usually not picked without thought, but the intuition that motivates some scientists is not always obvious to others. Luther Burbank was a plant hybridizer who created many hybrid fruits and vegetables during the early 1900s; a professional plant geneticist who was asked to work with Burbank and to organize his notes was unable to do so. Burbank achieved his results by intuition (one man's intuition is often another's common sense), hard work, and chaotic happenstance.

Still other scientists refuse to perform any manual operation until they have anticipated every conceivable outcome; the actual experiment, if they have thought matters through carefully, simply confirms what they have already "guessed." Guesses made by many persons (even scientists) are, of course, often wrong. Immediately after it was formulated, Einstein's theory of relativity was "disproved" by an experimental physicist at the University of Groningen in The Netherlands; Einstein ignored the disproof, and so does the rest of today's world. The "disproof" was itself erroneous.

Some scientists merely count, measure, or weigh. How many earthworms live in one cubic foot of garden soil? In the soil of an old pasture? Or in an apple orchard? Often scientists of this persuasion are reluctant to generalize their results: "My data," one will say, "apply only to my garden; your garden may be different."

One pervasive feature of scientists is their curiosity: they ask questions and pose problems for themselves. Nonscientists sometimes misunderstand the relationship between scientists and the problems they have chosen to solve. A naive and extremely religious typist once said of a famous evolutionary biologist, "He wouldn't have that terrible problem if he only believed in the Bible." A scientist without a problem is not a happy person. Now, not every scientist can think of problems for him- or herself; fortunately, however, others can think of problems faster than they can solve them. The latter are glad to share research problems with their less obsessed colleagues. In the end, every scientist has his or her "Why is that?," "How many of this?," and "How come?" questions to work on. It does not matter whether the scientist works in an easy chair, in the laboratory, on the

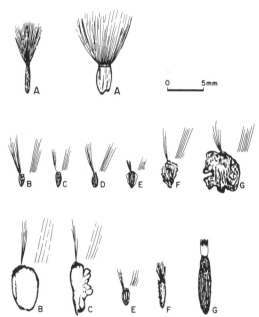

Figure 1.14. Composite photograph of honeybees visiting a small patch of lamb's ears (*Stachys olympica*). When the photographs were taken, two or three bees visiting a single spike were not an uncommon sight. Where did these bees come from? How did they manage to return home? How did they discover the patch in the first place? Accident? Odor? What evidence would suggest that bees tell one another about patches of flowers? (If, in answer to the last question, you find yourself reciting details about von Frisch's work and the bee's waggle dance, you are answering the wrong question: you are telling *how* bees communicate, not *if* they communicate.)

Figure 1.15. Filaments attached to seeds often aid in their dispersal by wind or even gentler air currents; the dandelion is perhaps the most familiar example. The two seeds illustrated at the top are typical of mainland species. The other seeds, which have fragile filaments (shown broken from the seed itself) or filaments reduced in either size or number, are from plants inhabiting small oceanic islands. Why should seeds develop dispersing mechanisms at all? Why are these mechanisms reduced on many small islands? These are the sorts of questions a biologist would ask if confronted with the above information for the first time. Furthermore, answers to the questions would be sought in natural—not *super*natural—causes. *(After Sherwin Carlquist)*

beach, or in the field. As long as one's curiosity is retained, as long as one's problem is approached in a rational manner that avoids seeking supernatural explanations, as long as solutions are proposed that can be subjected to confirmatory tests by others, one *is* a scientist using one or the other of the *many* methods of science.

NOW, IT'S YOUR TURN (1-3)

In what sense would you claim that the following persons are not scientists, or that they are not using valid scientific methods:

1. A Hindu microbiologist claims that the recitation of prayers over bacterial cultures increases the survival and growth of the bacteria.

2. A psychiatrist develops a means for analyzing individuals' personalities from physical measurements and three-way photographs; he then claims that no one other than himself has the ability to make such analyses.

3. An experimenter who, having said that if *A* is true, then *B* must follow, learns from his test that *C* (not *B*) is the outcome, but continues to argue that *A* must be true.

4. A graduate student who, while testing two varieties of wheat in order to compare their yields, discovers that he does not have enough fertilizer for all his plants. He then puts the fertilizer on what he believes is the better variety in order not to waste it on what he suspects is the inferior one.

Figure 1.16. Discussing one's research with visiting colleagues from foreign universities is stimulating for a scientist. Although visitors may be known from their published research papers, seminars are the mainstay of scientific communication. Seminars are given before departmental colleagues, at nearby colleges or research laboratories, and at national and international meetings of scientific societies. Are seminar notices posted on your departmental bulletin board? Note the variety of topics to be discussed by different speakers. Take time to attend a seminar or two, even though you think the topic might be too advanced for you to understand. *(Courtesy of Virginia Tech)*

A Manner of Speaking

In explaining that the present textbook does not emphasize biological terminology, we mentioned that the professional biologist—and the student who intends to make a career in biology—must learn the words that allow him or her to communicate with professional colleagues. Most beginning courses in biology require the student to learn more new words than do beginning courses in foreign languages. A hint as to what awaits a budding botanist is given in the following paragraph:

"Plicate ptyxis may be a synapomorphy for Arecales and Cyclanthales, while Pandanales has conduplicate–plicate ptyxis. The conduplicate–plicate ptyxis could also be regarded as a synapomorphy for Hypoxidaceae and Velloziaceae, but this is contradicted by seed characters. In general, when including ptyxis in an evolutionary estimation it is important to consider this character in conjunction with the general shape of the leaf."

The above paragraph can be crudely paraphrased as follows: "Palms and the plants used in making Panama hats are similar in having leaves that, in the leaf bud, are folded like a Spanish lady's fan; in contrast, screw pines have leaves that are folded along the midrib, with the upper faces of the leaf in contact. The screw pine leaf resembles that of star grasses and, thus, might suggest that these forms are related; seed characters, however, rule out this possibility. In estimating evolutionary relationships by means of the structure of leaf buds one should consider as well the general shape of the leaf."

Scientific Communication

This chapter's title refers to science as an "honorable enterprise." Although most scientists tend to work as individuals (team research is becoming more common, however), as a group they form what is recognized as the "scientific community." They exchange information, they repeat one another's experiments in order to verify the results obtained by others, and they constantly search for general explanations. In doing so, they bring diverse observations together into unifying syntheses. From naturalists' realization that nearly all higher animals develop from fertilized eggs came the more encompassing science of embryology. In recent years, embryology has utilized knowledge gained from genetic studies, and together these bodies of knowledge have yielded a new synthesis: molecular development.

Can a hermit who never talks or corresponds with another human being be a scientist? Certainly, a hermit who worked unseen and left no record of his studies would contribute nothing to that honorable enterprise called science. Furthermore, communication among scientists requires considerable precision in expression if each person is to understand what others have done. (At a testimonial dinner, one plant physiologist praised the guest of honor by saying that he was by far the best and clearest writer of all persons in their field: "When I do not understand one of Doctor S's sentences," the speaker exclaimed, "I know that Doctor S does not *want* me to understand that sentence.")

Information must be passed on accurately, and the data that are being reported must be those actually observed. Data recorded during the course of any experiment may not be those that provide an answer to the question a scientist *thinks* he asked. Nevertheless, data honestly and accurately reported do provide an answer to the question the scientist actually did ask. Chemical solutions that are stored momentarily in a refrigerator may absorb fumes of which the experimenter is unaware, but such fumes may have tremendous effects on the outcome of the experiment. The recipe for preparing a tissue stain may say, "Stir with an iron nail." The student who decides to be more "scientific" and instead uses a glass rod may not introduce enough iron ions into his stain to make it work properly. In this case, the recipe was precise; it failed, however, to explain why the use of an iron nail was important.

"Scientific evidence" has tumbled out of the laboratory and into television and magazine advertisements. A good deal of this popularized evidence involves intellectual sleight of hand. For instance, insurance company *A*, to convince the reader that its rates are lower than its competitors (*B* and *C*), claims the following: "In a survey of twenty-one cities, our rates were lower than those of both *B* and *C* in seven cities; our rates were second-lowest in an additional seven cities." Are you convinced? Nearly so? Notice, then, that if the three companies had identical average rates, one would expect each to be lowest in seven cities, second-lowest in an additional seven, and highest in the final seven. All three companies could have made the claim that *A* made as proof that it had the lowest rates!

Here is a second example of intellectual sleight of hand used in presenting "scientific" evidence: "In test after test, more doctors recommended *Asprino* for the relief of headache than *Buffero*." If doctors recommended the two pain relievers with equal frequency (on an average), then half of all tests would have a majority favoring *Asprino*, while the other half would show them favoring *Buffero*. Consequently, either manufacturer could claim that "in test after test" more doctors favored its brand over the competing one. Communication that is intended to mislead is not scientific communication; evidence that is presented by sleight of hand is not scientific evidence.

NOW, IT'S YOUR TURN (1-4)

1. To practice precision in writing, each student should describe some common object (a small rock would be fine). During class, place all the rocks on a table, and then exchange written descriptions. Each person should search for and retrieve the rock that matches the description he holds. If any stones are left unidentified, the describer and the seeker should together review the description in order to see what's at fault: the accuracy of the writing or the reader's failure to interpret correctly what was actually written.

2. A graduate student at the University of Michigan once said that she did not believe in evolution. "I do not believe," she said, "that frogs turned into birds." Try to compose a more precise statement expressing this student's skepticism regarding evolutionary theory. Remember, without precision, there is no opportunity to discuss scientific matters.

3. In studying the sexual behavior of flies, investigators often place males of one type with two different types of females; by discarding males after a stipulated time, successful matings can be scored later at one's convenience by dissecting the females and looking for sperm in each one's reproductive tract. Such experiments were first regarded as if they revealed a choice made by the males; later, the same tests were interpreted as revealing the female's preference. Describe why such tests are ambiguous with respect to who is choosing whom.

4. A thief, when he heard that the prosecution had two persons who had witnessed his crime, claimed that he could find ten who had not witnessed it. Why is negative evidence poor evidence?

A recent article in a national newspaper was headlined "Scientist confesses to fraud." The article stated that the scientist admitted that *one* of his published articles was based on fabricated data, but that a second, related article contained honest data, honestly reported. The flavor of the article, except for the sense that scientists should not lie, was what would be expected if a local car dealer admitted to selling a lemon: "Here's the refund for the lemon, but I do have some excellent cars over here for your consideration."

Science is an enterprise that is honorable, and its practitioners, scientists, are incensed by fraud. Scientists who have knowingly promulgated fraudulent data are generally excluded (blacklisted) from the fellowship. Why should this be? There are several reasons that come immediately to mind:

Fame. Anyone conversant with past observations in any field of research can make predictions about the possible outcome of future experiments that are made to test hypotheses, experiments that determine whether a given hypothesis will be discarded. Suppose someone fabricates data and reports that he (or she) has tested and, on the basis of these data, supports a given hypothesis. Other persons actually perform similar tests, expending a great deal of time and energy. If these tests confirm the guess made by the fabricator, he gets the credit. If they do not, he asks forgiveness for his fabrication and continues his "research." If forgiving were the custom, the fabricator would have everything to gain and very little to lose.

Time and energy. Upon reading of someone's experimental observations, dozens of other scientists can see the implications of those results for their own research. Each of these persons then continues to work, on the basis of the assumed accuracy of the reported observations. Perhaps, after two or three years' study, the truth emerges: the original observations were false. Because a scientist operates effectively only between the ages, say, of 25 and 65, a two- or three-year loss of time is a substantial loss; it is especially great if those two or three years fall within the five years of research during which a young scientist must convince older colleagues and university administrative officers that he or she deserves a tenured position. Persons who falsify results can ruin another person's scientific career.

Science is a self-correcting enterprise. Errors in experimental design, errors in reaching conclusions, and fraudulent attempts to promote conclusions that are based on make-believe data are all eventually detected and set straight. The misinterpretation of data is generally forgiven, albeit more gracefully by some than by others; scientists, after all, are human. Errors in reaching conclusions—"stupidity" describes the worst cases—are also forgiven. The publication or promulgation of fraudulent data is an act not forgiven by scientists—to the apparent mystification of many non-scientists.

Figure 1.17. A pothole in the Alaskan tundra caused by a construction vehicle. Because the permafrost lies just beneath the vegetative cover of the tundra, disturbances like these may remain for centuries. Off-road vehicles, which are used in desert areas of the American Southwest, also damage the plant life, so that it (and the animals dependent upon it) may require decades to recover. Few persons realize that harsh environments have their characteristic plant and animal communities, or that these communities are extremely sensitive to the changes that people and their vehicles cause in these environments. *(Courtesy of Orson Miller, Virginia Tech)*

Figure 1.18. Los Angeles was merely the first of a great many American cities that accepted the notion that every family was entitled to its own lot of ground and at least one automobile. What sorts of problems occur when millions of families demand this way of life? *(Photo courtesy of the Department of Housing and Urban Development)*

Value Judgments

If someone claims that dogs make better pets than cats, you may or may not agree. You may, for example, retort, "That, Sir, is your opinion." Unless "better pet" can be defined in a manner acceptable to you and all others concerned, there can be no rational discussion of the matter (although there certainly could be a heated argument!). Recall how a similar problem arose earlier with respect to identifying a "better" scientist.

The methods of science don't measure values. There is no scientific response, for example, to a suggestion that Los Angeles replace its trees with plastic ones, or that colorful plastic birds be placed within those plastic trees, or that Astroturf displace live grass for lawns. Or that single-family dwellings are better than apartments. Or that seashores without hotels are underdeveloped. These are matters of conscience or of personal values, which are based on personal points of view. Points of view, in turn, are influenced by convenience, nostalgia, and greed, and such feelings cannot be measured by the yardsticks of science.

The factual consequences of various acts, acts that in themselves seem to be subject only to value judgments, can often be measured and described objectively. Salt marshes and river-mouth estuaries are the breeding grounds for most commercially important species of marine fish; if the entire coastline were covered with landfill and converted into seaside resorts, the fishing industry (and more) would be destroyed. This is not a value judgment, it is a statement of fact. Similarly, suppose every family in the United States were to insist upon a suburban house and a half-acre lot. The demand for wood, for street and highway construction, for telephone and electric lines, and for water and sewage systems, as well as the total energy requirement, would be enormous. So would the loss of agricultural land. Thus, whether a person seeks a city apartment or a ranch house in suburbia may be a matter of personal preference, but the ultimate consequences of that choice can be expressed in measurable terms.

Although there are no scientific formulae by which one can directly evaluate the consequences of a personal preference, there are means by which the consequences of many otherwise unmeasurable actions can be evaluated. No textbook can dictate proper behavior; a textbook can, however, attempt to remove ignorance as an excuse for harmful behavior.

NOW, IT'S YOUR TURN (1-5)

1. A colleague claims that his ailing mother-in-law can visit the Adirondack Mountains in New York only by snowmobile; she cannot

hike, and highways do not penetrate the most scenic areas. Snowmobiles, however, disturb deer and other wildlife in regions where wild species must be left unmolested if they are to reproduce and continue their existence. How do you balance the disabled's right to enjoy nature against the right for wild plants and animals to perpetuate themselves?

2. Commercial hog yards in the Midwest once controlled parasites and other swine diseases by simply flushing off the topsoil from pig-pens into a nearby river (a practice that often caused problems for cities and villages downstream). The amount of topsoil, even in the Midwest, where prairie deposits were once forty or more feet deep, is limited. How do you balance today's expediency against tomorrow's dire need?

3. One of the arguments against exposing human beings to unnecessary X-rays or gamma radiation is that the consequences of genetic damage will harm some individual born at some future time. Many geneticists say that the welfare of a human being is of constant value, whether that person is born today or a hundred years from today. Most economists say that we constantly discount the future, even for ourselves. Other, more cynical persons ask, "What has the future done for me?" How much concern should we of today show for persons yet unborn, but who *will* be born if human beings are to continue their existence?

Figure 1.19. An aerial view of excellent lakeside farming land in New York state. The entire land area shown here encompasses 6000 acres. At the current U.S. standard of living of 5 acres per person, it will support 1200 persons. The three marked roads enclose a 500-acre area. Divided into quarter-acre lots, this area could house 2000 families or about 9000 persons. To feed these persons would require 45,000 acres, seven times the total area shown in the photograph. *(Courtesy of Cornell University)*

Problems Without Rational Solutions

Much of education consists of problem solving. (Most examinations are precisely that: What is the answer to this problem? What is the answer to that?) Some questions have factual or logical answers on which all knowledgeable persons agree: Columbus reached America in 1492. The Declaration of Independence was signed on July 4, 1776. Two plus two equals four. The answers to other questions are provided by consensus. For example, *free enterprise* is considered "right" in the teaching of American schools, but "wrong" in that of Russian ones.

Some problems have no logical solutions. This may sound heretical or idiotic, but it is true. An excellent example is the problem of a valuable resource that is shared by many persons. Each person can gain personal wealth by exploiting the resource shared by all. An essay by Garrett Hardin that appeared in *Science* during 1968 made thousands of persons aware of this problem for the first time. The essay describes how a commons (a meadow shared by many herdsmen) could be destroyed by overgrazing. Nevertheless, it always paid each herdsman to add one more animal to his herd because he collected *all* the gain accrued by that animal, but shared the concomitant loss with all the other herdsmen. Consequently, each herdsman, by doing what was profitable (that is, *logical* or *rational* in the language of economics), increased his herd until the commons and all who used it were destroyed.

This essay is relevant to us because much of the world is a commons. The air we breathe is a commons, but industries save money by dumping smoke and waste gases into the atmosphere; the gain to individual companies is large, while the losses are shared by everyone. Rivers, lakes, and oceans are commons; they are subject to exploitation and neglect. Industrial wastes can be profitably poured into the Mississippi River at St. Louis, for example, because the cost of removing them is borne downstream by the citizens of such cities as Memphis and New Orleans. Marine fishes and whales are commons that have been exploited almost to extinction. Scientific knowledge, because it is freely exchanged, is also a commons. Some persons (today, unfortunately, even more so than yesterday) argue that the education and training of scientists represent a needless expense. Rather than train scientists, they say, each nation should parasitize the findings of the scientists of other countries. Such is the irrepressible logic that leads to the destruction of the commons.

NOW, IT'S YOUR TURN (1-6)

That the destruction of a commons can occur as the result of carefully made calculations and rational behaviors on the part of every concerned citizen is a sobering thought. Be sure you understand the calculations with respect to cattle grazing on a common meadow. For example, convince yourself that an individual herdsman will profit by adding one more animal to his herd whether (1) he and nine other herdsmen have ten cattle each, and each animal is worth $100 or (2) he and the nine other herdsmen have fifty cattle each, and each animal (rather wretched, half-starved animals by this time) is worth $20. Having made this calculation, try your hand at proving the same effect for air and water polluters, marine fishermen, or technologists attempting to get the most scientific information at the least cost to themselves.

Human Social Ties

The logic of the commons cannot be refuted: property and resources shared by many persons and from which each can gain personal wealth will eventually be destroyed. Nevertheless, the commons of England, the public meadows upon which all local herdsmen maintained their cattle, lasted for 1000 years or more. The wells and other watering places in Africa south of the Sahara Desert were also commons that have been destroyed only during the past three decades. How can a commons last for centuries without being destroyed? By an irrational ethic! By the simple admission that every individual is a member of a larger community, and by often seeking what is best for the *community* rather than for the *individual*. The fatal logic that destroys the commons is the "rational" struggle to increase personal gains while sharing only partially in the total costs of one's actions.

Once more, a difficulty arises. By what right do biologists discuss social interactions holding communities together? Are these problems not best left for philosophers? Or sociologists? Or economists? Perhaps. Nevertheless, the consequences that follow from one's attitudes toward other persons bear upon the continued existence of humanity: in that sense, they fall within the scope of a humanistic biology.

At a tribal meeting held some fifty years ago in Tierra del Fuego, at the southernmost tip of South America, a group of Fuegians decided that the modern world was no longer a proper place into which to bring babies, so they resolved to bear no more children. Years later, a small item in the *New York Times* reported that the youngest surviving woman of the tribe had passed childbearing age (childless, of course)

Figure 1.20. Five Fuegian men and a Fuegian woman (*upper right*) as drawn by Robert Fitzroy, the captain of the *Beagle*—the ship on which Charles Darwin served as naturalist. The Fuegians' homeland, Tierra del Fuego, is an extremely harsh environment at the southernmost tip of South America (see map).

and, therefore, the tribe's fate was sealed. How should we look upon the self-inflicted extinction of a proud people? Is it a calamity? Would it be a calamity if all human beings entered into a similar compact? Once all human beings had disappeared, remember, the world would not care, because *caring* is a human trait.

Answers to these and similar questions must come from within each person. Biology is a natural science; it cannot serve as the basis for an ethic. It can, however, help in describing the consequences of differing attitudes. For example, economists (who are *social* scientists) assume that each individual behaves rationally, that each acts to maximize his personal status. Working on this assumption, calculations can be made that predict the flow of raw materials from mines and fields to factories, the fabrication and sale of manufactured goods, the distribution of prices and wages, and the accumulation of wealth by different segments of society. Of course, the underlying assumption is known to be incorrect. Nearly every professor of economics is himself an exception; rarely does a college professorship maximize an economist's status. Furthermore, we all know of persons who have sacrificed themselves—even their lives—for their families, friends, or strangers. Some of these have been doting parents, others are recognized as personal or national heroes. Technically, these persons have behaved irrationally. We admire and love them nonetheless. As we recall such irrational persons and their irrational acts, we also recall that it is *rational* behavior (as defined by economists) that leads to the destruction of the commons.

Human beings depend upon air, water, food, energy, and shelter for life. Because many of these essentials are commons that are shared by many other persons, we must admit that our lives, the lives of our neighbors, and those of our children and grandchildren depend upon irrational behavior. Our attitudes concerning others—even those who will not dwell on earth for generations to come—determine whether there will be a human biology in the future.

Personal preferences are often based on values that lie outside the scope of science. Nevertheless, the consequences of actions that are based on these values can be subjected to scientific scrutiny. If an examination of these consequences reveals that rational behavior will lead to destruction, many persons will choose to act irrationally. Or they will redefine the term "rational." Human beings live neither alone nor by bread alone.

NOW IT'S YOUR TURN (1-7)

A graph illustrating the growth of the human population is shown in Figure 1.21. Notice that for 9000 years the line is nearly horizontal,

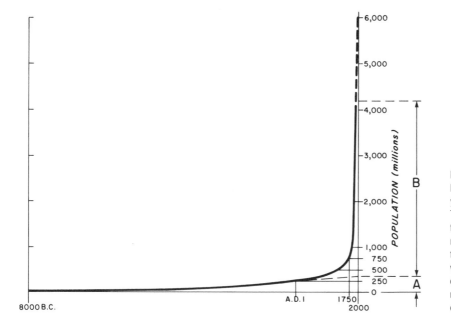

Figure 1.21. A curve showing the growth of the human population from a few millions some 10,000 years ago to more than four billion persons today. The growth has occurred primarily during the centuries following the Industrial Revolution. The segments labeled *A* and *B* were referred to by one author as measures of the contributions made to human welfare by the humanities and the sciences (technology), respectively. You are asked to consider whether numbers of individuals and human welfare are equivalent.

but during the 1600s it turns sharply upward and soon thereafter becomes nearly vertical. The upturn coincides with the Industrial Revolution—a period that marks the wedding of science and technology. The author of an article in the *American Scientist* some years ago divided the vertical scale as shown by the letters *A* and *B: A* represents (according to that author) the well-being of persons that can be accounted for by the humanities, while *B* represents the well-being that can be accounted for by the sciences. Recalling that the vertical axis represents "number of persons," discuss the bearing (if any) *A* and *B* have on the well-being of individuals, or on the quality of life. What assumptions did the author make? Are such assumptions justified in your opinion?

Science and Religion

Values that are not subject to measurement by scientific methods have been mentioned several times in the preceding paragraphs. These values are important because they determine what persons do, how they behave, and in what regard they hold one another. Biologists and other scientists are unable to comment professionally on values, but the consequences of given acts, as we have explained earlier, are another matter.

Many persons, including individuals in high public office, fear the growing influence of science and scientists. They fear that science will cause an erosion of public morality. Morality, in the eyes of these persons, is decreed from above, from a "higher power." Science, in contrast, is amoral. (Note that *amoral* is not the same as *immoral:* amoral means "without morals" whereas immoral means "wicked.") The laws of thermodynamics that deal with energy and its distribution in the universe offer no aid in establishing a code of ethics or guiding moral behavior. Neither do Newton's laws concerning the motion of planets, nor do Mendel's laws describing the transmission of hereditary traits.

However, eminent scientists are among the many persons who retain a belief in a Supreme Being whose laws govern the behavior of civilized persons. The Supreme Being may take different forms, though. The ordained minister or priest who serves Him sees Him in one form. And yet He was just as real to the scientist-philosopher Theodosius Dobzhansky, who said, "If mankind is meaningless, then my personal existence cannot be meaningful." Meaninglessness, to Dobzhansky, was a nightmare.

Other scientists feel differently. Jacques Monod, an outstanding molecular geneticist and Nobel Laureate, argues brilliantly that life arose by *chance.* Having arisen, however, it has undergone evolutionary changes that were dictated by *necessity.* Details of the evolving systems could not have been accurately predicted in advance, but the broad evolutionary pathways, the strategies for survival, and the devices for ensuring life's success on Earth might have been. Indeed, the search for life on other planets is conducted by the rules of human logic!

Monod claims that today human beings are able to discard the religions of the past (myths, in his words) and to erect an ethical system of their own making. The published reviews of *Chance and Necessity,* Monod's book, reveal that this proposal disturbs or even frightens many persons. It would remove responsibilities for decisions from a higher source and place them upon imperfect human beings. Many persons are not prepared to accept this burden. They believe that individuals should be responsible for how they *do* act, but the larger responsibility for how they *should* act, they suggest, should remain in the wisdom of the Supreme Being as revealed in His written word.

Ironically, the phrase "just a value judgment" would be the first casualty under Monod's suggestion. Values, if human beings were responsible for their own ethical system, would have to be measured (as we have suggested several times in the preceding sections) by their logical consequences. Perhaps human beings are not yet prepared to erect their own ethical systems. They willingly act to carry out the dictates of a Higher Being (many persons exhibit a pathetic obedience to "official" orders of all sorts) but they are often reluctant to promote their own values, to impose their values on others. The words "only a value judgment" can be used either to terminate an argument or to escape responsibility.

NOW, IT'S YOUR TURN (1-8)

1. The main difference between knowledge gained "scientifically" and knowledge gained by means of a religious experience is that the first is gained rationally and is subject to subsequent confirmation by others, while the second is acquired by divine revelation and is passed from one person to the other by persuasion, conversion, and faith. Explain, then, why scientific and religious "explanations" for the existence of the universe, of the Earth, or of life on Earth cannot easily be reconciled.

2. Contrast the ease with which the following differences of opinion can be resolved: (1) A Roman Catholic priest, a Lutheran minister, a Jewish rabbi, and an Iranian Ayatollah disagree on a theological point; and (2) a Chinese, a Brazilian, and a German embryologist disagree on the amount and source of variation in the developmental patterns of sea urchin embryos.

3. J. B. S. Haldane, one of the world's outstanding biologists, summarized one difference between science and religion as follows: In science, many persons make tiny (even trivial) observations until one day an enormous pattern is revealed; at that time all scientists pause and stand in awe at what they have accomplished. In religion, a person has a revelatory vision that is overwhelmingly awesome in its beauty and context; following his vision, the person is often moved to go forth and do good works. If you react to this summary, jot down your reactions in your journal.

Science, Scientists, and Society

Different persons look at science differently. Earlier you read that science is an assumed relation between human beings and the world about them that permits people to pose questions and through experimentation to seek answers. The scientist, the prober, looks upon science as a search for knowledge and upon himself (or herself) as a seeker of truth. Knowledge, most scientists will tell you, is value-free. It is neither good nor bad in itself—only the uses to which knowledge is put can be judged in this way.

Other persons, such as politicians and technologists (who include many scientists), regard science as a means by which knowledge is gained *for a purpose*. The purpose may be expressed as: bettering the lot of humanity, conquering hunger, making the world safe for democracy (or Communism), making the deserts bloom, curing cancer within our lifetime, or, as President Kennedy said in 1961, placing a man on the moon in this decade. No matter how the purpose is expressed, to most politicians and technologists science is a practical exercise that is undertaken for a practical purpose.

Figure 1.22. An aerial view (*right*) of the Fermi particle accelerator located near Chicago, Illinois. The multi-storied administrative building at the left of the accelerator, as well as the clusters of houses in the near distance, provide standards by which the size of this research facility can be judged. The interior of the circular tunnel (*left*) is lined by large electrical magnets, which provide the energy needed to accelerate subatomic particles to speeds approaching those of light. Research facilities of this size can be afforded only by large nations or by multinational associations. *(Photo courtesy of the Fermi National Accelerator Laboratory, Batavia, Illinois)*

Modern science is expensive. Modern physics requires subatomic particle accelerators so enormous that only the largest nations (or groups of nations) can afford them. Modern research in chemistry and biology costs a great deal of money. Today's astronomers are deeply involved with space probes that involve sending satellite laboratories to other planets. Radio-astronomy requires the construction of gigantic radio antennae to intercept radiation from galaxies far beyond our own.

Because research funds are provided by governmental agencies seeking practical results ("A dollar returned for each dollar spent"), little of science is entirely value-free. Even the most dedicated and least "practical" scientist operates within his or her culture. The problems a scientist chooses to solve are influenced by the prevailing culture. So are those private reflections that make these particular problems seem important. The salary that provides the scientist with food and shelter is forthcoming because the research done in return reflects the attitudes and aspirations of the larger society.

Albert Speer, who served as Adolf Hitler's Minister of Finance, has described the role of scientists in Nazi Germany during the 1930s and 1940s. His government actually encouraged scientists to regard their work as value-free. In this way the scientists were persuaded that they were not responsible for the atrocities that were committed throughout Europe on Hitler's orders. As an official of the German government of that era, Speer knew that the idea of truly value-free research is nonsensical. On the other hand, to gain needed knowledge he encouraged the German scientists in their self-delusion.

The example cited from recent German history is unique only with respect to the depravity of the uses to which the scientific knowledge

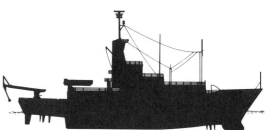

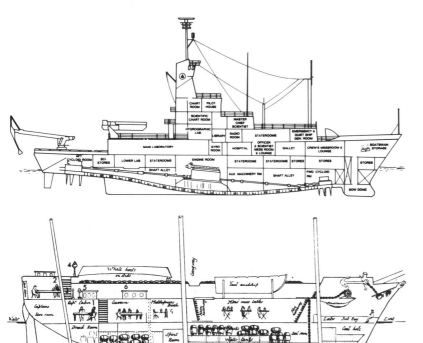

Figure 1.23. A comparison of the new 245-foot ocean-going research vessel *Knorr* and the 90-foot *Beagle,* on which Charles Darwin served as naturalist for five years. The *Knorr* carries a crew of 25 and 25 scientists; the *Beagle* carried 74 persons. The years spent aboard the *Beagle* observing plants, animals, and rocks throughout the world led Darwin to his theory of evolution through natural selection, a theory whose effects on human thought have not yet run their course. The *Knorr*'s crew discovered and photographed the sunken *Titanic,* lying two and one-half miles down in the North Atlantic. The silhouettes at the right show the comparative sizes of the two ships.

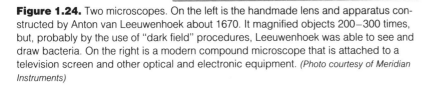

Figure 1.24. Two microscopes. On the left is the handmade lens and apparatus constructed by Anton van Leeuwenhoek about 1670. It magnified objects 200–300 times, but, probably by the use of "dark field" procedures, Leeuwenhoek was able to see and draw bacteria. On the right is a modern compound microscope that is attached to a television screen and other optical and electronic equipment. *(Photo courtesy of Meridian Instruments)*

We Wash Our Hands . . .

To carry out their planned extermination of the Jews (and of others who were considered unworthy of living), officials of Nazi Germany needed appropriately equipped concentration camps. The following excerpts are from letters written by German businessmen who competed for the opportunity to build and equip crematoria at Auschwitz, Dachau, and other concentration camps.

"To the Central Construction Office of the S.S. and Police, Auschwitz:

"Subject: Crematoria 2 and 3 for the camp.

"We acknowledge receipt of your order for five triple furnaces, including two electric elevators for raising the corpses and one emergency elevator. A practical installation for stoking coal was also ordered and one for transporting ashes."

"For putting the bodies into the furnace, we suggest simply a metal fork moving on cylinders.

"Each furnace will have an oven measuring only 25 by 18 inches, as coffins will not be used. For transporting the corpses from the storage points to the furnaces we suggest using light carts on wheels, and we enclose diagrams of these drawn to scale."

"Following our verbal discussion regarding the delivery of equipment of simple construction for the burning of bodies, we are submitting plans for our perfected cremation ovens which operate with coal and which have hitherto given full satisfaction.

"We suggest two crematoria furnaces for the building planned, but we advise you to make further inquiries to make sure that two ovens will be sufficient for your requirements.

"We guarantee the effectiveness of the cremation ovens as well as their durability, the use of the best material and our faultless workmanship.

"Awaiting your further word, we will be at your service.

Heil Hitler!
C. H. Kori, G.m.b.H."

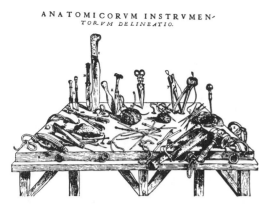

Figure 1.25. The surgical instruments used by Andreas Vesalius in performing the postmortem dissections that were the basis of his atlas of human anatomy, *De humani corporis fabrica* (1543). His dissections and the artistic renditions of them by Jan von Calcar are so fine that several of them appear as illustrations in subsequent chapters of this book.

Science and Values: Another Point of View

On at least two occasions we have claimed that science can provide information on how big, how hot, or how frequent certain things or events are, but that it has nothing to say about value judgments or about ethical values, generally. Many persons, including Dr. Jacob Bronowski (author of *The Ascent of Man*), would disagree. In his book *Magic, Science, and Civilization*, Bronowski refers to the possession of foresight and hindsight by human beings; in his terminology, each human being possesses a *plan*. One can ask of one's plan, "Did things take place like this, or did they not?" This question establishes a criterion of objective truth, "of making cognitive statements about which you could say they are true or they are false in a purely objective way." Later, he continues: "These open plans (i.e., rules of conduct or *values*) depend on our recognizing that other people and we are one—that we understand them through ourselves and ourselves the better through them. And for that reason it is basic to any ethical strategy . . . for carrying on one's life. . . . You cannot in fact carry on a scientific activity without just that sense of truth and dignity."

One can question whether any person can form an objective judgment about whether events happened according to plan. Reliance on one's senses can lead to the conclusion once expressed by a Yale University theologian: "May I venture to say that I radically deny the real existence even of all that can be seen by microscopical enlarging." One's senses can distort reality, or ignore it entirely.

was put under Hitler, in that it was used in the wholesale extermination of human beings—Jews, prisoners of war, and enemy civilians. Today, through national research projects, governments encourage researchers whose findings are needed for larger projects of which the individual scientist is often unaware. Thus, a plant physiologist who dreams of improving agricultural techniques finds to his dismay that, through federally funded research, he has contributed instead to the defoliation and destruction of another country's forests. A specialist studying the effect of whole-body irradiation on terminal cancer patients learns only later that his results were used primarily to estimate the usefulness of infantrymen who might be exposed to sublethal amounts of radiation from tactical nuclear weapons.

Scientists are neither lawmakers nor law-enforcers. They cannot compel others to do as they themselves might wish. Indeed, they are amazingly varied in their own views. For the most part, however, they are thoughtful persons. Many are educators. They all share a special responsibility for evaluating the uses to which scientific findings might be put. Each one must ask questions and insist on truthful answers. Each one must help the public to know the truth—the *whole* truth. Otherwise, new knowledge becomes just another commodity to be sold to the highest bidder; in that case, the claim that the store of human knowledge can never be too large would indeed be difficult to defend.

NOW, IT'S YOUR TURN (1-9)

1. More and more, the basis for Federal support of science is couched in the question: "What have you done for me today?" Scientists are told that for each dollar they receive in research support, they must provide a dollar's benefit to society. How do you view such "practical" constraints to scientific research? [You may be interested to know that Federal funds were denied to Dr. Milislav Demerec, who proposed during the early 1940s to search for a strain of *Penicillium* capable of producing much more penicillin than the then-known strains. Using private funds, Dr. Demerec carried out his proposed research successfully and discovered the high-yielding strain of *Penicillium* whose penicillin not only saved many of the wounded during World War II (and the later Korean and Vietnam wars) but is still the strain used in most penicillin production today.]

2. During the 1950s, unorthodox views on the genetics of populations were frequently cited as if they demonstrated that exposure to X-rays and atomic radiation was harmless. Today, Dr. Stephen Gould, who would like to view evolution in a somewhat novel light, is embraced—much to his chagrin—by creationists who cite his views as

anti-evolutionist. In your opinion, what should a scientist do when he finds his views being misquoted and misused by others? To what extent has your answer depended upon an example where the scientist and you happened to agree (or disagree)?

Summary

This introductory chapter has two main objectives: (1) to introduce science, one of mankind's truly honorable enterprises, to the student, and to relate this enterprise to other human endeavors and activities, such as value judgments, ethics, and religion; and (2) to emphasize to both the student and the instructor that education involves more than the acquisition of a specialized vocabulary, the memorization of today's trivia, and the adulation of grades as if they represent ends in themselves. An educated person internalizes knowledge and adjusts his or her behavior in a manner consistent with the possession of this knowledge. To become educated in this sense, the student cannot be a mere classroom spectator. The student's journal is a means of making biology an audience-participation experience.

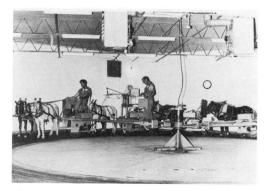

Figure 1.26. Studies such as the one shown here involving Shetland ponies have revealed that near-lethal radiation exposures have little effect on immediate work performance. The lead team sets the pace for all five teams; the animals are monitored for physiological stress over a period of three months. One member of each team is the unirradiated control; the other has received a near-lethal exposure. It appears that information of the same sort was gathered in the past by exposing terminal cancer patients to whole-body irradiation. *(Illustration from University of Tennessee–AEC Agricultural Research Laboratory)*

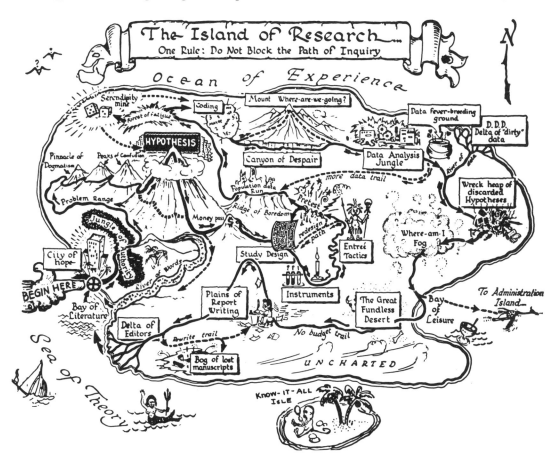

Figure 1.27. An example of the fun scientists poke at themselves and their colleagues. Science is an honorable enterprise as the text claims, but, for the serious practitioner, it is also an intellectually taxing one. Cartoons like the one reproduced here help both the authors of research papers and their readers maintain their perspective and their humor. *(Courtesy of E. Harbury and* The American Scientist*)*

Congress and the Financial Support of Scientific Research

Many of the complex interrelations between politics and science in a government-financed technological society are illustrated by the following remarks, which have been extracted from the *Congressional Record*, August 1, 1975. The names of both the speaker and other Congressmen have been deleted in order that personalities not distract from the facts and issues.

". . . the National Science Foundation actually receives some 25 to 28,000 research proposals per year, and last year the proposals were reviewed by about 34,000 reviewers, experienced scientists who are competent in the field of the proposals. They generated close to 120,000 reviews. Each proposal is also subjected to staff review, and, in the case of large sums of money, to review by the National Science Board. I think that we in Congress have better things to do than second guessing the NSF in areas where few of us have any expertise to begin with. We certainly could not go into detail on every proposal, and summaries and titles rarely provide an adequate basis for informed evaluation.

"This is not to say that we should abdicate our oversight responsibility. We must do all we can to insure that NSF acts wisely in allocating our money to scientific research. And the Science and Technology Committee, of which I am a member, is now conducting hearings into the process by which funding for NSF proposals is determined. I think that is where we should be concentrating our efforts: In making sure that the system works, not in trying to play scientist ourselves. . . .

"NSF each year undergoes intensive review of its policies, procedures, and specific program actions in hearings by the four congressional committees that have jurisdiction over it. It submits quarterly reports to the Congress; it supplies us with a daily list of every grant awarded and an annual report that lists every grant it has made, to whom made, for what purpose, and the amount. Does the Defense Department do the same?

"Mr. ***** cited a number of so-called crazy grants that NSF has made when he proposed his amendment to the NSF bill on April 9. I am particularly struck by his ridicule of a National Science Foundation grant to study the sex life of crabs. I think this is a beautiful illustration of what happens when we Congressmen take it into our hands to play scientist. Surely the Representative from Maryland's Eastern Shore must care about the future of the crabbing industry. Is he so focused on the prurient aspects of sex that he has forgotten that this is the biological means by which crabs create more crabs? Perhaps he would actually be quite upset if some Congressman objected to this grant and managed to prevent its being funded—especially if he knew that it concerned the consequences of minute amounts of oil pollution on the complex reproductive behavior of crabs. It has been found that the production of pheromones that govern this behavior is prevented when crabs are exposed to oil-soluble extracts of crude petroleum oil to a concentration of one part per hundred million. Under these minute amounts of pollution, reproductive behavior ceases. I wonder how many million dollars a year the crab industry is worth in Maryland? And I wonder what those 'ordinary people' who are employed in that industry, Mr. *****'s constituents, would say if they were told of the significance of that research?

"If we in Congress were to vote on funding for proposed research with titles like: 'Insect Chemoreception and Mechanism of Action of Attractants and Repellents,' and 'Pheromone Chemistry and Specificity in the Tortridae,' or 'Influence of Red Pine Hybridization on Olfactory Responses of Weevils,' or even 'Behavior Pattern of Solitary Hymenoptera,' we might decide these proposals sounded rather frivolous. Yet the research conducted under these research projects has led to the practical use of new, environmentally benign insect control techniques—which eliminate the need for DDT and other environmentally harmful pesticides.

"Or, if faced with a vote on esoteric proposals entitled 'UV Reflectance of Ice Clouds and Fronts Composed of Water and Ammonia,' and 'Mesospheric Trace Constituents with Rocky Probes,' and 'Low Level Vertical Ozone Distribution Determined by Inversion of Backscattered Ultraviolet Radiation,' we might find it difficult to believe in their importance to the world. But this research provided the groundwork for the discovery of the breakdown of the ozone layer, which we are all now so concerned about.

"Programs for teaching chimpanzees to use sign language have led to better understanding of ways to communicate with autistic and retarded children, and adults suffering from severe strokes. A grant entitled 'Legged Locomotion in Animal and Machines' has allowed the development of new artificial limbs and braces, as well as other new paraplegic devices. And studies of the 'Sex Life of the Screwworm' saved the U.S. cattle industry millions of dollars by eliminating a parasite that deposited eggs in the placenta of newborn calves.

"If these are the kind of 'crazy grants' for 'frivolous research' that have Mr. ***** so worried, I can only say that I hope we fund many more of them.

"Mr. Speaker, I urge the defeat of Mr. *****'s motion."

II
PERSONAL BIOLOGY

This part and Parts III and IV deal with biology at three levels of organization: the individual (or family members and friends who are to be treated as individuals), the community as an aggregate of individuals (many of whom are friends or acquaintances), and society, the political unit in which institutionalized behaviors and actions prevail. Throughout, biology is presented largely with reference to the human organism. A human-oriented presentation no longer requires an apology, because more and more contributions to modern biology are being made by investigators working with human cell cultures (and bacteria carrying fragments of the human genome), medical scientists studying the whole person, and geneticists analyzing human populations. Although some biological facts may be treated in a nontraditional manner, the important aspects of cellular and organismic biology are thoroughly covered in the following chapters.

Two reasons can be given for dwelling first on the individual when our eventual goal is to encompass societal problems. The first reason is that concern, like charity, begins at home. Human beings are self-centered, self-serving, and greedy, but these traits are tolerable—as long as the interactions and confrontations between and among persons are not irreversibly destructive. The five chapters of Part II can be used in a gratifyingly self-serving way by everyone. They provide a basis for attaining and maintaining good health. The ancient Greeks said "know thyself" and recommended "all things in moderation." The chapters on personal biology provide meat for those ancient sayings.

Personal health is not a trivial matter, nor is it likely to remain "personal" for long. The cost of health care has skyrocketed during the past twenty years. Those who are ill obtain medical care with increasing difficulty. Many persons now subscribe to the cynical view that the American patient is "doing better but feeling worse." The fault, however, does not lie entirely with the medical profession; it lies at least in part with the individual's abuse of his own health. Because of Medicare and other forms of health insurance, the cost of this abuse "is now a national, and not an individual, responsibility," according to John Knowles, former President of the Rockefeller Foundation. Dr. Knowles continued: "The individual has the power—indeed, the moral responsibility—to maintain his own health by the observance of simple prudent rules of behavior relating to sleep, exercise, diet and weight, alcohol, and smoking. In addition, he should avoid where possible the long-term use of drugs. He should be aware of the dangers of stress and the need for precautionary measures during periods of sudden change, such as bereavement, divorce, or new employment." He concluded his account by saying that the choice is individual responsibility or social failure. (Ironically, Dr. Knowles died shortly thereafter of cancer, a fate over which he had no control. No person can be held absolutely to blame for the physical ailments that impair or destroy his health.)

The second reason for dealing with the individual first is to set the stage for a later discussion of population size and resource utilization. Large populations require (*demand* may be more truthful) large quantities of food and water. Providing large numbers of persons with adequate food, water, and shelter requires large amounts of energy. The amounts of energy, food, water, and other necessities can be estimated accurately by simple multiplication:

the number of persons × the needs of each person.

If a person knows his or her requirements, that person automatically knows the basic physiological requirements of any group, whatever its size: 20,000, 1,000,000, or 5,000,000,000. Whether the environment can meet these calculated needs is another matter. The dependence of human beings on energy and on physical resources as individuals, as communities, and as societies is emphasized throughout the text because these needs provide a thread that joins the three levels of organization.

Finally, a comment must be made on the "wisdom of the body." Persons who regard the brain (that is, the cortex, or thinking part of the brain) as the body's seat of wisdom are wrong. Dr. Lewis Thomas, President of the Memorial Sloan-Kettering Cancer Center in New York City, explains the brain's limited wisdom in his book *The Lives of a Cell*. He writes: "If I were informed tomorrow that I was in direct communication with my liver, and could now take over, I would be deeply depressed. . . . Nothing would save me and my liver, if I were in charge. For I am, to face the facts squarely, considerably less intelligent than my liver. I am, moreover, constitutionally unable to make hepatic decisions, and I prefer not to be obliged to, ever. I would not be able to think of the first thing to do."

The counterpart to Dr. Thomas's genius liver appears in Chapter 5: the healing of a severe injury by the coordinated effort of the entire body. The wisdom of the body emphasizes the remarkable extent to which logical solutions to recurrent problems, even complex ones, can be built into and stored in genetic material (DNA) under the influence of natural selection. The accumulation of such genetic "knowledge" is, of course, the essence of evolution.

Although on one level the unique chemical structure of DNA and the uses to which this structure can be put may be regarded as evidence for the existence of a Supreme Being, a Master Mechanic of limitless technical ability, such a conclusion is unnecessary. The body's wisdom requires no phenomena other than those that will be encountered subsequently in this text: (1) random (or haphazard) changes in the structure (and associated function) of DNA, (2) the amplification of certain alterations by successful reproduction, and (3) the elimination of others by decreased survival (early death) and impaired reproduction. Genetic changes that provide adequate solutions (not necessarily perfect ones) to life's problems are successful; they are retained and multiplied by surviving and reproducing individuals.

Thus, natural selection has led to two systems, each capable of providing solutions to problems: (1) the DNA that runs Dr. Thomas's liver and leads to the healing of wounds and (2) the brain (whose physical structure is also based on information carried by DNA), which recognizes classes of similar problems and is capable of instantaneously solving almost any problem once it has been properly classified, as well as enabling Dr. Thomas to write engaging essays about his liver.

NOW, IT'S YOUR TURN (II-1)

Convince yourself that your body is in many ways more capable than your brain—in problem solving, for example. Fill a glass with water to within about one-half inch of its rim. Drop a pencil or some other small item on the floor near your feet. While standing with the glass of water held comfortably in front of you, look at the object on the floor, and stoop to pick it up—without spilling the water. Difficult? Probably not. The whole sequence of events could have occurred at nearly any party where one drops a fork, a napkin, or even a potato chip. Now, however, describe all the mathematical problems that confronted the body—problems that the body solves accurately and virtually instantaneously.

2

A PERSPECTIVE ON HUMAN LIFE

The human body is an amazing machine. It frequently operates trouble-free for years. Unlike most machines, it repairs itself. Like many machines, it can malfunction. Some malfunctions are caused by disease, others by external circumstances. Still others, hereditary disorders, are built into the original design. Of the latter, some bring the individual to an abrupt end during early development, while others allow the individual to survive early development, only to decline because of ever-failing health over five, ten, twenty, or even more years. Some malfunctions are characteristic of youth; others, such as heart attacks, strokes, and cancers, appear in later years. Finally, aging itself is debilitating; all machines—even our bodies—eventually wear out.

This chapter and the four that follow present two views of the organization and operation of the human body in health and in disease. In the present chapter, life is viewed dynamically over time—in growth, development, and aging. Subsequent chapters treat life as if it were static, held constant for at least a moment while we study the form and function, as well as the needs and problems, of the body. This division, of course, is arbitrary. All living things—human beings included—are open or dynamic systems into which food, oxygen, and water are poured, and out of which come heat, ashes (waste products), and (again) water. An individual organism is *never* static; it changes constantly, even during the shortest imaginable instant of time.

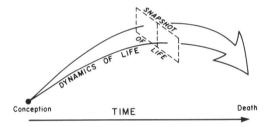

Figure 2.1. Two views of human life. The first view (*broad arrow*) encompasses the entire sweep of life from conception to death. The other view (*double rectangle*) is restricted to a single moment in midlife, almost like a snapshot—an unusual one revealing rapid physiological processes within the individual as well as the individual's anatomical details.

Have you ever had a serious illness? Since the age of ten, for example? If so, describe it while drawing upon your past biological training (i.e., high school) as much as possible. For example, if your illness was an infection, identify the infectious organism, the site of the infection, the danger it posed to you, and the means (antibiotics? natural resistance?) by which you recovered.

The Human Being from Birth to Death

The course of life resembles a trajectory extending from conception until death. Unlike a cannonball's trajectory, however, life's trajectory is not entirely predetermined by initial conditions (such as, for the cannonball, muzzle velocity, angle of elevation, and direction). Rather, the life of an individual more closely resembles the flight of a modern rocket: its trajectory can be changed in flight. The body has internal controls for growth and change, and methods for performing self-repair. Medicine provides additional corrective services throughout one's lifetime. Many diseases and even built-in genetic errors are corrected by pills, injections, dietary regulation, or surgery. Without this ability to correct and repair, the individual would not necessarily follow a "normal" course of development. For the moment, let's not get too involved with the meaning of "normal"; it is an elusive (and, in the context of politics or society, a potentially dangerous) word. Without internal and external corrective mechanisms, however, the individual might easily come to an early (but otherwise avoidable) death, or might suffer unnecessary pain and serious physical debilitation.

Each person begins life as a fertilized egg, passes through a series of prenatal developmental stages, is born, grows through childhood, adolescence, adulthood, and enters a period of middle and advanced age during which many bodily functions gradually fail. Finally, the person dies.

Each of us has a biased view of the trajectories being followed by those around us. First, we do not sense our own development, at least on a day-by-day basis. Age creeps up on us while we regard ourselves as fixed. A forty-year-old man attending a high-school reunion is shocked to see his aged classmates, since he still sees *himself* as he was when he was young. Each of us, then, sees his or her own life as more or less static.

Changes come in jumps: aging is not a uniform process. More importantly, each person tends to think of him- or herself as "normal." From this point of view, to a college student, babies and old people are different—that is, not quite normal. Persons of the opposite sex

Figure 2.2. A real-life, rural counterpart to the diagrammatic trajectory of life. Items of interest that can be seen in or inferred from this photograph include:

• The changing physical appearance of persons as they age.

• The pooling of many minds in arriving at a solution to common problems. The proximate cause of this gathering is the making of apple butter by old-fashioned methods; the conversations held here will touch on many other problems of local interest.

• Any consensus concerning local problems arrived at by the persons attending this function will not be static; the change in this consensus through time is an important facet of cultural evolution. *(Photo courtesy of Wayne Speer)*

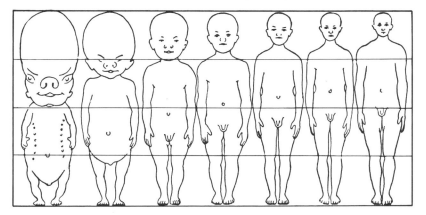

Figure 2.3. Changes in the proportions of the human body from fetus to adult. The head accounts for one-half the length of an early fetus, but it accounts for only one-eighth the height of an adult. The proportion of the legs is reversed: they make up only one-eighth of the length of a fetus, but one-half the height of an adult.

are often regarded as strange and different. They are different, of course, but they are not abnormal or inferior on that account. The entire spectrum of developmental states that is followed by individuals of either sex is normal.

Pre-conception Events

Germ Cells Events crucial for an individual's development occur even before the individual exists, before the union of the egg and sperm (germ cells) that starts the individual on his personal trajectory. The reproductive cells—sperm and egg—provided by the male and female parents differ markedly in structure (most obviously in size) but are remarkably similar in their genetic content—that is, in the contributions that they make to the offspring.

The Identification of the Genetic Material

The paragraph quoted below appeared in *An Atlas of the Fertilization and Karyokinesis of the Ovum,* by E. B. Wilson and Edward Leaming, which was published in 1895, five years before the "rediscovery" of Mendel's studies on inheritance in peas.

For a generation of students who have been exposed both in school and through newspapers and television programs to Mendelian genetics and to the advances made by molecular biologists during the past two decades, the quotation may come as a surprise. We should remember that the scientists of yesterday were not fools; on the contrary, they simply lacked information that has been accumulated since their time.

"These facts justify the conclusion that the nuclei of the two germ-cells are in a morphological sense precisely equivalent, and that they lend strong support to Hertwig's identification of the nucleus as the bearer of hereditary qualities. The precise equivalence of the chromosomes contributed by the two sexes is a physical correlative of the fact that the two sexes play, on the whole, equal parts in hereditary transmission, and it seems to show that the chromosomal substance, the *chromatin,* is to be regarded as the physical basis of inheritance. Now, chromatin is known to be closely similar to, if not identical with, a substance known as *nuclein* ($C_{29}H_{49}N_9P_3O_{22}$, according to Miescher), which analysis shows to be a tolerably definite chemical composed of nucleic acid (a complex organic acid rich in phosphorus) and albumin. And thus we reach the remarkable conclusion that inheritance may, perhaps, be effected by the physical transmission of a particular chemical compound from parent to offspring."

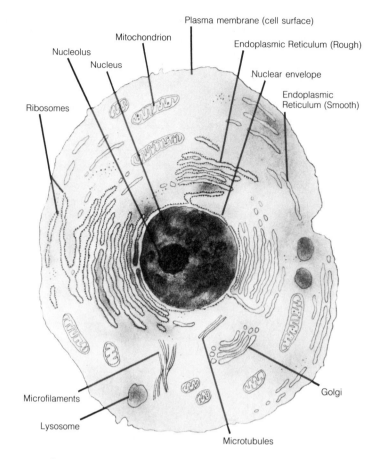

Figure 2.4. Generalized diagram of an animal cell. The organelles with which you may not be familiar and whose functions may not be obvious from the diagram are: ribosomes, the sites of protein synthesis; Golgi apparatus, for packaging proteins; lysosomes, small sacks of digestive enzymes; mitochondria, where high-energy adenosine triphosphate (ATP) is manufactured; microtubules, which serve as fibers and assist in transporting substances within the cell; and microfilaments, muscle-like filaments that "shape" the cell.

Germ Cell Structure The human egg is a large cell, visible even to the naked eye. As it matures in the ovary, it accumulates a storehouse of materials, some of which are transported into the egg from the outside, others of which are manufactured within the egg itself. The chemicals that provide the energy for early development are found within the mature but unfertilized egg. Similarly, many of the blueprints and much of the machinery for making new chemical substances are located in the egg, waiting to be activated at the moment of fertilization.

The role of the egg in initiating early development can be demonstrated in some animals simply by pricking the egg with a fine pin: stimulated by this artificial signal (normally, penetration by the sperm head would signal fertilization), the egg may divide and redivide until it has formed 1000 or more cells. In contrast to the size and complexity of the egg, the sperm is a small cell, containing scarcely any material except that which contains genetic information. Other than that, it possesses only the small store of chemicals that provide the energy for its mobility. The egg and the sperm meet only because the latter is mobile; the egg is essentially sessile.

The Germ Cell's Genetic Information Although different in size, the egg and sperm are similar in the amounts of genetic information they hand down from parent to child. In fact, the transmission of genetic material (deoxyribonucleic acid, or DNA as it is called for convenience) from one generation to the next among higher organisms occurs exclusively by way of the egg nucleus and the sperm cell (which is virtually a naked nucleus). We speak of DNA as being the genetic material as if this were common knowledge, and—if daily mention in newspapers and on television form a valid basis for judgment—indeed it is. This knowledge has not been common for long, however. DNA has been accepted as being the sole repository of genetic information among living organisms from bacteria to man only since the 1950s. The trail leading to its identification as *the* genetic material provides the main theme of genetics, one of the many branches of biology.

Genetic material (or, as we shall often say, DNA) has a function unlike that of any other substance found within living cells. Some of these other substances provide energy, some enable energy to be released or utilized properly, and some are structural molecules that account for the physical shape of the cell. DNA, however, is the repository for the information (i.e., the plans) according to which the cell is built, the information that provides for the replenishment of spent enzymes and other proteins in the individual cells, and the information that directs the orderly development of organs and tissues.

Molecules of DNA are long, threadlike fibers. In higher organisms (eucaryotes), they are packaged into chromosomes. Each chromosome is thought to contain a single, tremendously long DNA molecule. Some persons have even suggested that the DNA threads in different chromosomes might join at times so that the entire DNA complement of any cell might then be represented by a thread of truly gigantic length. The total length of the DNA in a human cell is about 170 cm (some six feet). By molecular standards, a single molecule of this length (or of even half this length, as would be found in an egg or sperm nucleus) is tremendous.

The quantity of DNA, unlike that of other cellular substances, is nearly constant from body cell to body cell, whatever the tissue, within any one species. Human beings, for example, possess some 3.25 billionths of a microgram of DNA per cell, while a mouse possesses 3.00 billionths of a microgram; a frog possesses nearly twice as

Figure 2.5. A diagram of the DNA molecule contained in the "head" of a bacteriophage. The molecule was released by an osmotic shock (exposure of the phage to distilled water). Its total length is about one-twentieth of a millimeter. Note how its length compares to the size of the phage that carried it (shown in the center). Also note that there are only two free ends—the DNA is a long, continuous filament. In contrast, the circular chromosome of a bacterium like *E. coli* is a single DNA molecule about 1 mm in length. Within each human cell, with its two sets of chromosomes, 23 chromosomes per set, are about two meters of DNA.

much, and the fruit fly only one-fortieth as much. Eggs and sperm of these organisms, however, possess almost exactly one-half as much DNA as the usual body cell. Furthermore, just before a body cell divides, the amount of DNA it possesses doubles; thus, when cell division occurs, each daughter cell comes to possess the normal amount of DNA.

The development of each individual is controlled by the DNA it receives in equal amounts from its male and female parents. Although early development can start and proceed without the presence of a sperm nucleus (even through 10–12 synchronous cell divisions that produce some 1000–4000 cells), a continuation of normal development almost always depends upon the possession of both maternal and paternal DNA.

Why is DNA from both parents needed? In part, it is merely a matter of quantity. There are moments during development when the quantity of DNA can limit the rate at which cellular reactions proceed. An ample supply of DNA at these times ensures (as well as the business of living can be ensured) that these reactions will proceed at the right pace. In addition, however, the developing embryo is less likely to encounter insurmountable errors if it develops according to directions received from two biologically independent sources. Mistakes do occur during the passage of DNA from one generation to the next, but an individual who receives two sets of instructions is not likely to inherit the *same* error from each of two parents, although the genetic information passed on by both parents may be incorrect in certain respects. Consider this textbook, for example: Would you have a better grasp of biology if you simply read it twice or if you read this book and a separate text one time each? Reading two textbooks could alert you to areas of disagreement and send you on a search for truth, a search that would not be undertaken if you were to simply read either textbook twice.

Putting Genetic Information to Use

Living organisms are unique in their molecular composition. Among their molecules, the most varied and important in cellular processes are the *proteins*. Proteins are large, complex molecules consisting of carbon, oxygen, hydrogen, and small amounts of sulfur. Protein molecules are so complicated that early biologists regarded the protein-containing fluid found within cells, *protoplasm*, as being identical with life. A textbook from the 1930s, for example, says: "Each cell consists of a mass of living substance, protoplasm, separated from the external milieu by a cell membrane."

During the 1950s, methods were developed that made the study of proteins not only possible, but also rather simple. Previously, proteins were known to be linear sequences of many (100 or more) amino acids. New investigative techniques have revealed that for a given protein the arrangement of these amino acids in every molecule is, with rare exceptions, identical. Once it has been made, a chain of amino acids can fold and twist upon itself (not for mystical reasons, but for thermodynamically economical ones) to form a complex three-dimensional structure. Furthermore, this folded chain can be "locked" into a stable three-dimensional figure by crosslinking certain amino acids; this is the function of the sulfur atoms. The sulfur atoms play the same role in maintaining the three-dimensional shape of the protein as do the half-dozen or so spot welds that maintain the shape of a chair that has been fashioned from a slender steel rod. The basic pattern of the protein molecule is its linear amino acid sequence, which determines what foldings and twistings are possible. That sequence determines whether the sulfur-containing amino acids will be at the proper sites for crosslinking; it is the eventual three-dimensional pattern of the molecule, of course, that determines whether the protein can carry out its proper function: transporting oxygen, synthesizing or degrading organic molecules (including other proteins), or protecting the body against bacteria or other foreign substances.

The building of a protein molecule so that each of 100 or more sites is occupied by the proper amino acid poses serious technical problems. There are twenty different amino acids. If the protein contains 150 amino acids in linear sequence, how can thousands and thousands of these protein molecules be built so that each of the 150 sites is occupied by the correct one of the twenty possible amino acids? The answer lies in the structure of DNA!

J. D. Watson on Biochemistry

"The growth and division of cells are based upon the same laws of chemistry that control the behavior of molecules outside of cells. Cells contain no atoms unique to the living state; they can synthesize no molecules which the chemist, with inspired, hard work, cannot some day make. Thus there is no special chemistry of living cells. A biochemist is not someone who studies unique types of chemical laws, but a chemist interested in learning about the behavior of molecules found within cells (biological molecules). . . .

"Until a few years ago, the chemists' understanding of the cell's large molecules, the proteins and nucleic acids, was much less firm than it is now. . . . Now practically all the important features of the protein myoglobin and the primary genetic material DNA are known. In both cases the chemical laws applicable to small molecules also apply."

In *Molecular Biology of the Gene* (1965)

NOW, IT'S YOUR TURN (2-2)

How many different sequences of amino acids do you think would be possible if any one of twenty amino acids could be inserted into any one of 100 sites? Make a guess, and jot it down in your journal. Next, make the calculation. For the first site, there are twenty possibilities. There are also twenty possibilities for the second site. For each of the first twenty, there are twenty additional possibilities for the second amino acid; therefore, there are $20 \times 20 = 400$ (or 20^2) possible combinations of the two amino acids. Each of these 400 combinations has twenty possibilities for the third site; this equals 20^3 possibilities. Clearly, the total number of possibilities for any number of sites equals 20 raised to that number; for 100 sites, this equals 20^{100}, or about 1.25×10^{130}—a number containing 130 zeros.

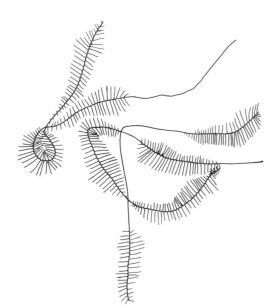

Figure 2.6. Fragments of a DNA filament isolated from an amphibian egg, and from which shorter strands of RNA project. The RNA is in the process of being copied (transcribed) from one of the two strands (not shown) of DNA. Short strands have just begun transcription, while longer ones have been transcribed over considerable distances; hence the featherlike appearances.

The DNA thread is a huge molecule. Variation in the pattern of DNA structure as well as variation in the environments within which individuals develop are jointly responsible for the uniqueness of individuals. Protein molecules are built according to information carried by the DNA molecule. Among these proteins are enzymes that build other cellular molecules—fats, carbohydrates, proteins, RNA, and even more DNA. The structure and physiology of an individual are ultimately dependent on its DNA.

Protein Synthesis Only a sketch of protein synthesis is provided here, since more details of DNA replication and the control of protein synthesis by DNA are found in Informational Essays.

Certain segments of the DNA molecule are responsible for specifying the arrangement of amino acids in protein molecules, and each type of molecule is specified by a particular segment of DNA. (These segments correspond to the genes discovered by Gregor Mendel and other early geneticists.) Protein molecules are not built directly from DNA. Instead, the DNA segments (genes) serve as patterns for building threadlike molecules of another nucleic acid (ribonucleic acid, or RNA). In higher (eucaryotic) organisms, whose cells posess nuclei, the RNA carrying genetic information for use in protein synthesis moves from the cell's nucleus to the cytoplasm (this RNA is known as *messenger* RNA, mRNA). The sequence of bases contained within the mRNA molecule directs the protein-making machinery of the cell so that the proper amino acid is inserted at each of the many sites of the newly synthesized protein. That the ability to direct the machinery (that is, the information about protein structure) resides in the mRNA was proven in the following way: mRNA from rabbit blood-forming cells was inserted both into bacteria and into frogs' eggs; both the bacteria and the frogs' eggs then made detectable amounts of rabbit hemoglobin—a protein that neither normally makes. This was the first experiment in a research area now known as "genetic engineering."

Genetic information, then, consists largely of instructions for making each of the thousands of protein molecules that are necessary for life. In addition, however, DNA molecules contain information about *when* protein molecules should be made. Molecular biologists are just beginning to understand the signals that lead to protein production. Within bacterial cells, for example, the enzymes (which are proteins) required for the utilization of lactose (milk sugar) are not made unless lactose is present in the culture medium. Even then, however, these enzymes are not made if the simple sugar glucose is also present. In the latter case, one can easily show that the bacteria in the culture reproduce rapidly until the glucose is gone, pause momentarily, retool their enzymatic machinery, and then once again multiply rapidly. The two-stage growth of bacteria in the two-sugar medium was observed in the early 1940s; the details of the genetic control mechanisms that are responsible for the retooling are still being studied—nearly half a century later.

The coordinated control of protein production in the different tissues of a developing embryo or even within an adult person poses problems even greater than the control within isolated cells like bacteria. In multicellular individuals, certain molecules—hormones, for example—trigger the production of RNA from DNA in specific target cells; thus these hormones control and coordinate the times at which proteins are produced. The basic problems, not the details, of human development will be understood when it is known *how* DNA controls protein synthesis with respect to both structure and time.

What Do Proteins Do?

Beginning with the discussion of sperm (from the male parent) and eggs (from the female parent), we have stressed the nearly equal genetic contribution of the mother and father to their newly conceived child. Genetic information has, in turn, been identified as the ability of DNA (1) to specify the amino acid composition of hundreds or thousands of protein molecules and (2) to regulate when and where within the developing embryo (within its individual cells, within its tissues and organs, and within the person who is finally born) these proteins are made. Just what is it, though, that proteins do?

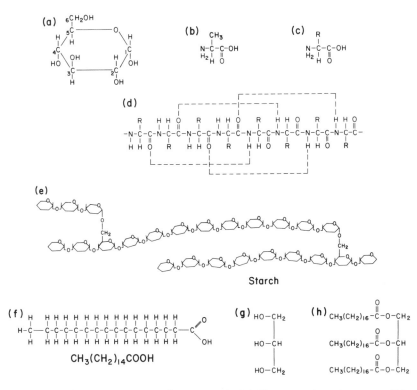

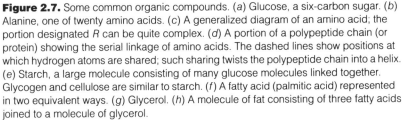

Figure 2.7. Some common organic compounds. (*a*) Glucose, a six-carbon sugar. (*b*) Alanine, one of twenty amino acids. (*c*) A generalized diagram of an amino acid; the portion designated *R* can be quite complex. (*d*) A portion of a polypeptide chain (or protein) showing the serial linkage of amino acids. The dashed lines show positions at which hydrogen atoms are shared; such sharing twists the polypeptide chain into a helix. (*e*) Starch, a large molecule consisting of many glucose molecules linked together. Glycogen and cellulose are similar to starch. (*f*) A fatty acid (palmitic acid) represented in two equivalent ways. (*g*) Glycerol. (*h*) A molecule of fat consisting of three fatty acids joined to a molecule of glycerol.

Putting Alien Genes to Use

Pioneer experiments using mRNA from one organism to direct protein synthesis by cells of another kind were performed a quarter century ago. Messenger RNA from rabbit bone marrow cells directed both frog eggs and bacterial protein-synthesizing machinery to make rabbit hemoglobin. Such experiments suggested that the genetic code is essentially universal.

Those early experiments have been enlarged greatly in scope and sophistication. Complete libraries of the DNA from human beings, farm animals, laboratory rodents (rats, mice, and hamsters), poultry, fruit flies, and bacteria are available for perusal. A "volume" in such a library consists of a small segment of an organism's DNA that has been inserted into a bacterial plasmid (a small, circular piece of bacterial DNA); thousands of volumes, each containing a different fragment, are required to ensure that nearly all of an organism's DNA has been packaged.

A casual review (of a single issue of a scientific journal) of studies made so far reveals that chicken genes are read (i.e., chicken proteins are synthesized) by mouse cells, bean genes are read by tobacco cells, human genes by yeast cells, human genes by rat cells, cattle genes by *E. coli*, and *E. coli* genes by the cholera organism. Nearly every organism can read genes from individuals of any other species.

The potential uses of these genetic libraries are far-ranging. For example, the DNA that specifies any human protein or proteinlike hormone can now be inserted into a bacterial cell. The bacterium will transcribe mRNA from that small segment of DNA and then translate the mRNA into the protein or hormone. That protein can then be harvested in virtually unlimited quantities.

Many persons who lack specific hormones (such as insulin) will, as the result of modern genetic technology, no longer need to rely on supplies from slaughterhouses. Currently, hormone-secreting organs of slaughtered animals are saved for pharmaceutical houses, which extract from them small amounts of hormone of uncertain purity, but in the future bacteria carrying the proper bit of DNA will be used to manufacture large quantities of hormone of known purity and of *human* origin.

The use of DNA specifying one or another of the many proteins found in the cell walls of pathogenic microorganisms allows those proteins to be manufactured by bacteria, such as *E. coli*. The manufactured proteins can then be injected into an animal, such as the horse, after which large quantities of the horse's antibodies can be recovered for use in human vaccines.

Table 2.1. The twenty amino acids used in protein synthesis, together with their one-letter and three-letter abbreviations

Amino acid	Abbreviation	
	One-letter	Three-letter
Alanine	A	Ala
Arginine	R	Arg
Asparagine	N	Asn
Aspartic acid	D	Asp
Cysteine	C	Cys
Glutamine	Q	Gln
Glutamic acid	E	Glu
Glycine	G	Gly
Histidine	H	His
Isoleucine	I	Ile
Leucine	L	Leu
Lysine	K	Lys
Methionine	M	Met
Phenylalanine	F	Phe
Proline	P	Pro
Serine	S	Ser
Threonine	T	Thr
Tryptophan	W	Trp
Tyrosine	Y	Tyr
Valine	V	Val

Many proteins are structural materials, the building blocks of life. Structural proteins are essential for all cells, but they constitute nearly all the material found in outer skin, hair, and in certain blood vessels. Numerically with respect to kinds, the vast majority of the many kinds of proteins found in living cells are *enzymes*, catalytic molecules that control the cell's biochemical processes or the metabolic processes that supply the cell with the chemical energy on which it runs. In these processes, complex molecules either from the body or from food are broken down, one step at a time, into smaller, less complex molecules, and some of these steps release the energy that is put to use for heat, motion, or other forms of work. Almost without exception, each of the metabolic steps is under the control of a separate enzyme. An enzyme cannot make a reaction occur that could not otherwise occur by itself (enzymes, that is, do not contradict the laws of thermodynamics), but the proper enzyme can make a reaction proceed at a rate millions of times faster than it otherwise would.

DNA and Proteins For each of the thousands of enzymes needed to carry out the chemical reactions within cells, there exists a corresponding segment of DNA that specifies its structure, as well as DNA capable of receiving signals saying when the enzyme is needed. These thousands of enzymes require more DNA than is required by the structural proteins that make up the bulk of the human body. In some cases, an enzyme may be used at only one moment during a person's

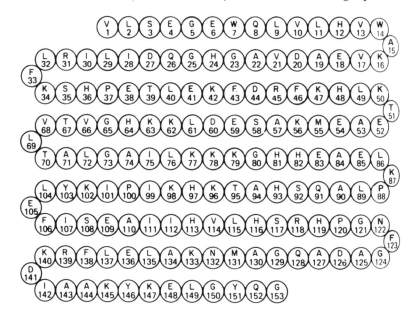

Figure 2.8. The amino acid sequence of whale myoglobin. The amino acid that occupies each of the 153 positions is specified by the three purine and pyrimidine bases that occupy a corresponding position in a portion of the whale's DNA. The three-dimensional structure of myoglobin, like that of other proteins, is determined by its amino acid composition.

lifetime and in only a few cells of the entire body; furthermore, only a few molecules of that enzyme may be needed even within those cells. Nevertheless, this enzyme must have its corresponding gene or it cannot be produced. Although each gene is tiny, nearly a meter of DNA is needed to spell out the chemical structure of all enzymes, and to specify when and where each will be made.

NOW, IT'S YOUR TURN (2-3)

1. Using both diagrams and words (your own), describe in your journal how chromosomes are transmitted accurately from one cell generation to the next during cell division and (without doubling the amount of genetic material) from parents to offspring. Be precise in your descriptions; make your diagrams clear.

2. A substance, E, is the end product of a series of enzyme-mediated reactions leading from substance A to B, from B to C, from C to D, and, finally, from D to E. The enzyme that converts A to B has a greater affinity for E than for A, its normal substrate; furthermore, once it is attached to E, the enzyme is unable to convert A to B. What is the effect of E's attachment to this enzyme on the production of E within the cell? (Hint; Assume that no E exists within the cell—will E be made? Now, assume that a great deal of E exists within the cell— will still more be made? Now think about intermediate situations, when some E, but not a great deal, exists in the cell.) What you have just worked out is known as *end-product inhibition*, one of the common types of feedback control found in living cells.

Prenatal Development

The Normal Pattern The human embryo follows a developmental pattern that has been programmed in human genetic material for millions of years. The program does not call at the onset for a tiny man or woman; instead it begins with the formation of a small cluster of cells. Then a neural tube is formed by an infolding of exterior cells along what will become the mid-dorsal line. The front (anterior) end of the neural tube becomes the brain, and the rest becomes the spinal cord. Development progresses from the anterior, or head, end of the embryo toward its rear, or posterior, end. As a consequence, to name one example, arms develop before legs.

As development proceeds from the early cluster of rather undifferentiated cells to the small, clearly humanlike fetus, groups of cells begin active differentiation, forming the brain stem, the heart, arms, legs, cerebral hemispheres, and various internal organs.

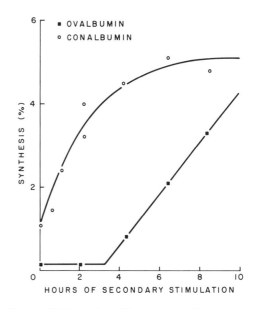

Figure 2.9. Induction of the synthesis of two proteins by chicken oviduct cells, following their experimental exposure to the hormone *estrogen*. Note that one protein (conalbumin) is made immediately, whereas the other (ovalbumin) is made only after a delay of about three hours. *(Copyright, Columbia University Press; reproduced with permission)*

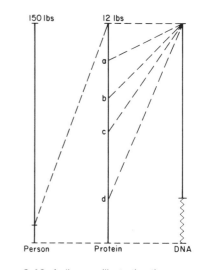

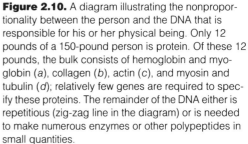

Figure 2.10. A diagram illustrating the nonproportionality between the person and the DNA that is responsible for his or her physical being. Only 12 pounds of a 150-pound person is protein. Of these 12 pounds, the bulk consists of hemoglobin and myoglobin (*a*), collagen (*b*), actin (*c*), and myosin and tubulin (*d*); relatively few genes are required to specify these proteins. The remainder of the DNA either is repetitive (zig-zag line in the diagram) or is needed to make numerous enzymes or other polypeptides in small quantities.

Figure 2.11. Stages in the development of the human embryo. Note that the magnification differs for each stage. (*A*) Embryo 27 days after fertilization (× 11); (*B*) embryo at 34 days (× 5); (*C*) 43 days (× 3); (*D*) 56 days (× 2). *(Copyright, Carnegie Institution of Washington; reproduced with permission)*

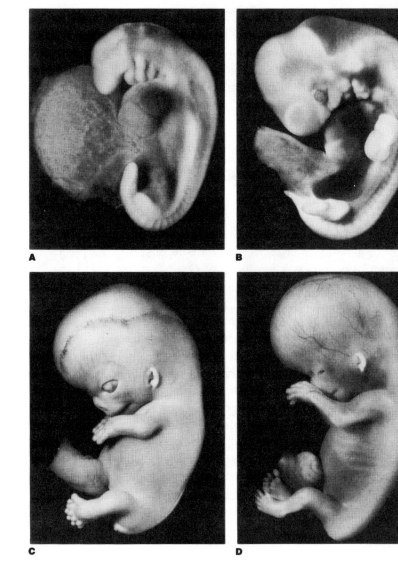

A

B

C

D

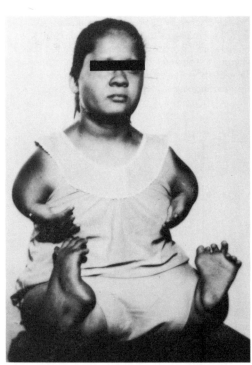

Figure 2.12. A photograph of a mature young woman whose arms and legs failed to develop as the result of intrauterine (prenatal) exposure to thalidomide, a sedative once widely prescribed for pregnant women. Thalidomide is only one of many substances capable of injuring unborn children. As knowledge concerning other harmful substances grows, so will the justification for legal redress by the injured child. *(Previously unpublished photo courtesy of Dr. F. M. Salzano, UFRGS, Porto Alegre, Brazil)*

Development can be said to be "normal" only if something different is available for contrast. The two preceding paragraphs described the "normal" pattern. But evidence of abnormal development is commonly found, and some causes are known. Therefore, if an expectant mother requires extensive abdominal X-ray exposure during the early months of her pregnancy, her doctor will often recommend that her fetus be aborted. Similarly, if she catches German measles (rubella) early in her pregnancy, her physician may recommend an abortion. A sedative once widely used in Europe—thalidomide—was prescribed for many pregnant women; its use resulted in the births of thousands of armless and legless babies.

Experimental Studies of Prenatal Development Why is the danger of an abnormal birth so high if the fetus has been exposed to radiation, German measles, or certain drugs, such as thalidomide? The answer can be found by studying animals, such as mice or chickens. The course of normal development in these animals is actually traced by noting the effects of harmful agents, such as radiation or certain chemicals.

X-rays can break chromosomal DNA so that, after cell division, one or both daughter cells may lack essential genetic information. If a cell lacks this information, it cannot function properly; it is an abnormal cell that may die. If female mice are exposed to X-rays shortly after mating, during the time that newly fertilized eggs are undergoing early divisions, many prenatal deaths occur among the embryos. Those embryos that survive, however, are generally normal. If, on the other hand, the pregnant females are exposed to radiation when their embryos are further developed, prenatal deaths are fewer, but a large proportion of the surviving mice are abnormal when born.

The difference in the effects of early and late exposure to X-rays has this explanation: When the damaged cells make up a large part of the embryo (during early cleavage of the egg), the affected embryo cannot survive; however, many embryos escape damage altogether. When the damaged cells are only a small part of the embryo, the affected embryo can survive, but, on the other hand, because each embryo now consists of a great many cells, few if any embryos escape totally unharmed by the radiation.

Timing the exposure of pregnant mice to radiation shows that the kinds of malformations change with time. Early exposures cause skull, brain, and vertebral column abnormalities. Somewhat later, the front legs are affected. Still later, exposure to radiation causes damage to hindlimbs, and after that to the pelvis and tail. Heavy damage occurs in those regions of the body where numerous cell divisions are occurring at the time of radiation exposure; the loss or misdivision of genetic information during cell division leads to descendant cells that lack the information they need for normal development.

Biologists experiment on mice because the knowledge they gain helps explain the damage done to human embryos by a variety of harmful agents. For instance, the virus that causes German measles damages the cells in which it reproduces. Viral damage, like that caused by radiation, is exaggerated in certain cells. German measles is most dangerous to embryos during the first three months of development, so it most commonly causes damage to the head, eyes, brain, and heart. Thalidomide, the sedative that was once prescribed for pregnant women, damages developing limb buds.

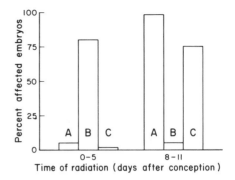

Figure 2.13. The contrasting patterns of deaths and physical abnormalities caused by exposure of developing mouse embryos to 200 rad or 400 rad of X-rays either immediately after conception or after considerable development had occurred. A = frequency of visible abnormalities; B = frequency of prenatal deaths; C = frequency of early postnatal deaths.

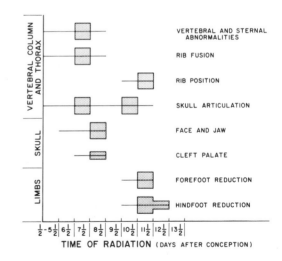

Figure 2.14. Times at which exposure of developing mouse embryos to X-rays causes different types of damage. Note that the hindlimb is radiation-sensitive beyond the sensitivity period of the forelimb, since, as is described in the text, development proceeds from the anterior toward the posterior end of the mammalian embryo. Features connected with the development of the central nervous system (skull and vertebral column) tend to have early sensitivity periods.

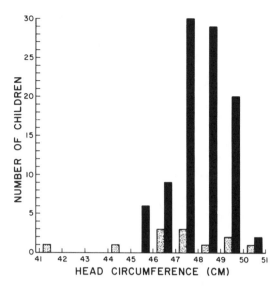

Figure 2.15. Head circumferences of children who were *in utero* at Nagasaki, Japan, during the spring of 1951. (The atomic bomb was dropped on Nagasaki in August 1945.) Control mothers (*black bars*) had been 4000–5000 m from the epicenter of the atomic bomb blast; exposed mothers (*cross-hatched bars*) had shown symptoms of severe radiation exposure after the bombing of Nagasaki and had been within 2000 m of the epicenter. Children of the control mothers had the larger, more normal head sizes.

Figure 2.16. An advertisement of the 1920s promoting the beneficial effects of radioactivity. Even today, some bottled waters sold in Europe list the amount of radioactivity in a promotional, rather than a negative, manner.

Control of Embryonic Development The developing embryo is under its own genetic control, but it is not independent of maternal influences. Most harmful chemicals that damage developing embryos, in fact, arrive by way of the mother. Such chemicals include common drugs, such as nicotine (in smoking mothers) and alcohol (in drinking mothers), as well as other chemicals, such as tranquilizers, stimulants, narcotics, and "street" drugs. The mother's hormones also affect the developing fetus; newborn babies of both sexes have overdeveloped mammary glands. The overdevelopment disappears a week or so after birth.

Problems During Development An embryo at an advanced stage of development can survive even severe damage because its respiratory and nutritional needs are met by the mother's circulation. Maternal blood provides both food and oxygen and removes carbon dioxide and other waste products. For example, to illustrate the type of damage that an embryo can survive, both rabbit and rat embryos continue to grow after their heads have been surgically removed, but surgery of this sort does have severe effects on other parts of the embryo, since the pituitary body of the brain is an important part of the embryo's development control system. Similar problems affect human embryos. As a result of genetic disease or prenatal accidents, human babies can be born with severe brain damage, including no brain at all. It is for fear of such abnormal births that physicians often recommend abortions when the embryo has been exposed to a known damaging agent.

Fortunately, embryonic development passes through several early critical points during which, if all is not going well, the embryo tends to abort spontaneously. Abnormal sets of chromosomes (from either parent) pose a severe threat to the embryo. Down syndrome, a rare genetic disease characterized by severe mental retardation and certain physical abnormalities, is caused by the presence of an extra chromosome corresponding to one of a normal pair of small chromosomes. From the frequency of children born with Down syndrome, one can calculate that five percent or more of all fertilized eggs should suffer from abnormalities similar to or worse than Down syndrome (these would be caused by extra chromosomes, or their converse, missing chromosomes). Infants with these predicted conditions are *not* found among newborn babies, however. Where are they? The answer is learned by a microscopic examination of tissues taken from spontaneously aborted fetuses. Many of these fetuses have abnormal chromosome complements. These severely damaged embryos do not complete their prenatal development; they die before birth. Down syndrome is one of the few chromosomal defects that are sufficiently mild (despite, in our eyes, the serious flaws of children suffering from Down syndrome) that they do not kill affected embryos.

Why does the last paragraph begin with the word "Fortunately"? Because spontaneous abortions relieve would-be parents of what otherwise would be difficult and heart-rending decisions. Despite terminology that has become commonplace in recent years, there are few, if any, pro-abortionists in this world. There are really only "anti-abortionists" and "anti-anti-abortionists." Although abortion laws have been liberalized throughout the United States, abortions are not pleasant to contemplate, to undergo, or to perform. The only reason these operations are done at all is that the consequences of *not* doing them promise to be worse than the abortion itself. For example, a damaged embryo might not be born as a functional human being, or it might suffer a slow, painful death, devastating to both it and its parents. Or the existence of a defective child could cause serious mental or physical harm to the mother (or father). Because the decision to abort an embryo is so difficult, it is indeed fortunate that abnormal fetal development so frequently leads to spontaneous abortion. Spontaneous termination of what would otherwise be an abnormal birth occurs in many animals. In most mammalian species, natural abortion prevents the investment of the pregnant female's effort and energy in bearing and giving birth to malformed offspring that would in any case be doomed to an early death.

Figure 2.17. Karyotypes of human chromosomes representing (*a*) a normal female, (*b*) a normal male, (*c*) an individual with Turner's syndrome (XO), and (*d*) an individual with Klinefelter's syndrome (XXY). Be sure that you recognize the differences between the chromosomal constitutions of normal males and females and those of the two aberrant syndromes.

Radiation Therapy of Yesteryear

X-radiation is a useful tool for studying embryonic development because it harms DNA (and, therefore, dividing cells), and because it can be administered in a known quantity, at a known time, for a known duration. It is, of course, the harm done by the radiation exposure that the experimenter studies; the abnormalities shed light on how normal development proceeds.

Radiation, however, has not always been regarded as harmful, despite the radiation-related deaths of such pioneer scientists as Roentgen, the discoverer of X-rays, and Marie Curie, the discoverer of radium. Health spas boasted (and to some extent still do) of the radium in, and the radioactivity of, their curative waters.

A 1926 issue of a magazine promoting good hygiene contained an article entitled "Methods of Removing Superfluous Hair." X-rays, the article claims, will cause hair to fall out and will prevent it from returning, but at the risk of wrinkled, scarred, mottled, or discolored skin or of developing warty growths. The article does not mention the risk of developing cancer ten to twenty years after the treatment. It does express the (vain) hope that a new and harmless X-ray will be developed. (A similar hope exists today concerning harmless tobacco!)

Health spas and beauty parlors were not the only places persons were deliberately exposed to X-rays. A 1925 article reports on the use of X-rays for treating whooping cough. After it was found that X-raying children stopped the convulsive cough of whooping cough, "attention was given to finding out just why the X-rays effect cures and improvements."

Enlarged tonsils could be controlled, according to another article of the 1920s, by surgery, by electric cautery and electric heat, or by X-rays. The latter provided a useful method in all cases in which an operation was not advisable. In the same article, X-rays were also recommended for elderly patients who were bothered by chronic sore throats.

Needless to say, the harmful effects of X-rays and other high-energy, penetrating radiation are now recognized, in that they cause gene and chromosome mutations and induce cancers. Frivolous use of X-rays has all but ceased. Superfluous hair is no longer removed by the use of X-rays; too many cancers were caused by such treatment. Shoe stores no longer have X-ray–activated flouroscopes for checking the fit of new shoes; too many children developed stunted and misshapen feet from use of such devices. Ringworm of the scalp is no longer treated by X-rays; too many thyroid cancers developed in the patients (mostly school children) as they aged. Finally, X-rays are no longer used to induce ovulation in women by the heavy-dose exposure of their ovaries; the germ cell, as you may have guessed, is the last cell that should risk the integrity of its DNA through exposure to ionizing radiation.

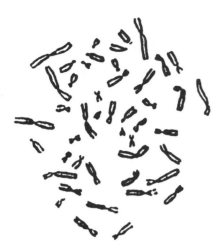

Figure 2.18. As suggested in the text, copy this figure, cut out the individual chromosomes, match them in pairs by size, and construct a karyotype like those shown in the previous figure. What is the sex of this unknown person? Can you spot any glaring genetic defect?

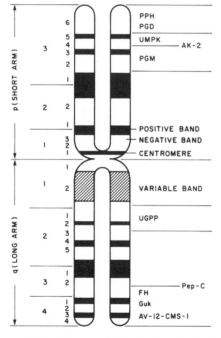

Figure 2.19. A diagram illustrating the detail with which human chromosomes are now being studied. The chromosome shown is the largest (Number 1). The positive (black) and negative (white) bands reflect the response of these chromosome regions to modern fluorescent dyes; the variable band reacts differently in preparations involving different persons. Note that twenty-four regions have been identified. Several gene loci have been located in these regions; these loci are responsible for specifying various enzymes (whose identifying symbols are not translated here).

NOW, IT'S YOUR TURN (2-4)

1. Figure 2.18 illustrates the full complement of human chromosomes (as they appear in a squashed and stained cell). Photocopy this figure, cut the individual chromosomes from the copy, and match them in pairs of what seem to be chromosomes of corresponding sizes and shapes. (Fasten a strip of transparent tape, *sticky side up*, on a sheet of typing paper. Place the chromosomes of corresponding size side by side on this tape. When all are in place, put another strip of tape sticky side down on top of the first strip and its adhering chromosomes. Transfer the lot—the *karyotype*—to your journal.) Were these chromosomes obtained from a male or a female? (Women have twenty-three pairs of chromosomes: twenty-two pairs of autosomes and a twenty-third pair of X chromosomes. Men have twenty-two pairs of autosomes and a twenty-third non-matching "pair" consisting of an X and a much smaller Y chromosome.)

2. Much discussion has gone on recently concerning when human life begins. Most of what one reads or hears could have been read or heard in Ancient Greece or even during Biblical times. Modern technology promises to change the ground rules of this discussion in ways not yet recognized by scientifically naive lawmakers. How would you react to the question, "When does a human life begin?" if you knew that a salve containing lanolin, some common salts, and two rather rare vitamins could cause the cells in most hair follicles (arms, legs, eyebrows, head, or wherever) to initiate embryonic growth? The main points of this question are to highlight how claimed attitudes toward "life" are often governed by the usually unmentioned matter of "sex," and to see how valuable an easily accessible embryo's life is if carrying it to term is removed from the concept of "deserved punishment" for an unwed mother.

3. Both spontaneous and therapeutic abortions have been mentioned in the text. The matter of abortion promises to be divisive for years to come. Try to clarify your own thoughts by considering the following:

a. A brain is not necessary for the near-normal physical development of a human embryo; with its oxygen and nutrients provided and its waste products removed by the mother's circulation, the anencephalic (brainless) embryo is much like an elaborate tissue culture that undergoes organized growth—not merely cell proliferation. (Recall that anencephaly is more extreme than "brain death," which in many states is sufficient cause for declaring an individual dead.)

b. What changes occur at the fusion of an egg and sperm that make the fertilized egg, now a diploid cell, something of moral value, whereas before fertilization both the egg and sperm are considered worthless and are continually being discarded by the millions and billions?

c. If fertilized eggs are to be regarded as persons in a legal sense, does the loss of a fetus for any reason become a legitimate matter for legal investigation in order to assess blame and, if warranted, impose punishment? How does one judge women's use of a fertility drug that is known to induce multiple ovulations but with a high risk of death for the resulting embryos? Or of embryo transplants, where only one fertilized egg is to be transplanted while the "stand-bys" are destroyed?

Postnatal Development

Postnatal development is something that we have all experienced by living through it. During growth, changes occur in the proportions as well as the size of the body. Changes take place in every part of the body.

NOW, IT'S YOUR TURN (2-5)

1. Try to locate your own or someone else's "baby" book in which a chart has been kept of the baby's weight from birth through the first year of life. Unless it happens to be your own, such a chart is not particularly interesting, but if a number of charts can be located and compared, an idea of the variation among infants can be obtained by plotting all the growth curves on one chart. As an exercise, you might take on the calculation of the mean or average weight at given weeks of age, some measure of the dispersal of weights about this mean, and (by an examination of the curves themselves) a determination of the extent to which large babies tend to remain large at a later time. (Question: At some point during childhood, the average height of children *must* be exactly half the average height of adults. This age is generally given as 2½ years. Does this mean that the adult height that will be attained by a given 2½-year-old child is two times his present height? If not, why not?)

2. In the late sixteenth century, Montaigne wrote: "I am in conflict with the worst, the most mortal, and the most irremediable of all diseases—kidney stones." He then continued, wondering how his father, who had died of kidney stones at age 75 (after seven years of torment), had transmitted the disease to his son (the only one of three children similarly afflicted) when the son (Montaigne) had been born twenty-five years before his father's first symptoms. In his words: "And he being then so far from the infirmity, how could that small part of his substance wherewith he made me, carry away so great an impression for its share? And how so concealed, that till five and forty years after, I did not begin to be sensible of it?" From what you have

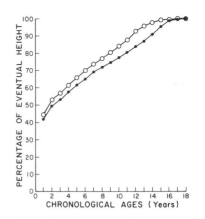

Figure 2.20. The percentage of mature height attained at different ages by 150 boys (*filled circles*) and girls (*open circles*).

Multiple Births: One Family's Saga

May 22, 1985 (First page headline)
SIX BABIES BORN TO CALIFORNIAN; 7th IS STILLBORN
"It's a neat experience." (father)

May 23, 1985 (First page headline)
SIX SURVIVING SEPTUPLETS WEAKEN BUT DOCTORS SAY THEY'RE "FIGHTERS"
"There is no impending death right now, but they are all critically ill." (attending physician)
The babies are "kicking around like polliwogs." (father)

May 24, 1985 (12th page headline)
FOUR OF SURVIVING SEPTUPLETS ARE SHOWING IMPROVEMENT
"We were not out to set any records." (father)

May 25, 1985 (First page headline)
SMALLEST SURVIVING SEPTUPLET DIES; DOCTORS GIVE REST 50:50 CHANCE

May 27, 1985 (8th page headline)
MOTHER OF 7 VISITS SURVIVING INFANTS
Prognosis "is hopeful, and for some of the babies quite good." (attending physician)

May 29, 1985 (14th page headline)
BABIES A TO G ARE NAMED ON WEST COAST
"There is no reason for me to think these babies won't have a full chance for survival and normal development." (attending physician)

October 9, 1985 (small news item)
PARENTS OF SEPTUPLETS SUE DOCTOR AND CLINIC
The parents of three surviving septuplets filed a malpractice and wrongful death suit today seeking more than $2.2 million from the doctor and clinic that gave the mother fertility drugs. . . . The three surviving children face a lifetime of medical problems including optic nerve damage, hernias, chronic lung disease, and heart damage according to the family attorney.

A Perspective on Human Life   **51**

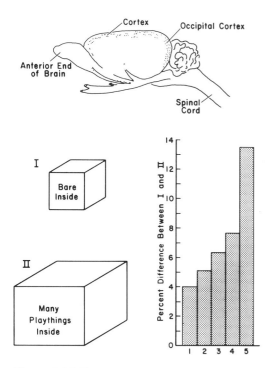

Figure 2.21. The structure of the brains of rats reared under simple conditions (I) differs from that of rats (matched for sex, age, heredity, and physical condition) reared under complex conditions (II) with many playthings. The bars of the graph show the degree of difference between the two groups of rats with respect to (1) total cortex, (2) ratio of cortex to rest of brain, (3) occipital cortex, (4) total protein in occipital cortex, and (5) size of nerve cells in the occipital cortex. Despite the larger size of the brains of the mentally stimulated group II animals, their brains were smaller than those of most wild-caught rats.

learned so far of DNA and its role in development, address an explanation to Montaigne, who concluded by saying: "He that can satisfy me in this point [i.e., provide a reasonable explanation], I will believe him in as many other miracles as he pleases."

Some Key Aspects Three aspects of postnatal development are of special importance today for understanding many of life's problems. The discussion that follows is limited to these three aspects. The first is the development of the brain in both pre- and postnatal life. The second poses a paradox. On the one hand, it is so important that it cannot be overemphasized, but, on the other hand, it is not the most important thing in the world as some persons seem to think: sex and sexuality! The third aspect of development to be discussed is aging. In today's society, aging is often looked upon as a disease. Yet old age is not a disease—it is a normal (although terminal) part of the trajectory of life.

On the Development of Intelligence The sins of the fathers, we like to believe, are not visited upon the sons. In the case of human beings, however, the economic circumstances of parents often have profound effects on the opportunities that are available to their children. Of all aspects of nongenetic inheritance (or *cultural heredity*), the worst is the mental damage that can afflict a child because of the poor nutrition of its mother.

There is a stage in embryonic development when the human brain grows rapidly in size. This stage of rapid growth begins somewhat before birth and lasts for a short while afterward. During this time when the brain enlarges, the individual nerve cells that form the brain grow *larger*. The *number* of cells is established earlier, during the development of the fetus, and this number depends mostly upon the level of nutrition provided to the fetus by its mother. The number never changes: brain cells, unlike other body cells, never divide.

A pregnant woman whose diet is inadequate (whether because of poverty, ignorance, or indifference) restricts the development of her child's brain by limiting the number of cells it contains. Just before and after birth, these cells will enlarge, but the deficient number can no longer be corrected. Because the damage caused by a mother's inadequate diet limits the mental ability of the child, the retarded child may suffer in turn from malnutrition, which may in its turn cause the birth of still more mentally retarded children. Cultural heredity in the form of environmentally caused mental retardation, then, is often transmitted from generation to generation as surely as is the family's DNA.

A large body of experimental evidence suggests that the postnatal development of the brain is related to the complexity of the environment in which the child develops. Experiments on rats and other animals have demonstrated this point. Baby rats raised alone have smaller cortexes than those raised with other rats of their own age in cages that are provided with numerous playthings. (Similar results have been obtained from studies of mice, gerbils, and monkeys.) Interestingly, wild-caught rats have cortexes that are larger than those of caged rats, even those that have been provided with manmade playthings. If results of this sort can be transferred to human development, it seems that a child whose environment is interesting and stimulating will have a greater brain development than will the child who is neglected and left alone in dull surroundings. Furthermore, "clever" playthings that are designed by adults for children but that fail to challenge the child's imagination may serve little purpose in stimulating a growing mind.

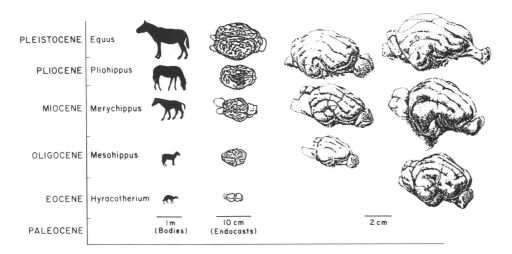

Figure 2.22. The evolution of brain and body sizes in horses (*left*) and of brain sizes in felines (*center*) and canines (*right*). Although this figure is not intended to suggest that catlike and doglike mammals preyed on the ancestors of the modern horse, it does illustrate the growing complexity of mammalian brains through time. In part, the increased complexity of the brain reflects the increased size and complexity of the animals' bodies; in part, however, the increased complexity of one type of mammal (predator or prey) matches that of the other. The increased complexity (that is, evolution) of all three series of brains in this figure reflects the action of natural selection on the genetic bases for the construction of complex nervous systems.

Sex and Sexuality Sex is an aspect of postnatal development that has cultural and emotional effects that trace back to early childhood. Small boys and girls are generally not dressed alike, nor are the reactions of adults to children of the two sexes identical. The main differentiation between the sexes, however, arises during adolescence.

In the trajectory of human development, the onset of sexual maturity, with all its manifestations, physical and emotional, completes the cyclic pattern of human generation. Germ cells mature in the reproductive organs of young men and women. The genetic informa-

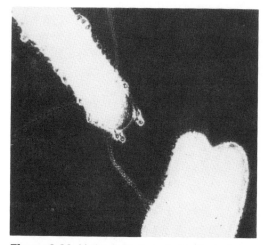

Figure 2.23. Mating in bacteria. In this illustration two bacteria (*E. coli*) are connected by a delicate filament through which DNA passes from the donor (male) to the recipient (female). To reveal the sex of these cells, males have been chosen that do not absorb a particular bacteriophage (and, hence, are resistant to it); however, phage particles can be seen absorbed to the surface of the recipient cell, the female. *(Photo courtesy of Charles Brinton, University of Pittsburgh)*

tion that each adult received at conception from his or her parents is now transferred into individual egg and sperm cells. The apportionment of this information, as we have already seen, is extremely precise. Each sperm or egg cell receives one complete set of chromosomes containing a unique combination of both maternal and paternal genes; that, you will recall, is the role of sex.

Of what use is sex? To find an answer to this question, first imagine a simple organism, such as a primitive bacterium, that has no sex. It has a bit of DNA that directs all of its life processes, and it reproduces by simple cell division that is synchronized with the replication of its DNA. The many individuals, with their identical bits of DNA, divide and divide again. Eventually they constitute a large population of individuals, each of which is identical to all others. If one needs the amino acid proline to grow and reproduce, all of them require proline to grow and reproduce. The sameness of all these individuals is a consequence of asexual reproduction.

Remarkably few organisms reproduce only asexually. The vast majority of organisms, including most bacteria, have a more complicated pattern of reproduction. In this pattern, the DNA of *two* individuals is brought together within a single organism. From this combined DNA, an amount that is suitable for one individual is selected for reproduction. Here is the key to the importance of sex: the DNA that is selected for the new individual represents a novel combination of DNA drawn in part from each of two contributors. This is the phenomenon that geneticists refer to as sex. (In contrast, the fascinating physical and behavioral maneuvers, from courtship to mating, are what most nongeneticists refer to as sex. It is not at all uncommon in science, as elsewhere, that one word may have several meanings.) Stripped of its frills, sex is the means by which one individual (offspring) can be provided with combinations of genes from two different individuals (parents).

At first glance, the ways in which bacteria exchange DNA may not seem particularly exciting, sexually or otherwise. Nevertheless, the ability to consummate the exchange has tremendous power in evolutionary terms. Sexually reproducing forms use the new combinations of genes in adapting to changing environments. They are much more successful in adapting and evolving than are those organisms that reproduce only asexually. Furthermore, populations composed of dissimilar organisms, each making slightly different demands on environmental resources, can attain higher densities than populations of genetically identical individuals.

In human beings, as in bacteria and most other forms of life, the original importance of sex was that it provided each generation with new combinations of genes. At first glance the aspects of sexual reproduction that most persons are aware of, and find most exciting—

Figure 2.24. The annual cycle of an insect (such as an aphid) that relies on asexual reproduction when food is plentiful but that switches to sexual reproduction as winter approaches. Each form of this insect requires a corresponding (operative) genetic program for its proper development. The signals that cause one or the other of these different programs to be activated come from the environment: a shortage of an amino acid, for example, can signal a coming shortage of food; short periods of daylight in the fall signal the coming of winter.

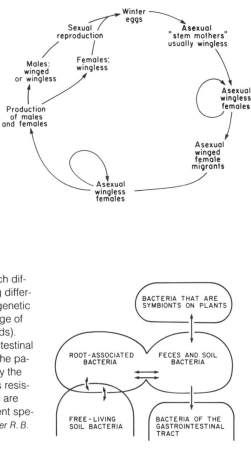

Figure 2.25. The extent to which different species of bacteria having different modes of life can still share genetic information by way of an exchange of infectious DNA particles (plasmids). After antibiotic treatment of an intestinal ailment, many bacteria living in the patient's colon will be found to carry the same plasmid—one that confers resistance to that antibiotic. Plasmids are even transferred between different species and genera of bacteria. *(After R. B. Davey and D. C. Reanney)*

those aspects that inspire poets, psychologists, and pornographers—seem to have little to do with DNA. Nevertheless, they are the mental and physical manifestations that are needed if individual men and women are to seek and find one another, fall in love, mate, and produce offspring. Physical development differs between men and women (a difference that is under genetic control). Their reproductive organs differ. As individuals of the two sexes mature, they develop contrasting sexual, emotional, and other psychological characteristics. Many of the beautiful and mysterious aspects of life, then, can be traced to the evolutionary advantage of a reproductive behavior that allows offspring to draw genes from two parents, rather than from one alone. The offspring then shuffle these genes once again when passing them on to the following generation.

Old Age and Death: The End of the Trajectory Human beings are unique in their acceptance of aged individuals who are well past the prime of life as active participants in community affairs. This acceptance of elderly persons is one of the key aspects of our humanity. The

Reproduction Without Sex

Sexual reproduction represents a gamble. Two adult individuals (even of bacteria, if one wishes to stretch the point) share DNA and, out of their pooled contributions, combinations of DNA are drawn and assigned to newly formed individuals—progeny. Some combinations may be better than the genetic information carried by either parent, while others may be decidedly worse. The parents, after all, have survived to adulthood and sexual maturity.

Many organisms do not gamble with reproduction: they reproduce asexually. The normal method by which bacteria reproduce is by simple cell division. *E. coli* divides every twenty minutes. Overnight, a single bacterium in a culture tube can produce a hundred billion progeny, most of which are genetically identical. Protozoans (paramecia and amoebae) and higher organisms, such as hydra, commonly reproduce either by simple fission or by budding.

Many plants reproduce asexually. The spider plant (*Chlorophytum*), which is commonly kept in hanging pots around the house, sends out stems that bud off small plants complete with leaves and roots. Any one of the buds can take root and produce a new plant genetically identical to its parent. Strawberry plants are another example; each plant sends out runners that produce new plants at frequent intervals.

Organisms that reproduce asexually avoid a gamble because they pass on to their offspring only the genetic information possessed by the original individual, which obviously worked (i.e., that individual survived). But just as gamblers often hedge their bets (bet against themselves), organisms that reproduce asexually nearly always possess the ability to reproduce sexually as well. Thus, asexual plants that have saturated their local area with runners can disperse their offspring by producing seeds as well. Aphids, small herbivorous insects, have a remarkable, finely tuned program of reproduction: when food is ample, wingless forms reproduce asexually; as the supply of food dwindles, asexual winged forms are produced; at the onset of cold weather, sexually reproducing individuals are produced. The vagaries of winter are faced by large numbers of genetically variable offspring of sexual origin.

Aging and Death

Is death necessary? Is life without death a real possibility, or an impossible dream?

No one can answer those questions, because we do not yet know the cause (or causes) of aging. Although the cause of an individual person's death is often obvious (disease, injury), we do not know why every human life always ends in death.

Even in single-celled life forms—both plants and animals—aging occurs. By dividing, these organisms produce more and more individuals. Eventually, however, reproduction slows down and stops altogether. In such cases, the organisms must undergo sexual reproduction—the exchange of DNA between individuals with different genetic backgrounds—in order to survive. Otherwise, the population simply dies out. (Not all single-celled organisms need sexual reproduction in order to survive.)

Almost all higher organisms exhibit aging. The signs vary from one species to another, but there is one feature in common in the aging of all organisms: the greater probability of death—from whatever cause—with advancing age. Nevertheless, some animals (amphibians and reptiles, for example) and some higher plants (such as redwood trees) have long natural lifespans.

Is there something about life that *requires* that aging and death occur? Why do they happen so predictably in most species, including our own?

There are many ideas concerning aging and death. Life is full of danger. Error in cell function, attack by viruses and other disease agents, and radiation damage may weaken the living system over the course of time, increasing the probability of death. Waste products may build up in the body and poison normal functions. Aging and death may actually be programmed into each individual's DNA. For instance, in the development of the fetus, the masses of cells that form the hands and feet are sculptured by the death of cells forming membranes ("webs") between the fingers and toes, so these become the nicely formed hands and feet of the newborn infant. This is a case of natural cell death, death that is essential for normal development. The periodic maturation of eggs in a woman's ovaries is turned off naturally at menopause. Human cells cultivated outside the body die after about fifty successive cell divisions. Natural death allows for the continuing replacement of organisms by new, genetically different individuals. The aging and death of individuals commonly aid the survival of the species.

younger members of the community, in contrast, share with all other forms of life the universal function of living things: reproduction. Among most animal species, individuals who are beyond reproductive age are not preserved. For example, once their reproductive role comes to an end, old grazing animals serve mainly to satisfy the hungry predators that continually stalk the herd.

Human beings, far more than other animals, have specialized in the development of intelligence. This specialization includes the formation of councils in which the intelligence of many persons is pooled: deliberative conferences are cultural heredity's analogue of sex. Within these councils, old and experienced persons have in the past had a special role: they were the members of the community who most likely had experienced rare happenings, events that occur no more than once or twice each generation. As long as knowledge can be transmitted by word of mouth, old persons are valuable as stores of accumulated experience. Ordinarily, wise men are not young men.

In Western societies, the special role of the wise old citizen has been diminished. Libraries and computerized data banks now store the experiences of generations. Furthermore, the vastly accelerated rate of change in modern societies has rendered useless many experiences gained fifty or more years ago. As a result, the contribution of the elderly to modern societies has become severely restricted.

In the United States, especially, the situation for most old persons has drastically worsened. Many take part in neither their families' lives nor the affairs of society. American families, many at the insistence of corporate employers, move from place to place at short intervals. Such continuous moving breaks family and community ties. Modern housing has also lessened the status of the old. For instance, the physical and social structure of suburban communities is determined largely by the economics of housing construction. As a rule, these communities make no provisions for the aged, who are shunted aside and concealed from view. In such a modern society, consequently, the trajectory of life no longer comes to a visible end; like a rainbow, it appears to fade away some distance above the ground. Unseen in the final weeks or not, each life does come to an end.

The Role of Death in Human Populations

Little can be said in strictly scientific terms about old age, except to emphasize that death is the natural termination of every human life. When many individual birth-to-death trajectories are combined, they constitute the age-structure of a human population.

Should we try to escape from aging and death? "Sorry," wrote one contemporary philosopher, "we're here for eternity." After referring to books, newspaper articles, television documentaries, and seminars urging people to accept death as a natural human experience, this writer exploded: "If dying is natural, then to hell with the tyranny of nature." He proposed, instead, that life-support technologies be directed toward achieving the "most liberating freedom of all"—physical immortality.

A fine, brief response to this impassioned plea for immortality was not long in coming: when death is abolished, birth is a disease. A population from which death was abolished would become old. And older. And still older. Birth could not be permitted because births without corresponding deaths would increase the population to an intolerable size. Therefore, at some moment a population lacking death would find itself condemned to one of two courses: first, to grow ever older, or, second, to surrender and let death return. Of the two, the second choice need not be the less pleasant.

Figure 2.26. Because human beings are not annuals like corn and zucchini, trajectories of individual lives overlap. In this case, the small boy and his great-great-uncle represent a span of four generations. *(Photo courtesy of Professor W. C. Johnson, Virginia Tech)*

NOW, IT'S YOUR TURN (2-6)

1. In a recent book, the authors point out (as have many earlier writers) that self-awareness is a unique and extremely important human trait. (When, for example, did you first become aware of yourself as an individual? In all subsequent years, have you ever forgotten who you are? Or thought for a moment that you were someone else?) Self-awareness, however, *must* be accompanied by death-awareness. Otherwise, the authors wonder, what will die if not one's self? Contrast this necessary relationship between self- and death-awareness with the growing emphasis on personal gratification (exaggerated self-awareness?) and the shunting of old persons to out-of-the-way nursing homes, where they die unseen (death unawareness?). Recall some relevant personal experiences bearing on both self- and death-awareness.

2. What is your (thoughtful) reaction to each of the following situations?

a. Having undergone amniocentesis and having learned that her child, if born, will be mentally retarded for life, the mother-to-be elects, nevertheless, to allow her pregnancy to proceed. Comment from the points of view of all persons who are, or will be, or think they are, concerned: the mother, the father, the doctor who was asked to perform the amniocentesis, the "town fathers" who must eventually assume custodial care of the child, and you as a concerned citizen.

b. Having learned that their unborn child does not exhibit the genetic disease that prompted their request for amniocentesis, a couple opts for abortion anyhow. The doctor believes that the choice is based on the now-known sex of the unborn child.

c. Having learned that one cannot legally destroy any viable aborted

Aging: The Roles of Research and Care

This chapter presents an account of human life from conception until death. However, at the chapter's finish the possibility of human immortality was raised, and the authors have clearly come down on the side of mortality: let death be the termination of each person's life.

Despite our (and, we believe, most persons') favoring mortality, we do not disapprove of gerontological studies. A great deal of federally sponsored research is devoted to the problem of aging. The purpose of this research, however, is not to achieve immortality, but to make life as pleasant and painless as possible for the elderly. The aim is to understand and to minimize the multitude of nonfatal but debilitating ailments that afflict the aged.

Other than human beings (and, perhaps, their relatives, the great apes), most organisms do not normally die of old age. (Exceptions are provided by human pets: cats, dogs, ponies, horses, and various birds.) In wild populations, aging individuals are displaced from the normal hierarchy, whether it be a nesting site for chickadees, a burrow in the case of gophers, or a position of prominence in a herd or pack, and death caused by natural enemies follows soon after.

Another Point of View

In the accompanying text, the authors state that in the matter of mortality versus physical immortality for individual human beings, they opt for mortality. At the first anniversary progress report on the then longest-lived recipient of an artificial heart (a man who, according to one account, "moves when he has to or is told to"), the implant surgeon reminded his audience that his patient had faced certain death the previous year. Commenting further on the patient's past and present physical states the surgeon added: "It is a lot better than the alternative—we lose track that he would have been dead that weekend."

A sound argument can be made that no one fatal ailment awaits aged human beings. The argument goes as follows: Assume that there were such an ailment. It then follows that there would now be a basis for natural selection to act, conferring individual resistance to that one ailment. As a result, the effects of that ailment would be postponed until they blended in and were often superseded by the effects of other ailments.

The conclusion that follows from the above argument is that human death involves a multitude of malfunctions. Geriatrics, that branch of medicine that is concerned with aging and the aged, must then deal with not one but a multitude of troubles—*per person*. Young doctors who perceive their mission as the saving of life are often confounded by the problems—and the attitudes—of the aged. The aging person's concern is not the avoidance of dying, but living in peace and happiness and dying with one's dignity intact.

Because each old person most likely suffers from several disabilities at once, caring for the aged is less the concern of the physician (who is as helpless as anyone else) and more the concern of the nurse and her colleagues—social workers, speech and physical therapists, paramedical personnel, and orderlies. Because the problems are numerous, the care provided must be correspondingly diverse. In the final analysis, terminal care is designed to relieve pain. As one experienced worker has said, "One *can* do that, to enable the patient not only to die peacefully, but to live fully until he dies."

fetus, a company manufacturing scientific equipment develops a "glass-and-chrome" artificial womb that is capable of maintaining embryonic development from the fertilized egg through the normal nine-month term of development. The company anticipates a tremendous, highly profitable market because, with their apparatus, *all* (spontaneously or induced) aborted fetuses are potentially viable in a legal sense, and hence must not be allowed to die. (In a recent Supreme Court ruling, one justice wrote that the Court had painted itself in a corner by using the term "viable" in an earlier decision. In the near future, every fertilized egg, whether in a woman's womb or in a laboratory test tube, will be viable if sufficient effort is made to sustain its development.)

3. The following two paragraphs outline an issue of some notoriety several years ago:

"The Governor of Colorado, Richard D. Lamm, had his heart in the right place when he warned that 'we really should be very careful in terms of our technological miracles that we don't impose life on people who, in fact, are suffering beyond the ability for us to help.'

"Speaking at a meeting of the Colorado Health Lawyers Association, Governor Lamm stirred widespread public criticism, apparently based on the misunderstanding of his remarks, when he said that 'we've got a duty to die, to get out of the way with our machines and our artificial hearts.' Later Governor Lamm said that he simply was urging that economically sound and sensible allocation of limited medical resources should preclude fruitless treatment of the terminally ill."

Prepare an essay for your journal in which you identify and analyze biological facts, factual consequences, issues that are relevant to science, and those that lie outside science. Refrain from personal remarks aimed at Governor Lamm; keep to the issues. Consult newspapers and magazines from March/April 1984 if you wish to read more about Governor Lamm's remarks.

INFORMATIONAL ESSAY
The Cell

Higher organisms, animal or plant, are collections of cells. These organisms arise from single cells—fertilized eggs. They grow and develop by means of cell division. As cells grow in number, they differentiate and form various tissues. Aggregates of tissues form organs. An adult human being (or ape, or mouse, or earthworm) is a functioning aggregation of cells that have become organized into tissues, organs, and organ systems whose cooperative interactions maintain each other, and hence maintain the individual.

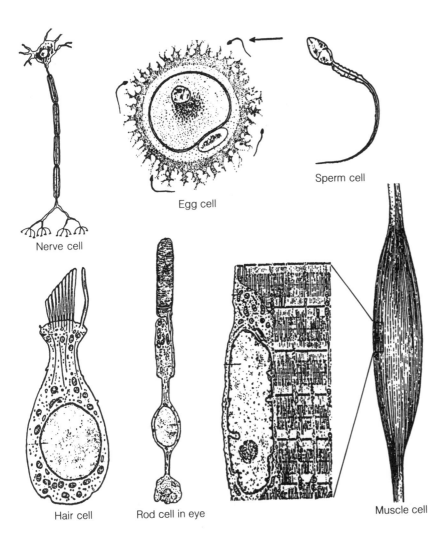

Nerve cell

Egg cell

Sperm cell

Hair cell

Rod cell in eye

Muscle cell

Figure 2.27. The variety of cells that make up the human body. The size and proportion of each cell type cannot be represented properly. For example, a nerve cell, like a strand from a spider's web, may be a fiber that is scarcely visible to the eye but two or more feet in length. Imagine the difficulty the cell theory originally encountered because cells exhibit such an array of structure. Nearly two centuries elapsed between the discovery of cells (in cork) and the acceptance of the modern view that all cells arise from pre-existing cells.

The term ''cell'' was first used not to describe a functional unit of life, but the nonliving skeleton of such units that can be observed in cork. Later, through the use of improved microscopes and stains, biologists came to see that all living organisms are composed of cells—many, many tiny cells. At one time they believed that cells arose by the transformation of noncellular fluids, but later that idea was rejected. Cells were seen to arise by the subdivision of previously existing cells. Thus, the prevailing view arose: all cells come from cells. Cells do not arise *de novo* from noncellular material.

The structure of different cell types varies tremendously, as the text describes for eggs and sperm. The lining of the mouth is a thin film; the cells that form this lining are themselves thin, flattened discs. Cells that make up the nervous system may possess elongated extensions one meter or more in length. Secretory cells may be shaped like small wine-bottle corks standing side by side; in addition, the two

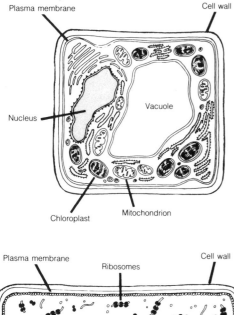

Plasma membrane — Cell wall

Nucleus

Vacuole

Chloroplast — Mitochondrion

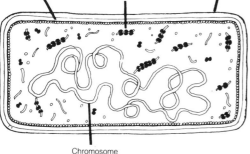

Plasma membrane — Ribosomes — Cell wall

Chromosome

Figure 2.28. Generalized plant (*top*) and bacteria (*bottom*) cells for comparison with the animal cell shown in Figure 2.4. Note that the bacterial chromosome is not surrounded by a nuclear membrane. Cells of green plants contain chloroplasts, the organelles within which photosynthesis occurs, and a liquid-filled vacuole that helps maintain the turgor of plant tissues.

Figure 2.29. A cell is not a simple bag of chemicals. This drawing shows a cell squashed thin. Part of a mitochondrion can be seen to the left and a section of endoplasmic reticulum on the right. Microtubules, which aid in the intracellular transport of materials, are shown along with a three-dimensional mesh of microfilaments, which give shape to the cell. The microfilaments contain actin, just as muscle fibers do. (*Photo courtesy of Professor Keith Porter, University of Maryland*)

ends may be differentiated, thus reflecting that the cell's secretion emerges from one end, not both.

Although the shapes and structures of cells may vary, certain general features can be recognized. From the onset, the presence of a nucleus has been considered essential in defining what is or is not a cell: a cell must have (or must have had until recently) a nucleus. Cytologically, the nucleus is seen as a dense body that stains differently from the rest of the cell or, if the optical properties of cellular material are the basis for observation, stands out as a body with its own characteristic refractive index. Sperm cells are essentially nuclei possessing a source of energy and filamentous tails that provide mobility.

Surrounding the nucleus is the cell's cytoplasm. Before protein chemistry advanced to its modern state, the cytoplasm was said to consist of protoplasm, a substance characteristic of living material. Biochemists once viewed the cell, especially the cytoplasm surrounding the nucleus, as a tiny bag filled with enzymes. Today it is known that the cytoplasm is largely an organized, membranous, reticulated structure. Small granules—ribosomes—are attached to the numerous layers of the reticulum, and they are the sites of protein synthesis.

Cells also contain small organelles: mitochondria in both animal and plant cells, and chloroplasts (green, chlorophyll-containing organelles) in certain cells of higher plants as well as in some one-celled organisms. The mitochondria are really the cell's source of energy; they produce the chemical (adenosine triphosphate, ATP) that is to the energetics of a cell what dry cell batteries are to a portable radio.

Surrounding the entire cell, separating it from its surroundings and thus delimiting it, is the cell membrane. In one sense, a cell membrane is like a person's skin: it separates the inside from the outside. The cell membrane is not a skin, however. First, a person's skin is composed of cells; the cell membrane is part of one cell. Second, skin—outer skin, at least—provides a barrier that protects the body within; in contrast,

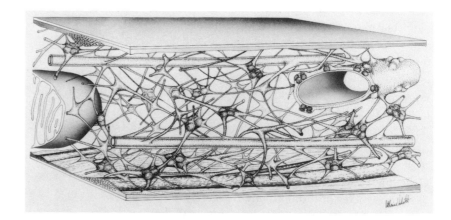

the cell membrane allows certain chemicals to enter the cell and others to leave. The cell membrane is an essential part of the functioning cell: holding together the structures and organelles that need to be near one another; allowing substances essential for the functioning of cellular machinery to enter; allowing poisonous waste products to leave; and, if the cell produces certain compounds that are destined for export and eventual use outside the cell, allowing these to pass out of the cell as well.

INFORMATIONAL ESSAY
Cell Division

All cells come from cells. No cell arises from noncellular material. These are well-established facts. On the other hand, genetic material representing the blueprints for the cell's activity, the cell's fate, and the cell's moment-by-moment self-maintenance ("housekeeping") comes from the individual's parents: half came from the mother's egg, the other half was brought in by the father's sperm. These are also well-established facts. The task before us is to explain how the precise apportionment of genetic material accompanies each cell division. This explanation is best grasped if presented in largely pictorial form.

The formation of germ cells can be illustrated easily by a simple analogy. Recall that the genetic material is packaged into bodies called chromosomes. Because both eggs and sperm contribute one set of chromosomes to the new individual, each cell of the human body contains pairs of corresponding chromosomes—23 pairs in all (one pair in

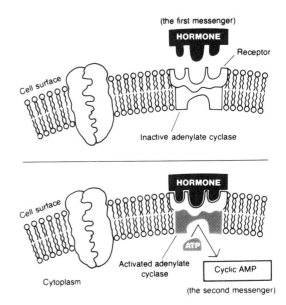

Figure 2.30. Communication through the cell membrane. The surface membrane of a cell is primarily a double layer (each layer one long molecule thick) of lipids—i.e., fatty material. Within the membrane are proteins (large, irregular structures in the diagrams) floating (as one person has said) like icebergs in a sea of lipids. The two diagrams of this figure illustrate how a hormone molecule, by coming to rest on a membrane receptor site *outside* the cell, can cause the machinery *within* the cell to react both promptly and properly.

Figure 2.31. The method of chromosomal apportionment in ordinary cell division (*mitosis*) and in the special divisions (*meiosis*) that must precede germ-cell formation.

In ordinary cell division (mitosis), each daughter cell acquires two complete sets of chromosomes identical to those of the parent cell. This is accomplished, glancing from top to bottom in the sequence at the right, by the duplication of each chromosome, by the movement of all duplicated chromosomes to an equatorial plane, and by the subsequent orderly passage of one member of each duplicated chromosome to each of the two daughter cells.

In reducing the number of chromosomes during germ-cell formation (meiosis), each primitive germ cell undergoes two divisions, while the chromosomes duplicate only once. The result is four germ cells (four functional ones in the case of sperm cells; one large functional one and three smaller nonfunctional polar bodies in the case of egg cells), each of which carries only one chromosome of each

original pair. Starting at the top left, homologous members of each pair of chromosomes come together in intimate contact. The configuration shown in the third diagram under "meiosis" arises from (1) the doubling of each chromosome and (2) an exchange of parts between chromosomes of maternal and paternal origin. The ability to make this exchange is one of the great advantages of sexual reproduction. The remaining diagrams show the two divisions and the consequent reduction in chromosome number.

If individuals are to arise by the union of two germ cells, each of which carries chromosomes, some mechanism for the reduction of chromosome number prior to germ-cell formation must exist; otherwise, the amount of chromosomal material per cell would double each generation. Continued doubling, generation after generation, would lead to the extinction of any species. The logic of this statement led to the postulation that a reduction–division exists even before anyone had actually observed the event occurring.

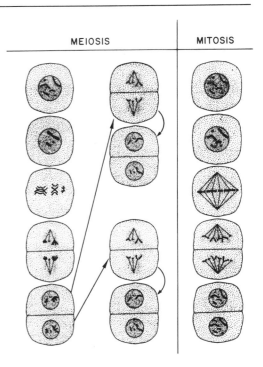

the male being a nonmatching, X–Y, "pair"). The genetic content of the different chromosomes is not identical; one chromosome cannot substitute for another. How, then, can we visualize the formation of germ cells that contain half of the normal cellular complement, one member of each pair?

Imagine a waiter in a fancy restaurant who is preparing the place settings at a table to be occupied by two persons. As he approaches the table, the waiter carries pairs of nearly every item constituting a place setting: knives, forks, salad forks, teaspoons, soup spoons, water glasses, wine glasses, dinner plates, salad plates, and more. Once the table has been set, each diner has before him one-half of the waiter's material: one knife, one teaspoon, one plate, and more. Diners do not receive merely one-half of the waiter's original load: two cups for one and two soup spoons for the other. Cups and soup spoons are not interchangeable; they do not serve the same function. The responsibility of the waiter setting a table for two is precisely that of the process (*meiosis*) by which the paired chromosomes of one cell are separated so that the resulting germ cells have only one member of each chromosome pair.

The process by which chromosomes are passed from a cell to its two "daughter" cells during normal cell division cannot be illustrated by an analogy based on silverware. The difficulty lies in the need for each chromosome to double, and then for the two halves to separate, with each half passing to one of the two daughter cells. Even before the chemical nature of genetic material was known, geneticists appreciated the necessity for this duplication and subsequent division of the chromosomes to proceed lengthwise; segments along the length of the same chromosome were known to be unable to substitute for one another. At the moment, we can only repeat that the DNA content of a cell doubles shortly before the cell divides; after the division occurs, each cell has precisely the original amount once more—not only in total weight, but also in the sense of a point-by-point duplication–division process involving the entire length of every chromosome.

INFORMATIONAL ESSAY
DNA—Structure and Replication

Many molecular biologists have forgotten (or are too young ever to have known) that the essential properties of genetic material had been specified by 1930: the material had to be able to duplicate itself precisely, it had to be able to undergo change (that is, to mutate), and the changed form had to be able to duplicate itself as faithfully as the original form. The last property was the most difficult to visualize: most molecules, one would imagine, either would fail to reproduce themselves if changed or, if they did manage to reproduce, would

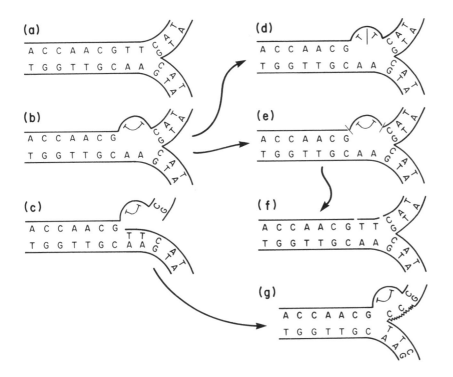

(a)

A C C A A C G T T
T G G T T G C A A

(b)

A C C A A C G
T G G T T G C A A

(c)

A C C A A C G
T G G T T G C A A
T T

(d)

A C C A A C G
T G G T T G C A A
T T

(e)

A C C A A C G
T G G T T G C A A

(f)

A C C A A C G T T
T G G T T G C A A

(g)

A C C A A C G
T G G T T G C

Figure 2.32. Diagrams illustrating how DNA manages to correct errors, thus ensuring (as far as is possible) that its replication is accurate.

Normal DNA replicates accurately (a) because the purines, adenine (A) and guanine (G), can only pair with the pyrimidines thymine (T) and cytosine (C) to form A–T and C–G crosslinkages. When the two strands of DNA are separated the purines and pyrimidines of each direct the construction of the new strands; thus, the pair A–T, when separated, becomes A– and –T. Each then specifies the missing partner, so that two identical A–T, A–T pairs are formed.

Exposure of DNA to ultraviolet (UV) light causes adjacent thymine molecules to interact, forming thymine dimers (b), which, if left unrepaired, cause replication of DNA to stop (c).

An enzyme that is activated by visible light can attach itself to thymine dimers, separate the two molecules of thymine (d), and (if the repair precedes replication) allow replication to proceed normally. (A proportion of bacteria that are exposed to UV light and subsequently kept in the dark die because their DNA cannot replicate. Exposure of UV-treated bacteria to visible light greatly reduces the otherwise expected mortality; hence, the enzyme is called the "photoreactivation" enzyme.)

In the absence of visible light, a thymine dimer can still be repaired through the cooperation of three enzymes. Sensing an error in DNA construction, an endonuclease will remove a portion of the dimer-containing strand (e). A polymerase can then reinsert matching purines and pyrimidines (f). As they are inserted, a ligase unites the bases into a coherent strand once more.

Finally, nearly all cells possess an SOS repair system. These are rather error-prone polymerases that ignore the dimer and insert bases at random into the corresponding spot of the new strand (g). Errors are made, to be sure, but DNA "replication" (and subsequent cell division) does occur. (Note that Cs have been inserted instead of As.) Those cells in which repairs were adequate survive and reproduce; the others die without leaving progeny cells. These different fates constitute natural selection.

tend to correct the change and revert to the original form once more.

The structure of DNA proposed by Watson and Crick in 1952 fulfilled the earlier specifications; if it had not done so, the search for the genetic material would have continued. DNA, whose structure has now been confirmed by a multitude of physical and chemical tests, consists essentially of two long filaments or fibers lying side by side and held in place by electrons shared by laterally projecting submolecules, purines and pyrimidines. Only two crosslinks give a proper fit for the matching filaments: adenine (a purine) with thymine (a pyrimidine), and guanine (a purine) with cytosine (a pyrimidine). In brief, A–T and C–G pairs form crosslinks holding the DNA duplex structure together. They form the rungs on what can be viewed as a twisted ladder.

The A–T and C–G rungs can occur in any order along the length of the DNA molecule. Whatever the order might be, however, if one were to separate the two fibers one from the other, and then use the "half-rungs" of each fiber to guide the reconstruction of the missing half, the outcome would be the construction of two identical DNA molecules. At a given spot along the original double-stranded molecule one might find a T–A rung; separated, one fiber possesses –T and the other –A. In guiding the reconstruction of the missing half of the DNA molecule, T demands A, to give rise to T–A once more, and

A demands T, also giving rise to T–A. Furthermore, if by some accident, G–C were to come to occupy the site we are discussing, it would reproduce itself as faithfully as the earlier (and correct) T–A pair. Thus, the essential properties of genetic material as specified by early geneticists are fulfilled by the chemical DNA.

INFORMATIONAL ESSAY
DNA—Protein Synthesis

Although early geneticists were able to specify certain essential properties for genetic material, they had little or nothing to say about the nature or origin of proteins. Nothing meaningful could be said about proteins until all proteins were recognized as being linear sequences of amino acids and, much later, all molecules of a particular kind of protein were recognized as being chemically identical—amino acid by amino acid, throughout the molecule's entire length. The latter realization coincided with the understanding of the structure of DNA provided by the Watson-Crick model. Together, the two types of information allowed the following type of reasoning:

DNA –A–A–T–C–T–G–G–A–A–A–C–G–T–T–C–G–A—
      –T–T–A–G–A–C–C–T–T–T–G–C–A–A–G–C–T—

Protein $aa_1.aa_2.aa_3.aa_4.aa_5.aa_6.aa_7.aa_8.aa_9.aa_{10}.aa_{11}.aa_{12}.\ldots$

The amino acid (aa) composition of a protein might be linearly related to the purine-pyrimidine sequence in a corresponding segment of DNA. Indeed, this view is correct; missing, however, is an important intermediary—messenger RNA (mRNA)—which reproduces the information contained in DNA (is *transcribed* from DNA) and carries it to the ribosomal machinery in the cytoplasmic reticulum, where the information is *translated* into the amino acid sequence of the protein molecule.

DNA –A–A–T –C–G–C –T–G–C –A–G–C –C–C–G—
      –T–T–A –G–C–G –A–C–G –T–C–G –G–G–C—

mRNA –A–A–U+C–G–C+U–G–C+A–G–C+C–C–G—

Protein — Asn — Arg — Cys — Ser — Pro —

This account has of course neglected a great deal of cellular machinery, in the form of ribosomal proteins, transfer RNA (tRNA, whose job it is to carry specific amino acids and insert them into their proper place within the growing protein molecule under the guidance of a three-letter code), enzymes, and energy sources.

SECOND LETTER

FIRST LETTER	U	C	A	G	THIRD LETTER
U	UUU⎫Phe UUC⎭ UUA⎫Leu UUG⎭	UCU⎫ UCC⎬Ser UCA⎪ UCG⎭	UAU⎫Tyr UAC⎭ UAA Stop UAG Stop	UGU⎫Cys UGC⎭ UGA Stop UGG Trp	U C A G
C	CUU⎫ CUC⎬Leu CUA⎪ CUG⎭	CCU⎫ CCC⎬Pro CCA⎪ CCG⎭	CAU⎫His CAC⎭ CAA⎫Gln CAG⎭	CGU⎫ CGC⎬Arg CGA⎪ CGG⎭	U C A G
A	AUU⎫ AUC⎬Ile AUA⎭ AUG Met	ACU⎫ ACC⎬Thr ACA⎪ ACG⎭	AAU⎫Asn AAC⎭ AAA⎫Lys AAG⎭	AGU⎫Ser AGC⎭ AGA⎫Arg AGG⎭	U C A G
G	GUU⎫ GUC⎬Val GUA⎪ GUG⎭	GCU⎫ GCC⎬Ala GCA⎪ GCG⎭	GAU⎫Asp GAC⎭ GAA⎫Glu GAG⎭	GGU⎫ GGC⎬Gly GGA⎪ GGG⎭	U C A G

Figure 2.33. The genetic code. A three-letter code that involves an "alphabet" of four letters provides for 64—and only 64—words; the table shown here is the complete dictionary. Three of the words signify "stop"; the enzymatic protein-synthesizing machinery terminates the protein (polypeptide chain) at this point. The word that specifies methionine (AUG) is the "start" signal; the amino acid methionine may or may not be enzymatically removed later from position one of the polypeptide chain. The remaining 60 words specify the other nineteen amino acids; there are, of course, a number of synonyms. This figure is intended to illustrate the simplicity of the communication system; it is *not* for memorization.

The Genetic Code and Mutation

The structures of DNA and of proteins were comprehended, if not discovered, nearly simultaneously. DNA, according to the Watson-Crick model, is an enormously long, double-stranded molecule within which certain nitrogenous bases (purines and pyrimidines) can be arranged in any sequence. Whatever the sequence, however, the molecule of DNA replicates itself faithfully; its two strands separate and, during the synthesis of new strands, each original one specifies precisely—base by base—the nature of the newly synthesized one.

Early chemical analyses had shown that individual proteins contain constant ratios of different amino acids; furthermore, if the rarest amino acid were assigned an integer value (1, 2, or 3, for example), the other amino acids took on higher, but still integer, values as well. Electrophoretic separation of the fragments of digested proteins, and chemical analyses of the resulting polypeptide fragments, revealed that the *sequence* of amino acids in a protein, not merely its overall composition, is precise.

That genes (that is, DNA) are involved in the synthesis of proteins (enzymes, in particular) has been known since studies were carried out on eye-color pigmentation in *Drosophila*; these studies led to the much finer studies of biochemical mutations in *Neurospora*, the pink bread mold. Out of the latter studies came a famous conclusion: one gene–one enzyme.

Given two linear macromolecules, DNA and (in its primary structure) protein, and the evidence that in some manner the one controls the synthesis of the other, many persons strove to find the key—or code—by which the control was exercised. Because proteins are built of twenty amino acids, whereas DNA is built of only four nitrogenous bases (adenine, thymine, cytosine, and guanine), one base cannot specify one amino acid. Two bases in combination could specify only sixteen amino acids, because the four bases can form only sixteen combinations. However, three bases can form sixty-four combinations; consequently, DNA words consisting of three bases could specify twenty amino acids—with words left over. These words (mRNA codons) are now known: three words ("stop" codons) specify the end of the growing polypeptide chain, one ("start" = AUG) specifies its beginning (and causes methionine to be inserted at site number one; the methionine may later be removed enzymatically), and the remaining sixty codons specify amino acids.

Given the means by which the sequence of purines and pyrimidines in DNA control, by way of intermediary mRNA, the sequence of amino acids in protein molecules, one can visualize the errors that might occur (i.e., mutations). A number of these errors are identified and illustrated in Figure 2.34.

Figure 2.34. A brief catalogue of gene mutations resulting from the gain (or loss) or substitution of base pairs in DNA. In each case, the change in DNA is identified (compare with the nonmutated form) and the modified mRNA is shown, as is the polypeptide whose synthesis would be directed by the altered mRNA. *(Copyright, Columbia University Press; reproduced with permission)*

DNA	mRNA	POLYPEPTIDE
NON-MUTATED		
A-T-T-C-T-G-A-C-T-T-G-C- T-A-A-G-A-C-T-G-A-A-C-G-	A-U-U-C-U-G-A-C-U-U-G-C-	Ile-Leu-Thr-Cys-
MISSENSE MUTATION : TRANSITIONAL , SYNONYMOUS		
A-T-T-[T]-T-G-A-C-T-T-G-C- T-A-A-[A]-A-C-T-G-A-A-C-G-	A-U-U-U-U-G-A-C-U-U-G-C-	Ile-Leu-Thr-Cys-
MISSENSE MUTATION : TRANSITIONAL , NON-SYNONYMOUS		
A-T-T-C-T-G-[G]-C-T-T-G-C- T-A-A-G-A-C-[C]-G-A-A-C-G-	A-U-U-C-U-G-G-C-U-U-G-C-	Ile-Leu-Ala-Cys-
MISSENSE MUTATION : TRANSVERSION		
A-T-T-C-T-G-A-C-T-[G]-G-C- T-A-A-G-A-C-T-G-A-[C]-C-G-	A-U-U-C-U-G-A-C-U-G-G-C-	Ile-Leu-Thr-Gly-
NONSENSE MUTATION		
A-T-T-C-T-G-A-C-T-T-G-[A] T-A-A-G-A-C-T-G-A-A-C-[T]	A-U-U-C-U-G-A-C-U-U-G-A	Ile-Leu-Thr STOP
FRAMESHIFT MUTATION		
[G]-A-T-T-C-T-G-A-C-T-T-G-C- [C]-T-A-A-G-A-C-T-G-A-A-C-G-	G-A-U-U-C-U-G-A-C-U-U-G-C-	Asp-Ser-Asp-Leu

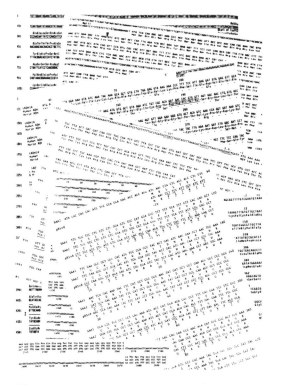

Figure 2.35. A montage of DNA sequences that were reported in several issues of a single biological research journal. The amino acids that correspond to the DNA strands are shown in many instances either by a 3-letter or by a 1-letter symbol for each amino acid. Note the lengths of the analyzed sequences: 4500, 2700, and 1700 base pairs are mentioned in three instances. Scientists can now obtain such sequences rather easily; however, they have not yet learned to *interpret* them, except for regions that are responsible for specifying the amino acid sequences of proteins.

For many years it seemed as if the nature of gene mutations had been fully understood: they consisted, it seemed, of base substitutions, base deletions, and base additions—the latter giving rise to "frameshift" mutations. All such changes would cause the protein-synthesizing machinery to go awry. Today, it is known that many mutations, especially in higher organisms, are caused by either the insertion of sizable pieces of DNA near or into gene loci or their removal. Little is known of the normal role of these transposable elements (*transposons*) except that they can at times control the transcription of mRNA at the gene locus. Because the orientation of an inserted DNA segment often determines the effect of that segment (one orientation leaves the gene "on"; the other turns it "off"), transposons could be intimately associated with the normal control of gene action during development. Indeed, such transposable elements have now been implicated in the origin of certain cancer cells—cells that undergo repeated, uncontrolled cell divisions.

INFORMATIONAL ESSAY
Amniocentesis

Hundreds of mental and physical abnormalities have genetic causes. Many of these can be detected early in pregnancy by the enzymatic or chromosomal analyses of fetal cells. The logic of these tests is that if the tests are performed early in fetal development, births of abnormal or potentially abnormal children can be avoided by induced abortions.

The actual procedures by which fetal cells are obtained from an unborn fetus, how these cells are multiplied in tissue culture flasks, and the types of tests that can then be performed using these cells are best understood by referring to Figure 2.36.

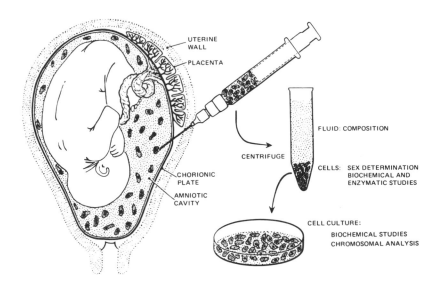

Figure 2.36. The basic steps in performing an amniocentesis. (1) Obtaining fetal cells and fluid by inserting a syringe through the mother's abdominal wall into the amniotic cavity, (2) separating the cells from the fluid, and (3) growing the fetal cells in tissue culture. Both the fluid and cultured cells can be subjected to a variety of biochemical or cytogenetic tests.

Having located the precise position and orientation of the fetus in its mother's womb, the doctor inserts a needle through the mother's abdominal wall, through the wall of the uterus, and into the amniotic cavity that surrounds the fetus. Some amniotic fluid, which contains embryonic cells, is withdrawn. The cells in the fluid, which have been sloughed off the developing fetus's skin and mucous membranes, are separated ("spun down") from the fluid by centrifugation and transferred into tissue-culture tubes. These growing cells, greatly outnumbering those in the original sample, are used for chromosomal analyses and biochemical tests. The amniotic fluid itself is also available for chemical tests. Chromosomal analyses include determining the fetus's sex, an examination for extra chromosomes, the detection of missing chromosomes, and the identification of chromosomal rearrangements (translocations or unusually large or small chromosomes).

The number of biochemical analyses that can be performed on embryonic cells continues to grow. Listed below are more than thirty disorders for which tests currently can be performed routinely, plus an additional dozen that will probably become analyzable in the near future:

Disorder	Affected Enzyme or Protein	Disorder	Affected Enzyme or Protein
Acanthocytosis	β-Lipoproteins (low density)	Orotic aciduria	Orotidine 5'-phosphate pyro-phosphorylase
Acatalasia	Catalase	Parahemophilia	Accelerator globulin
Afibrinogenemia	Fibrinogen	Pentosuria	L-Xylulose dehydrogenase
Agammaglobulinemia	γ-Globulin	Phenylketonuria	Phenylalanine hydroxylase
Albinism	Tyrosinase	Sulfite oxidase deficiency	Sulfite oxidase
Alkaptonuria	Homogentisic acid oxidase	Wilson's disease	Ceruloplasmin
Analbuminemia	Serum albumin	Xanthinuria	Xanthine oxidase
Argininosuccinic acidemia	Argininosuccinase		
Crigler-Najjar syndrome	Uridine diphosphate glucuronate transferase		

Disorder	Biochemical Manifestation
Favism	Glucose-6-phosphate dehydrogenase
Fructose intolerance	Fructose-1-phosphate aldolase
Fructosuria	Fructokinase
Galactosemia	Galactose-1-phosphate uridyl transferase
Goiter (familial)	Iodotyrosine dehalogenase
Gout	Hypoxanthine guanine phosphori-bosyl transferase

Disorder	Affected Enzyme or Protein	Disorder	Biochemical Manifestation
Favism	Glucose-6-phosphate dehydrogenase	Congenital steatorrhea	Failure to digest and/or absorb lipid
Fructose intolerance	Fructose-1-phosphate aldolase	Cystic fibrosis	Thick, viscous mucous secretion, high sodium content of all secretions
Fructosuria	Fructokinase	Cystinosis	Inability to utilize amino acids, notably cystine; aberration of amino acid transport into cells
Galactosemia	Galactose-1-phosphate uridyl transferase		
Goiter (familial)	Iodotyrosine dehalogenase		
Gout	Hypoxanthine guanine phosphori-bosyl transferase	Cystinuria	Excretion of cystine, lysine, arginine, and ornithine
Hartnup's disease	Tryptophan pyrrolase	Fanconi's syndrome	Increased excretion of amino acids
Hemoglobinopathies	Hemoglobins	Gargoylism (Hurler's syndrome)	Excessive excretion of chondroitin sulfate B
Hemolytic anemia	Pyruvate kinase		
Hemophilia A	Antihemophilic Factor A		
Hemophilia B	Antihemophilic Factor B	Gaucher's disease	Accumulation of cerebrosides in tissues
Histidinemia	L-Histidine ammonia lyase	Niemann-Pick disease	Accumulation of sphingomyelin in tissues
Homocystinemia	Cystathionine synthetase		
Hypophosphatasia	Alkaline phosphatase	Porphyria	Increased excretion of uroporphyrins
Isovaleric acidemia	Isovaleryl CoA dehydrogenase	Tangier disease	Lack of plasma high-density lipoproteins
Maple syrup urine disease	Amino acid decarboxylase	Tay-Sachs disease	Accumulation of gangliosides in tissues
Methemoglobinemia	Methemoglobin reductase		

INFORMATIONAL ESSAY
Atoms and Radiation[1]

ANATOMY OF THE ATOM: RADIATIONS

The atoms of the 92 elements known before 1940, and a few more discovered since then, can be regarded as consisting of three basic particles—electrons, protons, and neutrons. These particles are extremely small—10,000 times smaller in diameter than the atom itself. Moreover, the interior of the atom resembles, of all things, the solar system. In the center there is the atomic nucleus, compounded of protons and neutrons; the nucleus may be likened to the sun in the solar system. In the space around the nucleus, and spinning around it like the planets around the sun, are a number of electrons. The number of orbital electrons is characteristic for atoms of each element.

The simplest atom is that of the element hydrogen. A hydrogen atom has a diameter of about 0.000,000,005 of an inch. The center of

[1]Reprinted by permission from *Radiation, Genes, and Man,* by Bruce Wallace and Th. Dobzhansky, published by Holt, Rinehart, and Winston, New York.

the atom is occupied by the nucleus; the nucleus contains but a single proton, which is an electrically charged particle positive in sign. Rotating around the nucleus is a single electron. The electron is also electrically charged but its charge is negative, thus balancing the positive charge of the nucleus. The mass of the electron is, however, very tiny—about 1/1,840 of that of the proton. Most of the weight of the atom is thus concentrated in the nucleus. Atoms of other elements are more complex. Their nuclei contain varying numbers of protons and neutrons depending upon the element, and many electrons gyrate around the nuclei in fixed orbits. For each element the number of electrons balances the number of protons in the nucleus. Thus, an atom of one of the heaviest elements, uranium, has a nucleus compounded of 146 neutrons and 92 protons; 92 electrons spin around the nucleus in several concentric "shells." Atoms of other elements range in complexity, and in weight, between those of the lightest element, hydrogen, and up to and beyond uranium at the upper end.

If the properties of electrons are to be studied easily they must be removed from the atoms containing them. One of the early tools used for this purpose by physicists was the Geissler tube, a glass tube from which most of the air has been evacuated. Inside, at each end of the tube, is a metallic plate connected to the poles of a high-voltage source of electricity by a wire that pierces the glass. When the current is turned on, the tube gives off a glow, which is caused by the passage of a stream of electrons through the rarefied air. The electrons can be demonstrated to leave the plate connected to the negative pole and to jump across the tube to the plate connected to the positive pole. By coating the positive plate with a phosphorescent substance (zinc sulfide), one is able to discern the impact of the electrons as pinpoints of light.

In 1895 a German physicist, W. K. Roentgen, discovered that hitherto unknown radiations were coming out of Geissler tubes. The properties of these radiations seemed mysterious enough for them to be called X-rays—"X" designating the unknown nature of the rays. X-rays are invisible to the human eye, but they can be detected by photographic plates; strangest of all, X-rays easily penetrate screens of black paper and other materials completely opaque to ordinary rays of light.

The penetrating powers of X-rays made them important in ways which were not obvious to their discoverer and to many others. They became indispensable tools for the diagnosis of many diseases and for the treatment of some of them. Well-equipped hospitals, and even doctors' and dentists' offices, now usually include X-ray machines. Unfortunately, together with properties useful to man, these rays proved to have, when improperly used, some sinister ones as well. They can cause severe damage to the body and, as we have known since 1927, to the units of heredity, the genes.

X-rays are not streams of electrons escaping from the Geissler tube; electrons cannot pass easily through the glass or heavy paper. X-rays are a part of the radiation spectrum, of which visible light is one part and radio waves still another. The properties of radiation depend upon the length of its waves. As shown in the figure, radio waves vary in length from several kilometers to a few meters; microwaves (radar waves) from meters to millimeters; infrared (heat) waves from about a millimeter to $8/10,000$ (8×10^{-4}) of a millimeter; visible light occupies the narrow band from 8×10^{-4} to about 4×10^{-4} of a millimeter; ultraviolet from 4×10^{-4} to 10^{-5}; while X-rays extend from some 10^{-5} (soft X-rays) to 10^{-8} (hard X-rays) and to 10^{-9} (known also as gamma-rays). Still shorter wavelengths (those characteristic of cosmic rays) are known.

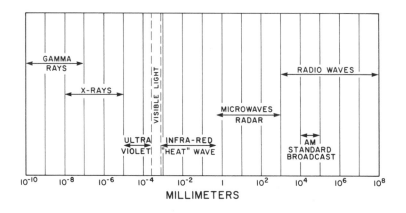

Figure 2.37. The spectrum of electromagnetic radiation, extending from cosmic (not shown) and gamma-rays to radiowaves. Note the rather narrow band within which all visible light (violet to red) falls. Ultraviolet and shorter radiation (X- and gamma-rays) can seriously damage genetic material (DNA).

Another basic fact of radiation physics is that radiation behaves as if it comes in discrete packages; these are called quanta. Each quantum of radiation has in it a certain amount of energy; the shorter the wavelength of radiation, the greater the energy contained in its quanta. Thus, a quantum of visible light has relatively little energy, quanta of ultraviolet light are more powerful, and those of X- and gamma-rays are more energetic still. This is why X-rays and gamma-rays are known as high-energy radiations.

EFFECTS OF RADIATION ON MATTER

When visible light falls on a material body, the light quanta are either absorbed in or reflected from the body. The energy carried in the quanta of visible light is, however, not sufficient to affect the structure or behavior of most atoms. Quanta of light resemble peas shot from a toy gun against a wall; the peas bounce off and do not go through the wall. Only some few substances are "phosphorescent." This means that when they are exposed to light, electrons are displaced in their atomic orbits and the substance gives off a glow (that is, emits visible light) for a short time after the exposure stops. The visible radiation is emitted as electrons return to their normal position. Other substances, such as silver bromide, which is used in photographic film, are light sensitive; their molecules undergo chemical changes because of the energy delivered to them by light.

X-rays, gamma-rays, and other high-energy radiations have quanta so powerful that they may perhaps be likened to bullets shot from a rifle. When such a quantum "hits" an atom, its energy is imparted to one of the electrons circling around the atomic nucleus. An "energized" electron may be ejected from an "excited" atom and fly at enormous speed into surrounding space, where it hits other atoms. Now, an atom that has lost one of its electrons has lost one unit of negative charge. It will, therefore, be a positively charged atom, or a positive ion. Such ions are chemically much more active than electrically neutral atoms; they will react easily with other atoms or atomic groups which are able to "share" an electron with them.

The electron ejected from an atom darts through the surrounding space, where it may collide with other electrons in neighboring atoms. These other electrons may in turn be knocked from their orbits. A hit by a powerful X-ray quantum thus produces a whole series of agitated electrons and ionized atoms. The disturbance caused by the initial hit gradually subsides as the energy of the absorbed quantum is distributed among more and more electrons. Finally, no one electron has enough energy left to displace other electrons, and the residue of the energy is dissipated as heat. By then, however, a number of electrons may have been displaced from their atoms; the atoms that have lost electrons are left behind as positive ions; the displaced electrons or the atoms that have picked them up are negative ions. Each displaced electron results in a pair of ions, one positive and one negative. This is why high-energy radiations are also called ionizing radiations.

Absorption of quanta of high-energy radiation by atoms is not the only cause of ionization. High-speed electrons, called beta-rays, exist within X-ray tubes; in fact, X-rays are produced when these electrons strike the metal in the positive electrode. Disintegrating atoms also eject electrons from their orbit at a high speed; beta-rays are, in fact, a characteristic emanation of disintegrating radium atoms. These rays have a wide range of velocities and, hence, of characteristic energies. Although more energetic beta-rays are known, most of those produced by atomic disintegration are capable of passing through only thin layers of aluminum or about 15 feet of air. They slow down rapidly as they collide with other electrons; their initial energies are rapidly dissipated by ejecting these other electrons from their orbits. Again, the result of irradiation by beta-rays is the formation of many pairs of ions.

Because they are simply rapidly moving electrons, beta-rays are electrically charged. Another type of charged particle is the alpha particle, a rather ponderous particle bearing two positive electric charges; it is, in fact, the nucleus of a helium atom. Again, velocities of alpha particles vary according to their source; a value of about one-twentieth the speed of light is not unusual. For all their weight, alpha particles are not efficient penetrators; a sheet of paper suffices to stop most such particles. However, in living tissue the extent of their penetration is not negligible, especially if their source is inside the body. Throughout the distance they do penetrate, their effect is tremendous; as they roar past and through atoms, they knock electrons out of their orbits at a tremendous rate. Not only electrons, but even whole atoms may be dislodged by the blows of these relatively massive particles.

Neutrons have effects of their own. Neutrons, it may be recalled, together with protons make up atomic nuclei. When atomic nuclei disintegrate, neutrons are quite frequently ejected. These may travel at a fast or slow speed. Since they have no electric charge, they can rush or drift right through the electronic orbits of other atoms. Eventually, they are "captured" by the nuclei of these other atoms. For reasons that are not completely known, certain numerical combinations of protons and neutrons are unstable; atoms having such unstable combinations undergo spontaneous nuclear disintegration.

This disintegration, or fission, may result in the ejection of additional neutrons. If, as in the case of uranium-235, the average number of ejected neutrons is larger than 1, there are more free neutrons after the disintegration of each atom than before. It thus becomes possible to initiate and sustain a chain reaction. This reaction may be slow and controlled (as in atomic piles) or almost instantaneous and of great violence (as in atomic, or fission, bombs). Electrons are also frequently expelled from unstable atoms as high-speed beta particles. Furthermore, the jostling of electrons in their orbits because of nuclear adjustments frequently results in the ejection of a quantum of gamma radiation. Gamma-rays are, as we have seen above, very powerful X-rays. Thus, the disintegration of radium atoms results in the release of all three kinds of radiation—alpha-, beta-, and gamma-rays. These rays are all ionizing, and the gamma-rays have, in addition, tremendous penetrating powers.

The disintegration of atomic nuclei poses a novel type of biological problem, since the atoms involved change from one chemical element into another. Thus, upon capturing a neutron, an atom of common sodium disintegrates into an atom of iron plus an atom of helium. The isotope (variety) of iron produced in this process is unstable and quickly changes into neon. It is obvious that such changes, if they occur in a living cell, may result in grave disturbances. Suppose that a physiological reaction depends upon the presence of sodium atoms; if some of these atoms suddenly change to helium and neon, the normal course of the reaction will be disturbed. The existence of a radioactive isotope of phosphorus poses a special problem for hereditary materials. The deoxyribonucleic acids (DNA), which are the most important constituents of the chromosomes, contain a great deal of phosphorus. If a radioactive variety of phosphorus becomes incorporated into chromosomal DNA, the affected molecule of DNA is doomed. Radioactive phosphorus exists for a few days only, and then changes into sulfur, an element entirely unsuitable for the chemical constitution of DNA.

The normal flow of life processes depends upon the presence, in the proper place and at the proper time, of certain chemical molecules. This is especially true of the processes of heredity. A change in a single gene-molecule in the sperm or in the egg cell may make the difference between a normal child and a pitiful invalid. High-energy, or ionizing, radiations disrupt the physical structure of atoms and initiate novel chemical reactions. The addition of a hydrogen atom here, its loss there; the ionization of a formerly uncharged atom; the production of hydrogen peroxide; the physical disruption of long, chain-like molecules—these and a dozen other effects or immediate after-effects of radiation may raise havoc with the cell's basic organization. In many instances, of course, the damage is done to an expendable substance and is repaired by the cell's built-in defense mechanisms. In many other instances, however, damage is done to a unique substance not easily replaced or repaired. Here the effects of radiation are serious. There is no more nearly unique, no more important, no more complex substance, nor one less capable of being repaired, than DNA—the chemical that conditions the hereditary processes and directs the manufacture of enzymes and proteins in every cell of the body.

MEASUREMENT OF RADIATION

The effects of radiation upon a living cell or a living body depend upon the amounts of energy the living substance absorbs from this irradiation. It is obviously important to measure the quantities of radiation applied and to relate them to the biological effects produced. Thus, exposure of the body to high-energy radiations may result in radiation sickness and in death; we must know, then, how much radiation of a given kind will be fatal to a mouse, or to a man, or to some other organism. We must also know how great are the amounts of radiation to which people are exposed from fallout products of testing atomic weapons, from X-rays used in medical practice, and from other sources.

The most useful unit in measuring exposure to X- or gamma-rays is the roentgen, usually abbreviated as "r." It is so named to honor the discoverer of X-rays. One roentgen corresponds to the amount of X- or gamma-radiation which produces about 2×10^9 ion pairs per cubic centimeter of air. This measurement is made fairly easy with the aid of an apparatus (dosimeter) which measures the leakage of electricity from a charged object. The amount of leakage depends on the number of ionized molecules; the greater the amount of X-rays, the greater the number of ions, and the greater the loss of charge as a result of leakage.

The number of ions produced by a given amount of radiation is, however, different in different types of substances. One roentgen of X-rays yields, for example, about 1.8×10^{12} ion pairs per gram of living tissue. The measurement of types of radiation other than X- and gamma-rays is more involved. The basic unit of the radiation absorbed is the "rad." The absorption of 1 rad delivers 100 ergs of energy per gram of matter. The relative biological effectiveness of 1 rad of alpha-rays is, however, approximately equivalent, under certain conditions, to that of 10 rads of gamma-rays. In practice, the biological effects of these radiations are measured by equating them with effects produced by a given dose of X-rays as measured in roentgens. Thus, we have a unit called "rem" (= roentgen equivalent for man). For measurement of very small amounts of radiation there are "millirads" (mrad, one-thousandth of a rad) and "millirems" (mrem, one-thousandth of a rem).

INFORMATIONAL ESSAY

Radiation and the Unborn Child

The cells most sensitive to radiation damage are those that are actively engaged in cell division. These cells are subject to damage because of the misdivision of damaged chromosomes during cell division and an inability of the cell's metabolic machinery to make or replace the many proteins that are needed in preparing for subsequent cell divisions. In adults, extremely sensitive tissues—blood-forming tissue, the lining of the gut, hair follicles, fingernails, and others—can be easily enumerated simply because there are so few of them. For young children the list of sensitive tissues would be extended to include the growing areas of various long bones, such as the arm, leg, finger, and toe bones. It is possible, for example, that many present-day Japanese adults are smaller today than they would have been had they not lived in Hiroshima during August 1945, when the first atomic bomb exploded. Persons exposed to radiation when very young may easily have suffered from retarded growth in later years because of radiation damage to their thyroid glands.

The calendar of events for the developing human embryo proceeds about as follows:

Day 0   An egg (weight: two-millionths of a gram) is fertilized by a sperm (weight: negligible).

Day 10   The embryo, now a mass of cells resembling a small raspberry, becomes embedded in the wall of the uterus.

Day 40   Major organ systems are now complete and the embryo, at least superficially, resembles that of nearly any other higher animal, from fish to mammal.

Day 60   Face complete; fingers and toes separated; tail diminishing. Total weight: 2 gm (about one million times the weight of the original egg).

Days 60–270   In the period between day 60 and birth, the embryo grows nearly 2000-fold in size; at birth the baby weighs about 3400 gm, or 7½ lbs.

An embryo does not grow uniformly. The head and brain are the first parts to form, and then development spreads posteriorly. The forelimbs (arms) take form, for example, before the hindlimbs (legs). Areas of differentiation are areas of rapid cell division. Consequently, if radiation strikes a developing embryo at a given moment, the most severely damaged portion is that which is differentiating at that instant.

During the first two weeks after fertilization, an embryo is merely a mass of dividing cells, with little or no organization in the sense of recognizable adult systems. Irradiation of the embryo at this time leads to its death or, should development continue despite radiation injury, to severe physical abnormalities.

Between the time of the embryo's implantation in the uterine wall and the fortieth day, when most organ systems are complete, even low levels of radiation can cause severe harm to one part or another of the developing child; the injured part is in each instance the one that was undergoing differentiation at the moment of radiation exposure. This effect, as was discussed in the text, allows the developmental embryology of an experimental animal to be studied by irradiating pregnant females at various times and noting the resulting abnormalities in their offspring. (For example, many embryos carried by pregnant women of Hiroshima and Nagasaki, the two sites of atomic bomb explosions, were killed by radiation exposure. Of the embryos surviving until birth, from one-fourth to one-half were mentally retarded.) The sensitivity of growing embryos to harm through radiation exposure is such that many governments now recommend a therapeutic abortion if the mother's abdomen has been exposed to as little as 10 roentgens of penetrating radiation.

3

LIFE: AN INSTANT-THICK SNAPSHOT

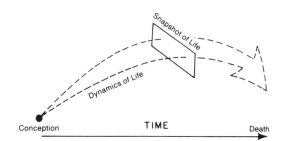

Figure 3.1. Two views of human life: the trajectory, which was the subject of the previous chapter, and the instant-thick snapshot, which is the subject of this and the following three chapters.

Most persons go through life paying no more attention to aging than many motorists pay to the countryside as they cruise along the interstate highways. Like the motorist, who is absorbed in thought and for whom the car radio and air conditioner create an illusion of constancy, most of us think of ourselves as unchanging. Let's look, then, at life—primarily, human life—in a brief moment of time when it seems to be stable, if not static.

To talk of life as static would be, of course, absurd. Life is not static. A living being is an open, dynamic system, albeit surrounded by some sort of recognizable boundary. Some materials pass into the interior, others are kept out. Some things are expelled from the interior, others are retained. Inside, a complex mixture of chemicals is maintained in a nearly constant composition, and doing so requires energy. Our snapshot, then, is illusionary, with no more reality than a scene in a discotheque revealed by a strobelight. Nevertheless, our illusion permits us to describe cellular systems whose moment-by-moment operation keeps us alive, thus giving all but one of our days a tomorrow.

The Life of a Bacterial Cell

The bodies of higher plants and animals, including the human body, are composed of billions of cells, most of which must live and function properly if the entire organism is also to live and function properly. As a whole, the multicellular organism has needs that lie beyond the total needs of its individual cells; nevertheless, many of its basic needs are dictated by metabolic demands that are common to many forms of life. These "housekeeping" tasks can be described adequately by considering a simple, well-known, single-celled organism, the colon bacillus. Each of us harbors an ounce or more of these bacteria in our large intestine; under these conditions, they probably grow more efficiently than other, possibly pathogenic bacteria, and in doing so render us a useful service. Here, then, we have an introduction to the ecological relationships between organisms—a subject considered again in Part III of this text.

The colon bacillus (*Escherichia coli*) is several hundred times smaller than most cells found in the human body. Its genetic apparatus consists of DNA in the form of a circular chromosome supplemented by small, nonchromosomal circles of DNA (plasmids) that carry only a few genes each. Plasmid-borne genes frequently confer resistance to one or another of the many antibiotics scientists have learned to obtain from *Penicillium* and soil fungi.

The bacterium illustrates several conditions that must be met if life is to exist. First, it has a membrane or cell wall that separates it from the outside world, since all living things, from simple cells to higher organisms, need protective boundaries. The exchange of chemical substances between the bacterium and the outside world takes place through, and to some extent with the active participation of, the cell wall.

The energy that keeps the cell running is obtained by the oxidation of sugars or other carbon-containing organic molecules. If only the beginning and the end of the chemical reaction involved in cellular combustion are considered, the reaction can be written as follows:

$$C_6H_{12}O_6 + 6O_2 \rightarrow 6CO_2 + 6H_2O$$
$$\text{glucose} + \text{oxygen} \rightarrow \text{carbon dioxide} + \text{water}$$

If the reaction took place in this manner, a great deal of heat would be generated—and lost without being put to use by the cell. Indeed, the cell would need to be fireproof.

NOW, IT'S YOUR TURN (3-1)

A simple experiment illustrating that sugar does indeed burn, and that the act of burning can be aided by intermediary substances, can be carried out with sugar cubes (sucrose) and cigarette ashes. Place a sugar cube on end in a small plate or saucer and attempt to light it with a match (or cigarette lighter). Roll a second sugar cube in cigarette ashes (which, obviously can no longer be set afire), place the cube on end in the dish, and attempt to set it afire. If all goes well, the second cube will burn readily—but it will probably leave a residue of unburned carbon.

Quantitative Arguments

Ernst Mayr, a world-famous ornithologist, evolutionary biologist, and philosopher of science, once criticized mathematical geneticists by asking: "But what, precisely, has been the contribution of this mathematical school to evolutionary theory?"

The late J. B. S. Haldane, one of those included in this criticism, responded in a manner that might be instructive for all persons who do not understand (and, therefore, fear or distrust) the quantitative formulation of problems.

First, Haldane denied that the mathematical theory of population genetics is impressive: "Our mathematics may impress zoologists but do not greatly impress mathematicians."

Second, citing the philosopher David Hume, he argued that algebra and arithmetic are the only sciences in which a chain of reasoning can be carried to any length and still preserve its exactness and certainty. (For example, in a ledger sheet of *any* size, the sums of the rows and those of the columns must add to one figure, the Grand Total.) Not only is algebra exact, it imposes an exactness on the postulates that are made before the algebra can start. This exactness is generally lacking in the case of ethical, political, or other arguments.

Here, then, is the crux of the matter: When one's thinking is so fuzzy that postulates and assumptions cannot be stated precisely, one's conclusions are likely to be wrong. If these postulates can be stated precisely, one's argument becomes mathematical *in appearance*—but not necessarily enough so to impress a mathematician. Our attempt to understand the origin of the many organic molecules found in *E. coli* and to compare *what is known* with what *appears possible* are not exercises in mathematics, but exercises in reasoning.

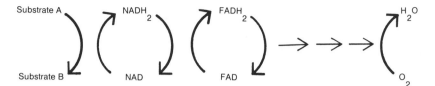

Figure 3.2. The cellular equivalent of burning sugars (or other carbohydrates) in order to recover energy. Instead of adding oxygen directly to Substrate A (a procedure that would be accompanied by a flame and a great deal of wasted heat), one of the cellular compounds (a coenzyme) removes two hydrogen atoms from Substrate A, converting it to substance B (NAD is converted to NADH$_2$ in the process). Then NADH$_2$ loses its two hydrogen atoms in turn to a second coenzyme (FAD, which becomes FADH$_2$), while NADH$_2$ returns to its original state, NAD. The two hydrogen atoms subsequently are passed through a series of reactions until they are eventually united with oxygen (H$_2$ + 1/2 (O$_2$) → H$_2$O, water). In the meantime, much energy that otherwise would have been wasted has been captured and used. Water-soluble vitamins are important components of this process.

In cellular respiration oxidation occurs not by combining organic compounds with oxygen, but by the chemically equivalent process of removing hydrogen. The removal of hydrogen atoms from molecules, like all organic chemical reactions, is mediated by enzymes. These enzymes, however, act in concert with smaller molecules, *coenzymes*. For any one reaction, the hydrogen atoms are transferred from one coenzyme to another; only in the terminal reaction are these atoms joined with oxygen to make water. The usable energy obtained by these enzyme-mediated reactions is a half-dozen times greater than the energy that would be obtained by direct oxidation.

For what purpose are carbon-containing molecules burned (oxidized) in cells? To build other compounds that constitute the cell itself. In all, there are thousands of such compounds—many, but not an infinite number of them. By harvesting many bacterial cells, breaking their cell walls, and subjecting the resulting soup to standard chemical analyses, a competent biochemist can identify hundreds of organic compounds. Where do these compounds come from? Not from the culture medium, because (for *E. coli*) it need contain only a simple sugar, such as glucose, some simple salts, including a source of nitrogen, and biotin, a common vitamin. Not from the original inoculum, because this may be a single bacterium whose total weight scarcely exceeds 10^{-12} g. The thousands of organic compounds that can be harvested from a culture of ruptured bacterial cells were made by the bacterial machinery. The proteins, the coenzymes, the nucleic acids, the fats and lipids, and all the intermediate metabolites, which either have functions of their own or are mere stepping stones to the synthesis of one chemical from another, are all made, one simple chemical at a time, through the catalytic action of enzymes.

The number of enzymes in a cell must at least approximate the number of compounds that exist in that cell. In the case of *E. coli*, the size of the chromosome is known: it contains about ten million nucleotide pairs. Thus, ignoring stretches of DNA that may serve regulatory or structural functions, the entire chromosome can *at most* code for some three million amino acids. If (and this is a rough estimate) the average protein were to contain 500 amino acids, a maximum of 6000 proteins could be synthesized under the control of *E. coli*'s DNA. Somewhat more than 1000 chemical compounds have been identified in *E. coli*; therefore, our snapshot of this cell, tiny as it is, is not yet fully developed—only about one-fifth so.

NOW, IT'S YOUR TURN (3-2)

The account of the activities of a bacterial cell given above assumes that nowhere within the cell is a mystical or "vital" force to be found: the chemical atoms of the living cell are those found in the nonliving world, no mysterious fluids or substances are encountered, the energy relationships are those that might be predicted by the use of a calorimeter, and the genes function in readily understandable, mechanical ways. This, of course, has not always been the predominant view of life; in the past, philosophers associated life with vital forces of one sort or another. Even some molecular biologists expected to encounter unfamiliar forces in the study of the gene. With references suggested by your instructor, trace the gradual elimination of vitalistic views from the science of biology.

Early Vitalism

Scientific discussion was common among the Ancient Greeks, one of the most notable of whom was Hippocrates (ca. 460–370 BC). The care that Greek scientists took in their observations laid the foundations for what is commonly regarded as the scientific method. One conclusion of Greek science influenced the beliefs about the body's constitution for several hundred years.

The early Greeks believed that all matter was made of four essential elements: earth, air, fire, and water. Each element was either allied with or opposed to another. For example, water was an ally of earth but a foe of fire. Moreover, each element was compounded by a pair of four "primary qualities": hot, cold, wet, and dry. The early Greeks also believed that living bodies contained four humours, each of which had a special relationship with one of the four elements. The relationship of the elements, their primary qualities, and the humours is shown in Figure 3.3.

A person's health or personality was believed to depend upon the amount and proportions of humour present. The humours (and the temperament they controlled) were blood (sanguine or ardent), yellow bile (choleric or sickly), black bile (melancholic or depressed), and phlegm (phlegmatic or apathetic). If any humour existed in excess, the person suffered or reflected this humoural condition accordingly. Modern scientists have long discarded this belief, but the words derived from it are still used in current expressions to describe a person's temperament. The effect was also seen in early literature, as for example in "The Nun's Priest's Tale" of Chaucer's *Canterbury Tales*:

> Certes this dream, which ye han met to-night,
> Cometh of the grete superfluitee
> Of youre rede colera, pardee,
> Which causeth folk to dremen in here dremes
> Of arwes, and of fyr with rede lemes,
> Of grete bestes, that they wol hem byte,
> Of contek, and of whelpes grete and lyte;
> Right as the humour of malencolye
> Causeth ful many a man, in sleep, to crye,
> For fere of blake beres, or boles blake,
> Or elles, blake develes wole hem take.
> Of othere humours coude I telle also,
> That werken many a man in sleep ful wo;
> But I wol passe as lightly as I can.

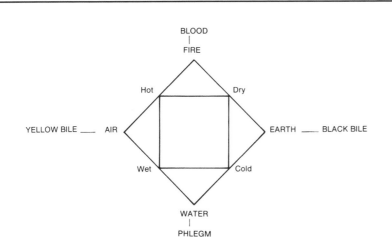

Figure 3.3. A diagram illustrating the Ancient Greek view of the four elements (fire, water, air, and earth), the four humours (blood, phlegm, yellow bile, and black bile), and the four qualities (hot, dry, wet, and cold). Although this view of the world has long vanished, its terminology still persists: we speak of phlegmatic or hot-blooded temperaments, for example. Note that, as with the genetic code, the four possible states characterizing each of the three factors (element, humour, and quality) yield sixty-four possible sorts of personalities.

Life: An Instant-Thick Snapshot   79

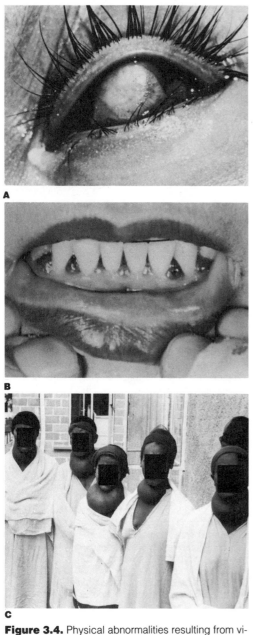

Figure 3.4. Physical abnormalities resulting from vitamin deficiencies and from an inadequate supply of iodine. A: Corneal scarring and blindness are a frequent result of vitamin A deficiency. B: A lack of vitamin C leads to *scurvy*, a disease that is vividly described in a marginal essay of the next chapter. C: Iodine, an element essential for the proper functioning of the thyroid gland, is lacking in many areas of the world, which are known as goiter belts. The persons shown here are suffering from enlarged thyroids (goiter). They are at a surgical clinic, where their thyroids will be removed. A small-town surgeon working in central Pennsylvania (a goiter belt) in the 1920s and 1930s reported that he removed an average of one thyroid a day. *(Photos courtesy of William Darby, M.D., Vanderbilt University)*

The Life of a Higher Organism

The bacterial cell, although hundreds of times smaller than most body cells, is an adequate model of the activities that must go on within all cells if they are to remain alive. Adequacy, however, does not imply identity. The cells of the human body cannot manufacture every needed molecular compound from a simple medium of sugar, salts, and biotin. There are amino acids that our cells must have but cannot make; appropriately, these are called essential amino acids. We obtain them for the most part from animal, rather than plant, proteins. Then there are *vitamins*. Their name suggests their importance: we are unable to synthesize them, but they are vital to good health. For the most part, vitamins are coenzymes—the medium-sized molecules whose attachment to enzyme molecules is needed for proper enzyme action.

A cell's inability to carry out a specific function can be viewed as a matter of economy: cellular machinery capable of carrying out various metabolic processes is not maintained if the processes are not necessary. The colon bacillus can synthesize every amino acid using nothing but a carbon source (sugar) and various salts. Mutant bacteria carrying defective genetic information may be unable to synthesize the amino acid histidine, for example. If these mutants are grown in a culture medium to which histidine has been added in appropriate amounts, they may grow faster than normal bacteria and even eliminate the latter from the culture. That is, if a small number of bacteria are transferred from old culture tubes to new ones (all containing histidine-enriched medium), the proportion of mutant bacteria will steadily increase, while that of the normal ones will decrease. Finally, the normal bacteria will be so rare that at some moment none will be included among the transferred cells; at that time the normal or wild-type bacteria will have been eliminated. The displacement of one type of bacteria by another as described here illustrates *natural selection*, the most important cause of evolution. To avoid confusion, the following point should be emphasized: natural selection does not *cause* the histidine-requiring bacteria to grow faster and displace the normal bacteria in the histidine-containing culture medium; the faster growth of the histidine-requiring organisms when they are provided with histidine and the consequent displacement (elimination) of the wild type *is* natural selection.

The cells of higher organisms have lost many of the metabolic abilities possessed by *E. coli*. Presumably, these abilities were lost as they proved to be unnecessary. This loss, in turn, freed DNA for the synthesis of enzymes needed for still other intra- and intercellular reactions and interactions.

For example, because of their size, higher animals—human beings included—have needs that lie beyond intracellular metabolism:

• The body is protected by skin and mucous membranes; it is supported by a skeleton that is moved by muscles.

• The digestive system brings fuels into the body; these fuels, combined with oxygen, furnish the energy needed for life.

• The respiratory system brings in oxygen and dispels carbon dioxide.

• The circulatory system carries oxygen and nutrients to the individual body cells, and removes carbon dioxide and other wastes. Because it touches all cells in the body, the circulatory system also serves as a communication network by transporting hormones and other chemicals.

• The excretory system removes most of the body's waste.

• Nerves and endocrine glands regulate and coordinate various parts of the body.

• The reproductive system, although not concerned primarily with the individual, is essential for the production of offspring—the next generation. Hormones produced by cells within the reproductive organs—ovaries or testes—have profound effects on the individual's characteristics.

The rest of this chapter is devoted to examining these systems. Before discussing each of them individually, we should note again that the need for them arose because higher plants and animals—multicellular organisms—must coordinate and regulate the actions of billions of individual cells. If we were to compare the body and its cells to a city or nation and its individual citizens, the various systems would compare roughly with city walls, systems of public transport, sewage disposal, and systems of communication. Finally, the sexuality of nations can be spoken of only in the most blatant symbolic manner: the city-states of Sparta and Athens of Ancient Greece could, if one were sexist, be referred to as *male* and *female*.

NOW, IT'S YOUR TURN (3-3)

The analogy between the body and its cells, on the one hand, and a nation and its citizens, on the other, is one that is often put forward with considerable seriousness by various persons—biologists and others alike. Evolution, it is claimed, has progressed from simple, self-reproducing molecules to primitive one-celled organisms, to more complex but still simple cell aggregates, to the highly organized multicellular plants and animals with which we are all familiar. The next logical step to this sequence, some persons argue (and even favor), is for society to become a true *supra-organism*, with individuals fulfilling the roles of automata, or of unthinking cells. Consider this possibility for a moment and comment in your journal. Be sure to extend your comments beyond your personal feelings concerning the supra-organismic nation. Whatever your personal views are, can you see why such a state, once undertaken, would or would not displace our present, rather loosely connected, societies?

Skin and Skeleton

Beauty is skin deep—or so they say. Certainly, the bases for our recognition of ourselves, and of our friends and acquaintances, are primarily features that are determined by the body's skin and its internal skeleton. The one sets the body's limits, while the other props it up from the inside. In between, unless a person is extremely emaciated, lies a contouring layer of muscle, connective tissue, and fat.

The skin has a complicated structure that reflects its many functions. Skin, one of the largest of all the organs of the body, serves to keep harmful organisms and substances outside the body; help control body temperature, by virtue of both the blood vessels it contains and the sweat it secretes; and eliminate wastes and metabolic byproducts, through its glands, which remove water, fats, and salts from the body.

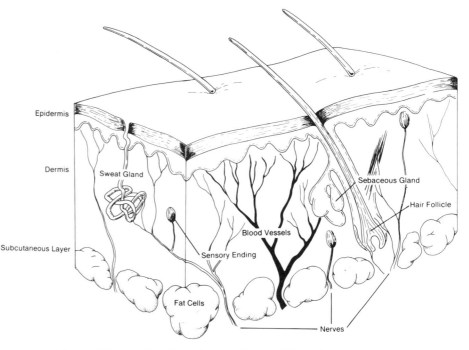

Figure 3.5. A small section of human skin. In addition to its obvious roles, the skin is also the body's source of vitamin D. Precursors of vitamin D, which occur in the skin, are transformed to vitamin D by ultraviolet radiation; the vitamin is then stored in the liver. This role of the skin has been cited in accounting for the pale skin of many peoples of Northern Europe, where sunlight (and ultraviolet light) is a scarce commodity. Vitamin D can be regarded as a hormone produced by the skin.

The skin has two layers—the epidermis and the dermis—which lie on a (third) layer of connective tissue and fatty cells. The third, or bottom, layer is the one that permits the skin to move about; if, as on the fingertips, skin cannot be displaced, it means that the underlying layer is reduced or absent.

The extreme outer layer of the epidermis consists of dead cells that are continually sloughed off (e.g., dandruff). Replacement of these cells comes from deeper dividing cells, which, as they are pushed closer and closer to the skin's surface, flatten, accumulate keratin (a tough protein), lose their nuclei, and die.

The dermis contains small blood vessels and houses the main portions of the various glands found in the skin. The latter consist for the most part of sweat glands, which are found virtually everywhere on the body, especially on the palms of the hands. Large sweat glands are located in the armpits and in the genital-anal region. The wax-secreting glands of the ear are thought to be modified sweat glands.

Each of the many hairs found on the body arises from a small follicle, and each follicle possesses a sebaceous, or fat-secreting, gland.

To a large extent, the pliability of one's skin is maintained by the moisture secreted by sweat glands and the fatty secretions of sebaceous glands. Sweat, of course, serves to lower body temperature should one's temperature rise, as during vigorous exercise. Since evaporation is a cooling process, the evaporation of sweat from the skin's surface cools the body's surface and the blood that circulates in or near the skin.

In passing, we might note that mammary glands correspond to enlarged sweat glands. Hormones that are produced at puberty cause

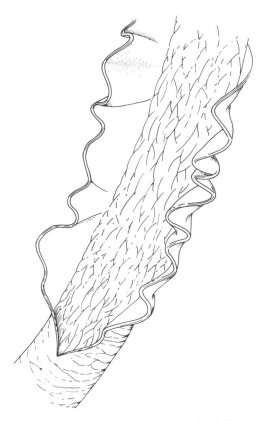

Figure 3.6. A small section of human hair, highly magnified. The outer cuticle has been laid back to reveal the core of the hair, which consists of dead cell membranes and nuclei. There are no nerves or blood vessels in the hair itself. Some years ago, a proposal was made in Congress suggesting that any human tissue that was once alive should be treated with appropriate respect. What is your opinion concerning shorn hair, dandruff, and fingernail clippings—not to mention the lining of the gut, which is continually being replaced? Where do you think appropriate respect might enter the picture? Why?

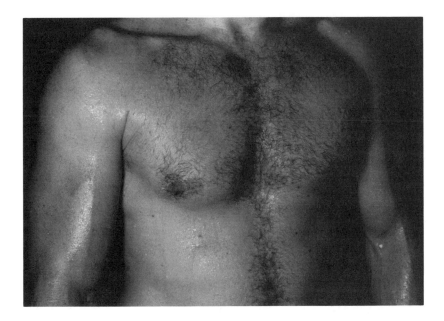

Figure 3.7. Sweat, an excellent coolant. The evaporation of three liters of water requires about 1.5 million calories. Without the loss of heat in evaporation of sweat (during strenuous exercise, for example), the temperature of a 70-kg person would rise 30°C—if death were not to intervene. To cool this efficiently, water must evaporate, not just drip from the body. Why? What does this imply about exercising while wearing plastic or other waterproof clothing? *(Photo courtesy of The Union Memorial Hospital, Baltimore, Maryland)*

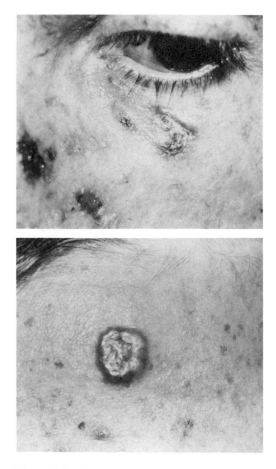

Figure 3.8. *Xeroderma pigmentosum*—dry, pigmented skin—a hereditary disease caused by a defective DNA repair enzyme. The ultraviolet rays in sunlight cause adjacent thymines on the same strand of double-stranded DNA to interact (see Figure 2.32). Normally, these thymine dimers are snipped apart, so that each thymine pairs with the adenine of the opposite strand once more. Some dimers are removed entirely: they and their surrounding purine and pyrimidine bases are then replaced. However, because they lack one or more of these repair mechanisms, persons suffering from *xeroderma pigmentosum* develop cancerous growths in areas of the skin that are exposed to sunlight; several of these cancerous and pre-cancerous areas can be seen in these photographs. *(Photos courtesy of J. Corwin Vance, M.D., University of Minnesota)*

these glands to enlarge in maturing females; they regress in males.

Finally, in addition to its biological roles, skin brings some color into our lives—often with important sociological effects. The deeper layers of the epidermis contain cells capable of manufacturing melanin, a brown pigment. Many persons react to ultraviolet radiation (short radiation near the visible spectrum; infrared or "heat" waves are long rays on or beyond the opposite end of the visible spectrum) by producing melanin; they *tan*. This phenomenon is especially apparent at Spring break, when many students at Eastern colleges take vacations on Southern beaches. There are persons, however, who are unable either to produce melanin (albinos are an extreme example) or, at least, to accumulate appreciable quantities of this pigment. Still others, blacks and other dark-skinned peoples, produce pigment whether or not they are exposed to sunlight.

Previously, we mentioned the ability of *E. coli* to assemble the enzymatic machinery that makes the cell capable of utilizing milk sugar, lactose. The wild-type bacterium, if exposed to both glucose and lactose, uses glucose first, then retools its battery of enzymes to use lactose when the supply of glucose is exhausted. *E. coli* mutants are known that cannot use lactose—ever. Other mutants produce lactose-digesting enzymes continuously, whether lactose is present or not. Human beings are much the same with respect to tanning: some cannot tan, even in the sun; others tan in the sun but lose their tan when not exposed to sunlight; and still others produce melanin and are pigmented even in the absence of sunlight.

The pigmentation of human skin is a purely biological phenomenon: ultraviolet rays stimulate the initiation of a series of metabolic activities within skin cells, and these activities lead to the formation of a brown or brownish-black pigment. On the other hand, the skin forms the surface by which human beings recognize themselves and their friends and acquaintances; unrecognized persons are strangers. Skin color plays a role in this recognition; for ages, skin color has served to separate one group of persons from another. To think of this separation as a simple biological problem involving enzymes and radiation would be to make a serious error. Racial discrimination, a discrimination based largely on skin color, is not a biological problem. Racial discrimination—racism—is for many persons an institutionalized response to a simple biological phenomenon. Consequently, racial problems are discussed in Part IV as a societal problem, rather than as a biological problem involving skin pigmentation.

NOW, IT'S YOUR TURN (3-4)

This journal entry is intended to stretch you a bit. Ultraviolet (UV) light damages DNA, but most organisms, from bacteria to human beings, possess enzyme-based repair systems that undo UV-caused genetic damage (remember, living things evolved in the presence of UV light). The UV damage, if it is not too great, affects only one of the two strands of DNA at any site along the DNA molecule; repair enzymes merely chew away the damaged strand and replace it (using the undamaged one as a model or template) with a new, undamaged (i.e., proper) strand. If cells containing damaged DNA are provided with tritium-containing purines and pyrimidines, the repair can be recorded on photograpic film as silver spots caused by the radioactivity of newly incorporated tritium.

Xeroderma pigmentosum is a genetic disease that affects individuals carrying two doses of a rare recessive gene. Sufferers of this disease develop widespread cancerous growths in skin that is exposed to sunlight; skin that is normally covered by clothing is not afflicted by these growths.

If tritium-labeled DNA precursors are injected into local patches of skin on normal persons and on those suffering from *xeroderma pigmentosum,* and if these persons are kept out of direct sunlight, microscopic preparations of skin samples (with photographic film superimposed on the tissue) show occasional heavily radioactive cells. These radioactive cells are those that underwent a cell division following the injection of labeled purines and pyrimidines, so the newly synthesized DNA was made using the radioactive chemicals.

If the same experiment is done, but the skin of both persons is now exposed to strong sunlight, many cells of the normal person show small amounts of radioactivity, but *there are no corresponding, weakly labeled cells in the skin of the person with xeroderma pigmentosum.* (Again, both types of person show occasional, heavily labeled cells that have arisen through recent cell divisions.)

In your journal, using the information provided here and differently colored pencils to represent unlabeled (normal) and labeled (radioactive) DNA strands, show how these observations have led persons to conclude that *xeroderma pigmentosum* is caused by a *defective repair system* for UV-damaged DNA. What evidence do you have that DNA replication occurs more or less normally? Can you suggest why even normal persons might develop skin cancer from overexposure to strong sunlight? (Hint: What repairs could be made if damage to DNA were very heavy, involving first one strand and then the other at short intervals?)

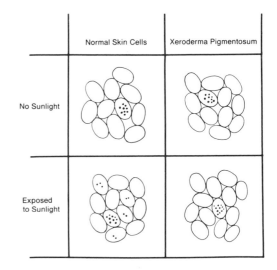

Figure 3.9. A diagram illustrating the outcome of an experiment in which small quantities of tritium-labeled (i.e., radioactive) DNA precursors were injected into the skin of normal persons and the skin of those suffering from the hereditary disease *xeroderma pigmentosum.* The treated skin areas of these persons were then either shielded from or exposed to bright sunlight. The black dots in the diagram represent silver granules in X-ray film that was exposed to small sections of skin from these persons. What evidence do you see in this diagram that sunlight damages DNA in normal persons? What evidence do you see that persons with *xeroderma pigmentosum* cannot repair damaged DNA? What evidence can you detect suggesting that chromosome replication during cell division is not affected by this hereditary disease?

The skeleton and its associated voluntary muscles form by far the greatest part of the entire body. The skeleton and its muscles (covered, of course, by skin) provided prehistoric artists with the models they needed for sketching not only human beings but also wild animals. Early anatomists gave excellent descriptions of bones and muscles.

Figure 3.10. The musculature of the human body as drawn in the mid 1500s by Jan von Calcar for Andreas Vesalius's *De humani corporis fabrica.* Note the arrangements of muscles that allow the operation of the elbow and knee, as well as the movement of the head.

They understood that the contraction of these muscles, pulling the tendons by which they were attached to the bones, caused all the body's movements. They also knew that the bones of the head, chest cavity, and spinal column provided protection for the vital structures located within them.

Despite appearances, the skeleton is not a fixed, stable structure like the framework of a house or the chassis of an automobile. For all its hardness, bone is in a state of constant change—that is, in a state of *dynamic* equilibrium. The bone's minerals—calcium, magnesium, and sodium—are laid down as salts within bone, taken up, replaced, and taken up once more in a continuing cycle. This cycle is regulated by the body's control system. When astronauts remain for long periods in the weightless conditions of outer space, seemingly unneeded minerals are reabsorbed from their bones and excreted. Astronauts return to earth with softened bones.

Bones serve as a reserve supply of calcium and other salts that can be drawn upon when needed. A nursing mother, for example, may produce calcium-containing milk only at the expense of her bones, which may be softened and weakened as a consequence. Adequate calcium in the nursing mother's diet can avoid this sacrifice on the part of bones. Aging persons often take calcium tablets to avoid *osteoporosis,* a weakening of the bones caused by the reabsorption of calcium.

The muscles—voluntary or striated muscles—that move the skeleton or hold it in place are highly specialized to contract with both speed and precision. They react only to messages conveyed by nerves emanating from the central nervous system. Muscles are actually dependent upon these messages; sever its nerve supply, and a muscle not only ceases to function, but also atrophies—it withers away. The loss of musculature is familiar to those suffering from paralyzing viral infections, physical injury, or inadequate blood supply to the brain (strokes).

Much muscular activity is voluntary; however, beyond voluntary or willed movements are the many activities under the direction of parts of the nervous system unrelated to consciousness. These movements include muscular contractions and relaxations related to posture, walking, and breathing. As I write these words, my mind composes ". . . walking, and breathing" but my fingers put them on paper with no willful interference on my part.

Voluntary muscles, together with the skeleton, offer protection for vital parts of the body. For example, very few major arteries lie just beneath the skin. Muscles also are the major heat-producing machine in the body. They store nutrients that can be drawn upon when needs arise; however, as numerous fasters in Northern Ireland demonstrated during early 1981, only 50–60 days of starvation can be sustained, even by otherwise healthy young men. All functions of muscles— movement, heat production, support, and the storage of nutrients—

require energy, so we shall turn next to those systems that provide the means by which energy is obtained by muscle cells: the digestive, respiratory, and circulatory systems.

NOW, IT'S YOUR TURN (3-5)

1. Much of the work of medical or biological engineers involves the fabrication of artificial joints or collections of artificial joints that form artificial, but functioning, limbs. If one does not think about the problems these engineers face, one not only will fail to appreciate their work but also will fail to appreciate the beauty of one's own body. For example, persons have an excellent sense of where their bodies are—that is, of the spaces their bodies fill. Earlier, you were asked to pick up a dropped object while holding a glass full of water. If that experience or the common experiences of slapping mosquitoes, brushing off flies and ants, and using your foot to scratch an itch have not convinced you, ask a friend to touch your arm, leg, or back in various places and observe how readily you can (1) describe the location, or (2) locate it with your finger. Using small stick drawings in which the arm and leg segments and the torso have fixed lengths, pose such problems as insect bites and draw the sorts of bodily contortions that are needed in order to scratch the offending spot.

2. Sensory receptors are not scattered uniformly over the surface of the body. Using a pair of dividers or a draftsman's compass, see how close together the two points can be placed before the tip of your tongue or a fingertip becomes confused and interprets the two points as one. Now, have a friend use the same pair of dividers to test the skin on your back. Are you ready to believe that points four to five inches apart may be interpreted as a single point?

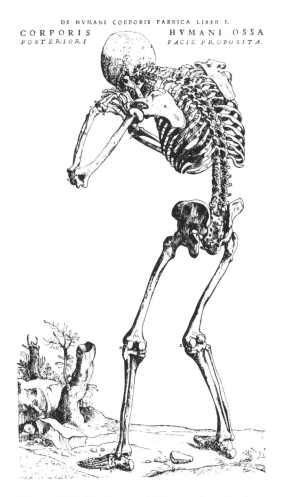

Figure 3.11. The human skeleton, as drawn by Jan von Calcar for Andreas Vesalius's *De humani corporis fabrica*.

The Digestive System

Details of the human digestive system can be illustrated more readily than they can be described in words—one picture, after all, is worth a thousand words—but a brief synopsis is given here. First, food taken into the mouth is softened by saliva and ground by chewing, then it is swallowed and passes into the stomach, where it is stored for a short time. While in the stomach, food is both mixed and subjected to chemical breakdown by enzymes and hydrochloric acid. As it softens, food passes from the stomach—a little at a time—into the small intestine. There, stomach acid is neutralized and the food is further broken down by a series of enzymes: proteins are broken down by proteases, starches by amylases, and fats by lipases. The

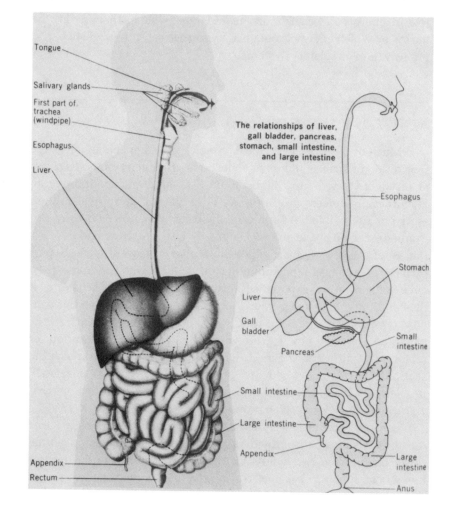

Figure 3.12. An overview of the gross anatomy of the human digestive tract from mouth to anus. Note that material lying within the digestive tract is not, strictly speaking, inside the body; only those substances that are absorbed through the lining of the gut truly *enter* the body. Note, too, that although waste material passes through the anus, the large intestine is not an excretory organ; the kidney is the main excretory organ of the body. *(Copyright, The Biological Sciences Curriculum Study; reproduced with permission)*

small molecules that result from enzymatic action are absorbed by cells in the wall of the small intestine and enter the bloodstream. Nondigested food (plant fibers composed of cellulose, for example) merely passes through the human digestive tract. As the undigested material passes through the large intestine, its water is absorbed. The rest of the waste material, as well as many intestinal bacteria, is eliminated as feces. Herbivores, such as cattle, horses, rabbits, mice, and voles, have special regions in their guts—enlarged stomachs, ceca (our appendix is a vestigial cecum), or large intestines—where bacteria digest cellulose and convert it into simple sugars. These animals either regurgitate and chew their cud (as cattle, sheep, and goats do); allow the fermented plant food to pass posteriorly, where the sugars and bacteria, with all the nutrients they contain, are digested (as in horses and other non-cud-chewing hoofed mammals); or eat a portion of their feces if the fermentation occurs in the large intestine or cecum, too late for nutrients to be reabsorbed without re-ingestion.

Most molecules that enter the body through the intestinal wall are picked up by blood circulating through the capillaries of the intestine and are carried directly to the liver, where they are delivered to liver cells by another set of capillaries. (A portion of the circulatory system that begins with capillaries and, without passing through the heart, ends in a second set of capillaries is known as a *portal* system; the one extending from the intestine to the liver is the *hepatic portal system.*) Substances that are *not* transported to the liver in this way are fats and some vitamins. These go directly to body cells by way of the lymphatic system; even they, however, may return to the liver for processing before they are put to use.

The liver is the chief food-processing organ of the body. Many substances arriving there from the intestine are not suitable for use in cellular metabolism. Liver cells, though, have enzymes that break these molecules down and rearrange their parts into new compounds that are suitable for the metabolic reactions of body cells. If the new compounds are not needed, they may be stored in the liver and released for use "on demand." Glucose, as we shall discuss later, is stored as a complex molecule, *glycogen*, when glucose is plentiful; glycogen is broken down into glucose and released into the bloodstream when glucose is otherwise in short supply.

Because the liver is the body's main chemical-processing organ, workers in industrial plants often suffer from cancer of the liver. Chemicals that enter the body by ingestion (or otherwise) eventually end up in liver cells for processing. Even though the original compound may not cause cancers, some of the enzymatic breakdown products may. Because these cancer-causing (*carcinogenic*) subtances are within the liver cells themselves, these cells may then lose their normal cellular control and begin dividing in an uncontrolled fashion. Cancers consist of such uncontrolled cellular growths.

Except for identical twins, no two persons are ever exactly alike genetically. Since liver enzymes, like all other enzymes, are protein molecules that are manufactured under the direction of information contained within DNA, the liver cells of some persons contain en-

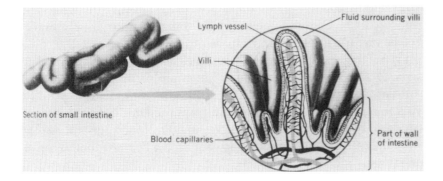

Figure 3.13. A cross-section of the wall of the small intestine illustrating the means by which its absorptive surface is enlarged, first by infoldings and then by means of the numerous fingerlike villi that project from the walls of the infoldings. *(Copyright, The Biological Sciences Curriculum Study; reproduced with permission)*

zymes that are lacking or nonfunctional in others. A practical ramification of this naturally occurring variation is that newly developed drugs and other pharmaceuticals must (or should) be introduced into general use with extreme caution; while some persons may take a particular drug with impunity, others may suffer severe liver damage for the lack of a particular enzyme.

NOW, IT'S YOUR TURN (3-6)

1. Many chemicals that cause cancer also cause mutations, especially mutations involving the introduction of an additional base pair into the DNA molecule. Conversely, chemicals that cause these mutations more often than not also cause cancer. These mutations are known as frame-shift mutations. You have already seen how sets of three bases in DNA (and mRNA) form the "words" that are translated into the amino acids of protein molecules. To see what the introduction of an extra base pair can do to a system that operates by these rules, read the following two sentences of run-together letters by introducing a space after each set of three letters (start at the left):

    (1) THECATANDTHEDOGATETHEBIGRAT
    (2) THECARTANDTHEDOGATETHEBIGRAT

The Ames test (named for Bruce Ames, the person who devised it) is a test that reveals whether mammalian liver cells synthesize any frame-shift mutagens as they break down certain complex substances, which themselves are neither carcinogenic nor mutagenic. The test cultures contain both *E. coli* and small enzymatic particles (microsomes) obtained from ruptured liver cells. The liver enzymes break down the original chemical (and even continue breaking down breakdown products). The newly formed chemicals are then picked up by neighboring bacterial cells, some of which may mutate.

The Ames test is sufficiently sensitive that it reveals a steady daily accumulation of mutagenic substances (these substances, remember, are also cancer-causing, or carcinogenic) in the urine of heavy smokers. The concentration of these substances drops each night while the smoker sleeps. Cancer of the bladder is common among heavy smokers, as one might expect from these observations.

Even if they are unaware of these Ames test results, few if any persons, smokers or nonsmokers, are unaware of the health hazards associated with smoking. With this observation in mind, explain in your journal your conception of the proper relation between knowledge and behavior. Your conception should include your attitude toward the assessment of risks. Keep your account precise, not vague or anecdotal.

2. Outline a possible solution to the following medical mystery: An employee at a dry-cleaning shop worked all afternoon cleaning a rug with carbon tetrachloride (CCl_4). After work, he joined some friends over a bottle of beer. As he left the bar, he collapsed on the sidewalk, turned greenish-brown, and very quickly died. [Hints: (1) Ethyl alcohol is metabolized in the liver; its first degradation product is acetaldehyde, a corrosive substance that is normally converted immediately by a second enzyme into acetyl-CoA, a tolerable form of acetic acid. (2) Carbon tetrachloride interferes with the function of the second enzyme, but not with that of the first.]

The Respiratory System

The first step in breathing is drawing air into the lungs. Oxygen from the indrawn air is picked up by hemoglobin in the red blood cells as they pass through capillaries in small terminal air sacs (alveoli) within the lungs. In all, there are about 300 million alveoli, each one of which is only a quarter of a millimeter in diameter. Because of their enormous number, the alveoli have a total surface area of about 75 square meters, or some 100 square yards. This is the moist surface through which oxygen enters the body—and carbon dioxide leaves it.

Breathing is carried out by muscles that move the chest walls and diaphragm under the influence of nerve impulses originating in con-

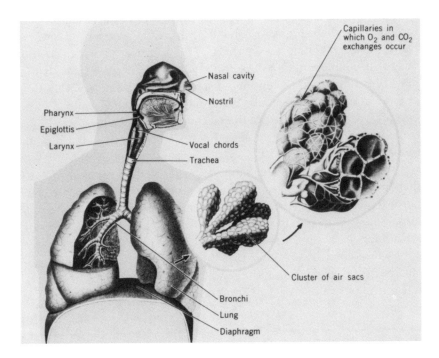

Figure 3.14. An overview of the human respiratory system. The cut-away of the right lung exposes the bronchial tubes and their subdivisions. A portion of the left lung has been enlarged to show, first, the clusters of air sacs and, second, the intimate relation between the capillaries and the individual air sacs (*alveoli*) that is needed for the exchange of CO_2 (from the body's cells) for O_2 (in the air). *(Copyright, The Biological Sciences Curriculum Study; reproduced with permission)*

Respiratory Failure

Respiratory failure can be dangerous to your health.

DROWNING

Approximately 7000 people drown each year in the United States. Drowning is one of the more distressing causes of death because it occurs most frequently among young children at play or among adults enjoying a holiday. In drowning, death is ascribed to asphyxiation (impaired respiratory exhange of oxygen and carbon dioxide). Actually, there are three different types of drowning:

- drowning in which no water enters the lungs, but asphyxia results
- drowning in which fresh water enters the lungs and hemodilution and hemolysis occur in addition to asphyxia
- drowning in which seawater enters the lungs, which results in hemoconcentration in addition to asphyxia.

Drowning without water inhalation results from a reflex laryngeal spasm—the larynx closes and does not reopen. In this case, the lungs remain "dry." Sometimes this phenomenon allows individuals who have been submerged for very long periods in extremely cold water to be revived, with no apparent brain damage.

BRONCHIAL ASTHMA AND EMPHYSEMA

Bronchial asthma is usually the result of an allergic reaction to substances in the air, especially plant pollen. The reaction causes localized edema (swelling) in the bronchiole walls, the secretion of thick mucus into their lumens, and spasms of the smooth muscle that lines the bronchiole wall. All of these effects make breathing difficult.

Emphysema is often preceded by bronchial asthma. It has become more prevalent in recent years, largely as a result of smoking. In emphysema, air flow through the bronchioles is blocked, and many of the alveolar cells are destroyed. These factors combine to increase the work of breathing, reduce the ability of the lungs to oxygenate the blood, and overload the right heart because of reduced circulation into the lungs (pulmonary arteries). These conditions may develop very slowly over many years, but eventually emphysema results in cardiac failure and death.

(continued on next page)

trol centers of the brain. The frequency of these pulses is tuned to the person's need at a given moment: fast, deep breathing during exercise or other exertion, and slow, shallow breathing when the body rests. These nerve impulses are largely involuntary; one does not breathe by conscious effort. In fact, one cannot stop breathing for more than a half or full minute (divers must train in order to hold their breaths for longer periods), nor can one breathe more than is necessary for more than a short time.

The entry of oxygen into the body can be disrupted in several ways. For example, one can choke, or, at high altitude, one can encounter too little oxygen in the atmosphere. Anemia—too little hemoglobin per red blood cell, or too few red blood cells per milliliter of blood—also causes oxygen starvation. Furthermore, exposure to carbon monoxide, cyanide, or other reactive compounds can produce hemoglobin complexes that are incapable of transporting oxygen. Finally, head injuries can stop the normal impulses that activate the chest muscles and diaphragm: breathing stops, and death may be nearly instantaneous.

NOW, IT'S YOUR TURN (3-7)

1. Can you describe the recommended procedures for assisting a person who is choking?

2. Hiccups can be "cured" by breathing into a paper bag or small plastic bag, such as the bag from a loaf of bread. The high level of CO_2 in the re-breathed air in the bag somehow alleviates the irritation that convulses the diaphragm. Try breathing into a bag. (*Do not put a plastic bag over your head:* hold the opening of the bag over your mouth and nose with your hands.) Notice the involuntary increase in both the rate of breathing and the depth of each breath that occurs within two or three minutes. This reaction is caused by CO_2. If lime water [a solution of $Ca(OH)_2$] were part of the test apparatus, breathing would remain normal *in appearance* because the CO_2 would unite with $Ca(OH)_2$ to form $CaCO_3$ and H_2O.

3. *Emphysema* is a severe, progressive respiratory ailment that afflicts many smokers, and others who are constantly exposed to airborne irritants. The walls between alveoli are destroyed, and the small air sacs fuse. Drawing upon your knowledge of mathematics, describe what happens to the surface area of the lung (through which oxygen enters the bloodstream) if air sacs one-quarter of a millimeter in diameter fuse to make air sacs one millimeter in diameter. Or, air sacs one centimeter in diameter. (Hint: The surface of a sphere = $4\pi r^2$; the volume = $\frac{4}{3}\pi r^3$.)

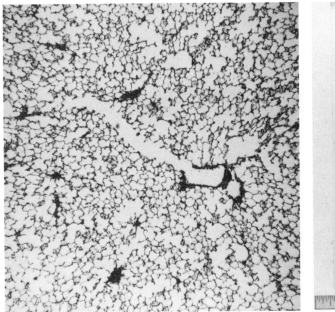 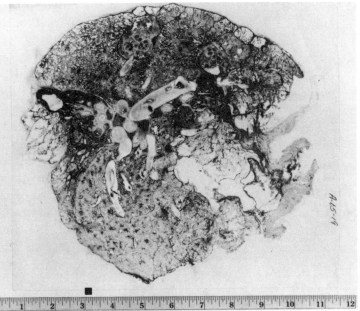

Figure 3.15. Emphysema. The lefthand figure shows a small section of a normal lung; approximately 1000 small air sacs are visible. The righthand figure is a section through a lobe of an emphysemic lung. Note that the small section of normal lung tissue (actual size shown by black square) could be easily fitted into even the smaller cavities of the diseased lung, which had become useless for respiration. *(Photo of normal lung courtesy of H. R. Steeves, Virginia Tech; emphysemic lung section courtesy of Gordon Madge, M.D., Medical College of Virginia)*

The Circulatory System

Figures 3.16 and 3.17 illustrate the main features of the circulatory system. The heart, of course, is the most familiar—by name, at least—structure of the circulatory system. Nevertheless, all parts of the system must function or the body as a whole does not, and the individual dies.

The heart is an amazing organ. Although it weighs scarcely 350 g (about 12 oz), it performs nearly 90,000 foot-pounds of work each day. Lying on an exercise bench, an athlete would need to lift a 90-lb barbell to arm's length (two feet) 500 times to equal one ordinary day's work for a heart.

Blood leaves the heart, under considerable pressure, by way of arteries. Arteries distribute blood throughout the entire body; they branch and rebranch until the smallest of them (arterioles) empty into the network of capillaries. Virtually every cell in your body lies on, or near, a capillary, which brings to it all its metabolic necessities and removes all poisonous waste materials. As capillaries reunite, they form first the small venules, then the larger veins, within which the blood returns to the heart once more.

Respiratory Failure (*continued*)

PNEUMOTHORAX

The lungs and the wall of the thorax are lined with a sheet of epithelial tissue called the pleural lining. Inspiration occurs when the diaphragm contracts because a partial vacuum is formed in the thoracic cavity and air rushes into the lungs. The vacuum forms, however, only if the pleural lining is intact. The pleural lining may tear under stress (such as a sudden jolt or labored breathing). The perforated pleura causes what is known as *spontaneous pneumothorax*. The lung, left or right, collapses and no longer functions. In minor cases, rest and re-establishment of the vacuum in the thoracic cavity is all that is needed. In major cases, surgery may be required to repair the pleural tissue.

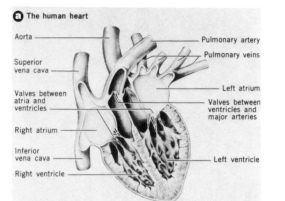

ⓐ The human heart

Aorta

Superior vena cava

Valves between atria and ventricles

Right atrium

Inferior vena cava

Right ventricle

Pulmonary artery

Pulmonary veins

Left atrium

Valves between ventricles and major arteries

Left ventricle

ⓑ The pumping cycle of the heart

Figure 3.16. The human heart and its operation. In the upper diagram, compare the thickness of ventricular walls and those of the right and left atria. Note, too, that the pulmonary artery leads from the right ventricle, while the aorta (the body's main artery) leads from the left one. The lower diagrams show (1) the relaxation of both ventricles, with blood entering them from the two atria; (2) the early contraction of both ventricles and the closure of the valves leading to the atria; and (3) the completion of the ventricular contractions, with blood going to the lungs from the right ventricle and to the body from the left. *(Copyright, The Biological Sciences Curriculum Study; reproduced with permission)*

Two pathways of circulation occur in mammalian bodies: The first "begins" at the heart, with the arteries that leave the heart carrying venous blood collected from all parts of the body. The blood passes through the capillaries that surround the alveoli of the lungs, and returns to the heart by way of a vein carrying what is now oxygenated ("arterial") blood. The second pathway also begins at the heart, leaving by way of the arteries that carry the now-oxygenated blood to capillaries that surround all cells of the body. The blood passes through these capillaries and is then collected to return to the heart by way of large veins. These two main circulatory pathways are supplemented by two pathways that do not "circulate" in the sense of making a circle: first, the hepatic portal system, which was mentioned earlier, begins with capillaries in the gut and ends with capillaries in the liver; second, the lymphatic system begins with blind capillaries, found in nearly all tissues (these collect blood serum that has seeped from the capillaries of the circulatory system and that has provided moisture

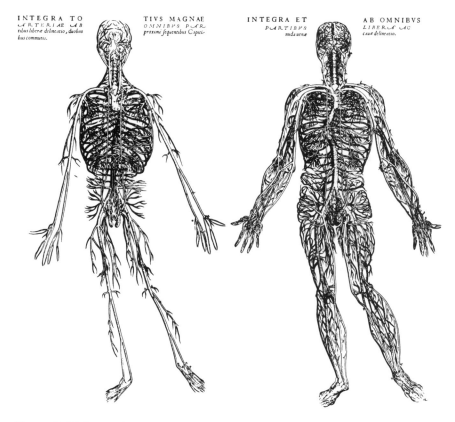

Figure 3.17. An overview of the body's arterial and venous systems, as revealed by Vesalius's dissections. Because capillaries were unknown at the time, depicting these two systems separately rather than as a single circulatory system was both reasonable and convenient. Note that veins lie close to the surface of the body and, therefore, the righthand drawing better reveals the body's contours.

for individual body cells), involves a series of ever-larger collecting veins, and then empties into the larger veins that lead back to the heart.

Blood is a complex mixture of water, nutrients, large molecules of many sorts, simple ions, waste materials, and special cells. Protein molecules in the blood have a variety of functions. For example, there are *antibodies*, protein molecules that can recognize and combine with many kinds of viruses, bacteria, or nearly any complex substance that enters the body. Still other freely circulating proteins are primed to undergo rapid chemical reactions should a blood vessel be cut or torn. They then transform flowing blood into a solid clot, which plugs the opening and stems the loss of blood. Other proteins combine with molecules that are in transport from one part of the body to another, in order to prevent chemical reactions that might occur before the molecules reach their target.

Vast numbers of cells are found in blood—some five or six million in every cubic millimeter. If, for example, in your imagination you take one of the o's in the printed word "blood" and rotate it to make a small sphere, that sphere contains about one-half a cubic millimeter. That small volume of blood would contain about three million cells!

Different blood cells have different functions. White blood cells attack and kill bacteria. Red blood cells (which in mammals have no nuclei) contain hemoglobin, the special protein that combines with oxygen in the lungs and releases it to body tissues. Hemoglobin is essential for large, active animals; very little oxygen can be absorbed directly into the blood plasma—enough for several seconds of human life at best. Even insects use hemoglobin (for example, the larvae of *Chironomus,* the midge fly, which live among rotting leaves at the bottom of quiet pools), thereby increasing the efficiency of oxygen utilization.

Because it continually circulates throughout the entire body, blood serves as a communication network. Normally, the nervous system would be regarded as the means by which various parts of the body are coordinated. Certainly, when a gallant man picks up a dropped handkerchief with one hand while watching and tipping his hat to a young lady, his body movements are coordinated and rendered graceful by nerve impulses generated by countless sensors and appropriate messages sent out in response by the central nervous system. On the other hand, messengers—hormones—from various glands in the body are poured into the bloodstream, which carries them to specific target cells in sometimes remote regions of the body. The desire on the part of a young lady to drop her handkerchief and the impulse of the young gallant to retrieve it are products not of split-second nerve impulses, but of months of hormonal communication between and among many body organs.

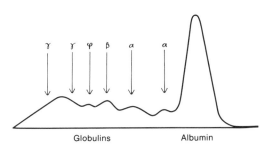

Figure 3.18. The separation of blood proteins in an electrical field by *electrophoresis.* Electrophoresis of blood is possible because in solution, some amino acids carry a negative electrical charge, whereas others are positively charged. A protein that consists of several hundred amino acids possesses a net electrical charge that equals the difference between the total numbers of positive and negative charges. According to its net charge, a protein will move toward either the positive pole (if its net charge is negative) or the negative pole (if its net charge is positive) when placed in a high-voltage electrical field. Thus, when blood proteins are placed in such an electrical field, they separate into characteristic bands (expressed as peaks, as drawn in this figure), which have been identified as albumin (the commonest blood protein) and the alpha-, beta-, and gamma-globulins. Among the latter are antibodies against such diseases as diphtheria, tetanus, scarlet fever, poliomyelitis, measles, influenza, mumps, smallpox, plague, cholera, and hepatitis. Persons who for genetic reasons are unable to make these proteins are extremely susceptible to infectious disease. Occasionally, someone who seems unable to shake off an infection will be given an injection of gamma-globulins that have been obtained from other (healthy) persons in the hope that, by chance, an effective antibody will be included in that injection.

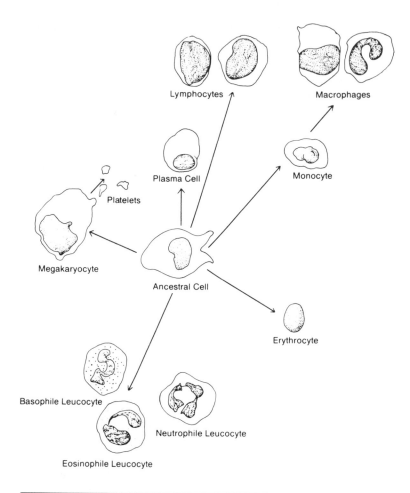

Figure 3.19. The many kinds of blood cells—white and red—arise from common ancestral cells located in bone marrow. Erythrocytes (red blood cells) transport oxygen, blood platelets are essential for clotting, lymphocyte and plasma cells are responsible for antibody synthesis, and leucocytes and monocytes (macrophages) engulf and digest bacteria and other foreign particles.

NOW, IT'S YOUR TURN (3-8)

1. The heart (specifically, the left ventricle) must exert considerable pressure in order to circulate blood through the body. Now, when the left ventricle, having expelled some 50–60 ml of blood in one contraction, relaxes and becomes refilled, blood in the arteries is still under pressure—just not as much as before. The two extremes in pressure are the *systolic* (maximum) and *diastolic* (minimum). The commonly used device for measuring blood pressure (the *sphygmomanometer*) consists of (1) a rubber bag, which, when wrapped around the upper arm and inflated under pressure, stops the flow of blood through the arm's main artery, and (2) a device for measuring the air pressure within the inflatable bag. Describe (with the help of diagrams) why, once the flow of blood has been stopped by the air bag, as the pressure is released, the first detectable pulse tells the examiner the systolic pressure, and why the cessation of an audible pulse tells the diastolic pressure.

Have you ever measured anyone's blood pressure? Maybe you will have the opportunity to learn how in this course. What is your blood pressure? How does it compare with that of your classmates?

2. The circulatory system is discussed here as if the passage of blood from the heart, through a capillary network, and back to the heart were well known. This was not always true. The first printed account of the circulation of blood was made by William Harvey in 1628. Since one of the stated purposes of this course is to encourage an appreciation of science, we should explore here how Harvey reached his conclusions.

Now, any military physician treating wounded soldiers could observe that arterial wounds, the bright red, spurting streams of blood, actually bled from one side only, the side nearest the heart. The corresponding observation for wounds involving large veins would show that they bled from the side of the wound away from the heart. Such observations ruled out the idea that blood simply surged back and forth in response to the beating heart. Not ruled out, however, was the possibility that muscles and other tissues contained two conflicting systems located side-by-side: one system that destroyed blood as it arrived from the heart, and another that manufactured blood and returned it to the heart by way of the veins. Harvey, in fact, spurned this possibility: he simply said that the blood that left the arteries was the same blood that was gathered into the veins and returned to the heart. What observation was needed to enable him to speak of a circulatory "system"? What instrument do you think was needed to make this observation?

3. As a practical exercise, enter into your journal evidence that you are familiar with pressure points for controlling arterial bleeding, as well as other first aid measures for the treatment of severe wounds, including nose bleeds.

The Excretory System

The kidneys have two main functions: they remove chemical wastes from the body, and they maintain the body's water balance. The kidney's excretory process requires two steps. First, small bunches of convoluted capillaries acting as filters (there are about one million of these bunches in each kidney) allow nearly one-fifth of the blood plasma flowing through them to pass into the small tubules of the kidney. These tubules are *outside* the circulatory system. As the filtrate passes down each tubule, the cells that make up the walls of the tubule resorb and pass back into the bloodstream much of the water and most of the metabolically valuable substances, such as sodium, chloride, calcium, magnesium, and glucose. Substances—notably waste products—that are not resorbed continue down the tubules, are collected in the ureters (tubes leading from the kidneys), pass to

Kidney Failure and Replacement of the Kidney

Although people normally possess two kidneys, either one of which can maintain good health, some unlucky people lose the function of both. For example, rapid (acute) kidney failure can follow certain streptococcal infections or the ingestion of substances that are specific renal poisons (carbon tertrachloride or mercury). Less rapid (chronic) failure can result from the gradual loss of glomeruli in disease, circulatory disorders, and old age.

As the number of functioning glomeruli decreases, substances normally lost from the body by way of urine tend instead to accumulate in body tissues. Salt and water accumulate, leading to edema. Of all substances normally excreted through the kidney, acids exceed bases; therefore, renal failure also leads to acidosis, the lowering of the pH of the blood despite the buffering capacity of blood serum. Nitrogenous wastes, notably urea, accumulate in the blood. High concentrations of urea and other substances normally excreted lead to uremia. The ultimate outcome of worsening acidosis and uremia is death.

Persons with malfunctioning kidneys can be aided by the artificial kidney, a machine that allows arterial blood from the patient to circulate in thin layers over cellophane, on the other side of which passes an equally thin layer of dialyzing solution. The solution has virtually the same concentration of ions as normal blood plasma, lacking only those ions that have accumulated to abnormal concentrations in the patient because of renal failure. The latter ions pass from the blood through the cellophane (which is porous to small molecules) and into the dialyzing solution.

The patient's blood is generally led from one of the arteries of the arm and is returned to the patient's body by a superficial vein in the arm. To prevent clotting in the machine, an anticoagulant (heparin) is added as the blood leaves the patient; to prevent the patient from bleeding, an anti-heparin substance (protamine) is added to the blood as it enters the vein.

The artificial kidney prolongs the lives of many persons who would otherwise die of kidney failure. Those whose problems are acute generally recover normal kidney function. The others must be treated once or twice a week for the remainder of their lives. Because more persons need the help of artificial kidneys than there are machines to aid them, decisions must be made concerning priorities. These agonizing decisions are generally made by committees of citizens much like yourselves: doctors, lawyers, ministers, and working men and women. They review the candidates, assess their futures, review their pasts, and attempt to evaluate each one's worth to society and society's debt to each. These fateful decisions need not be perfect, but they are better than assigning access to artificial kidneys through "first come, first served" or by chance alone. As organ donations by deceased individuals become more prevalent, reliance on artificial kidneys is reduced; the transplantation of kidneys is now a routine surgical procedure.

the urinary bladder, and are expelled from the body by way of the urethra. Nearly 100 liters (about 25 gallons) of blood plasma pass through the kidney's filters daily; only one liter (one quart) of urine is eventually formed. The other ninety-nine liters are returned to the blood. Because of the kidneys' efficiency in conserving the body's water and essential salts, most of us need not drink inordinate amounts of water; some diabetics, however, excrete large volumes of urine and must, therefore, drink correspondingly large amounts of water.

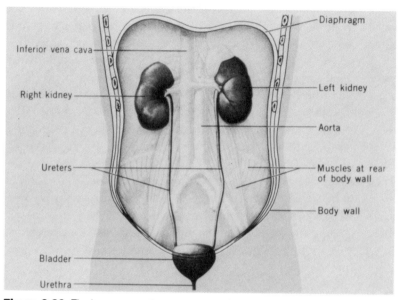

Figure 3.20. The human excretory system consists of two kidneys, the ureters, a urinary bladder, and the urethra, which drains the urine from the body. The adrenal glands, which secrete adrenaline (epinephrine), lie on the kidneys but are not part of the excretory system. *(Copyright, The Biological Sciences Curriculum Study; reproduced with permission)*

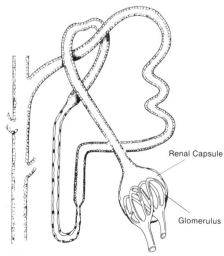

Figure 3.21. The basic functional unit of the kidney: the *nephron*. Arterial blood enters the glomerulus, which is enclosed by the renal capsule. Fluid from the blood seeps into the capsule, then through a long U-shaped tube that leads to a small collecting duct and, eventually, to the ureter. The U-shaped tube, however, is surrounded by a capillary bed; water and many useful molecules are reabsorbed from the kidney tubule into these surrounding capillaries. These blood vessels lead in turn to the renal vein, and the inferior vena cava, and the heart. Within the renal capsule surrounding the glomerulus, the fluid has the composition of blood serum; however, urine has much higher concentrations of salts than does blood, and its production requires energy. That is, the kidney performs *work*.

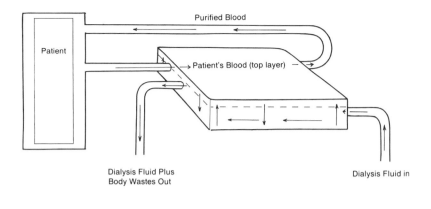

Patient

Purified Blood

Patient's Blood (top layer)

Dialysis Fluid Plus
Body Wastes Out

Dialysis Fluid in

Figure 3.22. A schematic view of a patient attached to an artificial kidney. Although the actual machine is much more complex, its principle is simple. Blood from the patient is compelled to form a thin film on a semi-permeable membrane. The dialysis fluid (regulated for pH, temperature, and chemical composition) is led into the chamber and is also compelled to form a thin layer on the bottom of the membrane. As described in the text, ions that have accumulated in the patient's blood in excess of normal levels pass through the membrane into the dialysis fluid as the two fluids are circulated in the machine. Subsequently, the blood, now cleansed, is returned to the patient and the contaminated dialysis solution is collected and discarded. The dialysis machine provides an example of counter-flow: the contaminated blood from the patient makes first contact with the dialysis fluid as the latter leaves the machine; the last dialysis fluid with which the blood comes in contact before returning to the patient is the fresh, uncontaminated fluid just entering the machine. As a result of this counter-flow, the blood returning to the patient is as purified of waste products as possible.

NOW, IT'S YOUR TURN (3-9)

1. The kidney, in the course of its functioning, takes one liquid, blood, and from it produces a second liquid, urine, in which the concentration of salts is four times greater than that in the first liquid, blood. If you divide a container by using a membrane through which both water molecules and salt ions can pass, the solutions on the two sides will equilibrate, i.e., they will become the same. If, on the other hand, only water molecules can pass through the filter, and the solution on one side of the filter is more concentrated than on the other, water will pass from the less concentrated to the more concentrated side, thus tending to make the *concentrations* on the two sides equal; in doing so, the level of water on the concentrated side will rise. How, then, does the kidney make water pass from the tubule (high concentrations of salts) into the bloodstream (low concentration)?

The answer to the above question is "work": the kidney concentrates urine by using energy to do work. Each of us does the same when we walk up stairs or lift a heavy weight. We act counter to expectations based on the second law of thermodynamics, which states that order decays, disorder increases, and the amount of useful energy declines: weights should not be lifted, we should walk down stairs, and water should pass from blood into urine, not the reverse. Describe in your journal how the second law of thermodynamics can be true *on the average* but not true in specific cases (like walking up stairs). Where does a person's energy come from that allows him (or her) to "defy" the second law of thermodynamics? What is the ultimate source of the energy that allows living organisms as a whole to circumvent this law?

2. The task facing the kidney is lessened by the arrangement of tubules shown in Figure 3.21. Engineers call this a counter-current. It can be illustrated most easily by reference to heat exchange between interlaced tubules. Make a simple diagram representing a stork stand-

Table 3.1. Hormones, their sources, and their effects[a]

Hormone	Secreted by	Effect
Neurotransmitter	Nerve cells	Stimulates (or inhibits) other nerve cells or muscle cells
Releasing factor	Hypothalamus	Causes release of ACTH, TSH, LH, FSH, prolactin, or GH by pituitary
ACTH (adrenocorticotropic hormone)	Pituitary	Stimulates adrenal cortex
TSH (thyrotropin)	Pituitary	Stimulates thyroid gland
FSH (follicle-stimulating hormone)	Pituitary	Stimulates maturation of gonads (male or female)
LH (luteinizing hormone)	Pituitary	Causes maturation of egg-bearing follicle in ovary, secretion of testosterone in male
GH (growth hormone)	Pituitary	Regulates growth metabolism
Prolactin	Pituitary	Regulates breast development
ADH (antidiuretic hormone) = vasopressin	Pituitary	Stimulates water absorption by kidney
Oxytocin	Pituitary	Governs sperm movement, childbirth, milk production
Parathyroid hormone	Parathyroids	Regulates calcium metabolism
Thyrocalcitonin	Thyroid	Regulates calcium metabolism
Thyroid hormone	Thyroid	Regulates rate of metabolism
Secretin	Intestinal lining	Stimulates pancreas
Gastrin	Stomach lining	Stimulates stomach to secrete HCl
Insulin	Pancreas	Regulates glucose metabolism
Glucagon	Pancreas	Regulates glucose metabolism
Epinephrine (adrenaline)	Adrenal medulla	Causes "fight or flight" reaction
Norepinephrine (noradrenaline)	Adrenal medulla	Increases heartbeat
Aldosterone Desoxycorticosterone	Adrenal cortex	Regulates kidney activity
Cortisol	Adrenal cortex	Participates in general metabolism
Androgen	Gonads	Bestows male characteristics; regulates sex drive
Estrogen	Ovaries (placenta)	Bestows female characteristics; regulates menstrual cycle

(*continued on next page*)

ing on one leg in a shallow pool of cold water. On the thin leg, draw an artery leading down and, in intimate contact with it, a vein leading up from the foot and back to the body. Label the body 110°F and the water (and foot) 40°F. Now, explain why *heat* does not circulate as *blood* does, and why the stork need not burn tremendous amounts of fuel to rewarm the blood that comes from its cold feet. (Not only is the principle of counter-current used by the kidney, as shown in the figure, but it is also used by the males of most mammals in maintaining the temperature of the testes substantially below their body temperatures.)

The Control System: Nerves and Glands

The respiratory, digestive, excretory, and skeletal-muscular systems of the body are coordinated with amazing precision; only by such coordination can the life of the individual be maintained. The digestive and respiratory systems bring in the food and oxygen needed for releasing energy and building structures within the individual cells; the circulatory system delivers the materials to the cells and removes wastes; and the excretory system clears the wastes from the bloodstream and discharges them from the body. If any part of the body expresses a special need, the other parts respond appropriately. Cooperation in the functioning of cells at different sites within the body depends upon controls exerted by the nervous system and the endocrine glands. Nerves and glands can be described separately, and they do differ considerably in how they perform; nevertheless, they are best thought of as a single *control system*.

The major human endocrine glands are shown in Figure 3.23. From these glands, special chemical substances—hormones—are secreted into and carried to other parts of the body by the bloodstream. Each circulating hormone affects only special cells that have the proper receptor—target cells. A good analogy to this system is the air over a town, which may teem with a multitude of television and radio signals (electromagnetic waves), but some persons can watch baseball, others can watch Johnny Carson, and still others, forgoing television, can listen to their favorite music on the radio, all because they have appropriate receivers.

The regulation of the release of the sugar glucose from liver cells is an example of hormonal action. The liver stores glucose in the form of glycogen (a starchlike polysaccharide) as it arrives from the intestine; glucose is released from the liver as requested by the body's cells. Almost all cells use glucose as an energy source. Because the brain is especially sensitive to the amount of glucose that arrives by way of circulating blood, a precise control of blood glucose is essential for the

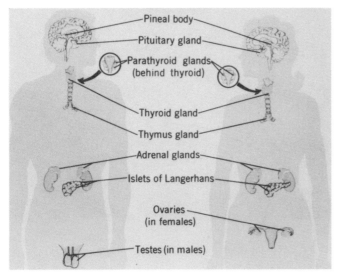

Figure 3.23. The major endocrine glands. At least one endocrinologist has claimed that *every* tissue of the body secretes at least one hormone and is the target tissue for at least one other tissue's hormone. Kidney cells, for example, secrete a protein, *renin*, if circulation to the kidney is impaired. Renin, an enzyme, starts a series of events leading to the production of *angiotensin*, a compound that raises the body's blood pressure. The glands illustrated here, however, are those whose functions are best known. *(Copyright, The Biological Sciences Curriculum Study; reproduced with permission)*

Table 3.1. (*continued*)

Hormone	Secreted by	Effect
Progesterone	Ovaries (placenta)	Regulates ovulatory cycle
HCG (human chorionic gonadotropin)	Placenta	Stimulates ovary to continue secreting progesterone
Relaxin	Ovary	Inhibits uterine contraction
Prostagladin	Gonads	Has general effects
Vitamin D	Skin	Promotes growth

a This catalogue of hormones is intended for reference, not for memorization. Notice the interplay among the different parts of the body: The releasing factors of the hypothalamus cause the release of hormones by the pituitary; the latter stimulate the adrenals, thyroid, or gonads, which have general effects upon the body, including the development of male and female sex characteristics.

proper functioning of the nervous system. This control is provided by hormones from the pancreas, adrenals, and other endocrine glands.

When a person is startled, frightened, or angered, the brain stimulates the adrenal glands (by way of nerves) to secrete adrenaline (epinephrine). This hormone has several effects, including stimulating the release of glucose by liver cells. The increased glucose in the blood provides the extra energy the body will need, whether the individual decides on fight or flight.

A second example of hormonal control relates to proper kidney function. The blood from which the kidneys remove waste products arrives at the kidneys by way of the renal arteries. Should the blood supply to the kidneys fall for any reason, kidney cells respond by secreting *renin*. Renin, a protein, reacts with a second protein that normally circulates in blood plasma to release a short polypeptide chain (eight amino acids in length), *angiotensin*. The events that follow the formation of angiotension include the constriction of small blood vessels and an increased output on the part of the heart as a reaction to the increased blood pressure. The rise in blood pressure, of course, forces more blood to enter the renal arteries, thus solving the kidney's original problem. However, if sustained for long periods, high blood pressure causes a set of problems of its own: The heart enlarges as a result of its extra work; small blood vessels in the brain may burst, causing brain damage; and the adrenals, stimulated by the high blood

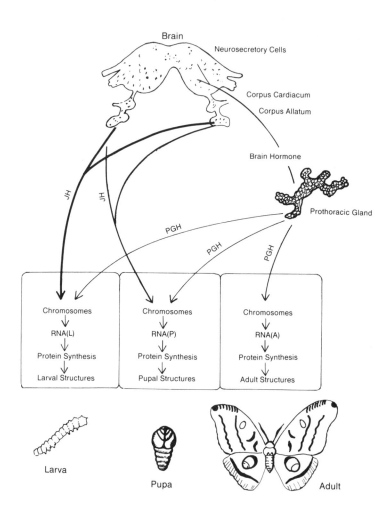

Figure 3.24. The control of protein synthesis in an insect (the Cecropia moth) by two hormones (JH, juvenile hormone, and PGH, ecdysone) so that the proper proteins—larval (L), pupal (P), or adult (A)—are made at each stage of life. Note that the ratio of JH to PGH determines which genes are activated, which mRNAs are synthesized, and, therefore, which proteins are made.

pressure, secrete a hormone that causes the kidneys to resorb more than normal amounts of sodium and water from its tubules, leading to *edema* (excess water in the body's tissues).

NOW, IT'S YOUR TURN (3-10)

1. When the dentist uses Novocain before filling a cavity, he injects both Novocain and adrenaline into the gum. The hormone constricts nearby blood vessels, thereby keeping the Novocain where it is needed. The next time you have a cavity filled, note your body's reactions to the Novocain injection. Not all of these reactions are caused by fright on your part; some can be ascribed to the injection of the "fight or flight" hormone, adrenaline.

2. Hormones act as messengers within the body. There are other organisms outside, however, which at times take advantage of these

circulating messages: eavesdropping, so to speak. Explain by way of hormones, for example, why fleas that normally stay near the head and neck of rabbits will, just before a pregnant doe gives birth, move to her posterior end, where they can jump onto and infect the newborn babies.

3. *Prolactin*, a polypeptide, is a hormone that is normally produced by the anterior pituitary gland; its normal function is to stimulate milk production in female mammary glands after birth. In the absence of prolactin, milk production ceases. Consider the implications of the following fact for gene control: certain types of cancer stimulate mammary glands to produce milk, even in males. (Hint: Recall how the sequence of amino acids in proteins and other polypeptides is controlled by DNA.)

The other part of the body's control system is the nervous system, whose anatomical details and circuitry are best illustrated by diagrams. Figure 3.26, illustrating the reflex action that accompanies the stubbing of one's toe, suggests the function of sense organs and identifies two kinds of nerves—*sensory* (carrying messages to the central nervous system) and *motor* (carrying messages from the central nervous system to other parts of the body). The most complex and dominant control of all parts of the body exerted through the nervous system is provided by the brain. Of course, we also identify the fascinating phenomena of the mind as a product of the brain's activity. Amazingly, through mental function, we can contemplate not only the outside world, but also our individual selves.

The seat of human intelligence—the ability to collect, store, and

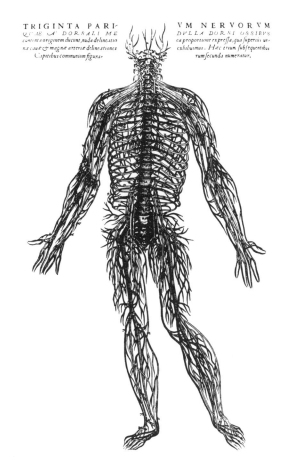

Figure 3.25. The major nerves of the human nervous system, as seen by Andreas Vesalius in the sixteenth century (very little has been added at this gross anatomical level since that time). The essential point here is that the nervous system is a communication network that extends to every part of the body. Missing in the illustration are the innumerable fine nerve fibers that go to literally every square millimeter of the body's skin and membranous surfaces.

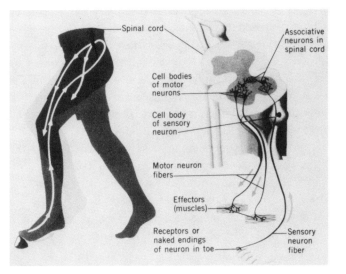

Figure 3.26. An example of a reflex action. The sharp rock is sensed by receptors in the skin of the toe. A sensory impulse travels to the spinal cord, where a motor impulse is immediately relayed to muscles in the leg. Either the foot is withdrawn, as illustrated, or it is flexed so that the body's weight is shifted to the heel. *(Copyright, The Biological Sciences Curriculum Study; reproduced with permission)*

The Human Mind Views Itself

A person defines himself using his nervous system and its images. Thoughts, perceptions of the world, and reactions to it are mediated and controlled by the nervous system. Yet even today there are persons who doubt that the mind and emotions are products of the brain. After a public lecture, a member of the audience complained to the speaker: "You concentrated only on matters of the mind, and ignored those of the heart!"

It is not surprising that the relationship between brain and mind is hard to grasp—or that many aspects of this relationship are still unknown. We cannot know our own brains, and the minds of other persons are revealed only sketchily by their words and actions. Injuries and disease, when they strike the brain, tell something about capabilities that are changed or lost. But the symptoms that emerge are bewildering. It is impressive that the Ancient Greek physician Hippocrates could write nearly 2400 years ago: "Some people think that the heart is the organ with which we think, and that it feels pain and anxiety. But it is not so. . . . From the brain and the brain alone, arise our pleasures, joys, laughter and jests, as well as our sorrows, pains, griefs, and tears."

Later Greek physicians swept these ideas aside. Their conclusions, embodied in the teaching of Galen, dominated medicine in the Western world for hundreds of years. They thought of a specialized air or spirit as central to nerve functions. This "air" flowed from the heart to the brain and all over the body. The brain seemed to have secondary roles—to cool the spirit, and to form mucus.

Even in the eighteenth century, the ideas of Franz Joseph Gall seemed revolutionary. Gall thought that different nervous functions were related directly to different areas—or "ganglia"—of the brain. Extending his theory, he made up a system termed *phrenology*, teaching that careful measurements of the outside of the head could reveal much about a person's intelligence and character. Because of the radical nature of his ideas, Gall was forbidden to lecture in Vienna. Later experimentation showed that, although he was wrong in many details, he was right in a most fundamental respect: Different areas of the brain are indeed specialized for particular functions. Special functions of the cortex, for example, can be identified, and there are special areas within the cortex that are concerned with vision, hearing, memory, and other processes that we recognize as functions of mind.

correlate information, and to draw upon this information to form mental images revealing possible *future* outcomes of this or that action—lies in the brain. More specifically, those abilities that are associated with intelligence seem to be associated with the outer layer (cortex) of the cerebral hemispheres. What makes the cortex of the human brain so special? In some other mammals, the *size* of the cerebral cortex is comparable to that of human beings: indeed, dolphins and elephants have cerebral cortices that are larger than those of human beings. The number of connections between cells, however, seems to be larger in human beings than in any other species. There are an estimated ten billion cells in the human cerebral cortex; each of these connects, on the average, with 10,000 other cells. Consequently, there are about 10^{14} connections in all.

The human brain makes possible speech, a phenomenon not duplicated in any other animal. In human speech, the noises that are uttered are not merely signals, but are substitutes for things and for relationships. Speech of this sophistication allows human beings to converse and to share experiences; information gained by talking is stored in an individual's memory as if it were a personal experience. Thus, the cumulative knowledge gained through thousands of years has been passed by word of mouth and by writing from one generation of brains to the next.

During the past century, the brain has been explored through the use of many techniques. Three important means for studying the functioning of the brain are (1) determining the correlation between persons' behaviors when they suffer from brain tumors or injuries and the evidence of damage revealed in autopsies; (2) noting the effects caused by applying small electrical shocks to the brain during surgery in people or experimentally in monkeys, cats, and rats; and (3) performing brain surgery—removing portions of the brain or cutting connections between different parts of the brain—and noticing changes in behavior or in sensory ability.

Epilepsy is a disease in which the person has unusual—and unwanted—movements and feelings. It is caused by uncontrolled activity of clusters of cells in the brain. Once these cell clusters start to misfire, activation may spread to more and more cells until the affected individual is in convulsions (an epileptic *fit* or *seizure*). The first symptoms sensed by an epileptic (a person afflicted with this disease) may differ from person to person. Some may sense a numbness in the foot, while others hear noises or see lights. In many epileptics, a damaged area of the brain has been found at autopsy. Furthermore, the symptoms of the disease in each case can be related to the particular part of the brain where the damage is located. For instance, the damaged area may lie on the side of the brain (in the temporal lobe). People with damage in this area were, when alive, epileptics with emotional rather than physical sensations at the onset of their attacks.

Motor area:
Muscles of limbs, body, and head (except for muscular walls of internal organs)
Premotor area:
Coordination of movements of the limbs and body. Impulses from the premotor area are passed along to the motor area.
Sensory area:
Touch and other tactile sensations
Speech:
Impulses are passed to the motor-speech area.
Motor-eyes (1):
Turning movements of eyes (and head); opening and closing of eyelids; dilation of pupils of eyes
Vision
Motor-eyes (2):
Turning movements of eyes (and head)
Hearing
Taste
Smell
Motor-speech:
Movements of the vocal cords

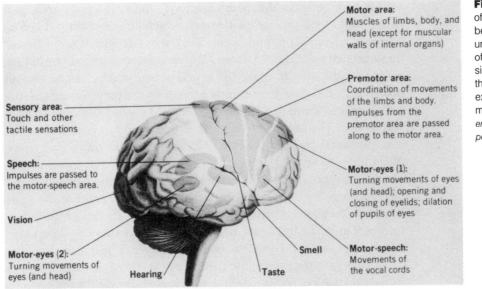

Figure 3.27. Some of the functional areas of the human brain. The simple relations between body structures and function are understood most easily; the shaded areas of the figure, then, govern such relatively simple functions. The unshaded portions of the cerebrum are those whose functions are extremely complex—that is, the regions that make us human. *(Copyright, The Biological Sciences Curriculum Study; reproduced with permission)*

Fear, dread, guilt, extreme depression or despair, and rage are some of the emotions experienced by epileptics at the onset of a seizure. Those exhibiting rage, for instance, tend to smash furniture, put their fists through window glass, or attack other persons. Others, with damage in the temporal lobe, have periods of automatic activity that they cannot recall afterwards. An example of automatic activity is walking along a street crowded with shoppers and other pedestrians and later being unable to recall doing so.

In order to locate damaged areas in the human brain, neurosurgeons sometimes stimulate it with fine electrodes and tiny electrical currents. These methods have also revealed a great deal about the brain's function. Surgeons can identify not only areas of the brain that (when stimulated electrically) make various muscles contract, but also those that give rise to sensations and various emotions, or which release old memories. The results found by electrical stimulation agree very well with those obtained at autopsy. In these and other ways, the brain and its many and complex control functions are being explored. There is still a long way to go, however, before the human mind understands itself.

Sympathetic
Parasympathetic
Cranial
Cervical
Thoracic
Lumbar
Sacral
Trachea
Lungs
Heart
Stomach
Liver
Kidney
Intestines
Bladder

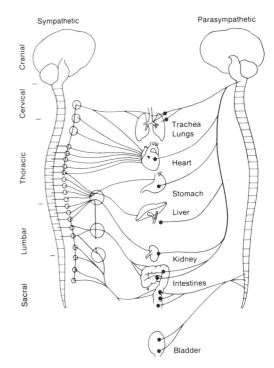

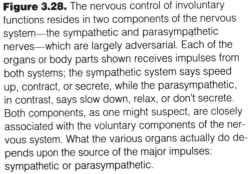

Figure 3.28. The nervous control of involuntary functions resides in two components of the nervous system—the sympathetic and parasympathetic nerves—which are largely adversarial. Each of the organs or body parts shown receives impulses from both systems; the sympathetic system says speed up, contract, or secrete, while the parasympathetic, in contrast, says slow down, relax, or don't secrete. Both components, as one might suspect, are closely associated with the voluntary components of the nervous system. What the various organs actually do depends upon the source of the major impulses: sympathetic or parasympathetic.

NOW, IT'S YOUR TURN (3-11)

You can perform several small experiments that reveal structural and functional peculiarities of the nervous system and its sensory apparatus.

1. For example, in the mammalian eye, the retina is covered by a network of blood capillaries, through which light must travel before an image is formed on retinal cells. Why, then, do we not normally see these capillaries? Because images that remain unchanged are not recorded by the eye. To make the capillary bed visible, punch a small hole in a 3″ × 5″ card, peer through this hole at a uniform surface (the ceiling or a wall will do), and move the card slightly but rapidly from side to side. The network of capillaries now becomes visible.

2. Not only blood capillaries but also a mat of individual nerve fibers lie in the path of light as it enters the pupil and forms an image on the retina. These fibers converge on one point (the *fovea*) and pass through the retina, continuing onward to the brain as the optic nerve(s). There are no retinal cells in the fovea. To demonstrate that these are indeed blind spots, draw two heavy dots about 3″ apart on a ruled paper or card. Note that when the paper is held at a particular distance before the eyes while the left eye is closed and the right eye focuses on the left dot, the right dot disappears, while the left remains clearly visible. The corresponding disappearance occurs if the role of the eyes is reversed: the left dot disappears. What about the ruled pattern? Is there a "hole" in the ruled paper?

3. Depth perception depends upon the slightly different views of the world that are provided to the brain by the right and left eyes. Artists can sometimes gain a better idea of how to depict a scene on paper by viewing it through only one eye; the mental image is then two-dimensional, as the painting or sketch itself will be. With an arm outstretched before you, and with one eye closed, walk across a room and try to touch a small object (a candle tip, the top of a soda bottle, or the end of a ruler that extends over the edge of a table) with your fingertip. You will be lucky to come within two inches of your objective.

4. Stereo diagrams are frequently used in scientific journals to illustrate the three-dimensional structures of complex molecules. With practice, no apparatus is needed to view these diagrams properly. To demonstrate this, use the stereo diagrams supplied in Figure 3.29. Hold the figures at a comfortable distance, and, while keeping them in focus, allow your eyes to relax so that they no longer converge on the page but at some point beyond. (Some directions say "Cross your eyes"; that is precisely what one should *not* do.) The two diagrams seem to move toward one another as your lines of sight become parallel; as they superimpose, the fused diagram stands out in three dimensions—with a less clear, two-dimensional diagram on either side.

5. The following illusion was devised by a one-eyed person who was never able to test his own predictions himself. The needed equipment is a piece of string about thirty-six to forty inches long, a baseball, softball, or tennis ball, and one lens from a pair of sunglasses. Having tied one end of the string around the ball to form a pendulum,

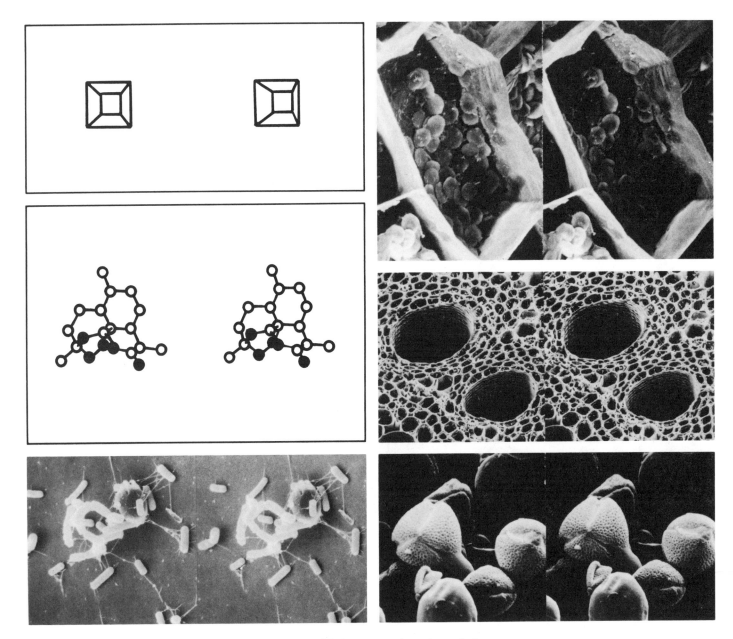

Figure 3.29. Stereoscopic—three-dimensional—vision is possible because each eye transmits its own view of the world to the brain. The two views differ, and the brain interprets the discrepancies in terms of distance. First, note that the paired drawings are not identical. Using the technique described in the text, practice fusing the small boxes in the upper left of the figure until you sense that you are peering into a square tunnel. Next, move on to working with the more complex figures. Proceeding counterclockwise from the boxes, the figures are: a stereo view of the structure of an antimalarial drug, *ginghaosu;* a collection of bacteria (*E. coli*); germinating pollen grains; ducts in plant tissue; and a dehydrated cell containing chloroplasts. The stereo view of *ginghaosu* is adapted from a view that was published in *Science;* it and most other scientific journals expect their readers to see these illustrations in three dimensions. *(Photographs copyrighted by The Biological Sciences Curriculum Study; reproduced with permission)*

one person stands with his back against the wall of the room, swinging the pendulum in a rather wide arc in front of him. The other person, standing ten to fifteen feet away, watches the pendulum with both eyes open, but with the smoked or colored lens held before one eye. What is the illusion seen by the viewer? How is it affected by holding the lens in front of the other eye? Can you diagram the viewer and the path of the pendulum and account for the illusion? (Hint: Less light enters the eye that looks through the colored lens and, consequently, the brain cells receiving information from that eye require a fraction of a second longer to process this information than do those cells receiving information from the other eye. The consequence is a double "image" in the brain that is then fused as if the eyes were viewing a three-dimensional scene.)

Reproduction

A survey of human reproduction can begin with a view of the structure of the sex cells, ova and sperm, and the organ systems that produce them. Still, the total process of reproduction involves far more than the fertilization of an egg by a sperm. In the wider sense, human reproduction includes all activities of people that are directed toward maintaining our species. As in all other higher vertebrates, the complicated activity of copulation is central to the act of reproduction. In many mammals the time at which copulation can occur successfully is brief. The female signals—by her activities, the coloring of her skin, or special scents—that she is at the stage during which eggs within her ovaries are made ready for fertilization. These signals attract the male, copulation occurs, and the resulting fertilization of ova starts the new generation. In people, the situation is different. In the reproductive cycle of females, whether young girls or adult women, an egg ripens in one of the two ovaries about once every 28 days. There is only a brief time during which the egg can be fertilized. Nevertheless, people do not restrict their sexual activity to this limited time. Usually, a strong emotional attraction exists between a man and a woman that is based on secondary sex characteristics, activities that are pleasant and rewarding to one or both, and customs of the society in which they live. The two are drawn together, and copulation may take place at any stage in the female's reproductive cycle. Therefore, sexual activity often occurs when there is no egg ready for fertilization in the female's reproductive tract. This seems like a "waste" to some scientists, who wonder at the striking way in which this aspect of human reproductive behavior differs from that of most mammals. One explanation for the special pattern in humans is that the strong tendency of males and females to stay together in stable

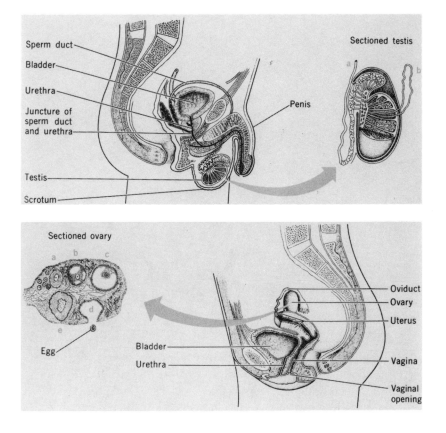

Figure 3.30. The reproductive system of the human male. The enlarged view of the testis shows (a) the sperm duct, which eventually joins the urethra, and (b) one of a great many sperm-producing tubules. *(Copyright, The Biological Sciences Curriculum Study; reproduced with permission)*

Figure 3.31. The reproductive system of the human female. The enlarged section of the ovary shows the egg in various stages of maturation within its follicle (*a–d*) and the formation of the *corpus luteum* (yellow body) (*e*) from the egg follicle after ovulation. Progesterone, a hormone secreted by the corpus luteum, prepares the uterus for the arrival of an embryo. Of course, the egg may or may not have been fertilized before arriving at the uterus. *(Copyright, The Biological Sciences Curriculum Study; reproduced with permission)*

pairs (husband-wife) is reinforced by the frequent sexual activity that is possible for both the female and the male. This stable *bonding*, as it is often called, is useful in the larger framework of human society, and certainly increases the chances for the survival and education of children born to the pair.

Once in a while, copulation leads to conception, that is, the fertilization of an egg by a sperm (occasionally more than one egg is fertilized, so multiple births may occur). An embryo is the result. If it is viable and succeeds in attaining attachment to the uterine wall, it begins a development that eventually leads to a new human being (Chapter 2). Changes in the mother's body, coordinated by hormones from the pituitary, ovaries, and placenta, result in an ever-increasing supply of nutrients to the growing fetus, and the efficient removal of its wastes.

With birth, new and specialized patterns of function and behavior, of both the newborn infant and its mother, explode into being. Under the influence of pituitary hormones, the mother's mammary glands begin to secrete milk, and they continue this secretion as long as the infant nurses. The infant sucks, cries, sleeps, and later begins to smile, make sounds, and cling to the adult body. The mother cares for the

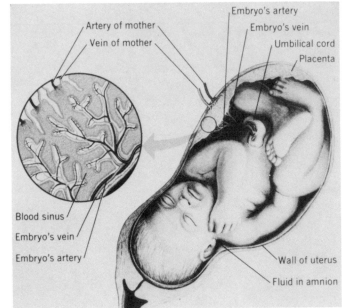

Figure 3.32. A view of an advanced human embryo in its mother's uterus. The insert shows details of the placenta, stressing the independence of the mother's and the baby's circulatory systems. Oxygen from the mother's blood is transferred through the placenta to the embryo; carbon dioxide (CO_2) is transferred in the reverse direction. Various waste products are passed into the mother's blood to be excreted through her kidneys; dissolved nutrients pass from mother to child. Chemicals other than nutrients can also be passed from mother to child; many are exceedingly harmful to the developing embryo. *(Copyright, The Biological Sciences Curriculum Study; reproduced with permission)*

infant, feeding it and keeping it warm and clean. She holds it close, carries it about, smiles at it, and plays with it. As the child develops, the mother (or other adult responsible for its care) and child continue to interact, to communicate, and to forever change their complex give-and-take pattern of behavior. Gradually, the child learns to sit up, stand, creep, walk, and talk. It becomes more and more independent. Much of its time is spent exploring, learning, and playing. Playing—which occurs in most infant mammals as well as people—is a useful behavior. It gives the young individual a chance to try out and to become skillful in many of the kinds of activities characteristic of adult life.

The adults who care for the child continue to watch out for it, and to see that it is fed and sufficiently warm (or cool) both by day and at night. The constant attention, interest, and loving interaction between adult and child are of great importance in the child's development. In particular, the development of the brain seems to be critically affected both by the physical care provided (nutrition, protection from injury, and so forth) and the social stimulation the child receives while still very young (Chapter 2).

This brief description of the activities leading from conception through childhood actually fits the ways and patterns of most human societies. Remarkably similar statements could also be made about the reproductive behavior of nonhuman primates, such as gorillas and chimpanzees—mammals that are relatively close to people in an evolutionary sense—and, indeed, about many other mammals as well.

In contrast, the care provided by human beings for their young differs from that of other animals in many ways. The great dependence of the young upon adults and the length of the period of this dependence are remarkable in people. A two- or three-year-old child, left entirely to itself, simply could not survive. Even older childen have little chance of growing to maturity outside the protection provided to them by caring adults. If deserted children do survive independently, they seldom develop socially acceptable patterns of mature behavior.

The total sweep of human reproduction defies the "snapshot" form that has been imposed on this chapter as a whole. The most fundamental aspect of the process of human reproduction is its flow through time, during the development of individual persons and from one generation to the next. The costly and complicated machinery of sexual reproduction is built into the biology of every normal person not for the sake of that individual, but for the survival of the species.

Summary

This chapter gives a view of the individual human being, a view contrasting with the overview that is presented in Chapter 2. In the overview, the person is presented as a changing and developing being, starting the path of life as a fertilized egg (a single cell) and passing through a series of stages to life's termination in death. In this chapter, the individual is "frozen" at a particular moment, like a single still frame from a motion picture film. From this point of view, the human being is shown as requiring the proper functioning of several systems—skin, skeleton, and muscles, and digestive, respiratory, circulatory, excretory, and control systems—in order to remain alive.

The individual is important in another respect, however. The human species survives from generation to generation through the successful reproductive (sexual) activities of individuals. Individuals are the units from which populations are made. If the effect of human populations on Earth is to be understood, a knowledge both of the needs (real and imagined) of individuals and of the total number of individuals on Earth is necessary. This chapter, taking a step toward describing the effect of people on Earth, has concentrated on the basic unit—the individual human being.

INFORMATIONAL ESSAY

Muscles

The tissue that moves the body, passes food through the digestive tract, squeezes secretions from glands, causes hair to stand erect, pushes blood through the body, and provides for many other types of movements is called muscle tissue. There are three types of muscles

Figure 3.33. Human reproduction entails more than sperm formation, ovulation, fertilization, and the coordinated hormonal control of a successful pregnancy. Years of loving, caring, entertaining, and educating must be invested in a growing child to ensure (as much as such things can be ensured) that it will become, in turn, a caring, loving adult. *(Photo courtesy of Virginia Tech)*

(skeletal muscle, smooth muscle, and cardiac muscle). We shall discuss skeletal muscle first because more is known about this type than the other two.

SKELETAL MUSCLE

Skeletal muscle is attached to and moves the skeleton. This muscle type is also often referred to as *voluntary* muscle because it is under direct control of the nervous system, and *striated* muscle because of the large contractile filaments that can be seen in each of its cells. Each skeletal muscle, such as the biceps, consists of many muscle cells running the entire length of the muscle. These cells are actually a *syncytium* or fused mass of several embryonic muscle cells and are called *muscle fibers* in a functioning muscle. The fibers may be very large (~100 microns in diameter and several centimeters long) and may contain many bundles of contractile filaments (several hundred to several thousand), called *myofibrils,* which also run the length of each fiber.

The myofibrils are responsible for the fiber's contractile ability. Each myofibril is composed of two proteins, *actin* and *myosin.* By chemical means, the myosin molecules pull the actin molecules together, thus causing the myofibrils, and hence the muscle itself, to contract. This account of muscle contraction is known as the *sliding bridge theory.*

Each muscle fiber is innervated by a motor neuron. The same axon, however, may branch to end on 100 or more muscle fibers. The axon and the fibers it innervates are called a *motor unit.* Skeletal muscle fibers will contract either completely or not at all—this is known as the *all or none principle* of muscle contraction. Because of this mode of action, it is more difficult to obtain a graded response from voluntary-striated muscles than from the muscle types found in such organisms as insects. Furthermore, striated muscle cannot maintain itself in a contracted state for long before fatigue sets in. How, then, does the body go about lifting a raw egg slowly, rather than quickly (and cracking it on one's face)? How is the body able to maintain a certain posture for long periods of time?

The answer lies in the number of motor units that are called into action at any one time. In the case of lifting the egg, the eyes send sensory impulses to the brain that tell it that the object is light. Only a few motor units fire. In the case of posture, sustained contraction is rotated between different muscle groups. No one group remains contracted until it completely exhausts its supply of ATP (adenosine triphospate, the "currency" of cellular energy).

SMOOTH MUSCLE

Smooth muscle is found in the walls of many organs, including the trachea, digestive tract, gall bladder, blood vessels, exocrine glands, bladder, vas deferens, and uterus. The cells are spindle-shaped and

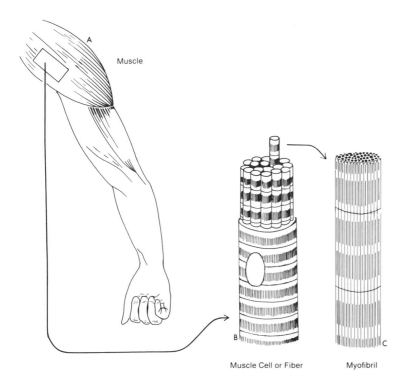

Muscle

Muscle Cell or Fiber   Myofibril

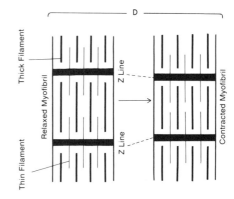

Figure 3.34. Muscle: an ever-closer look at its structure. The muscle of the arm illustrated in (A) is composed of muscle cells or fibers (B), which in turn are made up of myofibrils (C). The dark and light regions of myofibrils are responsible for the microscopist's term *striated muscle*. The sarcomere, the unit of contraction, extends from one Z (dark) line to the next (C); it is shown in detail in (D). When the muscle is relaxed, the thin lines are separated, scarcely overlapping the heavy ones. Contraction consists of bringing the Z bands closer together; this is accomplished (according to one model) by an interaction of actin molecules on the thin lines and myosin molecules on the heavy ones.

small (0.02–0.50 mm). Although these cells contain actin and myosin, the proteins are not arranged in the orderly fashion that they are in striated muscle.

Smooth muscle can sustain contraction for long periods of time without fatiguing. Smooth muscle also possesses other interesting properties. For example, the electrical signals that initiate contraction can pass from cell to cell; furthermore, these signals can arise spontaneously, without input from nerves or hormones. Thus, smooth muscles can react involuntarily or automatically. For example, smooth muscle reacts to stretching by contracting; this reaction occurs when blood shoots into the aorta from the heart's left ventricle. Contrary to popular belief, blood circulation through the body is not accomplished entirely by the heart; the smooth muscles of the arteries and veins also contract in a steady, rhythmic fashion to propel the fluid along. The role of the nervous system in controlling smooth muscle is one of modifying contractions, not initiating them.

CARDIAC MUSCLE

The news camera never sleeps, and the heart never rests. Cardiac muscle is unique in that it exhibits the useful characteristics of both striated and smooth muscle. It is striated, but it is not under voluntary control. It resists fatigue. It possesses even more automaticity than smooth muscle. If surgically removed from the body and maintained in a buffered saline solution with glucose, the heart will beat by itself for many hours. The heart also exhibits the *all or none* principle of striated muscle. However, the spontaneous rhythm is not the same for all parts of the heart. The heart's pacemaker, with its fast rhythm, regulates the other areas. As in smooth muscle, electrical impulses from the pacemaker travel from cell to cell without nerves. The innervation of the heart is still poorly understood; how the pacemaker operates in a continuous, spontaneous, rhythmic pattern is not known.

INFORMATIONAL ESSAY

The Lymphatic System

The most useful function of circulation occurs at the capillary level. The delivery of nutrients and oxygen in exchange for cellular waste products and carbon dioxide takes place here. No functional body cell is more than twenty to thirty microns away from a capillary. Most of the exchange and subsequent removal of waste products is carried out by the capillaries. The excess fluid surrounding cells, interstitial proteins (that is, proteins that are outside both cells and capillaries), and particulate matter, such as bacteria, cannot enter the capillaries. What happens to them? All this excess material enters the lymphatic system, an accessory route by which interstitial fluid, proteins, and bacteria are removed from the cells. The removal of these substances is a function that is essential for life.

Since lymph is merely tissue fluid, its composition varies from tissue to tissue. Some lymph will have high concentrations of protein, fats, carbohydrates, or salts, and some will have less. The excess tissue fluid drains away from the interstitial spaces and accumulates in channels that lead into lymphatic vessels. The vessels eventually come together into two large ducts that empty into the venous circulation near the bases of the jugular veins. Approximately 120 milliliters of lymph are accumulated, filtered, and returned to the bloodstream each hour in a resting person. During exercise, the rate of lymph flow may increase as much as five to fifteen times.

Like veins, all lymph channels and ducts have valves that ensure that lymph flows in one direction only; it does not surge to and fro. What drives lymph unidirectionally? Not the heart, but a series of "lymphatic pumps." In order of importance, these are: (1) contraction of muscles, (2) the movement of body parts, and (3) arterial pulsations.

The only way such things as particulate matter, intercellular proteins, and bacteria can enter the bloodstream from the tissue spaces is by way of the lymphatic system. But, in addition to maintaining the correct osmotic property around cells by removing these excessive proteins, the lymphatic system plays an important role in the body's resistance to infection. Central to this role are the lymph nodes. The lymph nodes are located intermittently along the course of the lymph ducts. Their function is to filter the lymph. The sinuses of each node are lined with highly specialized phagocytic cells, so that if any particle (bacterium or virus) or protein is recognized as non-self (i.e., foreign to the person's own body), it is phagocytized (engulfed and destroyed) and removed from circulation.

Sometimes the lymphatic system becomes overwhelmed by disease organisms; rather than retarding the infection, it now aids the spread. Cancer cells may spread—*metastasize*—through the lymphatic system from their points of origin to other sites within the body. Blood poisoning—*septicemia*—may result if the lymph vessels themselves become infected.

Relative to its importance, the lymphatic system is one of the least discussed systems of the body. The correct osmotic balance between cells and their associated capillary bed is essential for keeping those cells alive. The lymphatic system and its associated lymph nodes form the body's first line of defense against disease. It deserves better than to be equated with swollen tonsils and diseased adenoids.

INFORMATIONAL ESSAY
The Hemoglobins

The protein portion of human hemoglobin molecules consists of two pairs of polypeptide chains. These chains are designated *alpha* (α), *beta* (β), *gamma* (γ), *delta* (δ), and *epsilon* (ϵ)—the first five letters of the Greek alphabet. The bulk of all hemoglobin in the red blood cells of adult persons contains two alpha and two beta chains ($\alpha_2\beta_2$); a small fraction of adult hemoglobin consists of alpha and delta chains ($\alpha_2\delta_2$).

Adult hemoglobin appears in a person's circulating blood only about the time of birth. Before birth, the fetus utilizes fetal hemoglobin, which consists of alpha and gamma polypeptide chains ($\alpha_2\gamma_2$). For a brief instant during its early prenatal life, the embryo also contains hemoglobin that can be represented as ($\alpha_2\epsilon_2$).

The utilization of one sort of protein before birth, and the switchover to a second kind at the time of birth, raises several questions, one of which is physiological: Are there differences between fetal and adult hemoglobins that would account for the use of one before birth, and the other after birth? The answer appears to be Yes. Fetal hemoglobin has a greater affinity for oxygen at lower oxygen pressures than does adult hemoglobin. Thus, a pregnant woman's hemoglobin is oxygenated in her lungs, where the oxygen pressure (measured in mm of mercury) is about 100; the oxygen pressure in the maternal placental

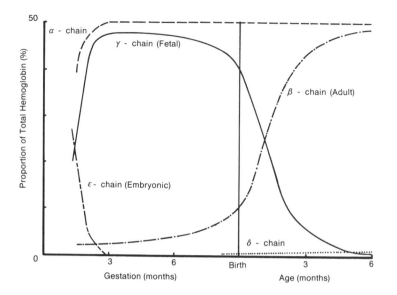

Figure 3.35. The appearance and disappearance of hemoglobin chains during human development. Because hemoglobin consists of two pairs of slightly dissimilar polypeptide chains ($\alpha_2\beta_2$, $\alpha_2\gamma_2$, $\alpha_2\delta_2$, or $\alpha_2\epsilon_2$), no polypeptide chain can constitute more than fifty percent of all hemoglobin. Note that the alpha chain is found in embryonic, fetal, and adult hemoglobins; the beta chain is primarily incorporated into hemoglobin at the time of birth and thereafter.

circulation is only fifty mm of mercury, one-half that in the lung. Because fetal hemoglobin has a greater affinity for oxygen than does adult hemoglobin, it can carry about twenty-five percent more oxygen than the latter. Furthermore, the amount of hemoglobin in a given volume of fetal blood is half again as great as the amount in the same volume of adult blood.

Other questions posed by the switchover from fetal to adult hemoglobin at birth are genetic: What genes are responsible for the synthesis of these different polypeptide chains? How is their activity regulated? The first of these questions has been answered thoroughly. There appear to be three regions of DNA that are responsible for the synthesis of the alpha, delta-beta, and gamma chains. The alpha region is separate from the others. The delta and beta regions are adjacent, although the beta region is by far the most actively transcribed. That they are adjacent is revealed by an aberrant form of hemoglobin in which the non-alpha chain is extra long and corresponds to the delta chain at one end and the beta chain at the other. An mRNA specifying such a polypeptide chain would be formed if a small interstitial region of DNA were missing: instead of reading d-e-l-t-a*b-e-t-a, the altered DNA reads d-e-l-t-e-t-a, because the intervening a*b has been lost.

The switchover from one hemoglobin to another in the developing embryo—first from embryonic to fetal, and then from fetal to adult hemoglobin—involves some sort of gene control, but just how that control is exerted is only poorly understood. Certain rare adult persons have red cells that contain only fetal hemoglobin, but these persons enjoy seemingly normal health. Molecular studies have shown that the DNA in hemoglobin-synthesizing cells of such persons has fewer methyl groups attached to it at the gamma locus than does that of persons with normal adult hemoglobin. Does that mean that methylation of DNA is a means for turning off gene loci? Perhaps. Whether or not methylation provides the means by which the hemoglobin loci are controlled, demethylation of DNA by chemical means does result in the production of fetal hemoglobin in adult mammals, including human beings. Tests have been made on persons suffering from thalassemia or sickle cell anemia (both extreme manifestations of defects in the beta chain); chemical treatment has induced the synthesis of fetal hemoglobin (and a suppression of the synthesis of the abnormal adult hemoglobin) in these persons. Because the fetal hemoglobin is not defective, such treatment promises to restore these individuals to normal health. Of course, much more needs to be learned about the duration of the induced switchover from adult to fetal hemoglobin and about the unwanted side effects of the chemotherapy.

The presence of different polypeptide chains in fetal and adult hemoglobins influences the observable types of genetic diseases caused by aberrant (mutant) chains. Most known hemoglobin disorders affect the beta chain. Sickle cell hemoglobin, for example, involves an amino acid substitution in the beta chain. The sickle cell mutation and others involving the beta chain have little or no effect on the developing embryo, which relies on fetal hemoglobin ($\alpha_2\gamma_2$). Conversely, mutations affecting the alpha chain could affect development of the embryo severely—even causing its death.

One final point can be made here, although it will recur at a later time. The amino acid sequences in the polypeptide chains of hemoglobin molecules are sufficiently similar to suggest that the responsible regions of DNA have arisen by gene duplication. In fact, there are at least three closely adjacent regions that code for recognizably different gamma chains in addition to the adjacent delta and beta regions and the separate alpha and epsilon regions.

In addition, corresponding alpha, beta, gamma, and delta chains are found among many different mammals. Consequently, one can trace not only gene duplications, but also the occurrence of these duplications during vertebrate (especially mammalian) evolution. This is just one type of evidence for evolution that molecular biologists have provided to their colleagues in evolutionary biology.

INFORMATIONAL ESSAY
The Pituitary-Hypothalamus Connection

That endocrine glands are controlled by the brain is easily forgotten. How is this control achieved? The interconnection of brain and endocrine glands is provided by the hypothalamus and the pituitary gland; even the close physical proximity of these two bodies suggests that they have a functional relationship. The pituitary gland, which consists of two parts, known as the anterior and posterior pituitary, is often referred to as the "master" gland, because it governs the secretion of many other glands. The hypothalamus, a part of the brain, is the most important control center of the body. Stimulation of the hypothalamus results in changes in behavior patterns with respect to such things as sex drive, pleasure, anger, fear, anxiety, stress, hunger, and thirst.

The hypothalamus sends chemical messages to the pituitary in two ways. In the first, the hypothalamus synthesizes an array of hormones that, when released into the bloodstream and carried (next door) to the anterior pituitary, cause it to release specific hormones. In the second, the hypothalamus secretes another group of hormones, but they travel down secretory neurons directly into the posterior pituitary, where they are subsequently picked up by the bloodstream.

All activities of the brain dealing with thought and feeling pass through the hypothalamus. Mental states, such as stress, anxiety, depression, and hostility, can be reflected in illnesses, such as heart attacks, ulcers, high blood pressure, obesity, and hives. Individuals often fail to appreciate the power of the brain to alter physiological states until they themselves have experienced the serious consequences.

INFORMATIONAL ESSAY
The Liver

Earlier, in contrasting the wisdom of the body ("wisdom" that is encoded in DNA) with the wisdom of the mind, we cited Dr. Lewis Thomas's claim that he is incapable of making hepatic decisions. Dr. Thomas's choice of the liver was no accident, since decisions regarding striated or skeletal muscle are not difficult: to contract or not to contract, that is their only question.

The liver, that three-pound gland lying on the right side of our bodies, largely protected by the lower ribs, has many functions; it stores and filtrates blood, it produces and secretes bile, and it is involved with nearly every metabolic system of the body.

The basic structural unit of the liver is an elongated cylinder of cells—a lobule—at the center of which is a small vein. This vein and the hundreds of thousands of central veins of other lobules lead through collecting veins to the main vein—the vena cava—leading to the heart.

Extending from the central vein in all directions are vanes composed of double layers of cells; these cells secrete bile into the space they enclose. The bile is collected in bile ducts, is led to the gall bladder for storage, and is released into the small intestine to mix with

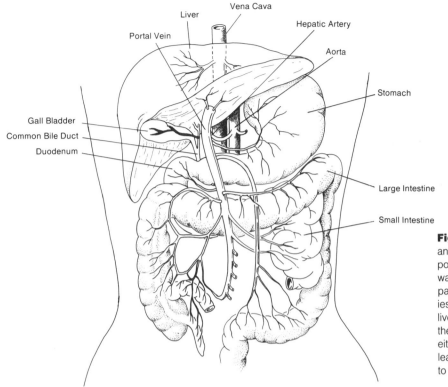

Figure 3.36. The physical relationship of the liver and the digestive tract. Nutrients and other compounds that pass into the blood vessels through the wall of the gut are transported to the liver by the hepatic portal system, a system that begins in capillaries (of the gut wall) and ends in capillaries (within the liver). Oxygenated blood enters the liver by way of the hepatic artery. All blood that arrives in the liver, either by the portal vein or by the hepatic artery, leaves the liver by way of the hepatic veins that lead to the vena cava.

food as it exits from the stomach. The main digestive function of bile is the saponification of fatty substances (saponification means "to make into soap"—hence, in short, bile plus fat equals soap).

Between the double-layered vanes are blood sinuses that receive blood both from the portal veins (of the hepatic portal system) and from branches of the hepatic artery. The latter, of course, supplies oxygen and nutrients needed to keep liver cells alive.

The blood sinuses contain large cells that are phagocytes. These cells, over and among which the portal blood from the intestine must

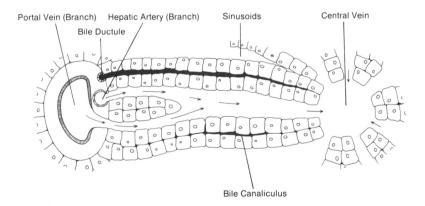

Figure 3.37. A functional unit of the liver. The arterial blood and blood arriving from the gut wall blend and together bathe long vanes of cells as the mixture makes its way to a collecting vein. The cells being bathed, in addition to carrying out complex biochemical processes on both nutrients and miscellaneous chemicals that have entered the body, secrete bile, which is gathered by small ducts and eventually sent to the gall bladder for storage.

Life: An Instant-Thick Snapshot

pass, can remove ninety-nine percent or more of the bacteria that might be in the blood. Blood in the hepatic portal system, arising as it does in the gut wall, might easily (as the result of a possible imperfection in the gut wall) contain bacterial contaminants. (Recall that the inside of the gut is *outside* the body.)

Surrounding the liver cells in each vane is lymphatic fluid. This fluid is collected in lymphatic vessels. The lymphatic glands through which these vessels pass are also extremely efficient at removing contaminating organisms.

The truly major function of the liver lies in the metabolic abilities of liver cells. (Later, this text covers at considerable length the conversion of glucose to glycogen, a stored carbohydrate, and the subsequent release on demand of glucose from this store.) Indeed, nearly all ingested molecules are altered in composition—either for use or for excretion—by liver cells. Numerous enzymes are needed virtually simultaneously to carry metabolic processes, once started, to completion. In one published summary covering sixteen systems of metabolic breakdown performed by liver cells, it was calculated that an average of nine enzymes must be activated in concert to handle each metabolic pathway. The largest number was seventeen, nearly twice the average.

In carrying out their genetically programmed metabolic functions, liver cells frequently encounter manmade organic compounds. In the process of degrading these for excretion, intermediate compounds that interact with the liver cells' genetic apparatus are frequently made. Cancerous cells are often the result, and liver cancers are frequent among industrial workers. Nature also takes advantage of the mammalian liver. The toxins of certain fungi are essentially harmless until cleaved by enzymes in liver cells; the cleavage products then wreak havoc in the liver—and with its owner.

NOW, IT'S YOUR TURN (3-12)

We have already discussed how radiation exposure damages dividing cells especially severely because of its effect on the orderly division of genetic material. We have also mentioned that the lining of the entire gut is continually being replaced; old cells are sloughed off into its lumen, only to be replaced by fresh ones arising through cell division. Does this information provide clues about why persons who have been exposed to whole-body radiation, such as the citizens of Hiroshima and Nagasaki or workers exposed during nuclear accidents, are in danger of contracting severe (systemic) bacterial infections? By what route might the invading bacteria enter the body?

INFORMATIONAL ESSAY

Feedback

Nearly all biological systems are controlled by negative feedback, in which a particular process automatically limits itself. *Homeostasis* is the term that emphasizes the ability of the body to automatically maintain its *internal* environment within narrow limits; negative feedback is the basis for that ability.

Although you are familiar with examples of feedback (recall "end-product inhibition"), one particular example might be discussed in detail. Consider, then, a person who realizes that he or she is in danger—perhaps of drowning. As the person perceives the danger, nerve transmissions related to that perception are picked up by the hypothalamus. The hypothalamus then secretes hormones that first activate the pituitary, which in turn secretes another hormone, *ACTH* (*adrenocorticotropic hormone*). ACTH enters the bloodstream and travels to the adrenal glands, where it initiates the production of *cortisol,* thus increasing blood sugar levels and raising the body's rate of metabolism. The body becomes agitated, tense, and anxious.

Second, but simultaneously, the hypothalamus stimulates the autonomic nervous system, which sends nerve impulses down the spinal column. Blood circulation to the digestive system is immediately restricted. Blood pressure is raised by the constriction of peripheral blood vessels. The same nerve impulses also go to the adrenal medulla, where *adrenaline* (*epinephrine*) is released. The extra adrenaline causes release of extra glucose, speeds up the heartbeat, and raises blood pressure still higher. As the danger recedes, the hypothalamus reacts by ceasing to secrete the hormones that aroused the body, and normalcy is restored.

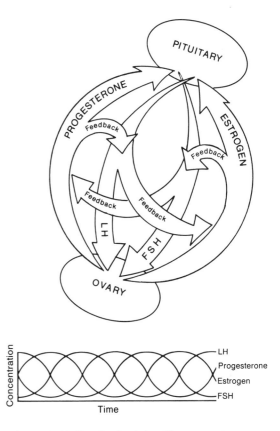

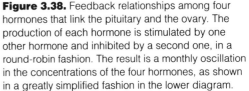

Figure 3.38. Feedback relationships among four hormones that link the pituitary and the ovary. The production of each hormone is stimulated by one other hormone and inhibited by a second one, in a round-robin fashion. The result is a monthly oscillation in the concentrations of the four hormones, as shown in a greatly simplified fashion in the lower diagram.

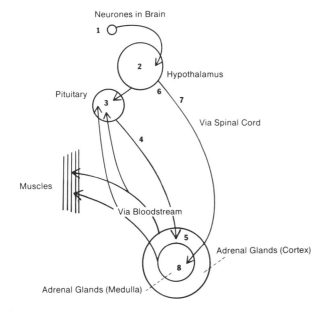

Figure 3.39. A schematic outline of the communication network that leads to stress. Nerve impulses resulting from stress or unpleasant sensations lead from the cerebrum (1) to the hypothalamus (2). The hypothalamus is stimulated to produce a chemical that travels either to the pituitary (3) or to the brain stem (6) and spinal cord (7). For its part, the pituitary produces ACTH, which travels by the bloodstream (4) to the cortex of the adrenal glands (5); here, ACTH stimulates the production of cortisol, which raises the body's metabolic rate. In the meantime, nervous impulses from the spinal cord reach the core of the adrenal gland (8) and cause the release of epinephrine (adrenaline), which leads to excess glucose in the blood, and of norepinephrine, which raises the blood pressure and speeds up the heartbeat. Both processes are continuously monitored by the pituitary.

Life: An Instant-Thick Snapshot   **121**

Feedback systems are sometimes positive, rather than negative. When the feedback is positive, secretions continue to increase, homeostasis is destroyed, and serious illness or even death can occur. One of the most serious, chronic problems of our society today is stress. (Stress also includes the reactions to stressful situations themselves, such as depression, anxiety, anger, and fear. One scientist has noted, "Our mode of life itself, the way we live, is emerging as today's principal cause of illness.") If we view stress as involving positive feedback, we can see how it contributes to illness.

Under continuous stress, the threatening situation that we perceive never passes or abates. The processes discussed earlier that bring the body to a state of alertness send signals back to the brain, where the signals are interpreted once more as danger signals. The hypothalamus reacts again by stimulating the pituitary and autonomic nervous system and the cycle is repeated again, and again, and again. This continuous feedback process eventually destroys the body's homeostasis, resulting in heart disease, heart attacks, ulcers, obesity, sickness, and perhaps even cancer. Recently it has been discovered that chronic stress inhibits the production of T-lymphocytes and macrophages, important components of the immune system.

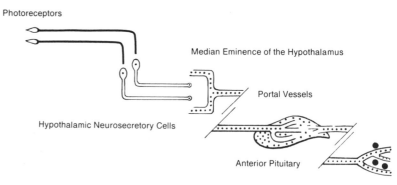

Figure 3.40. An example of hormonal control in an organism other than human beings. The white-crowned sparrow undergoes gonadal development, molts, and migrates in response to lengthening of daylight hours in the spring. The diagram shows that photoreceptors stimulate the hypothalamus, which in turn affects the pituitary, and, through the pituitary, the testes. This example illustrates a fundamental difference in the interpretation of facts by different sorts of biologists: Physiologists say that the reproductive behavior of this bird is controlled by hours of daylight (the photoperiod); naturalists say that light merely provides a signal that causes the bird to produce young at a place and time when food is plentiful. Both views are correct: one gives the *proximate* explanation, the other gives the *ultimate* one.

4

ENERGY FOR LIFE

What is your main occupation? Being a student? Training for one of the various athletic teams boasted by many colleges and universities? Studying? The job you have off campus? How about breathing and eating? Any job that requires only eight hours a day for five days each week—whether it be in a factory, shop, government office, or elsewhere—is less important by far than these last two. A person can stop work at any time; however, no one can stop breathing or go without eating for very long. The need for food and oxygen continues as long as one lives. True, most of us scarcely take notice as we fulfill these needs; nevertheless, they dominate our lives.

A newborn baby, not yet dry after birth, draws its first breath of air. With that first breath, it begins life as an independent being. From that first breath, breathing becomes an automatic, rhythmic activity that will continue for the rest of the individual's life. The breathing rate (breaths per minute) changes as the infant becomes a child and later matures as an adult man or woman. The rate at which this person breathes will also change with his or her physical activity. Recognizing that people vary, we can still estimate that between birth and seventy years of age, each person will breathe in some 250 million liters of air. From this air, the body will extract more than 15,000 kg (16½ tons!) of oxygen; it will discharge an even larger amount of carbon dioxide as waste—about one-third more carbon dioxide by weight expelled than oxygen taken in.

Feeding also starts early in an infant's life. Without example or training, the newborn baby sucks milk. It sleeps, and then wakes and cries until it is fed again. Thus, even the youngest human being is able to sense the need for food. It can also "measure" this need, because when its stomach is full, the baby stops feeding and falls asleep, only

Biological Rhythms

The beat of the heart and the rhythm of breathing are only two of countless examples of how life is organized in time. In fact, all known cell activities and processes exhibit changes with time. Some rhythms are fast, as in the case of the heartbeat. Others are slow, as in the monthly (menstrual) cycles of women of childbearing age. In many cases the observed cycle is about 24 hours long, and is coordinated with day and night. A well-known example of a 24-hour cycle is the alternation of states of alertness and sleep. Besides this cycle, there are many others known to have an approximately 24-hour period (the *circadian rhythm*, a term meaning *about a day*). The rates of cell division follow a circadian rhythm in the liver, the skin, the tissues that form blood cells, and elsewhere. So do the levels of cell enzymes and substances in blood, the sensitivities of body cells to attack by poisons and parasites, and many other aspects of life. The basis of these rhythms is unknown, but the rhythms of the human body (and similar rhythms in other organisms) have been shown experimentally to persist when the subject was deprived of clues to the actual (day or night) time in the outside world.

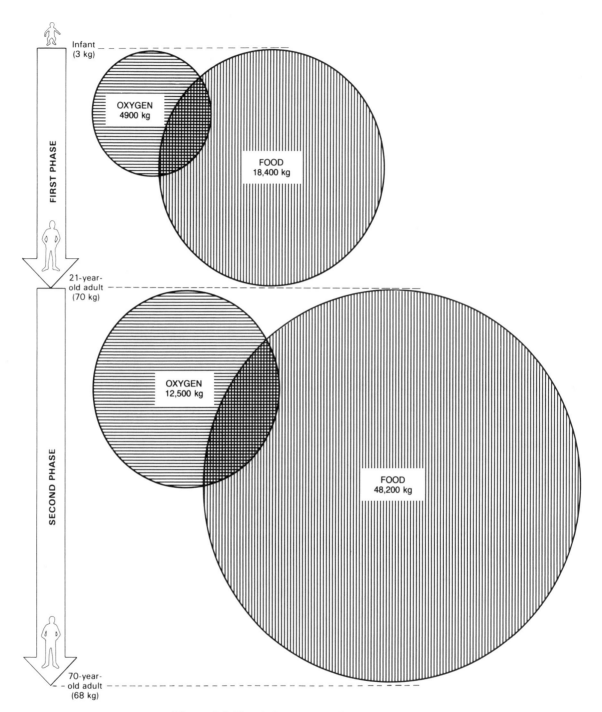

Figure 4.1. The relative amounts of oxygen and food that are required for a person (1) to grow from three to seventy kilograms and (2) to remain alive with no further growth for another fifty years. Note that the body's growth involves only a small fraction of the total amount of food consumed. Note, too, that an effective balance must exist between material going into and out of one's body; otherwise it would either swell to an incredible size or shrivel up and disappear. *(Illustration provided by The Biological Sciences Curriculum Study)*

to wake up again in need of still more food. As the infant progresses along its life's trajectory, feeding continues to be a main concern. By age seventy, this person will have consumed some 65,000 kg, or about seventy tons, of food—nearly 1000 times the body weight of an adult.

Supplying the body with oxygen (breathing) and supplying it with nutrients (eating) seem to be different activities. The ways in which the two differ can be easily listed: (1) Oxygen enters the body by way of the lungs; nutrients enter by way of the digestive system. (2) Oxygen combines with hemoglobin in the red blood cells and is carried directly to the body cells; food is changed both physically and chemically in the gut, is carried to the liver for further chemical processing, and may be stored in the liver before going to the body cells for use. (3) The supply of oxygen must be nearly continuous, because the human body cannot store an appreciable amount of oxygen; nutrients can be stored by the body, and during a fast a person may live as long as sixty days without eating.

Listing the above differences between breathing and eating obscures the common function they both serve. People, like all other living things, obtain energy from the world in which they live. Food contains chemical energy, and oxygen is essential for the release of energy in a useful form. Therefore, eating and breathing are two facets of a single fundamental task: obtaining the energy that is needed for life itself.

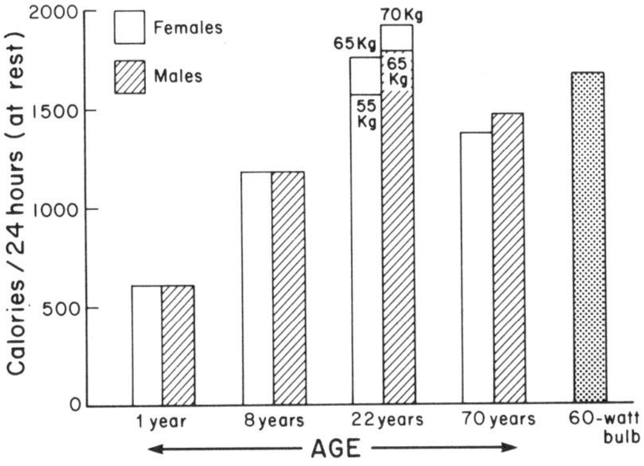

Figure 4.2. The number of calories needed each day by males (shaded bars) and females (unshaded bars) at different ages while at rest. Infants and children use fewer total calories than adults, but they use more relative to their body weight. Late in childhood, the caloric requirement for males becomes greater than that for females, and it remains greater thereafter. The boxes at the tops of the bars for 22-year-olds indicate the range of calories needed for persons of different weights: 55–65 kg for women, 65–70 for men. For comparison, the energy required by a 60-watt bulb burning continuously for twenty-four hours is shown. In its case, energy arrives as electricity and is emitted as light and heat.

NOW, IT'S YOUR TURN (4-1)

The text claims that the average person extracts 16½ tons of oxygen from the air during his or her lifetime. It also suggests that some 22 tons of carbon dioxide are exhaled. Where do the additional 5½ tons come from? (Hint: The atomic weight of oxygen is 16; that of carbon is 12.)

The Demand for Energy

How do oxygen and food provide energy? Why is energy needed for maintaining life? These and still other questions can organize our consideration of the vital role of breathing and eating in our lives.

How Do Food and Oxygen Function in Supplying Energy?

One way of answering the above question is to compare the body's machinery with manmade machinery. All machines draw on sources

Comparing People with Other Machines

People	Machines
Need chemical energy sources; only food can serve (by definition).	Need energy sources. Some are specialized (the automobile engine burns only gasoline), while others will accept almost any chemical energy source (a furnace burns wood, coal, paper, garbage, and all sorts of other substances, turning chemical energy into heat). Still other machines do not use chemical energy, but draw energy from movement, heat, or light (examples are windmills, the dynamos of steam or hydraulic plants, solar batteries, and solar heating plants).
Need oxygen to complete chemical reactions; energy is transferred into usable form.	Need oxygen only if they use a chemical energy source; no oxygen is needed by machines that run with heat, light, movement, or nuclear power.
Convert energy for many purposes—to build chemical substances (for growth and self-maintenance), for movement, and to supply heat.	Convert energy for special purposes; generally there is one main purpose for any one machine (energy for movement of an automobile, heat energy for a furnace, radiant energy for a light bulb).
Need energy to maintain the structures of enzymes and many other body parts. These structures break down if the energy supply fails; if they cannot be reformed, death occurs.	Do not need energy to maintain structure. If the energy supply fails, the machinery stops working, but when the right energy source is supplied once more, the machine starts up again.

of energy, often changing energy from one form to another and applying appropriate forms of energy to do particular kinds of work. For example, the automobile (auto=self, mobile=moving) engine burns gasoline. Air is drawn through the engine's carburetor and mixed with gasoline vapor, after which the oxygen and gasoline are combined (burned) in the cylinders, thus converting the chemical (stored) energy of gasoline—by way of the hot and explosively expanding gases of combustion—into the mechanical energy of engine parts (moving pistons). The energy released by combustion powers the movement of the wheels through the interplay of connecting rods, gears, transmission, and axles. Some energy is also changed to electrical energy in the generator. This energy either is used at once (for the car's lights) or is stored (in the car's battery). Although the hot gases exploding in the cylinder provide the power that moves the automobile, a large part of the chemical energy of gasoline is eventually wasted as unutilized *heat*.

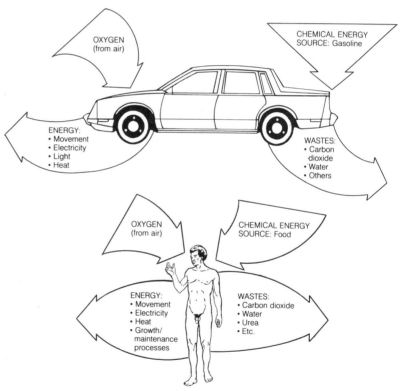

Figure 4.3. All machines, including the human body, use energy. In a person or an automobile, *chemical* energy in food or fuel is changed to mechanical energy—movement. Chemical energy in food is also retained in the body's materials as chemical energy. In a lightbulb, *electrical* energy is changed into light. All three machines—the body, the car, and the bulb—produce *heat* (also a form of energy). Usually heat is a waste product, but living organisms put it to use in maintaining body temperature. Heat cannot be considered a waste product of a furnace, of course. *(Adapted from an illustration provided by The Biological Sciences Curriculum Study)*

Many similarities exist between the automobile and the human machine. We also use oxygen in changing chemical energy into (to us) useful energy. Furthermore, the chemical reactions that occur within cells convert the chemical energy of food—one step at a time, not by way of an explosive blaze—into energy for movement, for producing electricity, or for heat. Our bodies, however, differ from automobile engines in one important way: much of our energy is expended on growth, self-maintenance, and self-repair.

The functioning of any machine depends on more than the available energy sources. First and foremost, the operation of the machine depends upon the way in which it is built—its parts, and how the parts are put together. The automobile engine is designed to use only one energy source, gasoline, although it can be modified to use alcohol or liquid propane gas. The engine can use its energy source only by mixing small quantities with air (or, rather, with the oxygen that forms about twenty percent of air) and burning it in a manner that allows the explosively hot gases to move the engine's pistons. In much the same way, special proteins—enzymes—are used by human beings as well as all other living organisms to convert chemical energy into utilizable energy. Furthermore, enzymes can use only acceptable nutrients as a source of energy; no starving person would be helped by gnawing on either bituminous or anthracite coal. Of all gases, only oxygen can be used in the conversion of energy. In fact, oxygen and food play their important parts in human life because our enzymes are adapted to their use—and to their use, only.

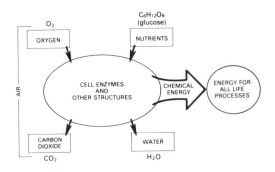

Figure 4.4. How the body obtains energy from food. The nutrient chosen as an example is glucose—a common ingredient of food, one that is part of table sugar and is the main sugar in human blood. It is the only energy source used by the brain. Within cells, oxygen and glucose enter into enzymatic reactions. As a consequence of those reactions, hydrogen from glucose combines with oxygen to form water, while the carbon atoms combine with oxygen to form carbon dioxide. Because these are enzymatically controlled reactions, a great deal of chemical energy (rather than heat) is made available for other life processes. *(Illustration provided by The Biological Sciences Curriculum Study)*

The Limits of Energy Supply

Oxygen Can't people live without oxygen? Might it not be replaced, even as some modern plastics have served as blood substitutes during medical emergencies? Most living organisms can obtain some chemical energy without utilizing oxygen, and some bacteria and even higher organisms use no oxygen at all. Even human muscles can obtain energy for short periods without oxygen. However, the total yield of energy from these alternative processes is low. In contrast to combustion involving oxygen, these alternative processes involve incomplete combustion. During this incomplete combustion, they produce acids that can cause serious physiological problems and that can be removed from muscle tissue only by the cooperation of blood, liver, and kidneys.

People, then, are dependent upon the availability of oxygen for most of their energy needs. Still, might some chemical *similar to* oxygen take its place in metabolic reactions? The answer is No. From laboratory studies, we know of other substances that may react in many circumstances as oxygen does: sulfur reacts with hydrogen to

Survival Without Access to Oxygen

Most of the oxygen in the mammalian body is combined with the protein *hemoglobin* in red blood cells, and with a related protein (*myoglobin*) in muscle cells. The total amount of oxygen in the body gives an upper limit to the *possible* time of survival if the supply of oxygen is cut off suddenly, as in drowning. Note, however, that the *real* time of survival is much shorter, because animals cannot actually use every molecule of stored oxygen. The figures given here are for a 70-kg man without special training; they are listed in terms of the volume (liters) of oxygen and also as minutes that the oxygen might last at an average rate of 0.24 l/min.

	Liters	Minutes
Oxygen in lungs	0.5	2.1
blood	1.2	5.0
tissues	0.3	1.2
	2.0	8.3

The real survival times of people cut off from oxygen supplies are very short. In contrast, marine animals have special adaptations (chiefly of the circulation) that make them far more able than people to stay under water. Here are some examples:

Species	Safe time to stay under water (in minutes)
People	
untrained	1
trained divers	2.5
Harbor seals	
(*Phoca vitubra*)	20
Sperm whales	
(*Physeter cetadon*)	75
Bottle-nosed whales	
(*Hyperoodon ampullatum*)	120

form H_2S, for example, much as oxygen does to form H_2O. However, the details of these other reactions differ from those that involve oxygen. Human enzymes are structured to function within such narrow limits that they will not accept oxygen substitutes. Oxygen, therefore, *is* essential for *human* life.

Foods The human body is also limited with respect to the chemical substances that can serve as its nutrients. Despite the thousands of enzymes their cells possess, human beings cannot use an unlimited range of food materials. A human being cannot use even all *possible* foods—that is, substances that serve as food for other organisms. People eat both plant and animal tissues, but they can digest neither cellulose, a complex sugar that forms the woody tissue in plants, nor hair, whose main component is protein. Because human beings do not have enzymes capable of breaking down cellulose or hair, these substances cannot be considered human food. In contrast, many protozoans can digest cellulose and the larvae of the clothes moth thrive on wool.

Why Is Energy Needed for Life?

What happens when the body's supply of energy fails is all too clear: when oxygen is cut off (say, by choking on a piece of steak), death is sudden. No energy means no physical movement. No energy means that enzyme molecules cannot be synthesized to replace those that are lost, nor can structural proteins be synthesized. Within minutes, in the absence of oxygen cellular changes are so profound that cells cannot be restored to life even if oxygen is provided once more. The body becomes cold, because heat is a form of energy and there is none. During starvation, the loss of energy is less abrupt but, in the long run, equally fatal. A child who does not eat, does not grow. Starving persons, children or adults, move about less and less, and they waste away as their own tissues are burned in lieu of food. Energy is needed both to keep a living organism in repair and to provide the power needed to keep it running.

NOW, IT'S YOUR TURN (4-2)

The following data, copied from the nutritional information printed on boxes, cans, or jars of food, will give you a chance to use your high-school algebra. How many calories does one gram of each substance (protein, carbohydrate, and fat) provide? Are the data consistent throughout?

Food	Grams per serving			Calories per serving
	protein	carbohydrate	fat	
Salad dressing	0	2	7	70
Whole milk	8	11	8	150
Peanut butter	8	5	17	190
Breakfast cereal	4	20	2	110
Canned soup	2	20	2	110
Jelly	0	4	0	25
Cereal (#1)	4	25	1	120
Cereal (#2)	3	20	4	130

Figure 4.5. *Hunger.* Käthe Kollwitz, 1923. This woodcut of hunger—bloated stomach, skeletal limbs, and skull-like features for the child; empty breasts and pronounced ribcage for the mother—has been chosen to illustrate hunger because it illustrates the general features of starvation. Kollwitz specialized in illustrating the social ills of Central Europe, especially Germany and Austria. However, today, as these words are being written, hundreds of thousands of sub-Saharan African children have either died of hunger or are starving. Hunger is a problem that confronts all living organisms—human beings included. It need not be illustrated with children who are recognizably of any nationality, because hungry children are European, Asiatic, African, American Indian, and Latin American.

How Much Energy Does One Need?

"How much?" is an important question. How does anyone know how much air to breathe, or how much food to eat? From birth, every person must judge how much energy he or she requires at every stage of growth. Furthermore, no two people have exactly the same requirements, and requirements change systematically throughout life as well as from moment to moment as activities change. Human beings use energy at a high rate; only bacteria, birds, and other mammals compare with human beings in their energy-related "high cost of living."

The Control of Breathing and Feeding

The question, "How much energy does one need?" has two answers. The actual energy needed can be measured accurately by any one of several laboratory procedures, as Figure 4.6 shows. However, a second answer is provided by the body itself as it measures its needs for oxygen and food and as it adjusts its rates of breathing and feeding. These adjustments are initiated by control systems within the brain that are implemented by endocrine glands. The brain and endocrine glands are part of an overall control system that gives rise to "the wisdom of the body," a phrase made popular by a great American physiologist, Walter Cannon.

The regulation of breathing rate depends in part on nerve endings that respond to the amount of carbon dioxide and oxygen in the blood. The nerve endings are located in blood vessels; indeed, there are nerve endings in the brain itself that are especially sensitive to changing concentrations of carbon dioxide. When the blood concentration of carbon dioxide rises, or that of oxygen falls, messages are carried to the respiratory centers of the brain. These centers react by speeding and deepening breathing movements. In addition, a decrease in carbon dioxide concentration depresses respiration. These responses occur almost instantly, and they are triggered by even the

Measure change in
TEMPERATURE

Measure decrease in volume of OXYGEN

Measure amount of RADIATION from neck

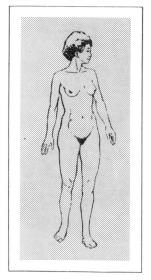

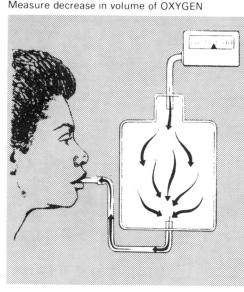

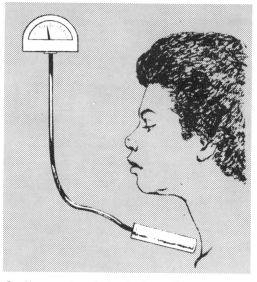

A. How much heat does the person give off?

B. How much oxygen does the person use?

C. How much radiation is given off by radioactive iodine in the thyroid gland?

Figure 4.6. Measuring the amount of energy a person uses. Several procedures can be used to measure the energy needed to maintain a human being (or other organism) for a given period of time. (A) The heat that the body produces can be measured. (B) The amount of oxygen consumed can be measured. Because oxygen is used only to oxidize carbohydrates, the amount used can be expressed as energy released. (C) The activity of the thyroid gland can be monitored by measuring the amount of iodine (radioactive in this case) that accumulates; the amount accumulated is related to the energy the body has used. *(Illustration provided by The Biological Sciences Curriculum Study)*

slightest changes in gas concentration in the blood. The same control systems aid the body when there are long-term conditions that interfere with an adequate supply of oxygen. For example, loss of blood or anemia (caused, perhaps, by insufficient iron in the diet) decreases the number of red blood cells (or hemoglobin) carrying oxygen to body cells; the control centers cause compensatory changes, such as rapid breathing.

Feeding is also closely tuned to the body's demands. Human beings,

Figure 4.7. Balancing food intake against energy use is an extremely complex affair, involving:
- The supply of food (Enough? Too much?)
- Food supply to body cells (Digestion? Transport? Liver? Hormone levels?)
- Physical circumstances and activities (Growing? Pregnant? Surrounding temperature? Diseases?)
- Control centers (Glands functioning properly? Social pressures? Stress?)

Whether one gains or loses weight depends on whether the energy entering the body exceeds (gain) or is less than (loss) the energy that is expended. *(Illustration provided by The Biological Sciences Curriculum Study)*

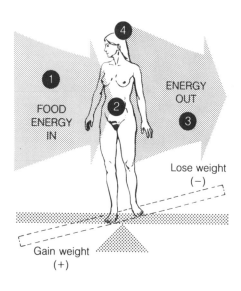

like other living things, have an amazing ability to control the amount they eat so that their bodies are kept in balance, with repair just equaling breakdown: energy is supplied as energy is needed. Imagine, for example, a moderately active adult woman who eats a mixed diet that yields about 2500 kilocalories (abbreviated kcal) each day. Her activity and diet are in balance; she neither gains nor loses weight. Now, if she should take in 100 kcal *less* each day, she would lose about ten pounds in a year. Notice, though, that 100 kcal is the amount supplied by a single apple, an onion, or a muffin. People's eating habits are, as a rule, extremely accurate, in the sense that what is eaten corresponds to what their bodies need. Little or no conscious thought goes into maintaining this balance.

The ability to feel a need for food and to measure the correct amount from a hodge-podge of choices (meats, tubers, green vegetables, cereals, fleshy fruits and watery ones) is of utmost importance. Important as it is, though, the physical basis for this ability is not fully understood. The major control centers are known to be in the brain. Nerve cells in the hypothalamus play an important role, but they may be controlled by still higher centers in the brain. Damage in, or disease of, the hypothalamus can cause both persons and laboratory animals to eat abnormally. Depending upon the location or type of injury, there may be a complete loss of appetite or, in sharp contrast, an uninterrupted urge to eat. Either way, the individual's life is placed in jeopardy by either self-imposed starvation or gross obesity.

The questions about eating never seem to end: How does the brain know how much food is needed? It may receive and interpret signals carried by blood from the rest of the body. Such signals could be the actual concentrations of various nutrients, such as sugars, fats, or amino acids. Hormones are an alternative possible means for signaling the brain; a promising candidate is insulin, which is manufactured

Table 4.1. The daily activities and energy requirements of two young men, similar in all respects except that one (A) is an office clerk, while the other (B) is a coal miner

Measure	Man A	Man B
Hours of the week spent		
sleeping	56	58
awake but not working (walking, standing, washing, dressing, etc.)	62	65
working	50	45
Energy use—kcal per week		
sleeping	3,810	3,690
awake but not working	9,800	10,900
working	5,710	11,870
TOTAL	19,320	26,460

Now, it's your turn:
The study revealed that A ate fifteen percent less food than B. Are the data consistent with this finding?

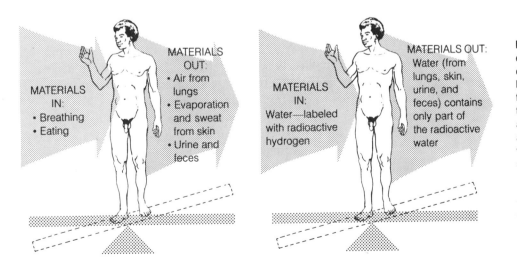

Figure 4.8. The balance of materials that enter and leave the body is an extremely complex matter. Unlike manmade machines, living bodies keep themselves in repair. In this process, food not only provides energy to keep the body's machinery running, but also provides replacement parts. Water balance, too, is complex. In experiments in which people drank radioactive water, only a small amount left the body immediately; the remaining labeled water molecules displaced others and took their places within the body, only to be displaced in turn by water that was drunk during the coming days. *(Illustration provided by The Biological Sciences Curriculum Study)*

MATERIALS IN:
• Breathing
• Eating

MATERIALS OUT:
• Air from lungs
• Evaporation and sweat from skin
• Urine and feces

MATERIALS IN:
Water—labeled with radioactive hydrogen

MATERIALS OUT:
Water (from lungs, skin, urine, and feces) contains only part of the radioactive water

Scurvy

People have always suffered from diseases caused by deficiencies in their diets. In the past, long, severe winters were especially dangerous times. Sometimes, without anyone's knowing the cause of a disease, a cure was found by chance. For example, European explorers who spent months at sea often suffered from *scurvy*, the "sailor's disease." The disease was described vividly by Jacques Cartier, who led an expedition from France to the eastern coast of Canada in 1535. In Cartier's words, during the winter, his men were attacked by a horrifying disease:

"Some did lose all their strength, and could not stand on their feete, then did their legges swel, their sinnews shrinke as black as any cole. Others had all their skins spotted with spots of blood of a purple colour: then did it ascend up to their ankles, knees, thighes, shoulders, armes, and necke: their mouth became stinking, their gummes so rotten, that all the flesh did fall off, even to the roots of the teeth, which also almost fell out. With such infection did their sicknes spread itself in our three ships, that about the middle of February, of a hundred and ten persons that we were, there were not ten whole, so that one could not helpe the other, a most horrible and pitiful case . . . there were already eight dead, and more than fifty sicke and, as we thought, past all hope of recovery."*

The American Indian natives helped Cartier and his men, advising them to take daily the "juice and sap of the leaves of a certaine tree." Their recovery was miraculous, and in time they returned to France.

The disease from which Cartier and his men suffered, *scurvy*, is a condition caused by a lack of vitamin C (ascorbic acid). This vitamin is present in large amounts in most fruits and vegetables, but more than 200 years elapsed after Cartier's experience before limes, oranges, and other citrus fruits were routinely included among the food supplies on ships during long voyages. Vitamin C, itself, was not discovered until 1932.

*Reprinted from Keevil, J. J. 1961. *Medicine and the Navy: 1200–1900.* E. S. Livingstone/Edinburgh.

in the pancreas. Temperature may serve as a signal. Whatever the signaling mechanisms, it is certain that the brain receives these signals and integrates them with a great deal of additional information in deciding whether, when, and what to eat. Otherwise, diners would be less concerned than they often are about the taste, odor, and appearance of the food they eat. Who has not enjoyed second and even third helpings at Thanksgiving and other holiday dinners?

Material for Growth, Maintenance, and Repair

To think of food only as a source of energy would be wrong; energy alone cannot sustain life. Food must also supply the body with specific chemicals. A growing child must accumulate physical matter: new cells, more cells, and more noncellular matter must be laid down if the child's body is to enlarge. From birth until death, the body is in a continual cycle of building, breaking down, and building anew. Enzymes, RNA molecules, and other molecular structures of the body's cells function only for a limited time, after which they are replaced. Small accidents and even minor injuries call for the replacement of dead and dying cells or for the rebuilding of body parts. Normally, these routine processes go unnoticed; however, let someone stop eating for any reason, and they become all too evident. The starved body simply withers away.

Food Quality

To be coldly practical, one eats in order to supply the body with necessary chemicals. Most of these chemicals eventually serve both for energy and for the construction of needed cellular molecules. An adequate diet *must* contain certain (essential) amino acids, vitamins, some fats, certain minerals (sodium, potassium, calcium, phosphate, iron, and zinc, to name a few), and, of course, water. Otherwise, neither health nor (ultimately) life can be maintained. The need for special or essential substances reflects the limited armory of enzymes manufactured by human cells. Our repertoire is limited. We can make some, but not all, of the amino acids required for protein synthesis; those that we cannot make must be present in the food we eat.

Vitamins are chemicals that are required in extremely small quantities. Our cells cannot make them, but they are essential for routine metabolic reactions. Some vitamins are soluble in water (these are the ones that are often thrown away when water is drained from cooked vegetables); these water-soluble vitamins are either cofactors or parts of cofactors for enzymes. Many enzymes, that is, do not function unless a second, small molecule is attached to them; these small molecules are the cofactors. Oil-soluble vitamins may or may not be cofactors. They are needed in small amounts, however, for whatever purpose they serve, and our cells are unable to manufacture them.

Minerals, of course, cannot be manufactured by the body; they must be ingested with the food we eat. Calcium and phosphate for bones must be in the food we eat or our bones will be soft and may be deformed. Hemoglobin contains atoms of iron; the food we eat must contain an adequate amount of iron or the amount of hemoglobin in our red blood cells will be too small. Trace amounts of still other minerals are essential for good health.

A Dietary Problem: Pellagra

Knowledge about vitamins and other essential components of food has grown enormously during the past fifty years, but it is still incomplete. Much of this knowledge—like much of our understanding of our body machinery—was gained from tragic events. The case of *pellagra* is presented here to show how luck, observation, and experimentation combined to advance our understanding of human nutrition.

Unlike many diseases afflicting human beings, pellagra does not have a long recorded history. Only during the eighteenth century was this "new" disease described by a Spanish physician. Somewhat earlier, it had been recognized in Italy, where it was named "pellagra," meaning "rough skin." In this serious, often fatal, disease, the skin reddens, develops a rash, and becomes encrusted—especially after exposure to the sun. The mouth and throat are inflamed. Afflicted persons have a poor appetite, and what they eat is poorly digested. Eventually those afflicted grow weak and often show symptoms of mental depression, epilepsy, and even insanity—all symptoms reflecting an involvement of the nervous system.

During the eighteenth and nineteenth centuries, because the disease spread rapidly in Italy, Spain, and other countries both in Europe and the Americas, many thought that it was contagious, that perhaps it was caused by an undescribed bacterium. On the other hand, at times it showed a familial pattern—affecting some family members but not others; perhaps, then, it was a hereditary disorder not previously described. Most persons who suffered from pellagra were poor, and many were confined to institutions, such as prisons; hence, many suspected that it was caused by spoiled or even contaminated grain.

During the nineteenth century and early decades of the twentieth century, thousands of persons died of pellagra each year. It was an especially serious problem in the southern United States. Dr. Joseph Goldberger, a U.S. Public Health Service physician, set out in the early 1900s to identify the cause of this disease. He noticed among other things that pellagra was common among the children in one particular orphanage, but not among the orphanage staff. Thus, the dis-

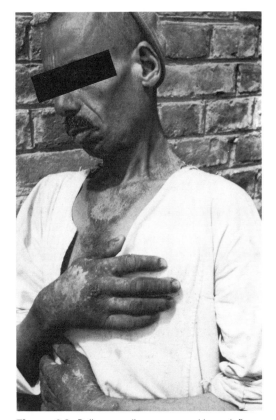

Figure 4.9. *Pellagra*, a disease caused by a deficiency of nicotinic acid (one of the B vitamins). The disease is common among persons whose diet emphasizes corn products, since corn contains little tryptophan, an amino acid that human beings can convert into nicotinic acid. The extensive lesions on the hands and chest of the patient shown here are reminders that the ultraviolet light in sunlight causes cellular damage (especially to DNA). Cells lacking nicotinic acid cannot muster the energy required for performing necessary repairs. Contrast this patient's plight with that of the *xeroderma pigmentosum* patient (Figure 3.8), whose cells have the necessary energy but lack needed enzymes. *(Photo courtesy of William Darby, M.D., Vanderbilt University)*

ease did not appear to be contagious. The main staple of the children's diet was corn (maize). Dr. Goldberger first guessed, and then proved through studies on prison volunteers, that pellagra was caused by a corn diet. Corn lacks some important nutrient. That nutrient, he showed, could be supplied by yeast extract: pellagra could be prevented or cured by adding yeast extract to a diet of corn.

By 1930, the symptoms of pellagra were known to result from a lack of nicotinic acid (niacin, one of the B vitamins). This vitamin is present in yeast, meats, eggs, and some other foods, but corn is exceptionally low in this vitamin and in tryptophan, an amino acid that human cells can transform into nicotinic acid. Corn, however, is a plant that produces high yields of grain for food; it is often the main food staple for poor persons. The connection between corn and pellagra explains the sudden appearance of the disease in Europe: corn (maize) became a European crop only after the discovery of the New World. Thanks to the present understanding of pellagra, the disease is rare in the United States. It is now "caused" by ignorance, food faddism, or conditions (such as alcoholism) that affect normal nutrition. Unfortunately, pellagra still rages elsewhere in the world—in Egypt, South Africa, and India, for example.

A fair, but not complete, understanding has been reached concerning the role of nicotinic acid in the development of the symptoms of pellagra. Nicotinic acid is the coenzyme of many enzymes (dehydrogenases) that are involved in obtaining energy from food; it is, in fact, the molecule that serves as the hydrogen acceptor. Without nicotinic acid, then, the energy-transforming mechanism of every cell in the body is more or less shut down. We saw in a previous chapter that exposure to sunlight causes damage to the DNA of the cells of the skin, and that to repair this damage requires energy. Hence the exacerbation of pellagra's skin lesions by sunlight. The inflammation of the mouth and throat and the poor digestion (often accompanied by intestinal bleeding) reflect the constant replacement of old cells in the intestinal tract by new cells: cell division requires energy. Lesions in the central nervous system reflect the actual death of cells; these cells require more energy than most other body cells, and, after a person is born, they never divide. In the absence of nicotinic acid and of the energy this cofactor makes available, the brain and central nevous system can only go downhill, since cells of the central nervous system, once dead, can never be replaced.

NOW, IT'S YOUR TURN (4-3)

Many traits "run in families." Pellagra is (or, was) such a trait. Hookworm was another. The "love of the sea" and "wanderlust" are also two familial traits. Later we shall discuss genetic disorders that are passed from one generation to another by way of genes and chromosomes. In the meantime, think of traits in your family or in friends' families, such as owning a business, holding a professional degree, owning a farm, or any other characteristic that might be looked upon as possibly "running" in the family. What is the basis for the occasional similarities between parents and offspring?

Intake vs. Needs

If one's diet provides enough energy and the variety of molecules needed for good health, one's body may remain stable and vigorous for long periods of time. However, this fine balance can be disturbed. Furthermore, such disturbances constitute major health problems. Ironically, these disturbances (the tragedies referred to in the previous section) often help us understand normal functions. A good deal of nutritional research is governed by the saying, "You never miss the water until the well runs dry." Only when things have gone wrong do we understand why earlier they were going right. With respect to the food we eat, the obvious problems are (1) too little, (2) too much, (3) an improper balance of needed nutrients, and (4) the presence of harmful contaminants.

Too Little Food

Millions of hungry and starving persons live in today's world— mostly in developing countries, but also among poor people everywhere. The amount of food available to these individuals does not provide the amount each needs. In addition to poverty, illness and disease can cause hunger. Some persons have enough food but still go hungry because intestinal parasites steal from them and interfere with normal food absorption. Some fever-causing diseases increase the body's need for energy; if these needs are not met, the body wastes away. In advanced cancer, the cancerous cells rob normal cells of needed nutrients. (The terminal aspects of malignancy, however, often reflect the inability of cancer cells to perform duties normally performed by body cells—for example, the termination of many cancers is a fatal hemorrhage caused by the cancer's destruction of the wall of a major blood vessel.)

Too Much Food

Overeating leaves its mark: the excess intake is stored in the cells of fatty tissue. Worldwide, overeating is a relatively rare disorder. Only in affluent societies or among affluent persons is overweight a common problem. For many fat persons, natural control systems are unbalanced by circumstance. A great deal of food is available at all times, in tempting forms, and with colors, tastes, odors, and textures so pleasing that it is eaten—hunger or no. In many human societies, eating is a social activity rather than a mere biological act. Businessmen arrange luncheon conferences; Washington politicians have working breakfasts; hostesses strive to arrange rich, tempting dinners in their homes. Alcoholic cocktails enhance appetites; furthermore, alcohol is metabolized before other nutrients, which, having proven to be unneeded, are then stored in fatty tissue. Even when their appetites fail, custom and courtesy make people eat. Also, lack of physical activity, if not accompanied by decreased food intake, makes persons fat. Finally, diseases that interfere with control mechanisms of the brain or hormonal messages can result in obesity.

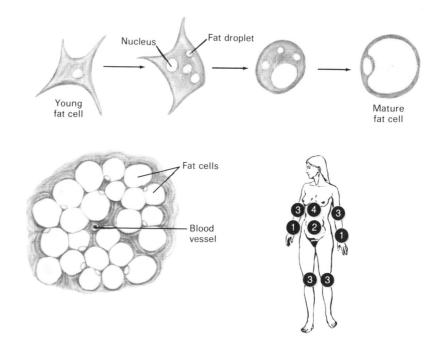

Figure 4.10. The storage of fat in the human body. (*Top*) A young fat cell looks much like any other cell. However, it accumulates fat droplets in its cytoplasm that grow and coalesce until the cytoplasm forms a thin layer around one large droplet of fat. (*Lower left*) Fat cells are organized into fatty tissues. Blood vessels running through these tissues can deposit more fat or withdraw fat for needed energy. (*Lower right*) Fatty tissues are located on the body at places where they can serve secondary functions: (1) under the skin (for insulation), (2) in body cavities (as packing material), (3) at the joints (as packing), and (4) in the liver. Too much fat—obesity—is regarded as a major health hazard, and it is difficult to overcome. One theory holds that at the start of a diet, fat cells inform the body's control centers that they are being consumed and, by so doing, intensify the dieter's urge to overeat once more. *(Illustration provided by The Biological Sciences Curriculum Study)*

Of course, within limits, stores of fat are a useful and normal part of the body structure. Unfortunately, though, large stores of fat place severe physical strains on the body. Fat persons, just because they are heavy, need more energy to move around than is normally required. The extra strain affects every part of the body, not only the heart, but also the muscles, bone, and joints—especially the conformation of the legs. In affluent societies, diseases of the heart, arteries, and supporting tissues are common problems, frequently caused by excess weight.

Needed Materials in Improper Balance

Serious problems may arise if one's food does not contain enough of the many essential chemicals and substances, such as vitamins, essential amino acids, and minerals. The lack of even one substance—the dietary bottleneck—is equivalent to undereating, no matter how much other food is actually consumed. Disease can result from a single vitamin deficiency. Pellagra, as we have learned, is caused by a lack of tryptophan or its metabolic derivative, nicotinic acid; scurvy is caused by a lack of vitamin C. Sometimes people eat much more than they should in an effort to get the proper amount of scarce but essential substances. Such improper diets can expose these persons to attacks by viruses, bacteria, and various parasites.

Harmful Contaminants

Food is extremely complex chemically. Many food products contain substances that are harmful—even poisonous—in large quantities. These substances are often present in harmless, trace amounts, but become harmful when the food within which they are found is eaten to excess. There is, for example, an organic compound in cabbage that, for those persons living almost entirely on a diet of cabbage, interferes with the proper functioning of the thyroid gland. Milk is a fine food for normal infants and for most (Caucasian) adults; however, milk contains a sugar (lactose) that some babies and many adults cannot digest properly. This sugar can be broken down by bacteria in the digestive tract; consequently, milk can cause diarrhea and other digestive problems in many adults.

The matter of the chemical composition of food is sufficiently important that, when plant breeders create a new variety of a food crop, nutritional experts are asked to verify that the chemical composition of the edible portion has not changed appreciably. Some fruits, leaves, and seeds contain deadly poison, as do certain fungi. The leaves of rhubarb are poisonous and cannot be eaten. Apple seeds are poisonous, as are peach pits. Alfalfa is a popular forage crop for dairy cattle; concentrated into food pellets, however, it may cause fatal cyanide poisoning. Thousands of potentially harmful or poisonous compounds are known to exist in animal and plant tissue; the list has recently been extended by the use of hormones, antibiotics, and pesticides in agricultural practices.

Food that may otherwise be healthful can be rendered dangerous through spoilage. The bacterium *Salmonella* can cause diseases that range from "food poisoning" to the dangerous, often fatal, typhoid fever. Nearly 20,000 persons became ill and dozens died in the Midwest recently after drinking milk contaminated with *Salmonella*. *Clostridium botulinum* bacteria (botulinus bacteria), living under anaerobic conditions in contaminated sealed cans of meat (tunafish, for example) or proteinaceous vegetables (beans or mushrooms), produce a short polypeptide, *botulin*, that is one of the most toxic substances known. Fortunately, botulin is destroyed by the heat of cooking.

Long before historic times, people learned to preserve food. Bacterial growth is slowed by high concentrations of salt or sugar, by drying, by sterilization, or by cold. Modern chemistry has added to the means by which spoilage is retarded. Exposure of food to high levels of radiation is one technique that has not yet been adopted; what new compounds might be produced by the irradiation of others is not well known. The chemicals BHA and BHT are widely used food additives; their presence in food must be acknowledged on the label. Most persons believe that trace amounts of these chemicals are harmless; other persons are not that confident, and they prefer "natural" foods that contain no artificial additives.

NOW, IT'S YOUR TURN (4-4)

Enter into your journal your complete dietary intake—meals and snacks—for one week. Calculate your caloric intake using a calorie-count table, which can be found in almost any "book of knowledge" or "information" almanac.

The Ultimate Source of Oxygen and Food

Human beings, like all animals, are dependent upon food and oxygen for their lives. They cannot manufacture most of the substances that life requires; by definition, they cannot manufacture *essential* foodstuffs, such as essential amino acids. What, then, is the ultimate (original) source of the sugars and starches, the amino acids, the fats, vitamins, and other food materials? And, what is the source of oxygen? The answer to both questions is the same: green plants. Green plants release oxygen, synthesize basic food materials, and transform solar energy into stored chemical energy. All other animals, and even nongreen plants (fungi), such as mushrooms and toadstools, owe their continued existence to the alchemy performed during photosynthesis by green plants: $H_2O + CO_2 \rightarrow CH_2O + O_2$.

Plants as Food

Green plants, *green* because of the chlorophyll that their cells contain, trap the energy of sunlight (solar energy) and use it to make the energy-rich chemical molecules of which their cells and structures are

Figure 4.11. Air-drying of fish, in Italy (*top*) and Alaska (*bottom*). Long before human beings recorded their history, they learned to preserve fish and meat by drying the flesh in air, exposed to the sun if possible. Air-dried fish and meat, unlike that preserved in salt (or in brine), retain water-soluble vitamins; consequently, air-drying helps stave off diseases caused by vitamin deficiencies. (*Bottom photo courtesy of the Alaska Division of Tourism*)

composed: cellulose and lignins (woody material), sugars, and amino acids. The entire process is called *photosynthesis*—*photo* for the trapping of light, *synthesis* for the construction of new chemical compounds through the use of light energy. The raw materials for photosynthesis are carbon dioxide from the air, and water and certain minerals from the soil. In carrying out photosynthesis, plants give off oxygen (and some heat) as a byproduct.

If there were no green plants, life as we know it could not exist on earth. Scientists know in considerable detail how photosynthesis is carried out, but they are not yet able to duplicate the plant's efficiency. They cannot yet make complex foodstuffs starting only with water and carbon dioxide, two simple and extremely stable compounds. A high government official, in commenting on his department's support of scientific research, once said, "I am not interested in why plants are green." He *should* have been interested; his own existence depended on the fact that they are green. Persons who appreciate the role that plants play in their lives sometimes display bumper stickers reading: "Have you thanked a green plant today?" They should also thank the green plants of yesterday, because the gasoline they buy for their car—all oil and coal, in fact—represents solar energy trapped and stored by green plants of the distant past. If all green plants on earth were to die, so would all animals, including human beings. Death would occur relatively soon, because large quantities of extra food are not lying about unused in today's world.

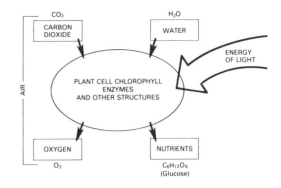

Figure 4.13. A diagram of photosynthesis. Unlike the process shown in Figure 4.4, in which energy was released by burning glucose, photosynthesis has a net *input* of energy, in the form of radiant energy. Within the chloroplast, carbon dioxide and water (with the help of the incoming energy) are combined to yield glucose and oxygen. The amount of energy consumed, if this were a perfect world, would equal the amount stored in glucose; actually, somewhat more is required. Other organic molecules—fats, amino acids, higher sugars and starches—are assembled by drawing upon the energy that has now been stored in glucose. *(Illustration provided by The Biological Sciences Curriculum Study)*

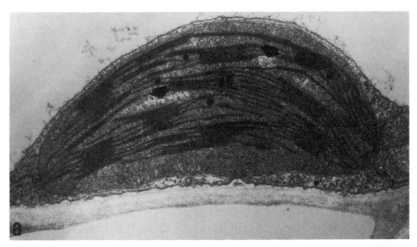

Figure 4.12. A chloroplast, the organelle within which photosynthesis occurs. This chloroplast is one of many that would be found in the chlorophyll-containing cells of green plants (see Figure 2.28). Chloroplasts, like mitochondria, possess their own DNA (*arrows*) that is separate from (but cooperates with) the DNA in the cell's nucleus. *(Photo courtesy of D. A. Stetler, Virginia Tech)*

Energy for Life 139

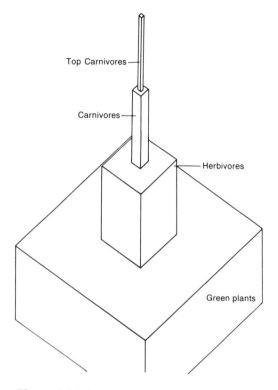

Figure 4.14. A generalized diagram in which the volumes of the solids represent the energy residing in organisms at various levels of the food chain. The energy contained in green plants has been captured from solar radiation. A rapid attrition of this initial store occurs as one moves up the food chain toward the top carnivores (carnivores that eat other carnivores). The energy initially captured does not simply vanish; it is radiated into outer space as heat from all sorts of organisms.

(labels on figure: Top Carnivores, Carnivores, Herbivores, Green plants)

Animals as Food

Although solar energy, trapped and converted by plants, is the ultimate source of the chemical energy in food, most persons do not eat plant tissues only. Some animals, *herbivores*, eat only plants, while others, *carnivores*, eat only other animals. Human beings are *omnivores:* we eat both plants and animals. The animals people eat are generally herbivores, although carnivores may also be eaten. Animal foods generally contain higher percentages of protein and fat than do plant tissues. Many amino acids that are essential in the human diet are more common in animal than in plant proteins. As a result, it is easier for human beings to get both energy and cellular materials from meat and milk than from plants. Furthermore, many persons prefer the taste and texture of meat.

When solar energy that is trapped by plants is passed on to human beings by way of an intermediate animal, an herbivore, much of the original energy is wasted. An herbivore, for example, requires energy for living and for moving about; furthermore, like all machines, living beings are wasteful—a great deal of potentially useful food energy passes through each individual unused and is lost in feces. A useful approximation can be stated as follows: a consumer stores only ten percent of the energy that is available to him, and makes that ten percent available to those who consume him in turn.

The loss of energy that accompanies energy's passage from one form of life to another explains why there are so many gnus, zebras, and gazelles on African plains and so few lions, cheetahs, and hyenas preying on these herbivores. It also explains why human societies differ so widely in their use of plants and animals as food sources. Where populations are large and food is scarce, plants must provide the bulk of all food eaten; hence the rice-based diets of India and China. Inhabitants of affluent countries, where the land is not (yet) crowded and where more food per person is available, tend to eat larger quantities of animal products, such as meat, eggs, and milk.

NOW, IT'S YOUR TURN (4-5)

1. Merely saying that only ten percent of the available energy is passed on from one trophic level to the next (from plants to herbivores, or from herbivores to carnivores) does not convey the rapidity with which this energy is lost. To understand it better, consider the following: Suppose a given area can support ten kilograms of plants. These plants are eaten by insects. These insects are eaten, in turn, by small birds. The small birds are preyed upon by sparrow hawks. How many "grams of sparrow hawks" can be supported in this area? A pair

of sparrow hawks and their offspring may weigh a total of one kilogram. Calculate how many grams of plants would be needed in an area to support this one family of sparrow hawks. Express this amount in pounds or tons.

2. The following information was included in a figure illustrating the role that eelgrass of the coastal wetlands plays in the maintenance of marine and land life. The original figure is not reproduced here. Try your hand at making a figure that will accurately illustrate the following:

Organism	Total amount (tons)
(1) Eelgrass	24,000,000
(2) Ducks and geese (grazers on eelgrass)	5,000,000
(3) Small marine herbivores	800,000
(4) Fish herbivores	5,000
(5) Predatory crustaceans (see 3)	50,000
(6) Predatory fish (see 3)	25,000

Based on the ten-percent rule, does it seem that ducks and geese live entirely on eelgrass? How about the marine predators in categories 5 and 6?

Limits to Food Supply: Plants

Plants seem to grow everywhere—in vast fields and forests; in rivers, lakes, and oceans as aquatic plants; and even as weeds in city streets. Despite appearances, plants are sensitive living creatures, with their own special needs. These needs put strict limits on *where* plants can grow, and on *how much energy* they can store.

Plants grow only where adequate supplies of the energy (light) and raw materials (carbon dioxide, water, and minerals) that they need are to be found. Other conditions must also be met: Temperature can be neither too high nor too low, else the plants are damaged. Plenty of oxygen must be available for the cellular metabolism carried out by the plant's tissues. The air must be relatively free of pollutants, such as smog, sulfuric and nitric acids, and ozone. The soil and water must not contain poisons, or even high levels of common salts. Vast areas of the earth either lack or have limited quantities of one or the other of the raw materials needed by plants. Other areas are too hot, too cold, too wet, too dry, or otherwise unsuited for plant growth.

Water and minerals commonly limit plant growth. Deserts are too dry for most plants. Aquatic plants growing in the ocean, although immersed in water, tend to grow near the shore, where minerals from land are deposited by streams and other runoff waters. Air pollution and soil poisons, to continue the litany, have destroyed and are de-

Nuclear Winter

Tens of thousands of inhabitants of Hiroshima and Nagasaki, the only two cities so far exposed to atomic explosions, died of burns, concussion, or radiation sickness. The horror of such devastation was largely responsible for terminating World War II and, in the opinion of many, for preventing the start of a World War III.

Weapons analyses and tests since the 1940s and 1950s have concentrated on the blast, heat, and radiation effects of ever-larger bombs—atomic bombs in the early years, followed quickly, however, by even more powerful nuclear devices.

More recently, a previously overlooked effect of nuclear explosions has been examined in detail by climatologists and others. Prompted by data relayed by instruments situated on Mars—data revealing a substantial drop in surface temperature that accompanied the occurrence of a gigantic dust storm—both American and Soviet scientists have tried to assess the likely effect of the dust and smoke of a nuclear war on the world's climate, and on its plant and animal life. The climatological outcome of a nuclear war has been dubbed the "Nuclear Winter."

The analyses, because of the many variables and their complex interactions, are not easy to carry out; some of the country's largest computers have been used in these studies. One can say, however, that even the most optimistic analysis predicts devastating results. Agriculture throughout the northern hemisphere would be at a virtual standstill. A cooling (not necessarily a freezing) trend at the wrong moment during the growing season can prevent the formation and maturation of wheat grains; this grain equals bread for much of the world's population. The Great Plains of the United States provide food for Asians, Africans, and Europeans as well as Americans.

Photosynthesis would be drastically reduced by lowered temperatures, lowered light intensities, and increased levels of ultraviolet radiation. This would be true both for agricultural crops and for the plants that sustain wildlife. Plants and animals are adapted to function under normal day–night cycles of light and darkness, and under annual temperature cycles. Darkness resulting from clouds of dust, smoke, and smog caused by nuclear explosions would interfere with food gathering, thus dooming many species. Frost or near-freezing temperatures occurring unexpectedly in midsummer would doom many others.

The consequences of nuclear war, therefore, extend far beyond the instantaneous deaths of millions of persons through torn, shattered, and burned bodies and through lingering radiation sickness. Beyond lies a period of many years during which human survivors as well as many "wild" plant and animal species would struggle for continued existence. The weedy species of the world, including cockroaches and rats, may be the eventual winners.

Figure 4.15. Many regions of the earth's surface are unsuited for plant life and, as a result, unsuited for animal life as well. Two barren regions are illustrated here: at the top is a view of a mountain range in Antarctica; at the bottom, a view of Death Valley, California. Despite its barren regions like Death Valley, some persons have calculated that the United States has 3,000,000 square miles for raising food crops (3000 miles from east to west times 1000 miles from north to south). Such calculations are simplistic and false. *(Bottom photo courtesy of J. A. Moore, University of California at Riverside)*

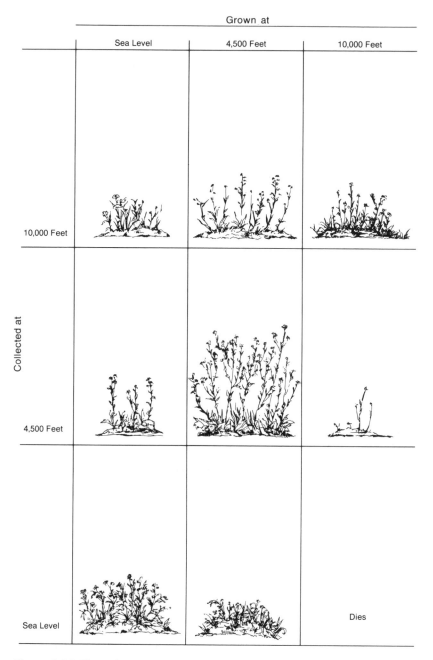

Figure 4.16. The regions on earth that support plants differ considerably in their overall climatic conditions and, as a result, support different species of plants. They may also support different varieties of the same plant species. In time, local races (or varieties) of plants become especially suited for life under the prevailing local conditions. Shown in the figure are the growth patterns of three individuals of the same plant species that were collected at three altitudes (600 feet, 4500 feet, and 10,000 feet) in the Sierra Nevada Mountains. The plants were separated into three (genetically identical) clones each, and then were transplanted at the three altitudes. The differences observable among plants in the same horizontal row are caused by the environment acting on genetically identical plants; the differences among plants in the same vertical column are caused by the different heredities now growing in nearly identical environments.

stroying many species of plants. Other than trees, few plants are found in the dark recesses of dense evergreen forests, because light is needed for photosynthesis. Polar regions are devoid of most common plants because these regions are too cold.

Because of the above limits on plant life and growth, less than one-tenth of the ice-free areas of our globe are suitable for growing food crops. Because there are so many persons living today and because so many of these persons are hungry, much work is being done in an effort to increase both the area that is suitable for food crops and the productivity of those areas that are suitable. The task is formidable. For example, the Soviet Union, despite huge successes in nuclear physics, rocketry, and space exploration, and even with the expenditure of strenuous effort over the past half-century, has failed to increase its grain production. In the United States, as well as in other areas of the world, land suitable for growing food crops is lost through both erosion and industrial and commercial exploitation.

On the other hand, tidelands have been reclaimed and turned into farmland. Fertilizers, added to mineral-poor soils, supply the chemicals whose lack previously limited plant growth. The removal of salt

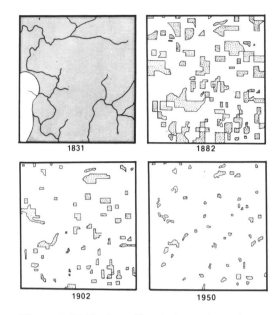

Figure 4.17. The loss of forested areas in one township within the state of Wisconsin over a century or more. Some areas of the globe cannot support certain types of plants because the environment is harsh or lacks some attribute that is essential for the plant's growth and reproduction. Some regions are denuded of plants (especially forests and prairies) by human activity. Today, perhaps more widespread than the effects of chopping and plowing is the enormous devastation of forests caused by acid rain—rain containing sulfuric and nitric acids as a result of air pollution.

Figure 4.18. Suburban housing poses one of the greatest threats to prime farmland both in the United States and abroad. In the eight years from 1967 to 1975, about 6.5 million acres of prime farmland was lost to urbanization and another 1.5 million acres to water projects. Near metropolitan areas this loss means a loss of fresh fruits and vegetables of local origin; those that are not grown locally have often been selected for useful characteristics (size, shape, resistance to bruising, resistance to spoilage) rather than flavor or palatability. *(Photo courtesy of the U.S. Department of the Interior, Bureau of Reclamation, Stan Rasmussen, photographer)*

Table 4.2. U.S. and world production of agricultural products (in millions of metric tons: 2200 lbs/ton)

Product	Production	
	U.S.	World
Wheat[a]	76.0	453.4
Oats	8.4	47.0
Corn	208.3[b]	438.8
Barley	10.4	160.3
Rice[a]	8.4	410.9
Soybeans	55.3	87.3
Edible vegetable oils	11.4	41.5

NOTE: The numbers shown in this table are tremendous. It has been estimated that ten percent of all photosynthesis occurring on earth occurs in cultivated plants.

[a] Nearly one-half of the caloric intake of all persons on earth, combined, is provided by wheat and rice.

[b] Corn produced in the U.S. is used primarily in the production of animal products: beef, pork, poultry, dairy products, and eggs.

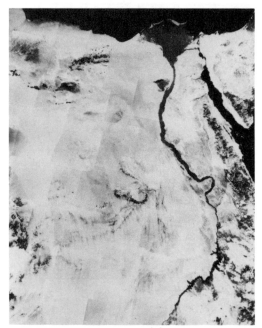

Figure 4.19. A satellite view of Egypt that reveals the narrow band of farmland bordering the Nile river and forming the delta at the river's mouth. Despite their pressing need to feed approximately fifty million persons, the Egyptians find it *economically* beneficial to convert the delta into housing developments and industrial complexes. *(Photo courtesy of the General Electric Company, Space Systems Division)*

from seawater can produce the fresh water needed by plants. Pollution control should save many food plants from damage and from diseases to which weakened plants are susceptible. However, these are not easy solutions to the problems posed by limited plant growth. Each solution has a cost and, furthermore, creates new problems of its own. For example, the destruction of tidal marshes sacrifices animal protein (clams, oysters, shrimp, and marine fish) for plant protein. The salt from desalinization plants is generally dumped into the ocean once more, killing much valuable animal life in the immediate area. The manufacture of fertilizers demands either natural gas or electricity, commodities that are already in short supply.

Limits on Food Supply: Animals

Plants do manage to live and grow, despite serving as food for a host of animals and plant pathogens. The temperate-zone landscape, after all, is green for much of the year. However, food crops, especially those crops that are grown in pure stands over scores or even hundreds and thousands of acres (*monocultures*), encourage the growth of huge populations of rapidly breeding insect herbivores. These insects constitute major agricultural pests and, as such, impose a limit on our food supply.

Limits on the supply of animal foodstuffs (meat, eggs, and milk, for example) are also set both by factors that limit plants and by other factors that are important for the good health of animals. When animals—cattle, sheep, hogs, or poultry—are raised for slaughter, they need a moderate temperature (indoor or outdoor), balanced diets, adequate supplies of pure water, and sanitary living quarters to dis-

Figure 4.20. The extent of soil erosion in the United States. Listed for each state is the average number of *tons* of topsoil lost per acre for that state in 1977 through water and wind erosion. Normally, topsoil is replaced at a rate of 1.5 tons per acre; any loss that exceeds that figure is considered serious because agricultural production cannot be sustained. More than one-half of all cropland in the United States (500 million acres) has either been permanently damaged or is in serious need of conservation treatment.

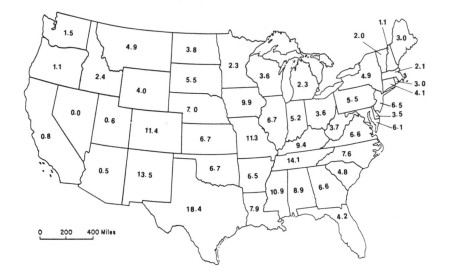

courage parasites and disease. Because they need food and oxygen, we are returned once more to discussing plants. Rearing animals for food is an expensive undertaking in modern agriculture. On a single-family plot, chickens exchange eggs for the table scraps that would otherwise be wasted, and the family cow produces milk (and eventually veal or beef) in exchange for grass that is unsuited for human consumption; agriculture on this scale is a profitable energy-generating occupation. In contrast, modern agricultural industry (large-scale production) is an energy-consuming, not an energy-generating, occupation. It is estimated that, today, five calories of oil are required to put one calorie of milk on the consumer's table. The calories in meat, on the average, are about equal to the petroleum calories that were used in raising the animals and getting the meat to the consumer. Thus the rich diet of affluent nations using the procedures of modern agriculture is subsidized by the world's supply of oil and coal. With an increasing world population and with an increasing desire on the part of all peoples to become affluent, the supply of food for human beings must be examined not on a local, individual-by-individual basis, but on a global scale, starting with the influx of energy from the sun and the flow of this energy through living things.

NOW, IT'S YOUR TURN (4-6)

A city of about a million persons in a developing country is situated within convenient walking and bicycling distance of a large, shallow freshwater lake. Children and other family members take 50 million pounds of fish from this lake each year, enough to provide one pound of meat for every inhabitant every week. A fertilizer plant that is built to aid the nation's agricultural output pollutes the lake, so fish are no longer available to the city's dwellers. Assuming that the average yield of wheat in the country is 2000 pounds per acre, calculate the number of acres of wheat that must be grown to replace the *animal* protein that has been lost (use the ten-percent "rule" in going from plant to herbivore).

Summary

Food and oxygen supply the substances and the energy human beings need for keeping fit, for self-maintenance, for growth, and for physical activity. The energy needed by individuals can be measured, but it differs from moment to moment, from one person to the next, with sex, age, activity, and state of health.

Oxygen is needed to release chemical energy, that is, for combus-

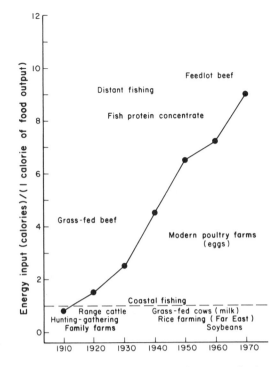

Figure 4.21. The relation between the energy that is invested in food production and the energy that is harvested. Below the dashed line are forms of agriculture that are energy-producing; above the dashed line are practices that are energy-consuming. The latter must be subsidized by energy provided by oil, gas, or electricity. Superimposed on the general chart is a graph showing the increased need for energy in agricultural practices in the United States over a period of sixty years.

Table 4.3. Where does the sunshine go? The following information has been adapted, at third hand, from an article of that title. The data pertain to Ottawa, Canada (or the northern United States).

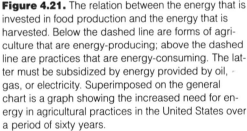

Measure	Calories/ cm^2	Car mileage from energy per acre
1. Top of atmosphere	228,000	5,500,000
2. Growing season	163,000	4,000,000
3. Incident on crop	109,000	2,600,000
4. Available for photosynthesis	5,500	134,000
5. Stored in plants (avg)	100	2,350
6. Made available to man by farm animals (avg)	10	250

tion. Its supply to body cells is controlled by the rates of respiration (breaths per minute) and circulation (pulse). Ideally, the food that is consumed should exactly counterbalance the body's needs. Natural hunger and eating patterns normally maintain this balance with remarkable precision. Nevertheless, neither the quantity nor the quality of food eaten by many persons (including the world's poor, but not limited to them) is adequate. Natural regulatory mechanisms may be upset by social behaviors and by stress, grief, and many other emotional factors.

On a global scale, there is a balance between the use and replacement of both oxygen and food in the world. The basic equation can be written with a double-headed arrow:

$$CO_2 + H_2O \leftrightarrow CH_2O + O_2$$

The tissue cells of both plants and animals require oxygen, but the supply of oxygen in the air is continuously replaced by oxygen produced during photosynthesis. The overall balance of available food depends upon the primary production of foodstuffs by plants, and on its use by fungi, animals (including human beings), and certain microorganisms. Later chapters of this book deal with matters concerning this balance: the limits of resources *versus* the increasing demands made upon these resources by human beings.

INFORMATIONAL ESSAY
Equilibria

In the preceding text, great stress is placed on the accuracy with which nutritional needs are met by one's food consumption. If, for example, one were to eat a single biscuit more each day, one's weight (all else remaining unchanged) might increase ten pounds in a single year. Conversely, if one were to eat one biscuit less, one's weight might decrease some ten pounds.

The awe in which we regard this precision is lessened, however, when we recall that in the one instance the person would not gain ten pounds per year for years on end, nor in the other, lose ten pounds per year over many, many years. In each case, the person's weight would stabilize at a new—higher or lower—level, and thereafter remain as stable as before. There are, in brief, innumerable stable weights, each of which is in equilibrium with the individual's food intake. These are examples of stable equilibria.

Over all, equilibria can be of three sorts: stable, unstable, and (for the lack of a better term) indifferent. A *stable* equilibrium is one in which, following a slight disturbance, the original state is restored, as we just saw. A marble resting at the bottom of a soup bowl is another example. If the bottom of the bowl is smoothly concave, the marble will rest at the lowest point. Flick the marble with your finger, and it will roll about until at last it returns to the precise spot at which it rested initially. A second example is the playground swing that, when not in use, hangs vertically. After normal use, it once again returns to its equilibrium position.

The stability of a stable equilibrium applies only for the area immediately surrounding the stable point. Flick the marble too hard, for example, and it will fly out of the soup bowl; it will then come to rest elsewhere. Play too hard with the swing, and its ropes may become entangled with its supporting rods. As another example, a brick has three pairs of stable positions. When standing on end, a brick is stable. When nudged gently, the brick tips slightly but then returns to its upright position. Nudged too hard, the brick tips over onto one of its other sides. The most stable position for a brick is, of course, resting on its largest side; in that position it will resist strong disturbances.

An *unstable* equilibrium is one in which a body is at rest, but the slightest nudge will send it toppling into a new position. A yardstick balanced on end is an example: it may be possible to balance it for a fraction of a second, but then it topples. If the yardstick initially is balanced on the edge of a table, when it topples it may fall either to the floor or merely to the tabletop; the place at which the yardstick comes to rest is independent of its initial instability. This is true even though a yardstick resting on a tabletop possesses potential energy no longer possessed by one lying closer to the center of the earth—on the floor. The brick mentioned earlier has twenty positions in which it might be at an unstable equilibrium: balancing on any one of eight corners where all three sides meet or on any of the twelve edges where two sides meet.

An *indifferent* equilibrium can be represented by a billiard ball on a level pool table, or by an automobile with its brakes released resting in a level parking lot. Nudge either the ball or the car so that it moves a short distance, and it will remain at rest in the new spot. There is no tendency for either the ball or the car to return to its former location, nor is there any tendency to move from the new one. Both ball and car are indifferent to their locations, provided that the tabletop and parking lot are level.

Let's return to the woman whose weight is stable but who then decides to eat one extra biscuit each day. Under this nudge, her weight increases, but if at any time she decides to eliminate the extra food, her weight will return to its former amount. The formerly constant weight does, in fact, represent a stable equilibrium. The woman's weight does not represent an indifferent equilibrium, because if an extra biscuit becomes a permanent feature of her diet, she merely stabilizes at a new and higher weight. The work entailed in carrying the extra weight around all day—walking up stairs or getting out of chairs, the bed, and the bathtub—eventually consumes the calories contained in the extra biscuit. The woman has then reached a new, *stable* equilibrium. In contrast, a constant force applied to an unbraked car in a flat parking lot will tend to make the car roll faster and faster.

Biogeochemical Cycles

Our biosphere has the remarkable property of cycling minerals and other materials. This cycling involves not only metabolism by living organisms, but also a series of strictly abiotic chemical and physical reactions between solar (radiant) energy, the atmosphere, the geologic crust of the earth (lithosphere), and water. The movements of matter between the large reservoirs of living tissue, atmosphere, lithosphere, and hydrosphere are termed *biogeochemical cycles*.

THE HYDROLOGICAL CYCLE

The hydrological cycle involves the movement of water between the atmosphere, living organisms, and large reservoirs on and in the earth's crust. The hydrological cycle is the heart of atmospheric heat transfer and provides a vehicle for the movement of other material in biogeochemical cycling. Living organisms are mostly water, and water constitutes much of the rest of our environment. Its scarcity is a major limiting factor in the distribution of organisms, while in superabundance it is responsible for floods and erosion. Figure 4.22 summarizes the basic hydrological cycle, but a few points may be added:

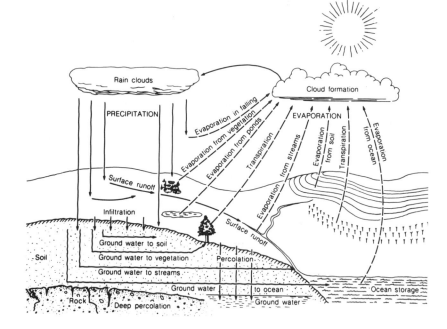

Figure 4.22. The water cycle, a cycle that operates on solar energy. Water enters the atmosphere through evaporation from the oceans or from land-based water, including water that passes from the soil through living plants (transpiration). It returns from the atmosphere as rain or snow, falling either on land or at sea. Water falling on land may seep into deep pockets of water or into the ground-water tables, or it may collect in streams and rivers, which empty into the oceans (surface runoff). A freshwater lake is an unstable, temporary entity caused by an interruption of the flow of a stream.

More than eighty percent of the energy that is absorbed from the sun's rays is used to evaporate water.

Nearly ninety-five percent of the earth's water is chemically bound in rocks and does not cycle.

Of the water that does cycle, most occurs in the ocean, polar icecaps, glaciers, and ground water. Ground water is an important source of fresh water, but one that is being rapidly depleted as the result of industrial and municipal needs and is being polluted by improper waste-disposal practices.

Water cycles rapidly between the earth's surface and the atmosphere. If all the water vapor in the atmosphere were to be suddenly and completely condensed, the resultant liquid water would cover the earth to a depth of 2.5 cm. The *average* rainfall for the earth is about 81 cm; therefore, the water vapor in the earth's atmosphere falls as precipitation and is regenerated by evaporation from surface water more than thirty-two times a year!

THE CARBON CYCLE

The carbon cycle is involved in the flow of energy through food chains because living things store energy in, and obtain it from, carbon compounds. Carbon in the form of ATP is the universal currency for exchanging or expending energy. The carbon cycle actually consists of two cycles: one a long-term cycle, the other a short-term cycle. The short-term cycle is commonly called the *food chain* or *food web*. The accompanying diagram illustrates the short-term cycle. Atmospheric carbon dioxide is transformed into glucose by photosynthesis; the glucose and more complex carbon compounds are then passed to the primary consumers (herbivores). The carbon is subsequently passed on to one or more secondary consumers (predators). Throughout, carbon is returned to the atmosphere as carbon dioxide through the respiration of both plants and animals, and through the fungal or bacterial decomposition of dead organic matter.

In the long-term cycle, carbon in the form of carbonate rocks (limestone), fossil fuels, or forests may persist unchanged for centuries, millennia, or even longer before returning to the atmosphere as carbon dioxide. The increased use of fossil fuels and the increased burning of forests, particularly tropical forests, has raised the carbon dioxide content of the atmosphere in the past few decades. An increase in atmospheric carbon dioxide may cause what has been called the "greenhouse effect." Carbon dioxide in the atmosphere operates much like the glass panes of a conventional greenhouse. The glass allows the passage of short-wavelength radiation from the sun, but reflects the infrared radiation (heat) that is radiated upward from the earth. Perhaps because of the greenhouse effect, the earth's surface temperature has been rising. If the rise in temperature continues, the polar icecaps will melt, causing ocean surfaces to rise. Geologists have

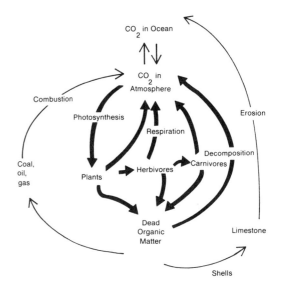

Figure 4.23. The carbon cycle, a cycle that, because of the role of photosynthetic plants, also operates largely on solar energy. The inner group of heavy arrows illustrates the short-term carbon cycle: carbon dioxide is withdrawn from air by plants and, through the magic of photosynthesis, is combined with water to make glucose and more complex organic substances. Plants enter the food chain (herbivores and carnivores), each member of which puts carbon dioxide back into the atmosphere. When they die, both plants and animals decompose, thus returning even more carbon dioxide. Hard shells of calcium carbonate (from clams, oysters, and small diatoms) accumulate as chalk and limestone. Material not fully decomposed accumulates as fossil fuel; the erosion of limestone and the burning of fossil fuels completes these two components of the long-term carbon cycle.

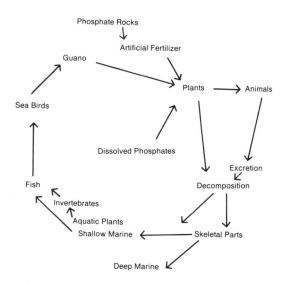

Figure 4.24. The phosphorus cycle, a two-part cycle (long-term and short-term) similar to the carbon cycle. The short-term cycle operates within the food chain, with animal feces serving to provide phosphorus for plants. Dissolved phosphates enter the short-term cycle through plant metabolism. The long-term cycle involves loss of phosphorus to the deep ocean, the formation of phosphate ores, geologic upswellings, and erosion of those ores to form dissolved phosphates. Mining and the manufacture of artificial fertilizer accelerate the cycle. Dense ocean-front population centers, such as most of Long Island, accelerate the loss of phosphorus when private septic tanks are replaced by community sewage plants that dispose of their wastes at sea.

predicted a rise in the oceans of two to twenty centimeters in the next decade. Additional points can be made regarding the carbon cycle:

Carbon dioxide is one of the rarer gases in our atmosphere (0.03–0.04%).

Carbon occurs not only as organically fixed carbon (carbohydrates, fats, proteins, and nucleic acids, for example) but also as carbonate (CO_3) in marine and freshwater shells, in the secretions of some algae, and in coral reefs. Carbonate may enter the long-term cycle as limestone (calcium carbonate—$CaCO_3$). Incompletely decomposed algae of past ages accumulated on the bottoms of ancient bogs and contributed to the formation of today's fossil fuels; the oil-shale beds of Wyoming are an example. Coal represents the carbon remains of plants that accumulated in swampy areas millions of years ago.

THE PHOSPHORUS CYCLE

Phosphorus is an important mineral for living organisms. The phosphorus cycle differs from the others because its source is the lithosphere. As rocks erode and weather, phosphorus becomes available to living organisms in the form of phosphate salts. Farmers provide their crops with phosphorus in fertilizer consisting of animal manure, guano (sea-bird feces), or crushed phosphate ores. Phosphorus must be used first by plants; only then can it enter the food chain for use by other organisms. Phosphorus eventually returns to the ocean and the lithosphere through waste products that are carried by inland rivers or deposited in the soil. Additional points about the phosphorus cycle are:

Because of its reactivity, the movement of phosphorus through the ecosystem is more complicated than that of many other elements. In certain types of soil, phosphorus atoms may be so tightly bound that plants are unable to absorb them.

Phosphorus may be absorbed against a concentration gradient; this is *active absorption*.

Living organisms recycle phosphorus within their cells and tissues; it is excreted in feces.

Marine protozoans mineralize phosphate compounds.

Wind-borne dust particles provide needed phosphorus in some localities.

THE NITROGEN CYCLE

Nitrogen is still another element that is essential to life. It is present in amino acids (and proteins), nucleic acids, vitamins, and plant pigments.

Elemental nitrogen (N_2), the most abundant gas in the atmosphere, cannot be used by most living organisms. Organisms known as *nitrogen fixers*—bacteria, fungi, and blue-green algae—"fix" nitrogen by combining it with hydrogen to form ammonia. Some ammonia is excreted and then oxidized to nitrite (NO_2), which may be oxidized, in turn, to nitrate (NO_3). These oxidative steps are collectively known as *nitrification:*

free nitrogen $\rightarrow$ ammonia $\rightarrow$ nitrites $\rightarrow$ nitrates.

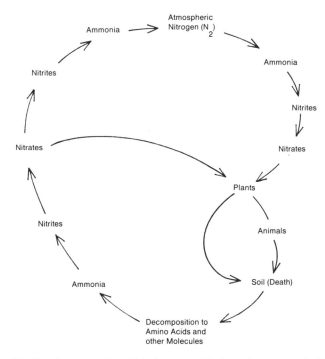

Figure 4.25. The nitrogen cycle, which, because of its importance to agriculture, is continually manipulated by human beings. Artificial fertilizers are made largely from atmospheric nitrogen. Proposals have been made in the past to destroy denitrifying bacteria so that fertilizer (ammonia) will not be broken down into elemental nitrogen. (What would happen to the nitrogen cycle—or any cycle—if one step were to be prevented?) Currently, molecular biologists are attempting to create plants capable of utilizing atmospheric nitrogen directly.

A series of organisms exist that can reduce nitrate to ammonia and, next, to elemental nitrogen. This process is known as *denitrification.*

Depending upon the fertility of the soil, anywhere from 0.1 to 20 g of nitrogen per square meter may enter food chains by nitrogen fixation each year. The high fertility of alpine lakes is directly related to the alder trees and their symbiotic nitrogen-fixing bacteria. The continued fertility of rice fields results from the blue-green algae that grow in the paddies during much of the year. Unlike phosphorus, nitrites and nitrates are soluble in ground water and do not adhere to soil particles. As a result, these compounds are easily carried into

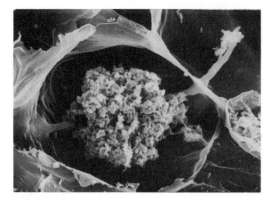

Figure 4.26. Nitrogen-fixing bacteria, *Rhizobium meliloti*, in one of thousands of small cells of a nodule on an alfalfa root. It is from such bacteria that genetic engineers hope to isolate DNA carrying the nitrogen-fixing genes and their regulatory sequences. They hope then to insert the entire genetic package into plant cells. The engineered plants, these biologists hope, would then be able to utilize free nitrogen (N_2) obtained directly from the atmosphere. *(Photo courtesy of Ann M. Hirsch, Wellesley College)*

ground water; nitrogenous fertilizers have caused serious pollution problems in some agricultural areas.

The overall rate of protein synthesis depends upon the local availability of nitrogen fixers. In many areas of the world nitrogen fixers are the bottleneck to good nutrition. Protein deficiency causes a disease known as *kwashiorkor*, which is particularly severe—often lethal—among children. Many genetic engineers believe that the solution to poor nutrition lies in the insertion of bacterial nitrogen-fixing genes into crop plants, thus allowing our food plants to fix atmospheric nitrogen directly, without reliance on microorganisms or fertilizers. Others argue that the number of human beings is the source of the nutrition problem and, if this number continues to expand, no manipulation of food plants will provide a lasting solution to world hunger.

Here are some additional points concerning the nitrogen cycle:

Nitrate is produced in the atmosphere by lightning and moisture.

Nitrogen compounds (primarily NO_3), having passed through the food chain, are excreted as ammonia and related compounds, such as urea.

Fixed nitrogen compounds leave the food chain directly as nitrogen oxides when forests and grasslands are burned.

Bacteria and fungi, when decomposing organic matter, convert nitrogen to NH_2 and then to ammonia.

INFORMATIONAL ESSAY

Alternatives to Oxygen

Combustion, as we normally think of it, is the union of some substance or element with oxygen. To produce a fire, we learned in lower grades, one must have a fuel, oxygen, and a temperature sufficiently high to cause ignition. Because air is only twenty percent oxygen, substances burn more slowly in air than in pure oxygen. A red-hot iron wire simply cools in air; plunged into a flask of oxygen, the same wire flashes and sparkles as it burns spontaneously. Human flesh burns with a flame in pure oxygen; three astronauts were lost through combustion in the pure oxygen within their space capsule.

Cellular metabolism does not involve combustion. The carbohydrates, fats, and proteins that we eat do not combine directly with oxygen. Instead, hydrogen atoms are removed from various organic molecules (hence there is an important class of enzymes known as dehydrogenases); these hydrogen atoms are passed from substance to substance until they are eventually combined with oxygen to form water. Having lost hydrogen atoms, the organic molecules gain an ever higher proportion of carbon atoms, which, too, are finally combined with oxygen to form carbon dioxide. Thus, oxygen is needed in the final analysis to combine with hydrogen and carbon to produce the two extremely stable products of combustion: carbon dioxide and water.

The slow release of energy through the metabolic system of dehydrogenation can be compared to the slow release of energy from a watch spring by the small oscillating ratchet wheel, the one that produces the characteristic tick-tock of watches and clocks. Winding a clock takes considerable energy. Anyone who has accidentally released the energy of a wound spring in his fingers while dismantling a clock knows how startling and painful this sudden release may be. Released second by second in an orderly way, the same energy is employed for the useful purpose of rotating wheels and moving the hands of the clock. And so it is with the energy of nutrients. Each bit of energy that is released by dehydrogenation is used to insert nearly as much (but not *as much*, otherwise we would be discussing perpetual motion) energy into another compound, where it is held ratchet-like until ready for use. Its release is used to perform useful work, such as transcribing RNA from DNA, replicating DNA, or translating mRNA into new protein molecules.

At times, oxygen is not available, and activity, such as muscular activity, proceeds with an insufficient oxygen supply. Microorganisms sometimes live in the absence of oxygen; indeed, some may live only

Figure 4.27. A black smoker—water at nearly 400°C emerging from a vent at the bottom of the sea, about two miles down, just off the tip of Lower California. The black color is caused by an iron sulfide precipitate. Bacteria use the hydrogen sulfide (H_2S) emerging from these vents as a source of energy that permits them to combine carbon dioxide and hydrogen atoms into carbohydrates. A *chemical* energy source (H_2S) plays the same role for these bacteria that sunlight (photo energy) plays for green plants. As plants do in normal communities, the chemoautotrophic bacteria form the basis for complex undersea communities of mussels, clams, crabs, tubeworms, and other exotic fauna. *(Photos courtesy of the Woods Hole Oceanographic Institution)*

anaerobically (as opposed to *aerobically*—in the presence of air). In anaerobic metabolism, however, the final products are not carbon dioxide and water, but alcohol or lactic acid. These substances still possess a great deal of energy (recall that alcohol can run automobiles, and that food can be cooked on sterno or alcohol lamps); thus the amount of energy gained anaerobically is less than that gained aerobically.

The secret of gaining energy through metabolism is having some element or compound at the end of the process that is capable of accepting the hydrogen atom (as oxygen does in forming water) or the carbon (again, as oxygen does in forming carbon dioxide). Various organisms—primarily bacteria—use substances and elements other than oxygen for this purpose. Nitrate (NO_3^-), nitrite (NO_2^-), sulfate (SO_4^{--}), and ferric (Fe^{++}) ions as well as carbon dioxide (which is converted to methane, CH_4) and sulfur (S) are used by one or another group of bacteria. Furthermore, it seems that newly discovered deep-sea organisms that live near hot-water vents at the bottom of the oceans either use sulfur or rely on bacteria that use sulfur for the bulk of their energy.

Organisms that utilize sulfur in place of oxygen produce, of course, H_2S (hydrogen sulfide) instead of H_2O. At the bottom of the ocean, this may not be important because the hydrogen sulfide, although toxic, is dissipated into an effectively unlimited sea. In freshwater ponds and lakes that are hidden beneath the icecap of Antarctica, however, the situation differs. Divers investigating the primitive algae inhabiting such enclosed bodies of water must be extremely careful: hydrogen sulfide can be absorbed through the skin and, in high enough concentration, can interfere with the oxygen-transport mechanism of the body's red blood cells—as does the more deadly carbon monoxide (CO) of automobile exhausts, gas ranges, and gasoline lanterns and stoves used by campers.

5

WATER AND HEAT

Adequate supplies of food and oxygen are required for the maintenance of life, but they are not enough—they are *necessary*, but not *sufficient*. There are other conditions that, if not met, are fatal. Water must
be available to all forms of life; from the lowest to the highest, all creatures are aqueous in construction. Life can exist, too, only in a narrow
range of temperatures. This chapter looks at water and temperature
from the point of view of human life.

Temperature

Heat is a form of energy. Temperature, the aspect of heat that we
measure with a thermometer, reflects the state of agitation of the
molecules or atoms of a substance. Temperature does not reveal the
amount of heat in a physical body because the amount of heat also
depends upon a characteristic of the body known as its *specific heat*.
The specific heat of iron, for example, is 0.10; that of liquid water is
1.00. If two grams of water, one at 10°C and the other at 90°C, are
mixed, the result would be two grams of water at 50°C. If a gram
of water at 10°C is poured over a gram of iron whose temperature is
90°C, the water-iron combination would possess a temperature of only
17.3°C. Thus, at a given temperature, iron possesses only one-tenth
the heat energy that water does at the same temperature.

The addition of heat to an object causes an increase in temperature:
a pan of cold water is placed on a hot stove to bring it to a boil (100°C
or 212°F). The reverse is also true: warm gelatin can be placed in snow
(or in the refrigerator) in order to cool it so that it will gel. Thus, temperature determines the direction in which heat will flow. The flow of

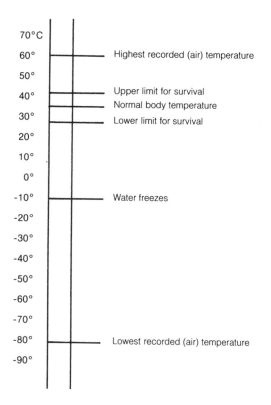

Figure 5.1. The range of temperatures encountered on the surface of the earth, the freezing point of water (0°), normal body temperature, and the upper and lower limits for human survival. The limits for survival fall well below (and above) the air temperatures that persons frequently encounter. These external temperatures are not fatal because our bodies have means for controlling internal temperatures.

heat is from an object at high temperature to an object at lower temperature. These, then, are the basic ideas that we shall use in discussing heat, temperature, and life:

• Heat is a form of energy.

• Heat can be transferred from one object to another. When an object is heated, its temperature generally rises. The exceptions that are especially important in the case of water involve changes of state: the change from water at 0°C to ice at 0°C requires the loss of 80 calories of heat for each gram (or milliliter) of water; the change from water at 100°C to steam at 100°C requires the addition of 539 calories per gram. [One calorie is the quantity of heat needed to raise the temperature of one gram of water one degree Centigrade. The "nutritional" calorie is really a kilocalorie—the amount of heat needed to raise one kilogram (liter) of water 1°C.]

• Heat flows, either slowly or rapidly, from the warmer object to the cooler one. Insulation, whether in a house or in a ski jacket, is material through which heat flows slowly. On the other hand, copper and aluminum are plated on the bottoms of steel pots and pans because these metals conduct heat rapidly, thus distributing the heat evenly.

• The temperature of the body and particularly of its individual organs and organ systems is of great physiological importance. In part this is because the structures of cells and cellular molecules are sensitive to temperature. Temperature also affects the chemical processes that occur within cells. As is true for chemical reactions generally, high temperatures speed up cellular chemical reactions, while low temperatures slow them down. If the temperature is too low, vital processes may not proceed at a rate sufficient to maintain life. On the other hand, when the structure of proteins has been altered—by denaturation, as in the cooking of an egg—vital processes cease abruptly.

Temperature and Maintenance of Life

What are the limits of temperature that are consistent with life? The answer depends upon the organism. Some insects and spiders live on ice. Some bacteria and algae thrive in hot springs, even at temperatures near the boiling point of water. Human beings, however, can survive in only a narrow temperature range. The brain, for example, normally varies in temperature between 36°C and 39°C. Consciousness is lost at 32°C; death results at 26°C. At the other "extreme," should the brain be heated to 44°C even briefly, convulsions, coma, and death occur. The narrow range of tolerable internal temperatures, when compared with the wide range of external ones, suggests that the body possesses means for conserving heat in cold surroundings and for avoiding heat gain in hot ones.

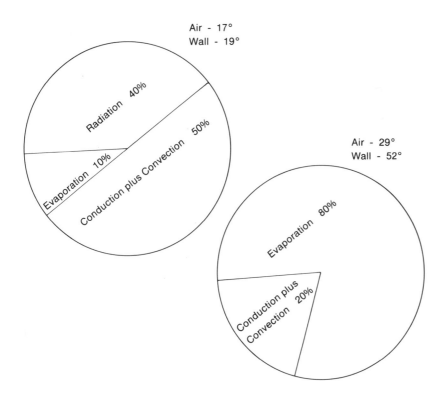

Air - 17°
Wall - 19°

Radiation 40%

Evaporation 10%

Conduction plus Convection 50%

Air - 29°
Wall - 52°

Evaporation 80%

Conduction plus Convection 20%

Figure 5.2. The amount of heat gained or lost in exchange with the environment and the means by which it is gained or lost depend upon many factors: Is the sun shining? What is the air temperature? Is the wind blowing? What is the relative humidity? How is the person dressed? Is the body dry? The diagrams show some experimental results obtained about twenty-five years ago. In each experiment, a volunteer sat naked in a small room where the air was dry; the temperature of both the air and the walls of the room was experimentally controlled. The body temperature of each volunteer remained constant; thus, heat formed by body cells was dissipated at the same rate at which it was created. The means by which heat was lost—conduction, radiation, and evaporation—varied considerably depending upon the experimental conditions.

In most parts of the world, especially at night, human body temperature exceeds that of the environment. Consequently, the body as a rule continuously loses heat. Because cellular metabolism continually generates heat, a balance between heat production and loss is possible. Indeed, such a balance normally exists; body temperature normally varies by only 1°C or 2°C. Within this pattern of general stability, body temperatures tend to be higher in the afternoon than in early morning.

For persons living in hot environments, the problem of maintaining a balance is different: the body tends to gain heat from the environment, while cellular metabolism generates even more heat. Human beings can and do live in hot environments because, as we shall see, the human body possesses means by which heat can be lost.

Physiological Responses to Heat Loss

Imagine a nude man escaping from a burning house on a cold winter morning. Upon his meeting the frosty morning air, heat flows from the man's warm skin to the air (*convection*) or *radiates* into space; these losses cause his body temperature to drop. Following the slightest initial changes, his body makes compensating adjustments automatically. Blood vessels in his skin contract, causing him to become

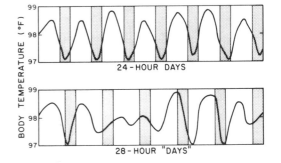

Figure 5.3. Body temperature varies through the 24-hour daily cycle. Usually it is highest in the afternoon and falls in the early morning hours. This normal cycle can be shifted somewhat to fit the timing of day and night (after a transcontinental plane ride, for example), but it cannot be lengthened or shortened much. As the graphs show, a volunteer lived in an underground cave for several weeks under two regimes: 17 hours light/7 hours dark (a 24-hour day) and 19 hours light/9 hours dark (a 28-hour day). The upper graph shows that his body temperature varied as expected while he was on the 24-hour day. However, during six 28-hour days, a low temperature was sometimes recorded during the light period and a high temperature sometimes occurred at night. Compared to the upper graph, the lower curve is conspicuously erratic.

pale because his surface circulation is restricted. Blood circulation to his muscles, internal organs, and brain becomes limited. When blood ceases flowing through his skin, that organ becomes an insulating blanket that helps prevent heat loss.

There are still other adjustments: the man may shiver; in fact, he may even jump up and down, wave his arms, and slap himself. Shivering and physical exercise increase muscular activity and increase the production of heat by muscle cells. The production of heat tends to counterbalance heat loss. If the man were unable to gain shelter at a neighbor's home, he would use other devices for self-protection. His behavior would change. For example, he might locate a sheltered nook where he could huddle, thus reducing the exposed area of his body.

Changes in circulatory pattern, cellular activity, diet, and fat stores are the body's natural defenses against falling temperature. Such defenses, however, can cope with moderate heat loss only. In meeting the much greater stresses that are commonly encountered, human beings have devised some of their most useful inventions: clothing, shelter, and fire. Using these inventions, people have successfully colonized some of the coldest, windiest, and most hostile areas of the earth. Vast numbers of other organisms are better equipped than human beings in their individual capabilities to endure cold; none, however, normally lives in the wide range of environments that are inhabited successfully by people.

Physiological Responses to Heat Gain

The human body is even more sensitive to rising temperature than to falling temperature. If the temperature of the brain reaches and remains at 41°C to 45°C through high fever or exposure to heat, the result is permanent brain damage or even death.

As in the case of exposure to cold, the body possesses natural defenses against moderate temperature rises; they are roughly the reverse of those mentioned above. What happens, for example, to our nude man when friendly neighbors take him in from the snow and cold? Within seconds, circulation of blood in the skin is restored and his skin becomes hot and flushed. The blood in his skin now loses heat to the air, provided that the temperature of the house is lower than that of the man's body. Sweat glands are activated; as the water in sweat evaporates, body heat is lost, because evaporation is a cooling process. In extremely moist air, sweat may run off the skin without evaporating; sweat that drips does not remove heat from the body. This explains why it is dangerous to exercise in sweat-proof clothing: if the body cannot get rid of extra heat, its temperature may rise, causing heat prostration or heat stroke.

NOW, IT'S YOUR TURN (5-1)

An internationally famous marathon runner collapsed shortly after completing each of two races. The causes of his collapse in the two instances were completely different: the first collapse resulted from *heat stroke*, the second, from *hypothermia* (abnormally low body temperature). From what you have learned about the physical characteristics of water in this chapter, explain how a grueling 26-mile race can leave a runner in one or the other of two completely different pathological states. [Hint: If your explanation is correct, you should also be able to explain the weather conditions (temperature and humidity) on the days of these races. In fact, both days were extremely hot. What do you think the humidity and wind conditions could have been?]

Long-range adaptations to an increased heat load are also possible. Sweat formation is enhanced by an increase in the total volume of circulating blood. The circulation of blood in the skin is repatterned to facilitate heat loss. Natives of hot climates tend to move slowly and to avoid needless exercise; they also tend to eat less than persons living in cold climates.

When suffering from too much heat, few people remain passive. They fan themselves; open windows if they are indoors; move into the shade if they are outside; find water to moisten their faces, arms, and body; and do other things that tend to remove body heat. These behaviors reflect "common sense" or, as scientists would say, are "adaptive." Human beings have also responded to problems posed by heat with a series of inventions: fans, air conditioners, refrigerators, and artificial fountains represent a few. Finally, in the tropics clothing and housing are designed to provide shelter from the sun and to allow free circulation of air as an aid to evaporation.

NOW, IT'S YOUR TURN (5-2)

1. Have you ever spent a period of time in a climate considerably different from that in which you normally live? Recall in your journal the contrasts in your behavior, dress, diet, and housing under the different climates, pointing out the way in which each of the patterns is, or was, adapted for the prevailing conditions.

2. In tropical islands that are under U.S. protection, governmental agencies have often provided sheet-metal huts to replace the primitive shelters normally used by the natives. Would this change in housing necessarily represent an improvement?

Control Systems

The adaptive reactions of the body that protect it against temperature changes are activated by sensitive control systems that can be compared with the thermostats that regulate temperatures in buildings. Special nerve endings in the skin, as well as certain nerve cells in the brain, are acutely sensitive to temperature change. They respond to rising temperature (heat response) and falling temperature (cold response) with increased activity. These sensory structures are, in fact, thermometers. They send information to the major control center in the brain, which then activates appropriate responses. If this system were a home thermostat, it would turn on either the furnace or the air conditioner. In the human body, the control center sends out nerve impulses that activate sweat glands, muscles (for shivering, if for nothing else), and all other responses needed to warm or cool the body. When the body is forced to remain for long periods in an extremely hot or an extremely cold climate, adjustments occur in endocrine glands and elsewhere throughout the body. These bodily adjustments reinforce the immediate or short-term reactions.

Water and Life

The Need for Water

About two-thirds of the human body is water. This water is not a stable or static component of the body; it is taken in and expelled continuously. The rate at which water is lost depends upon many variables, such as physical activity and the temperature of the surroundings. When a person works and produces a great deal of heat, some of it is lost by evaporation of water from the skin and lungs. High temperature also leads to sweating and water loss. An average adult loses some five liters of water (five to eight percent of the total body water) every twenty-four hours. Does this mean that a person would lose all body water in twelve to twenty days if he refused to drink any? No, because the body would slow its rate of loss, chiefly by reducing the amount of urine produced. Dehydration leads to a decline in the volume of circulating blood; eventually circulation, respiration, and nerve function fail. Death comes long before all water is lost from the body. Depending upon the temperature and other conditions, persons trapped without water have survived from about two to eighteen days. Normally, of course, water that is lost from the body is replaced very accurately by water from food or water that is drunk.

Can There Be Too Much Water in the Body?

To see if blood could be diluted by drinking water, the English biologist J. B. S. Haldane once downed about six liters of water in six hours.

Figure 5.4. Water turnover—that is, the balance of water entering and leaving the body—varies from one animal species to another and, of course, depends on external conditions as well. In frogs (and most other amphibians), the soft, moist skin is easily penetrated by water molecules. Frogs lose water so rapidly in dry air that their survival depends on immediate access to water. Snakes, like other reptiles, lose water much more slowly than frogs, partly because their skin is less permeable to water and, in addition, because their kidneys prevent much water loss during excretion. Such differences in water exchange account for different ways of life: amphibians cannot live in the dry desertlike areas that are home to many snakes and lizards. *(Sketch from* The Animal Kingdom*)*

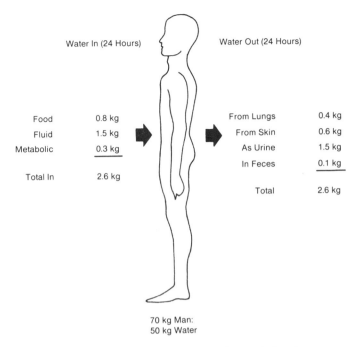

Water In (24 Hours)   Water Out (24 Hours)

Food	0.8 kg	From Lungs	0.4 kg
Fluid	1.5 kg	From Skin	0.6 kg
Metabolic	0.3 kg	As Urine	1.5 kg
		In Feces	0.1 kg
Total In	2.6 kg	Total	2.6 kg

70 kg Man:
50 kg Water

Figure 5.5. In comparison with other vertebrates, human beings have average rates of water exchange. For someone weighing 70 kg, loss of water through the lungs and skin may vary from one to five (or even more) liters per day. The flow of urine may range from less than one to five (maybe as many as twelve) liters per day. Persons with access to water generally drink enough to balance water losses, so that the total water content of the body varies by only 0.2% in any 24-hour period.

Figure 5.6. Many organisms are much better adapted to a lack of water than human beings are. Some animals can survive without access to free water. Mealworms and flour beetles, for example, live and grow on dry grain and flour. Desert rodents often live entirely on water that is contained in their food (roots and seeds) or that is produced by the metabolic breakdown of sugars and fats. Most such animals make no attempt to control their bodies' temperature by evaporation. In addition, they possess extremely efficient kidneys and water resorption regions in their guts. The koala shown here is a native marsupial of Australia; it will drink water if it is available, but seemingly can survive without water. *(Sketch from* The Animal Kingdom*)*

The amount of urine he produced increased greatly. A liter and a half of urine was passed during one of the six hours; that was Haldane's record performance. Throughout the experiment, however, Haldane was unable to detect any change in the amount of water in his blood, although blood carried the excess water from his intestine to his kidneys.

An experiment such as Haldane's shows that, in healthy persons, the amount of water in the body is regulated precisely. Too much water is seldom a danger to health. Nevertheless, disease and environmental conditions can upset the balance. When things go wrong, a person may drink too much water and suffer from "water intoxication." Its symptoms include the following: muscular weakness and cramps, mental confusion, loss of coordination, convulsions, and coma. Drinking a great deal of water (or beer) after heavy sweating may induce water intoxication even in healthy persons. The symptoms of water intoxication are caused, then, by the dilution of those body salts that remain after severe sweating.

The Control of Water Balance

Because the body is largely water, and because a great deal is taken in (and lost) every day, one might expect a great deal of "play" or

Figure 5.7. An artist's sketch of a small oasis in the interior of Egypt. Here one senses the role that temperature (hot days, cold nights) and water (both its presence and absence) play in controlling the types of plants and their physical distribution, the pattern of human life, including the nature of the house, and the types of domesticated animals to be found. In the left foreground is a water wheel, which is operated by a water buffalo. The water wheel provides water for the irrigation of the small plots of grain and vegetables.

variation in the body's water content. This, however, is not the case. The sensations of hunger and thirst closely balance intake with loss through urine, sweat, and respiration. The accuracy with which body weight is controlled depends on more than an accurate intake of food; it also depends upon the accuracy with which water intake corresponds to water loss.

How does the body sense a need for water? And how does it measure this need? Nerve cells in the "drinking" center of the brain, located close to the "feeding" center, sense the concentration of dissolved substances in the blood. The drinking center also receives information from the mouth, especially about the flow of saliva. Other sense organs located in blood vessels both near the heart and in the kidneys send additional information to the brain about the quantity and composition of the blood. These items of information are interpreted by the brain: if water is required, a search is begun; if not, drinking becomes an unpleasant chore. Of course, as with most human activities, drinking behavior is more complicated than this account would indicate: after all, Haldane did, by his own choice, choke down six liters of water. The cortex of the brain plays its part as well. Persons do have customs, memories, wishes, and ideas about drinking: Drink at least eight glasses of water each day! Have another cup of coffee! To your health! Down the hatch!

The search for water can become the dominant, compelling occupation of dehydrated, thirsty persons as well as for other higher animals. Because the compulsion to drink is so great and the relief from thirst so specific, no mammal or other animal attempts to establish exclusive control over watering holes in the arid regions of Africa and Australia; desperation on the part of those animals not allowed free access would quickly make a watering hole an undefendable resource.

On the other hand, people normally do not drink a great excess of water; they drink water and other watery fluids in moderation. Once it is ingested, water enters the intestine, from which it passes into the bloodstream. As part of the blood, the water is distributed throughout the body, where it joins the fluids already in and around cells. If a need existed, the arrival of water now satisfies that need. If there were no need, the excessive dilution of body fluids activates specific nerve cells in the brain. These cells act in turn on the pituitary gland. An antidiuretic hormone stored in the pituitary gland (*vasopressin*) controls the function of the kidneys in removing water from the body. The hormone stimulates the kidneys to resorb a sizable fraction of water that has filtered through the capillary walls into the renal tubules. Lack of this hormone allows excess water to pass down the tubule and into the urinary bladder, from which it is expelled from the body. In response to excess water in body fluids, the amount of vasopressin released from the pituitary is reduced. The pituitary stimulation of kidney resorption subsides, and urine flow increases. In this way, the extra water is drained from the body.

NOW, IT'S YOUR TURN (5-3)

1. In most hospitals, the patients, especially young children, are monitored with respect to fluid intake and loss so that they will avoid the risk of dehydration. The monitoring is not difficult; it requires only care and record keeping. Record in your journal an inventory of water intake and loss (urine) for several days (a weekend perhaps). It is unlikely that your intake and loss quantities will balance. Attempt to list for both intake and loss some items that have escaped measurement. (Hint: Weigh yourself carefully upon going to bed and upon rising. Is there a systematic, overnight change? How would you explain it?)

2. There seems to be a contradiction in the text: Despite drinking six liters of water in a period of six hours, Haldane was unable to detect any dilution of his blood, yet excess water was being lost as urine. Later, however, the text mentions water intoxication that can result from drinking too much water, and speaks of the *dilution* of the body's salts. Reconstruct the fate of water as it goes from the gut into the body (the circulatory system) and then leaves the body by way of the kidneys as urine. Need Haldane's observation and water intoxication be contradictory?

Figure 5.8. A sub-Saharan village being engulfed by sand. Although water may seem to be available in limitless quantities to some persons, it is an endangered (even nonexistent) resource for others, as in this village, where the advance of dry desert sand threatens to obliterate all forms of life. *(Photo courtesy of the Church World Service)*

Water Supplies

In the past, water has been freely available to most persons. Of course, there were dry areas, such as deserts, where water was to be found only at an occasional watering hole or oasis. Elsewhere, however, ponds, lakes, streams, rivers, and ground-water aquifers provided ample quantities of water. Towns and cities arose near these sources of water because good health, as well as transportation and industry, has always required a supply of water. However, much of the earth's water cannot be used directly by human beings. Ocean water, the bulk of all water, contains too much salt. Many lakes and streams are contaminated with salts, mineral acids, alkalis, poisonous materials, bacteria, or animal wastes. In recent times, industry has competed with both human beings and other forms of life in its use of water; lakes and streams have been polluted by both chemicals and excess heat. Modern agriculture also demands huge quantities of water. Water that is directly suitable for human use is currently in short supply.

Figure 5.9. A deadly trade-off involving cattle (food), vegetation (food for cattle), and water. Because they require water, in arid regions cattle are concentrated near sources of water (upper photograph). Soon, however, they will have stripped the surrounding area of vegetation, thus destroying their source of food (lower photograph). *(Photo courtesy of U.S. Agency for International Development)*

Table 5.1. Water use

A. Water use by individuals for nonindustrial purposes

Use	Gallons per day
Drinking and washing	1–2
Washing dishes	1–4
Laundry	8–9
Bathing	20
Toilet	24
TOTAL	54–59

B. Total water use in the U.S. (in billion gallons per day) for a variety of purposes

Use	1965	1980	2000 (estimated)
Rural homes and agriculture	108	135	153
Urban homes and business	27	41	75
Industry	55	75	128
Steam-generated electricity	330	525	892

How much water does one person need? Personal needs vary from one individual to the next, depending upon how much water is lost as urine, sweat, and tears. An adult man or woman living in a moderate climate, without heavy activity or physical stress, drinks about two liters of water a day. Water, though, is needed for cooking: rice, beans, cereals, and many other common foods cannot be digested properly unless they have been softened by being boiled in water. Water is used for bathing and for laundering clothes because it is a remarkable solvent, capable of removing many unwanted or unpleasant substances. It is also used for flushing human wastes out of our homes and into the community sewage system. Some data on water use are given in Table 5.1.

Quality, as well as *quantity*, of water is crucial for good health. All natural water supplies contain bacteria and other small organisms. The numbers and kinds of these contaminating organisms that are tolerable in drinking water are limited; without proper safeguards, a public water system can become a deadly dispenser of disease. "Soft" water contains little calcium or other minerals; "hard" water, on the contrary, is rich in dissolved minerals, particularly calcium. Recent studies suggest that some circulatory diseases are especially frequent in localities where drinking water is relatively soft. Minerals that are often present in hard waters may be helpful in maintaining health; it is no accident that health spas have been associated with mineral springs for centuries. Alternatively, soft water may be more likely than hard to contain unprecipitated traces of harmful substances. Our drinking water may have subtle effects on our health that we never think of as we drink it. The biological effects of chlorine, for example, are virtually unknown; nevertheless, it is the water purifier of choice in every municipal water station worldwide.

NOW, IT'S YOUR TURN (5-4)

Very few people have any idea about the amount of water that falls upon the United States annually, or what happens to it. Undertake a project for your journal in which you discuss the water cycle, from evaporation through precipitation (as rain or snow), runoff into the oceans, and back to evaporation. For the United States, cite as accurately as you can the figures for water that is usable for agriculture and for hydroelectric power. One item for which you will be unable to obtain precise data will be the number of times the same (river) water is reused before it returns to the sea; much of the water used in New Orleans, for example, was urine in Pittsburgh and St. Louis.

Summary

The ability of the human body to function normally depends on moderate environmental temperature. The range of body temperature that permits good health and is consistent with survival is narrow, perhaps 4°C or 5°C. External temperatures may vary anywhere within a range of one hundred degrees or more. Both short- and long-term controls regulate the human body and human behavior, thus permitting survival despite erratic external temperature fluctuations.

Water is necessary for life. Each person is about two-thirds water, and as much as eight percent of the body's water can be lost every day. This loss must be balanced by water gained through eating and drinking. The body's normal water balance is maintained by the delicate interplay of nerves, hormones, and behavior.

Temperature and water are two major components of an environment that determine whether living organisms (including human beings) can survive.

INFORMATIONAL ESSAY
Heat Transfer

Any object can lose heat in three, or maybe four, ways:

• Conduction: Heat flows *from* the object to a colder object with which it is in contact, or *to* the object from a second, warmer one.

• Convection: Flowing air (or water) can either bring heat or remove it.

Figure 5.10. The UN helps provide clean water. For most persons in the United States, the notion that a piped spring or a community pump is an enormous step toward sanitation and the control of disease is incomprehensible. On the other hand, before becoming too smug, recall the August 18, 1983 *New York Times* headline that read: "Don't drink the water in Akron, Ohio." Although the quantity of water in the United States is generally not limiting (the arid Southwest may prove to be an exception), insidious pollution of ground water and rivers threatens water quality. *(Photo courtesy of WHO; Paul Almasy, photographer)*

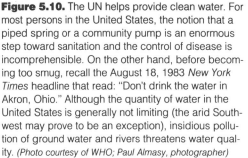

Figure 5.11. A diagram illustrating the three ways by which a solid object resting on a surface can gain or lose heat: conduction, convection, and radiation. If the object were moistened with water, *evaporation* of the water would provide a fourth means by which heat is lost.

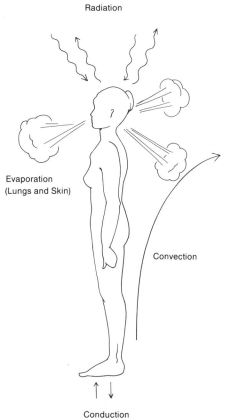

Radiation

Evaporation
(Lungs and Skin)

Convection

Conduction

Figure 5.12. The four ways by which a human being (a heat producer) could exchange heat with his or her surroundings if standing in the open, unclothed.

• Radiation: Heat leaves an object as infrared (heat) rays and radiates to a colder recipient (the blackness of outer space, for example) or arrives at the object from a warmer heat source (radiator or heater).

• Evaporation of water: Any object with a moist surface may lose heat by evaporation (539 calories per gram of water evaporated).

Human bodies are physical objects; they sometimes gain heat from their surroundings, but at other times they lose heat. Human beings are also heat *sources* because of their cellular metabolism.

People control their heat balance and body temperature by "natural" and "artificial" means. Natural controls can be described as follows: (1) Radiation, conduction, and convection are altered by the flow of blood in the skin; (2) evaporation is changed by increasing or decreasing the flow of sweat; and (3) heat production is changed by body movements, shivering, and increasing cellular activity.

Artificial controls include clothing and shelter (which reduce all forms of heat loss) and radiators or air conditioners (which adjust the temperature of the immediate environment).

Heat exchange sometimes exceeds the body's control system. The result is either hypothermia (a fall in body temperature) or hyperthermia (a rise in body temperature). Unfortunately, these extreme conditions are not uncommon; too often, they result in an individual's death.

INFORMATIONAL ESSAY
Water, a Special Substance

In 1913, L. J. Henderson, an American physiologist, wrote a book entitled *The Fitness of the Environment*. He was impressed by the characteristics of water and of carbon dioxide that make these substances suitable for life. Although he admitted that living organisms undoubtedly evolve as Darwin had suggested a half-century or more earlier, he argued that the environment could not have evolved. Consequently, when he considered all the characteristics of the environment (especially of water) that make life possible, he thought that the probability of seeing so many special properties should be exceedingly small. Although he offered no solution, he obviously entertained the notion that the environment was designed especially to support life. The weakness of Henderson's position was that "life" should have been cited by its full title: "*life as we know it.*" Life as we know it is analogous to a bridge hand that has been dealt to an individual player: once it has been dealt, it is possible to calculate that the probability of the player's getting precisely that hand equals $\frac{1}{52} \times \frac{1}{51} \times \frac{1}{50} \times \ldots \frac{1}{42} \times \frac{1}{41} \times \frac{1}{40}$ (13 terms) or 1 in approximately 4×10^{21} deals. That is, if the environment had not possessed the properties that impressed

Dr. Henderson, neither life as we know it nor Dr. Henderson's book would exist; acknowledging that we all do exist, however, it then becomes meaningless to calculate the combined probabilities of the features that make existence possible. None of this detracts, however, from the properties of water that are important for life.

The first of these properties is water's specific heat, which is considerably higher than that of most substances. As a consequence, large bodies of water (oceans, lakes, and rivers) change temperature extremely slowly. Therefore, it is unnecessary for aquatic organisms to adapt to highly variable temperatures. The moderating influence affects land organisms as well. The marine climates of the poles and the equator differ by 35°C; continental climates differ by 60°C or more. Ocean currents and atmospheric winds (weather patterns) can also be ascribed to the high specific heat of water.

Water has high latent heats of melting and evaporation. When water is cooled to 0°C, 80 additional calories per gram must be removed in order that it be converted into ice at 0°C. The effect is that an ice-water mixture maintains the same temperature (0°C) despite the gain (less ice, more water) or loss (more ice, less water) of considerable heat. To transform one gram of water into steam requires 539 calories; the effect is that the temperature of boiling water remains at 100°C—additional heat merely reduces the quantity of water. Stated differently, the evaporation of one gram of water removes 539 calories, or, as expressed earlier, evaporation is a cooling process.

At 0°C, the freezing point of water, many chemical reactions still occur with considerable speed. Apart from this consequence of a high freezing point, the latent heat of melting serves again as a moderating influence. A washtub full of water placed in a vegetable cellar is generally sufficient to prevent the freezing of vegetables during winter months.

Sun shining upon a body of water, because of water's transparency, warms a layer of considerable depth. Evaporation and the heat required for evaporation—539 calories per gram—maintain water temperature nearly constant; only a small fraction of the solar heat absorbed goes toward raising the temperature of the water. The rest goes into evaporation. Latent heat of evaporation, then, serves to moderate the temperature both of the earth and of individual organisms.

Water conducts heat much better than most liquids do: thirty times faster than alcohol, for example. The consequence of this property is that temperatures within individual cells are equalized; one point does not boil while another nearby point remains cool.

An especially important feature of water as far as the earth's ecology is concerned is that it expands upon freezing, even upon cooling below 4°C. The densest water, that which is 4°C, sinks to the bottoms of lakes and oceans. If water contracted as it froze, ice would sink to

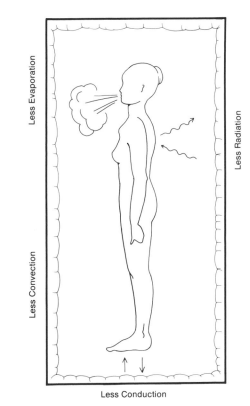

Figure 5.13. The role played by a shield (clothing or shelter) in altering the heat exchange between a person and his or her surroundings.

Table 5.2. Water is nearly a universal solvent. The following is a list of organic acids that have been identified in a urine sample; many other substances—organic and inorganic—have not been included in this list. A perusal of this list should engender (1) an awe for the physical properties of water that make it suitable for sustaining life and (2) an appreciation for the wizardry of cellular enzymatic machinery. Recall that the origin of all these chemicals lies in the equation $CO_2 + H_2O \rightarrow CH_2O + O_2$, the reaction carried out during photosynthesis.

Lactic acid	Pyroglutamic acid
Glycolic acid	2-Hydroxyoctene-2-oic acid
Diaminopropionic acid	2-Ketoglutaric acid
3-Hydroxyisobutyric acid	4-Hydroxybenzoic acid
Propylene glycol	4-Hydroxyphenyllactic acid
Sulfuric acid	Suberic acid
Phosphoric acid	Aconitic acid
Benzoic acid	Citric acid
Succinic acid	Isocitric acid
Fumaric acid	Dihydroxyphenylpropionic acid
Tartronic acid	4-Hydroxy-3-methoxy mandelic
Galactonic acid	acid
Adipic acid	5-Indole carboxylic acid
3-Methyladipic acid	Uric acid
2-Hydroxyglutaric acid	Stearic acid

the bottoms of lakes and seas. In time the large bodies of water would come to be large bodies of ice, and only thin surface layers of water would melt each summer—much as only the surface of the arctic's permafrost regions melts.

Additional properties of water to which Henderson refers are (1) its unequaled ability as a solvent, (2) the ionization of most substances in solution, and (3) its surface tension, which is higher than that of most liquids except liquid mercury. Water's ability as a solvent has repercussions not only on a global scale (an estimated 3.5×10^{16} tons of table salt are in solution in the oceans) but also on the scale of individual organisms (literally scores of chemical compounds are known to be contained in human urine). The matter of surface tension is important both in allowing water to reach the tops of tall trees (or to move through the small tubules of any plant) and in delivering water to the surface of soil from underground sources.

Darwinian fitness, according to Henderson's view, required that a fit organism inhabit a fit environment. The origin of life—granted that life could arise and continue in existence—virtually guaranteed that the reciprocal relationship would be found by anyone who took the time to look carefully.

INFORMATIONAL ESSAY
Survival at Sea

The oceans provide a favorable environment for vast numbers of organisms. Sea turtles and snakes, many birds, and mammals, such as seals, walruses, and whales, live their entire lives in the oceans. The oceans also harbor a great variety of marine invertebrate animals, including jellyfish, crabs, lobsters, octopuses, and giant squid; marine plants range from microscopic algae to giant kelp.

A human being at sea is, despite contrary appearances, in constant danger. The water is cold; it contains little oxygen; and it is too salty to drink. Human beings lack those characteristics of marine organisms that allow them to live at sea.

Marine organisms must be able to tolerate a cold environment. Some have body temperatures that are identical with the water in which they live, and their metabolic rates are exceedingly slow. Birds have waterproofed feathers that provide excellent insulation, whereas sea mammals are generally large, with thick layers of fat beneath their skins.

Reptiles, birds, and mammals that live in the oceans breathe air, just as human beings do. Those organisms that obtain oxygen from seawater have large surface areas—gills, for example—through which the oxygen can pass.

The salinity of seawater poses a problem for most marine life. Most invertebrates have body fluids whose salinity is about the same as seawater; the cells of these organisms function at this high concentration. Sharks have body fluids that also match seawater in overall salt concentration, but a good deal of the ionic strength of sharks' blood plasma and body fluids is provided by urea—a substance that would cause uremic poisoning in human beings. Marine fish, reptiles, birds, and mammals have special cells—in gills, kidneys, or special salt glands—that expel high concentrations of salt from the body, thus retaining the water and maintaining their body fluids at low salt concentrations.

When a person drinks seawater, many of its salts are not even absorbed; severe diarrhea results, with an accompanying loss of body fluid (and an increased thirst). Salts that are absorbed increase the salt concentration of body fluids; these salts are filtered out by the kidneys and excreted in urine—together with extra water. Consequently, a person lost at sea and suffering from severe thirst loses more water by drinking seawater than he gains; his chance of survival is no greater than if he were in a desert without water.

To survive at sea, human beings have drawn upon their culture—their intellect. They build ships, diving bells, submarines. They use insulated wet suits to prevent heat loss. Bioengineers are even pondering now how artificial gills might be constructed. To obtain drinking water from seawater, life rafts are equipped with evaporators operated by solar heat; as we have just discussed, the human body cannot cope with salty drinking water.

6

WHEN THINGS GO WRONG

The preceding three chapters are concerned with the essential requirements for health, even for life itself: food, oxygen, water, and an appropriate temperature. These requirements are examined in terms of extreme conditions: too much, too little, too high, too low, too wet, too dry, too hot, and too cold. The main theme has been a human body in balance, a human body seen briefly during an instant in time. We have considered the body, a challenge, and the body's response to that challenge; we have examined how the body recovers from one or another of several disturbances. In this chapter, we shall watch the body while it makes the best of bad situations. We shall see what happens when things go seriously wrong. To cover all contingencies would, of course, require a library of medical texts. That is not our intention. Rather, we shall pick and choose from among several illustrative examples in the hope that an understanding of these malfunctions will lead to a better understanding of normal life processes.

Why Do Things Go Wrong?

Do things go wrong? Of course! Nearly everyone reading this book has at one time or another been seriously ill. Most of us have had days when we felt "out of sorts." The obituary page of the daily paper is largely a testament to things that have gone wrong.

Why do things go wrong? There is no easy answer to this question, even when Why? is interpreted as How come? Many factors must operate properly if things are to go "right" for a living organism; the failure of any one of these is enough to make things go "wrong." Just as an automobile trip can be interrupted by a blown-out tire, a blown gasket, a broken piston rod, or a ruined transmission, so can a person's life be terminated by any one of many different failures.

The phrase "How come?" was used above in the sense that biological or medical scientists can attempt to pinpoint the specific malfunction and to identify the faulty mechanisms involved. In this role, the scientist is a model for the popular medical examiner, Dr. Quincy. The phrase "Why?" implies a different question, an inquiry into "For what reason?" or "Why me, Oh Lord?" A once-distraught mother wrote to her local paper saying how God had explained all to her, and how happy she had become: scarcely had she buried her 10-year-old daughter who died of a massive stroke when the family's pet horse died of a previously unsuspected congenital defect. Now, her daughter could ride her favorite horse in heaven. The reasoning is not scientific; not one claim in this woman's entire account can be subjected to test. Nevertheless, shame on those who would shatter her faith. Shame, too, on those who would in the name of science substitute instruction in such faith for instruction in science. A search for underlying purposes—purposeful explanations—does not fall within the scope of science. Here, we can only describe the faulty functioning of our body's machinery, or the role of external circumstances in overwhelming our body's machinery.

NOW, IT'S YOUR TURN (6-1)

A recurrent theme has surfaced here once more: the distinction between science and religion, whose answers to questions are obtained solely by revelations or other religious experiences. This theme *should* recur, because—despite the advanced level of technology American society has achieved—most of our citizens, educated or not, have little grasp of what is, and what is not, science.

Considerable time has now elapsed since you made your first entry in your journal; review what you wrote then. Would you alter any of your original thoughts and ideas? Alternatively, scan any of the major news magazines in order to learn what sorts of problems are confronting the U.S. Congress or your own state assembly. Which of these lie completely outside the scope of science? Which require action based on scientific data? Are there any for which you believe scientists should authoritatively decree solutions? (Foretelling problems is difficult, but among those of the past that would have been suitable for a discussion in your journal are abortion, equal rights for women, the further development of nuclear arms, the use of federal lands, air and water pollution, ownership of minerals lying on ocean beds, wartime

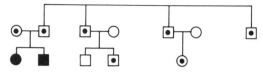

Figure 6.1. The transmission of thalassemia within three families where the fathers (squares) were brothers. In the first family, two children were anemic (solid symbols), and tests revealed that both the mother (circle) and the father had some abnormal hemoglobin—that is, both carried a mutant gene causing abnormal hemoglobin (shown as solid dot). Additional tests then revealed that two other brothers (uncles of the affected children) were also carriers of the mutant gene; in their cases, however, their wives carried only genes for normal hemoglobin. One child in each of the other two families had inherited the mutant gene from the father. A fourth brother (also a carrier of the mutant gene) was unmarried at the time of the study.

use of defoliants, federal support for remedial education and Head Start programs, and the introduction of Biblical theories into science curricula. There is not likely to be a shortage of corresponding topics in the future.)

Physical Malfunction: Genetic Diseases

Genetic material (DNA) serves a function completely unlike that of any other cellular substance. DNA carries instructions that tell each cell *how* to make every protein molecule the cell might need, and *when* to make each one. These proteins are of two sorts: *structural* proteins, which give each cell and its subcellular inclusions their characteristic shapes, and *enzymes*, protein catalysts that control each step of every metabolic reaction. The need for a different enzyme for each one of an enormous number of metabolic reactions demands a considerable length of DNA.

In an earlier chapter, we learned that the bulk of the DNA of human beings—like that of most plants and other animals—is located in small bodies called chromosomes, and that each individual possesses two sets of chromosomes, one set having been obtained from each parent. A result of having two sets of chromosomes rather than a single set is that, between the two, each individual is likely to possess at least one *correct* (or "normal," or "wild-type") gene at each of the thousands of gene loci. (Note that the introductory phrase "A result of" is the scientifically correct one; the phrase "The purpose for having . . ." would have been nonscientific.) During the replication of DNA, copying errors are made at infrequent intervals. Rare as these errors (or *mutations*) are, they accumulate with time—over numerous generations—so that any one set of chromosomes is likely to contain a number of recessive mutant alleles among the thousands of loci. The mutant alleles of either chromosome set—paternal or maternal—are so infrequent among all loci that only rarely does any individual carry two mutant alleles at the same locus. These rare individuals, however, are provided only with *incorrect* information by that locus; hence, they suffer a more or less debilitating genetic disease.

As an example of a genetic disease, we shall consider *thalassemia*, a disease in which hemoglobin molecules are improperly constructed. The disease is common among persons whose ancestors lived in the Mediterranean region: Italy, Sicily, Greece, and nearby countries.

Hemoglobin, as we have learned, is the protein that forms the bulk of each red blood cell; indeed, it is responsible for the red color of these cells. Oxygen is carried from the lungs to all cells of the body in combination with hemoglobin; carbon dioxide is picked up in the body's tissues and carried to the lungs mostly in blood plasma, but also in combination with hemoglobin. Interestingly, it is the union of oxygen with hemoglobin in the lungs that displaces carbon dioxide from the plasma and allows it to be exhaled; in the tissues, it is the loss of oxygen from hemoglobin that allows blood plasma to absorb carbon dioxide. Hemoglobin is clearly one of the body's essential constituents.

Too few red blood cells or too little hemoglobin in each cell causes diseases known collectively as *anemias*. Too little iron in one's diet, for example, limits the amount of hemoglobin each red blood cell can synthesize, since hemoglobin is an iron-containing substance. A lack of certain vitamins can lead to anemia, because without their cofactors, the enzymes leading to hemoglobin synthesis cannot function properly. A severe injury followed by considerable loss of blood can cause a temporary anemia. Finally, genes that specify the amino acid sequences in the protein portions of the hemoglobin molecule exist in many (usually very rare) mutant forms; the erroneously constructed (hemo)globin molecules may function improperly, thus giving rise to anemia.

In *thalassemia*, the blood-forming tissues of the bone marrow are too slow in manufacturing one of the two polypeptide chains that are eventually joined to make hemoglobin molecules. As a result, the red blood cells are thin, misshapen, and deficient in hemoglobin. White blood cells that normally check the red ones, destroying those that are abnormal in shape, work overtime in persons suffering from thalassemia: the badly formed red cells are rapidly destroyed, thus making a bad situation worse. The growth of thalassemic children is stunted by chronic ill health. Fetuses that have inherited mutant genes for thalassemia from both parents may die before birth; a child who inherits the defective gene from one parent only, however, may have normal health. There are actually many genetic diseases that go by the label *thalassemia*; sorting them out and determining the molecular basis of each disease is one of the occupations of molecular biologists who are interested in human genetics.

Another mutation that affects hemoglobin gives rise to *sickle cell anemia*. The effect of the mutation is well known: it substitutes the amino acid *valine* for *glutamic acid* at position 6 in a polypeptide chain (the β-chain) of 150 amino acids. The second polypeptide chain (the α-chain) of 141 amino acids is unaffected. The one amino acid substitution in a total of 291 sites gives rise to an abnormal hemoglobin that forms needlelike crystals when oxygen tension is lowered (as it is in all tissues of the body). Once more, the individuals who suffer greatly are those who have received one sickle cell mutation from each parent (*homozygotes*); individuals who carry one normal and one mutant allele (*heterozygotes*) are essentially normal in health. An accompanying figure lists some of the compensatory changes (dilated heart,

Figure 6.2. The distribution in Italy, Sicily, and Sardinia of the mutant gene that is responsible for thalassemia. The regions where the mutant gene reaches its highest frequencies coincide with the regions in which malaria was once prevalent. Indeed, it is now known that thalassemia and other inherited diseases of hemoglobin give their carriers a greater resistance to malaria than that shown by persons having only normal hemoglobin. *(After J. V. Neel)*

Table 6.1. The incidence of sickling red blood cells in various populations[a]

Population	Total	Number affected	Percent of total
British soldiers (in Africa)	568	0	0
Americans (White)	2621	1	0
Mexicans	239	0	0
Brazilian Indians	1379	0	0
Brazilian Negroes	1536	143	9.3
Colombian Negroes	489	46	9.4
Central American Negroes	3000	246	8.2
Americans (Black)	10,858	1154	10.6

[a] The test of British soldiers stationed in Africa was made to determine whether the sickling phenomenon was caused by an infectious agent.

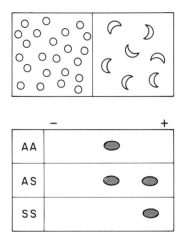

Figure 6.3. Schematic diagrams of normal and sickling red blood cells (top). The sickle shape is caused by crystalline accumulations of hemoglobin molecules at low oxygen tension; the hemoglobin needles distort the membrane surrounding the red blood cell. The lower diagram shows the migration of hemoglobins through a gel matrix in an electrical field. Under a convenient set of conditions (pH and salt concentration), normal hemoglobin moves toward the negative pole, whereas sickle-cell hemoglobin moves toward the positive one. Most persons (*AA*) and those individuals suffering from sickle-cell anemia (*SS*) possess but one type of hemoglobin; carriers of the sickling gene (*AS*), however, have both types. Each allele is responsible for specifying one type of protein molecule, right or wrong.

for example) the human body makes in a futile effort to undo the extensive damage that traces to this substitution of one of 291 amino acids in a single protein.

NOW, IT'S YOUR TURN (6-2)

1. In Chapter 2, you were asked to write in your journal an explanation for the question posed by Montaigne: "How did my father pass on to me an affliction from which he was to suffer only decades after my birth, and which is only now afflicting me?" The preceding paragraphs summarize the extent to which gene action has been (and will be) covered in this text; now is the time to review your reply to Montaigne, and to improve on it if you see inaccuracies or ambiguities.

2. Choose one disorder from the following list of genetic disorders, and (using references provided by your instructor) prepare for your journal a report in sufficient detail to convince your instructor that you understand (1) the role of genetic material (DNA) in "causing" hereditary diseases, (2) the basic biochemistry or physiology of the trait in question, and (3) the relationship between the biochemistry (or physiology) and the affected individual's phenotype (that is, his physical appearance or the clinical symptoms of disease).

Albinism	Tay-Sachs disease
Phenylketonuria	Alkaptonuria
Hemophilia	Favism
Porphyria	Gangliosidosis
Lactose intolerance	Maple syrup urine disease

Physical Malfunction: Improper Controls

Normal development and good health depend on the precisely controlled regulation of interrelated metabolic processes by nerves and hormones. Let any aspect of these controls fail, and the individual's development or physiology goes awry.

Literally hundreds of chemical reactions are sufficiently well known to serve as examples for the point to be made in this section, but as our example we have chosen glucose (blood sugar), one of the major nutrients for many forms of life. Glucose is distributed in the human body in blood by way of the circulatory system. Its concentration in blood is precisely controlled. Glucose enters the bloodstream through the intestinal wall and is carried to the cells and tissues, where it is used as a source of energy. If there is more glucose in the blood than is immediately needed, some is taken up by the liver and stored in the form of glycogen (a starch) or is converted into fat (by liver cells and,

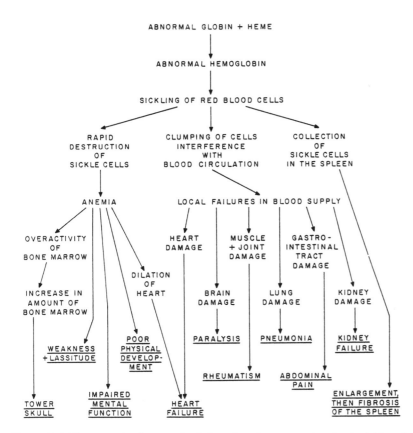

Figure 6.4. The cascade of events that follows from the possession of only sickle-cell hemoglobin. Some abnormalities stem from anemia; these would be characteristic of anyone suffering from a permanent deficiency of hemoglobin. Other abnormalities arise because of the tendency for misshapen red blood cells to become entangled and to clog small blood vessels. As you examine the diagram, be sure you understand the reason for each item listed. For example: What destroys sickle cells? *(Figure used with permission of J. V. Neel and The University of Chicago Press)*

to some extent, by fat cells), which is stored in fat cells. Later, when cells need glucose again, it can be drawn from these stores. The storage and withdrawal of glucose (and fat, when needed) is controlled through interactions of the nervous system and several endocrine glands.

One hormone, *insulin,* is manufactured by certain cells (the B cells) of the pancreas. B cells measure changes in the level of glucose in the blood. When the level of glucose rises, insulin is secreted and is carried in the bloodstream to all cells of the body. Insulin stimulates the uptake of glucose by the cells; it also stimulates liver cells to take up glucose and transform it into the storable starch, glycogen. After glucose has been taken up by body cells and changed into glycogen by the liver, the blood level of glucose falls once more. The B cells of the pancreas, in addition to being stimulated by changes in the level of

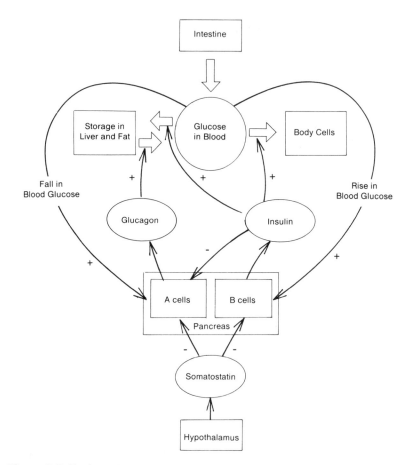

Figure 6.5. The interrelations of the hypothalamus, pancreas, and intestine (expressed through chemical messengers, *hormones*) in controlling the level of glucose (fuel) received by the body's cells by way of the bloodstream. Plus signs (+) indicate stimulation; minus signs (−), inhibition. An account of this feedback network is given in the text.

glucose in the blood, are stimulated directly by a hormone released by intestinal cells as ingested glucose passes through the intestinal wall and into the bloodstream.

Other pancreatic cells (the A cells) manufacture another hormone, *glucagon* (this hormone is also manufactured by cells in the walls of the stomach and intestine). The A cells, like their counterparts the B cells, also measure changes in the level of glucose in the blood. When the level falls, glucagon is secreted into the bloodstream and is carried to the liver and fat cells, where it stimulates the release of glucose, causing the level of glucose in the blood to rise.

The A and B cells of the pancreas work in opposition to one another in maintaining normal blood glucose levels. They are further coordinated in the sense that the glucagon of the A cells (and stomach and intestinal wall cells) stimulates the release of insulin by B cells;

once released, insulin tends to inhibit the release of glucagon by A cells. There are, of course, still other controls (nerves, hormones from the hypothalamus, pituitary, and adrenal glands) that work in regulating the level of blood sugar. Furthermore, each interaction that we have mentioned is accomplished by a half-dozen or more enzymatic steps that have been ignored here. (Similarly, one can say that hitting the A key of a typewriter results in an "A" being printed on the typing paper; this statement ignores the complex system of levers and electrical solenoids that is responsible for the proper working of the struck key.)

In many persons the control of blood glucose levels fails; this failure can lead to the disease *diabetes mellitus*. In this disease, blood glucose is too high, and is actually excreted by the kidney into the urine. The high blood glucose level reflects the abnormally low rate at which cells pick up glucose; to operate properly, then, the glucose-starved cells burn amino acids and fats. Eventually, blood vessels and nerve cells become damaged. Glucose is the *only* source of energy that can satisfy the brain and the cells constituting the retina of the eyes. Diabetes can cause blindness; in addition, death results if the disease is left untreated. The number of persons suffering from diabetes appears to be increasing: diabetes is the fifth most common cause of death in the United States.

For centuries physicians knew that diabetics have sweet (that is, sugar-containing) urine. Little else was known about the disease. During the early 1900s, however, two Canadian physiologists, William Banting and Charles Best, discovered the hormone *insulin*. The discovery of this hormone improved the treatment of diabetes tremendously. When insulin (prepared largely from the pancreases of slaughtered sheep or cattle) is injected into a diabetic, the blood sugar falls and many symptoms of the disease disappear. The life of diabetics is prolonged by insulin treatment.

If insulin injections aid those suffering from diabetes mellitus, does that mean that the disease is caused by the failure of pancreatic B cells to secret insulin? Not always. A lack of insulin is only one of many possible causes of the symptoms that are grouped under the name *diabetes mellitus*. The problem can lie with nearly any other part of the control system: with the supply of a hormone we have not yet mentioned, *somatostatin*; with abnormal A cells that secrete too much glucagon; or with B cells and A cells that can form hormones but that fail to react to the incoming signal provided by changes in the level of blood sugar.

Diabetes mellitus is becoming a common disease in the United States. No one knows why, however. Some persons think that insulin treatment has allowed persons carrying defective genes to reproduce, whereas previously they died before having children (this is a popu-

Figure 6.6. Persons of unusual appearance from the world of the circus. Rather than regarding these photographs with the awe and amazement of the usual side-show customer, try to determine the physical (including genetic) basis for each abnormal appearance. What might determine the elasticity of skin? How is the normal distribution of hair regulated? What malfunction of growth hormone might seriously affect body size? *(Photos by permission: Circus World Museum, Baraboo, Wisconsin)*

lation problem better discussed in Part III). Doctors and even nonprofessionals are now alert to the symptoms of diabetes, so the high frequency of its diagnosis may reflect the large number of persons who can recognize the disease. Some persons have suggested that viral diseases (of which people have more, now that bacterial diseases are largely controlled) may lead to glandular malfunction. Others prefer to point to modern sugar- and fat-rich diets, and to argue that the body's control system has been overwhelmed. Toxic chemicals in the environment have also been blamed. The more one learns about a disease, the more problems one encounters. This is the usual fate of scientists; it keeps them in business.

The greatest concentration of persons for whom the proper control of development had gone berserk were once found in circus side shows; they were commonly called freaks, although nearly all were decent, intelligent, and loving persons possessing both families and friends. With the help of references provided by your instructor, prepare an explanatory account for your journal of any one of the following old-time curiosities:

The Elephant Man	The Siamese Twins
The World's Tallest Man	The Lion-Faced Man
The Bearded Lady	India Rubber Man
Tom Thumb	Half Man–Half Woman

Physical Malfunction: Cancer

Cancer is the second most common cause of death in the United States today. It is responsible for months or years of pain, sickness, and anxiety, not only for those who are afflicted, but also for their families and friends. Cancer, however, is not a single disease. Many different ailments are called cancers because they share various features. In all such diseases, certain cells multiply out of control, without reference to the normal organization of tissues. These cells often lose their normal structure and the ability to carry out the reactions that are characteristic of normal tissue cells. For example, a cancerous brain cell possesses some, but not all, of the features of a normal brain cell. The same is true of skin cancers: the cancerous cells only partially resemble normal skin cells. Cancerous cells grow rapidly, out-competing other cells, drawing on the body's nutritional and metabolic resources, physically pressing upon their neighbors, and generally disrupting the normal structure and function of the affected tissue. Some cancer cells lose the normal cell's "sense of identity"; these cells become detached and are carried by lymph to distant parts of the body, where they start new, secondary growths (*metastases*). Brain cancers remain within the skull, where they invade nearby nerve and endocrine centers; death follows the destruction of vital control centers.

The origins of cancer are not understood; there may well be several origins. One important feature is the loss of normal genetic control over cell division. Because of the vulnerability of DNA to chemical damage (the substitution of one purine or pyrimidine for another) or physical damage (actual loss or gain of paired purine and pyrimidine bases), it is not surprising that chemicals (formaldehyde, mus-

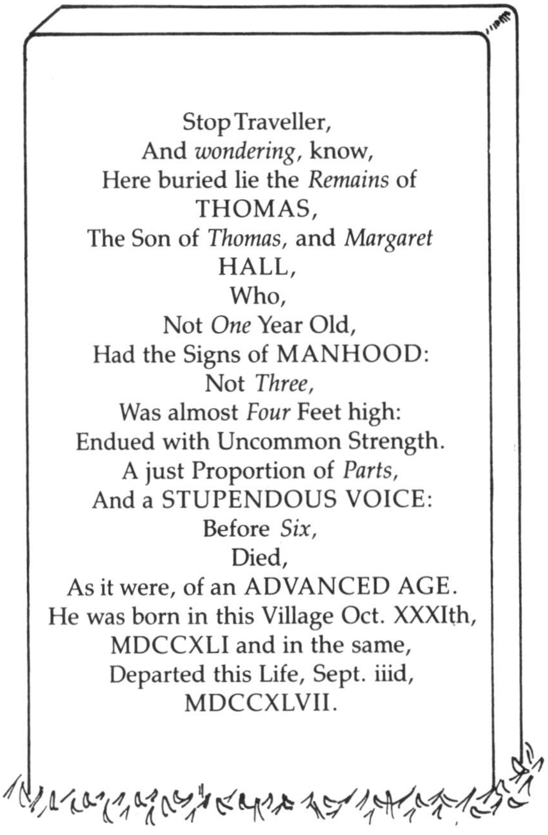

Stop Traveller,
And *wondering*, know,
Here buried lie the *Remains* of
THOMAS,
The Son of *Thomas*, and *Margaret*
HALL,
Who,
Not *One* Year Old,
Had the Signs of MANHOOD:
Not *Three*,
Was almost *Four* Feet high:
Endued with Uncommon Strength.
A just Proportion of *Parts*,
And a STUPENDOUS VOICE:
Before *Six*,
Died,
As it were, of an ADVANCED AGE.
He was born in this Village Oct. XXXIth,
MDCCXLI and in the same,
Departed this Life, Sept. iiid,
MDCCXLVII.

Figure 6.7. An epitaph found in a village cemetery in Britain. From other records it appears that, before dying, Thomas showed signs of old age (wrinkled skin and baldness) and advanced senility. Thomas suffered from a hormonal disorder (precocious virilism); with hints provided by the epitaph attempt to identify the source(s) of his abnormal characteristics.

Figure 6.8. County-by-county maps of the relative death rates for two common cancers: cancer of the bladder and cancer of the stomach. (Such maps are based on county medical records and death certificates. Because there are 3056 counties within the continental United States, the national map shows patterns with remarkable detail.) *Bladder cancers* tend to occur with high frequency in certain industrial areas (Northern New Jersey, Chicago, Detroit, and Los Angeles). Many of the industries in these areas manufacture or use organic chemicals. The workers in these industries are especially prone to cancer of the bladder. The bladder is often the primary site of cancer in persons who are heavy smokers. (Recall that organic chemicals, no matter how they are introduced into the body, pass through the kidney and are then stored in the bladder before passing from the body in urine.) *Cancer of the stomach* has a geographical distribution different from that of bladder cancer. Individual metropolitan areas are heavily shaded, thus confirming a commonly held view that this type of cancer is an affliction of the poor. In addition, high rates of stomach cancer are found in rural areas inhabited by people of certain nationalities: Russian, Austrian, Scandinavian, German, and Spanish-American.

The compilers of these maps (and many similar ones for other types of cancer) believe that the environment (by way of pollutants and diet) causes more than eighty percent of all cancers. *(Photos courtesy of T. J. Mason, National Cancer Institute; copyright 1975 by AAAS)*

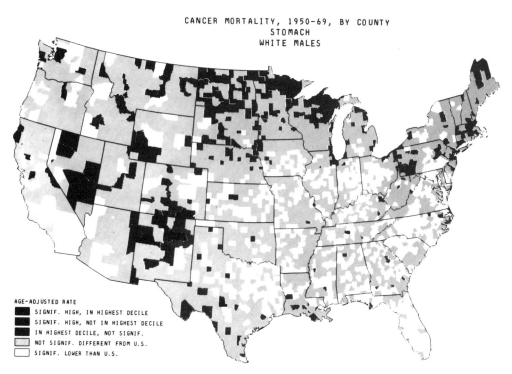

CANCER MORTALITY, 1950-69, BY COUNTY
BLADDER
WHITE MALES

AGE-ADJUSTED RATE
SIGNIF. HIGH, IN HIGHEST DECILE
SIGNIF. HIGH, NOT IN HIGHEST DECILE
IN HIGHEST DECILE, NOT SIGNIF.
NOT SIGNIF. DIFFERENT FROM U.S.
SIGNIF. LOWER THAN U.S.

CANCER MORTALITY, 1950-69, BY COUNTY
STOMACH
WHITE MALES

AGE-ADJUSTED RATE
SIGNIF. HIGH, IN HIGHEST DECILE
SIGNIF. HIGH, NOT IN HIGHEST DECILE
IN HIGHEST DECILE, NOT SIGNIF.
NOT SIGNIF. DIFFERENT FROM U.S.
SIGNIF. LOWER THAN U.S.

tard gas, vinyl chloride, epoxies, and various steroid compounds) and physical agents (ultraviolet light, X-rays, gamma rays, and other ionizing radiations) can cause cancer. The effect of these agents has been established for both human beings and experimental animals. There are cancers, however, that are known to be caused by viruses. Furthermore, since it is known that viruses can insert themselves (or DNA complements of themselves if they are RNA viruses) into chromosomal DNA, one need not be surprised to learn that some forms of cancer seem "to run in families." Individuals differ in their susceptibility to cancer and, furthermore, live under varied conditions. Nevertheless, recent statistics show clearly that cancers are most common where environmental pollutants that are known to cause cancer are highest.

Stressful Environment: Ultraviolet Radiation

Radiation is a special form of energy. The amount of energy that would warm a cup of tea only a few degrees, if delivered to the human body as penetrating radiation—X-rays or gamma rays—will cause death.

Radiation of any sort—visible light, ultraviolet radiation, X-rays, gamma rays, or other ionizing radiations—affects the human body for one reason: radiant energy is absorbed by cellular molecules, which are physically altered as a result. For instance, visible light acts on chemicals in the retinal cells of the eye. Because of the energy in the light rays, these chemicals undergo changes that activate nerves to the brain. If the messages reach the cerebral cortex, the person "sees the light." Meanwhile, back at the retina, the altered chemicals have already returned to their original state. This is a short-term chemical effect of radiation.

Ultraviolet rays are one of the physical causes of cancer; this fact emerged in our earlier discussion of *Xeroderma pigmentosum*. In addition to the damage it causes to DNA, ultraviolet light induces a darkening or tanning reaction in most people's skin. Tanning is an adaptive response of the cells in the skin evoked by exposure to a potentially damaging radiation; the brown pigment formed by the skin reduces the amount of ultraviolet light that penetrates to the deeper, still-reproducing (i.e., dividing) cells of the skin. Most suntan lotions also absorb ultraviolet radiation, thus shielding the skin from full exposure.

The ability of different persons to respond to the ultraviolet radiation in sunlight varies: some (notably Anglo-Saxons and others from Northern Europe) scarcely respond, other than turning beet-red and forming freckles; other persons of European descent (as well as Japanese and Chinese) become dark-skinned during the summer but are

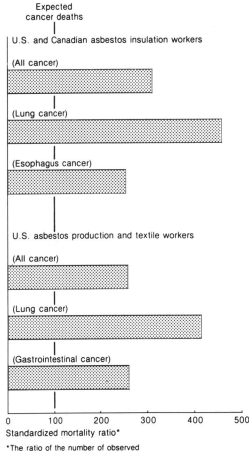

Figure 6.9. The increased incidence of cancers among persons working with asbestos, whether in its production, in installing it for insulation, or in manufacturing various textiles.

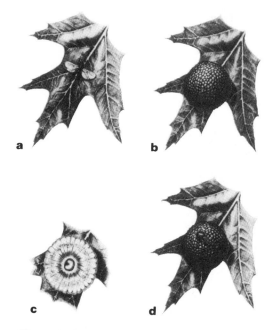

a   b

c   d

Figure 6.10. Like animals, plants can suffer uncontrolled, disorganized cellular growths (cancers). As these pictures show, some insects have the ability to induce reorganized growth of plant cells—galls. The gall wasp (*a*) lays an egg in the vein of an oak leaf; chemicals introduced with the egg cause the plant's tissue cells to divide abnormally, producing the gall (*b*), within which the developing wasp larva feeds and grows (*c*). Eventually, an adult wasp emerges from the gall (*d*). If one fails to observe the initial egg-laying, one might (as some early biologists did) conclude that wasps arise from galls by spontaneous generation. To cite a second example of tumor induction, a gall-producing bacterium actually inserts part of its DNA into the plant's chromosome, thereby taking over the course of development of the plant cells within the resulting gall. *(Modified with permission of The Biological Sciences Curriculum Study)*

otherwise quite pale; still other persons (mainly from tropical regions of the world) have perpetually dark skins. The genetic control of tanning, to repeat an earlier account, is analogous to the genetic control of lactose utilization by bacteria: some bacteria are unable to use lactose under any circumstance; others use it only if (1) lactose is available and (2) glucose is absent; still others are enzymatically able to metabolize lactose at all times, whether or not lactose is present in the growth medium.

Individual variation in the nature of control systems often reflects the presence of benefits the organism can gain through choosing one or the other of conflicting possibilities. In the case of lactose-utilizing bacteria, energy would be wasted synthesizing the enzymes for lactose metabolism if the sugar were not available for use. On the other hand, if lactose is available, bacteria that are unable to use it for growth and reproduction may be overrun and displaced by those that can. In the case of tanning, the damage that ultraviolet light causes to DNA is counterbalanced by its ability to stimulate the cells of the skin to secrete vitamin D. In this role, skin is an endocrine gland and vitamin D a hormone. Tanning reduces the quantity of vitamin D that is produced. Thus, in the far north, where for many months of the year there is little sunlight, obtaining an adequate supply of vitamin D poses a greater problem than does skin cancer. Farther south, tanning is adjusted according to the amount of sunshine each person receives. Finally, in the tropics, the production of sufficient vitamin D by the skin is not as great a problem as is the possibility of acquiring an ultraviolet-induced skin cancer.

Stressful Environment: Limits to Endurance

The limits to people's ability to endure stressful environments are largely unknown. The ability varies from one person to another, and from one situation to the next. Here are some of the known facts:

• The sensitivity of persons varies with age; the very young (including unborn fetuses) and the very old suffer most from toxic chemicals and radiation exposure.

• Any disease or physical stress increases a person's sensitivity to noxious chemicals.

• Two or more toxic substances, acting together, may be far more dangerous than either acting alone (they have *synergistic* effects).

• Radiation and toxic chemicals also interact, each making the other more dangerous.

• The effects of either radiation or toxic substances may not be seen for years, but this does not mean that the effects, when seen, will be slight.

Individual persons are not standardized like automobiles, vacuum cleaners, and toasters. They differ genetically. They differ in age. They differ in their individual histories of past illnesses and accidents. And they differ in their physical surroundings. As a consequence, a multitude of suggestions for, and reactions to, the control of environmental pollution have been made. Some of these suggestions are made on the basis of economic greed: one industry-based medic has suggested publicly that it might be less expensive to provide the aged and those suffering from respiratory illness with air-conditioned housing than to prevent air pollution. Still other suggestions are based on a poor sense of the future: if I am not hurt by exposure today, then I have not been hurt at all. This attitude led one New York State governor to volunteer to work briefly in a state office building found to be contaminated with PCBs (polychlorinated biphenyls), known cancer-causing agents. Perhaps one should not be surprised that most persons whose positions require that they compute risks for risk-benefit analyses do not live in those neighborhoods where residents are exposed to the risks: the ones who perform the calculations reap the benefits; others assume the risks. The hazards, that is, do not often test the limits of endurance for those who create them, but instead test those of other people, including innocent bystanders.

NOW, IT'S YOUR TURN (6-4)

The preceding section mentioned a New York governor who offered to work temporarily in a PCB-contaminated office building, in order to demonstrate, apparently, that PCBs are not dangerous. (He subsequently apologized for his foolishness!) To prove that the transport of nerve gas through Georgia was not dangerous, a former governor of that state offered to ride atop one of the loaded tank cars. A U.S. Ambassador to Spain, in order to prove that an accidentally dropped A-bomb had not contaminated one of Spain's resort beaches, took his wife and children swimming in the endangered water. One cabinet officer took his wife and children to a small Aleutian island to witness an underground nuclear blast. Each of these acts gives rise to *anecdotal data:* that the ambassador, cabinet officer, and their family members were unaffected or that, had they carried out their proposals, the two governors might not have been harmed, proves nothing. Nothing is proven either by anecdotal data concerning intelligence, race, or the poor; anecdotes can be found to bolster either side of any controversy.

Watch for the use of anecdotal data in either a national or local controversy (political figures, such as presidents, congressmen, and mayors, frequently cite them). Point out the flaws in the example you choose to cite, and describe data that you believe would be more accurate and, therefore, more trustworthy.

How the Body Reacts

Fortunately, the body has defenses that permit it to deal with many of the problems it encounters. (Only problems that are out of control are classified as diseases: genetic, bacterial, viral, or environmental.) Many problems lead to the same *symptoms*—fever, cough, and skin rash—and, indeed, symptoms are reactions of a body in trouble. Symptoms are the signs of a body defending itself.

The facets of self-defense are many, wide-ranging, and powerful. Many persons survive to old age with little or no "outside" help; this has always been so. Today's high life expectancy has been achieved by lowering the death rate of infants, not by lengthening the life of the elderly: threescore years and ten is still descriptive of old age in the United States, as it is among Australian aborigines. Defense mechanisms extend from the molecular and cellular levels to alterations in gross behavior. Enzymes, for example, degrade (digest) many foreign organic chemicals that enter the body; these chemical reactions take place largely in the liver. Should the amount of the foreign chemicals increase, the concentrations of enzymes also increase. Certain enzymes metabolize alcohol. The amount of these enzymes increases in the livers of those who drink regularly; as a result, confirmed drinkers must imbibe more and more in order to attain the desired level of intoxication.

Antibodies, complex protein molecules that are manufactured by specialized white blood cells (plasma cells), serve as another powerful molecular defense. Any individual human being (or other vertebrate) can form hundreds or thousands of different antibodies during a lifetime. The collection of antibodies carried by each individual depends upon his or her lifetime experiences with respect to infections and other invasions by foreign substances (by blood transfusions, for example). No two individuals possess exactly the same antibodies.

Antibodies are not made in large quantities in advance, merely waiting for the proper foreign substance to come along. They are actually made in quantity only in response to the presence of the foreign substance. The segment of DNA that specifies the antibody protein is unstable: several repeated loci are involved, each of which, by means of internal inversions of small segments of DNA, can assume a variety of purine and pyrimidine sequences. Each plasma cell, instead of producing an antibody identical to that of all other plasma cells (in the way, for example, that an individual's red blood cells come to have

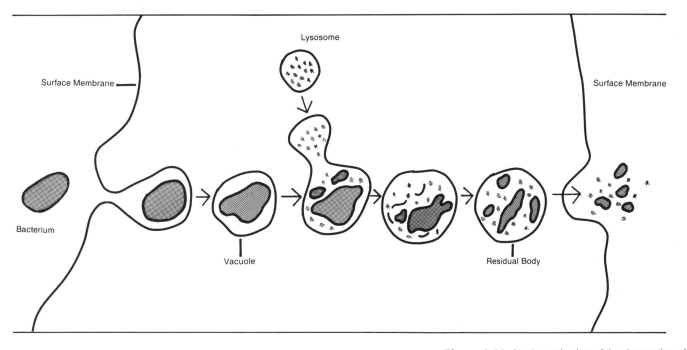

Bacterium

Surface Membrane

Lysosome

Surface Membrane

Vacuole

Residual Body

Figure 6.11. A schematic view of the destruction of a bacterium by a white blood cell. A *lysosome* is a small package of proteolytic enzymes. When its contents are emptied into the vacuole containing the engulfed bacterium, the bacterial wall and its contents are destroyed. The remnants are then ejected from the cell. In a similar process, a lymphocyte (B cell) encountering a potentially cancerous cell can attach to its surface momentarily and inject lysosomes into the suspected cell, which dies within a short time.

identical hemoglobin molecules), produces its own private version of antibody. Among the billions of plasma cells of the body, a number will have antibodies that are likely to react with nearly any conceivable chemical. Once the antibody on the surface of a cell has interacted with a foreign substance, that cell is stimulated to divide. Its daughter cells divide repeatedly, forming hundreds or thousands of cells, each of which synthesizes and secretes the *same* antibody—the one capable of reacting with the foreign substance that initiated the division of the original cell by reacting with its private antibody.

Antibodies provide protection in several ways. They "coat" the foreign substance, the *antigen*. Such antibody–antigen combinations are a signal that something is wrong; they are ingested by white blood cells that digest the entire complex, including the invading substances. The common feature of the foreign substances is that they are *alien*: they are not *self*. Proteins and other substances that are normally found in circulation do not invoke antibody build-up.

Still other systems of molecular defense exist. For example, cells secrete special proteins, *interferons*, that block the reproduction of viruses. Specialized cells (phagocytic cells) found in the bloodstream, in the lymphatic system, in the liver, and in intercellular tissue spaces ingest and destroy not only foreign subtances but also damaged, abnormal, or dying cells of the body itself.

What to do when things go wrong? Let's review some of the body's

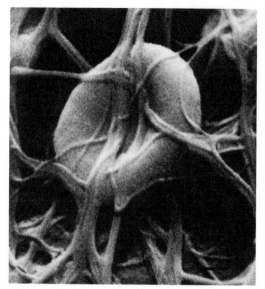

Figure 6.12. An electron micrograph showing a red blood cell caught within a mesh of fibrin threads in a blood clot. Most people take blood clotting for granted: blood clots form when one is cut, they do not form otherwise. However, blood clots that form within the circulatory system are life-threatening. About ten steps are required to convert circulating blood proteins into fibrin threads; these steps are set in motion by trauma—the destruction of cells in the blood vessel wall. Because each step involves the interaction of protein factors, each step can go wrong as the result of gene mutations. Persons who lack clotting factors or who possess nonfunctioning factors of any sort are said to suffer from *hemophilia*. Unlike other people, hemophiliacs do not take blood clotting for granted; even a small wound may be life-threatening. *(Courtesy E. Bernstein and E. Kairinen, Gillette Research Institute, Rockville, Maryland. Copyright 1971 by the AAAS)*

reactions to a minor but nontrivial gash you have sustained in an automobile accident. As you read the text, ask yourself how many of the defense reactions you would remember to mobilize if you had to control each step consciously. (Keep in mind Dr. Lewis Thomas's statement, cited earlier, that he would not want to take control of his own liver's decision-making responsibility.)

The accident results in a gashed hand. Blood flows from the cut, but it soon stops because of changes in the blood that lead to clotting. Pain reminds you to hold your injured hand carefully, protecting it from further injury.

In response to chemicals released by damaged tissue cells, where the skin is broken, bacteria can enter the body. White blood cells, including phagocytes, also move into the wounded area. These cells release proteins that tend to fix invading organisms in place, thus preventing an infection of the entire body (*septicemia*, or blood poisoning). They also release enzymes that destroy invading bacteria. Phagocytic cells clean up damaged tissue in the wounded area.

The chemicals released by damaged cells also stimulate the flow of blood, bringing the oxygen and nutrients that are needed both for energy and for the construction of new structural proteins. Indeed, immediately after an injury, repair processes start. New tissue fibers are laid down. Skin and other cells divide and move in, restoring both the cut skin and the damaged tissues. Your wound, in all likelihood, will be healed within days of the accident. Damaged tissue has been repaired; infection by bacteria and viruses prevented. The accident is a thing of the past. Or is it?

Despite appearances, the consequences of the accident will linger on. Stimulated by hormones, blood-forming tissues increase their activity until blood lost during the accident is replaced. Plasma cells will continue for some time to make antibodies against the kinds of bacteria that entered the wound. A scar will remain on your hand for the rest of your life. If you learned from the accident, you will drive more carefully in the future. Reactions to the injury, then, involve short-, intermediate-, and long-term responses at all levels of molecular, cellular, and tissue structure, and at the level of behavior as well. Except for the matter of behavior, all of these reactions are controlled immediately by chemical messengers and nerve impulses, and ultimately by the complement of DNA you received from your parents—and they from theirs, and so on for generations throughout millions of years.

NOW, IT'S YOUR TURN (6-5)

Women who have never contracted German measles but are exposed to them during early pregnancy are often given shots of gamma-globulin that has been obtained from pooled blood serum drawn from many individuals. Do you know why the exposure of pregnant women to German measles causes concern? Why do you think many doctors suggest that young girls be deliberately exposed to German measles in their preteen years? Recall that some physicians recommend abortion if the pregnant woman *contracts* German measles; explain the possible role of gamma-globulin shots for women who have been *exposed*, but who have not yet contracted measles. Why are the shots prepared from pooled serum?

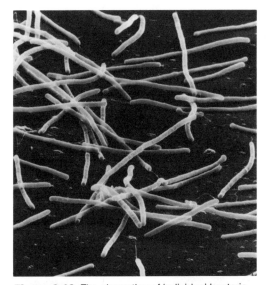

Figure 6.13. The elongation of individual bacteria, *Pseudomonas aeruginosa*, after a two-hour exposure to an experimental antibiotic. Normally, the length of these bacteria is approximately twice their diameter. This bacterium, the green-pus bacillus, can cause ear and mastoid infections and septicemia (blood poisoning). The elongation of cells that are exposed to X-rays and antibiotics appears to be a defensive reaction associated with the SOS mechanism of DNA repair (see Figure 2.32). Cell division is suppressed. Within the elongated cell, DNA replication and recombination are occurring. As a consequence, the probability that an intact (and genetically correct) bacterial chromosome will be included within a normal-sized bacterium when cell division resumes once more is greatly increased. *(Photo courtesy of M. E. Bayer, M.D., Institute for Cancer Research)*

Summary

Human health is constantly endangered by (1) abnormal functioning of the body itself, (2) unfavorable external environments, and (3) attack by other organisms. In combination, these threats are often more serious than would be expected from a knowledge of any one alone. Because human beings are complicated organisms, no two being alike, the causes of disease may be hard to identify. Human beings, in common with other higher forms of life, possess many life-preserving mechanisms. These include enzymes, antibodies, and interferons; protective cells among the white blood cells; the capability for self-repair (although human beings cannot *regenerate* lost limbs, as salamanders can); and patterns of response involving all body systems and coordinated by nerves and endocrine glands.

INFORMATIONAL ESSAY
Antibiotics

From accounts presented in the text, it may seem that the living cell functions only because of information that is encoded in DNA, that is transcribed (in RNA), and that is translated (into enzymes and other proteins) when and where needed. Transcription and translation need to be precise, of course, but in this imperfect world things do go wrong, and cellular machinery has the leeway and "wobble" needed for making the best of less than perfect circumstances.

When things go wrong, especially when an individual goes to the doctor and is found to have an infection, chemicals known as *antibiotics* are often prescribed. Antibiotics have been called "miracle drugs." When they first appeared on the market, seemingly fatal infections were instead cured, often overnight. The effectiveness of antibiotics suggests that they have extremely precise modes of action; to borrow a gross anatomical phrase, antibiotics seem to "go for the jugular vein" on a molecular level.

Indeed, at the molecular level there are many vulnerable points,

Table 6.2. The modes of action of some commonly used antibiotics. The precise modes of action of these "miracle" drugs qualify each of them for the old medical dream title: magic bullet.

Mode of action	Drug
Inhibition of cell-wall synthesis	Penicillin
Inhibition of DNA synthesis	Macromycin
	Tubercidin
Inhibition of mitochondrial action	Rotenone (an insecticide)
Inhibition of protein synthesis	Chloramphenicol
	Colicins
	Cycloheximide
	Erythromycin
	Neomycin
	Primaquine
	Puromycin
	Streptomycin
	Tetracyclines
Inhibition of RNA synthesis	Actinomycin
	Ethidium bromide

molecular "jugular veins." *Penicillin*, for example, interferes with cell-wall synthesis in bacteria, a step that effectively blocks their reproduction. Microbial biologists make use of this effect in searching for organisms sensitive to the effect of penicillin. They add both penicillin and a poison that kills growing cells to the nutrient medium in a cell culture. All cells resistant to penicillin continue to grow in the culture—and are killed by the poison. By diluting the poison, and replating the cells on normal medium, the microbiologist obtains colonies of bacteria whose parents were sensitive to, and hence unable to grow in the presence of, penicillin. Mammalian cells do not have cell walls in the sense that bacterial cells do; consequently, penicillin is a splendid antibiotic for use with human infections, as well as with those of our pets and our work animals. Bacteria are generally sensitive to penicillin; we are more or less immune to it.

The same "immunity" does not apply so readily to those antibiotics that act by suppressing protein synthesis. *Chloramphenicol*, for example, is an extremely powerful antibiotic that stops protein synthesis. Unfortunately, when a person uses it for any length of time, it stops that person's protein synthesis, too. An alarming side effect of chloramphenicol treatment is the destruction of the marrow (red blood cell–producing tissue) in long bones. A second effect—an increase in cancer among chloramphenicol-treated patients—might stem from the antibiotic's interference with both protein synthesis and the synthesis of short polypeptide chains: these short chains are often the chemical signals that control gene action.

Actinomycin and *ethidium bromide* are antibiotics that interfere with the transcription of RNA. Actinomycin, for example, slips between adjacent guanine-cytosine pairs and then reacts with the G-C and C-G nucleotide pairs it encounters on either side of its newly found position.

Macromycin, one of the antibiotics frequently used in the treatment of bladder infections, inhibits DNA synthesis. *Rotenone*, an insecticide rather than an antibiotic, acts by inhibiting the normal function of mitochondria.

Many of the substances that are effective in killing bacteria are normal products of fungi. Penicillin is synthesized by certain *Penicillium* species. Chloramphenicol, erythromycin, neomycin, streptomycin, and the tetracyclines are synthesized by various species of *Streptomyces*, a common soil fungus. Presumably, antibiotics are part of the armament that fungi use to protect their nutritional resources from use by bacteria. On the other hand, if pure antibiotics are added to soil they are quickly degraded.

To the chagrin of clinicians, bacteria have evolved defenses against antibiotics. Some of these defenses—genetic defenses encoded in DNA—spread extremely rapidly through various species of bacteria,

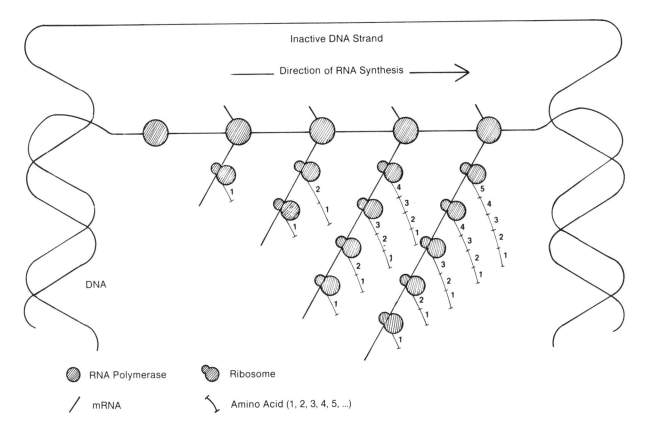

Inactive DNA Strand

Direction of RNA Synthesis

DNA

RNA Polymerase   Ribosome

mRNA   Amino Acid (1, 2, 3, 4, 5, ...)

Figure 6.14. A generalized account of the transcription of mRNA from one strand of DNA in a bacterium, the attachment of ribosomes to the mRNA strand, and the translation of the information carried by mRNA into the amino-acid sequence of growing polypeptide chains. This scheme illustrates points at which things might go wrong for a bacterium; these points are the targets for antibiotics and other medicines. Note that each growing polypeptide chain begins with amino acid 1 (actually, methionine) after which 2, 3, 4, 5, and still others are added in sequence.

especially those inhabiting the colon. The genes that confer resistance to antibiotics have proven to be located not on the chromosomes of the bacteria, but on small circular pieces of DNA that (1) control their own (rapid) rate of replication and (2) can pass from one individual bacterium to another without either individual reproducing. The latter process is *infective resistance*. At one time, scientists wondered how these circular pieces of DNA (*plasmids*) came to possess only the DNA needed to specify the enzymes conferring resistance. That mystery was solved when plasmids were discovered in stored bacterial cultures that predated the earliest human use of antibiotics. Bacteria have been squabbling over nutritional resources with fungi for millions of years, fungi have had time to invent antibiotics, and bacteria have developed and perfected their countermeasures. Human beings and their physicians have only recently arrived on the battle scene.

INFORMATIONAL ESSAY
Immunoglobulins (Antibodies)

The human body (and that of any other vertebrate) can form antibodies against virtually any large alien molecule that invades the body or is injected into the bloodstream by accident. These large molecules include those that make up the outer membranes of viruses, bacteria, and other microorganisms. The antibodies are to be found among the globulins of the blood serum, specifically among those proteins known as gamma-globulins.

As are other proteins, antibodies are assembled under directions carried ultimately by DNA, but which are transcribed into mRNA for actual translation into the proteins that are the antibodies. The cells in which antibodies are synthesized are known as plasma cells—cells descended from bone marrow lymphocytes.

The synthesis of antibodies has in the past posed problems for both immunologists and geneticists. First, they are extremely specific with respect to the antigen (the foreign molecule) with which they will combine to form a precipitate. For example, blood serum from animals into which antigens X and Y have been injected can be treated with X, which will remove anti-X antibodies while leaving anti-Y antibodies unaffected, or it can be treated with Y, which will remove anti-Y antibodies while leaving anti-X antibodies unaffected.

Despite their specificity, the structure of antibodies—like that of all proteins—is specified by genetic information residing in DNA. For each antigen there is a gene whose purine and pyrimidine bases spell out the corresponding sequence of amino acids in an appropriate antibody. One method by which the specificity could be accomplished would be to have a gene—that is, a segment of DNA—corresponding to every known antibody, a gene capable of making an antibody for every conceivable antigen (of which there are thousands). Because it is difficult to imagine genes making antibodies against antigens not yet encountered in their organism's evolutionary history, one would imagine that during the course of evolution all sorts of chemicals had been encountered at one time or another, and the successful production of an antibody had, in each case, been recorded as one more gene in the immense battery of antibody genes.

However, if there were so many antibody genes, the problem of activating them would still remain. Were there lymphocytes in which this one or that one of each of the thousands of genes was activated? If so, these cells would lie in wait for the appearance of the proper antigen, and then react by dividing repeatedly and synthesizing the proper antibody. Alternatively, the chemical antigen, having touched one lymphocyte, might in that way cause the lymphocyte to identify the proper gene among the thousands it possessed, and to call upon this gene to start transcribing mRNA and synthesizing antibody. In this case the antibody would be made in response to a specific antigen. All lymphocytes in this case would be similar, but each would possess the ability to respond to whatever antigen it encountered; in the earlier scheme, there would be thousands of different types of lymphocytes.

The correct answer seems to be as strange as the two incorrect guesses we have proposed here. Generally, antibodies are composed of four polypeptide chains, two heavy (long) chains and two light ones. Within the genome, there are several duplicated loci capable of making heavy chains, and several capable of making light ones. Within each of these, there is a region that can be occupied by a variety of amino acids.

How can a segment of DNA give rise to different sequences of amino acids? The extreme case—one found in a few viral genes—is a shift in the reading of the mRNA molecule. Suppose a segment of mRNA has the following sequences of purine and pyrimidine bases:

$$*A\underset{\wedge}{-}U-U-G-G-C-A-G-U-C-U-A-U$$

Starting at *, this section would specify

Ile·Gly·Ser·Leu·

whereas starting at the caret produces

Leu·Ala·Val·Tyr·.

Viruses (and to a somewhat lesser extent, bacteria) have managed to use the same stretch of DNA to specify two different proteins with two different functions by the "simple" device of shifting the reading frame.

Another means by which the same DNA can be made to produce two dissimilar proteins is by rotating a section so that it reads one way in some cells, and another way in others. For example, consider the following section of DNA:

A T:C C A A A G G:A T C
T A:G G T T T C C:T A G

If the section between the dotted lines were excised and rotated 180°, the DNA would read

A-T-C-C-T-T-T-G-G-A-T-C
T-A-G-G-A-A-A-C-C-T-A-G.

mRNA transcribed from the lower strand of DNA in the *original* diagram would read

A U C C A A A G G A U C

and the sequence of amino acids it specifies would be

Ile·Gln·Arg·Ile.

mRNA transcribed from the lower strand of DNA in the *rearranged* lower diagram would be

A-U-C-C-U-U-U-G-G-A-U-C.

The corresponding sequence of amino acids in this case would be

Ile·Leu·Trp·Ile.

Instead of Gln and Arg, the lower polypeptide contains the two amino acids Leu and Trp.

The variation in antibodies can be traced to manipulations of DNA that were undreamed of even twenty years ago. Corresponding but dissimilar parts of different DNA segments can be brought together and then transcribed as if they were a single gene in the "classical" sense. Segments of DNA can be flipped over so that the DNA strand that is transcribed can take on a variety of base compositions. If these maneuvers (one is tempted to say, "shenanigans") are performed willy-nilly in every one of millions of lymphocytes, then millions of different antibody molecules can be constructed. Here, then, is the source of the many different antibodies.

Antibody proteins protrude from the surface of the lymphocytes that make them. Now, suppose one of these cells encounters an antigen that interacts with (becomes attached to) the surface antibody. This cell divides and redivides to form a large number of daughter cells, each of which secretes large numbers of antibody molecules— all identical, all capable of attaching to the type of antigen originally sensed by the parent cell.

Our molecular knowledge of antibodies has been aided by persons suffering from a blood cancer, *myeloma*. Myeloma consists of the uncontrolled multiplication of plasma cells—but not *all* plasma cells. One particular cell divides in an uncontrolled fashion, and all its daughter cells secrete the type of antibody specified by the one antibody gene. The concentration of this protein in the bloodstream becomes so high that it is excreted in the afflicted person's urine. All of this protein (called Bence-Jones protein) secreted by one person has the same amino acid composition. The Bence-Jones proteins of different myeloma patients have different amino acid sequences in the variable regions of the light and heavy chains (hence the term *variable* region); otherwise, the antibody molecules are constant in composition.

As an incidental comment, we might note that myeloma patients come to possess more and more lymphocytes that produce but one type of antibody. Consequently, they become more and more susceptible to bacterial infections. A minor operation, such as the removal of a small skin growth, may prove to be a major challenge to someone who has myeloma. Infections occuring during what should be a routine healing are sometimes controlled only by a continuous application of antibiotics. Much worse, of course, is the fate of infants who, because of genetic defects, are unable to synthesize *any* antibodies. These babies, who generally have very brief lives, are said to be *agammaglobulinemic.*

Acquired Immune Deficiency Syndrome (AIDS)

During 1981, hospitals on both the East and West Coast recorded an increasing incidence of an otherwise rare cancer, Kaposi's sarcoma, among young male homosexuals. This cancer is rare because it occurs primarily among organ-transplant patients who have undergone treatment designed to suppress the immune system. Without such suppression, transplanted organs would be promptly rejected by the recipient's blood.

Kaposi's sarcoma was only one affliction noted among the homosexuals who were arriving at hospitals for treatment. Many suffered from an unusual pneumonia caused by *Pneumocystis carinii.* Others suffered from yeast (*Candida*) infections, or thrush, of both the mouth and rectum.

The organisms involved in these cases are considered *opportunistic:* they lurk, threatening all persons with infection, but actually invade or infect only persons whose immune system has been impaired. With increasing numbers of similarly afflicted persons being reported, the designation AIDS (Acquired Immune Deficiency Syndrome) was assigned to cases exhibiting the above symptoms and meeting several other defined diagnostic criteria. This discussion of AIDS deals with aspects of the human and mammalian immune systems that are relevant to common features of AIDS.

THE IMMUNE SYSTEM

If a foreign substance of considerable size (molecular weight of 8000 or more) finds its way into one's bloodstream *for the first time,* it will eventually encounter white blood cells (lymphocytes) whose surface molecules (proteins) will interact with it. This interaction stimulates the reacting cells to proliferate rapidly, forming enormous numbers of cells of that particular type. Should the excited, interactive cells be of one type (B lymphocytes), large quantities of the surface proteins (antibodies) are released into the blood serum; if they are of another type (T lymphocytes), a large number of similar interactive cells are produced. Either way, the novel foreign substance becomes attached either to free antibodies or to interactive cells and is removed from the bloodstream. These particular antibodies (and corresponding T cells) may remain in the body, in circulation, for years, ready to react immediately should the foreign substance reappear.

How can there be a cell or cells capable of interacting with *any* in-

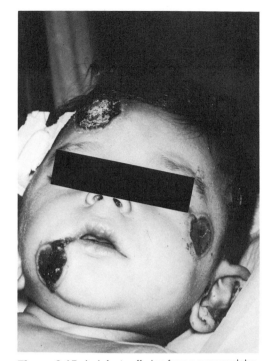

Figure 6.15. An infant suffering from agammaglobulinemia, a lack of antibodies. The ulcers that are visible on the baby's chin, forehead, and temple are caused by attenuated bacteria in what should have been a routine preventive vaccination. In this case, the vaccination proved to be fatal. (*Courtesy of Vincent A. Fulginiti, M.D., from D. Bergsma (ed.):* Birth Defects Compendium. *New York: Alan R. Liss for the National Foundation—March of Dimes, 1979)*

vading molecule? At one time, persons thought that during the millions of years of evolution, every conceivable substance had entered the bodies of primitive animals, and a gene capable of making an appropriate antibody had arisen by chance and had been preserved by natural selection. Such a system would, of course, demand an enormous number of antigen-producing genes. This notion has been discarded. It appears, instead, that the DNA comprising the few loci that are responsible for producing antibody proteins can, during the earlier development of lymphocytes, be shuffled, rearranged, inverted, and otherwise altered in different cell lines. Thus, among the enormous number of lymphocytes in the human body (about 2000–2500 per cubic millimeter or $1.0–1.5 \times 10^{10}$ in all) there are many (100,000 or more) different clones of cells, whose members carry identical antigenic surface proteins. The reaction of a foreign substance with one or the other of these many potential antibodies is a matter of chance—chance made greater, however, by sheer numbers.

SELF AND NONSELF

The surface proteins of lymphocytes are capable of recognizing virtually any foreign substance. What, then, prevents those substances (including protein) that normally occur in a person's bloodstream from triggering an antigenic reaction? How do lymphocytes recognize one's "self"?

The answer to this question lies in the "processing" that lymphocytes undergo in the thymus gland or other regions (fetal liver and, perhaps, bone marrow) shortly before birth. If a foreign substance is injected into the fetus of a laboratory animal at this time, throughout its life that individual will not form antibodies against that particular substance. On the other hand, removal of the thymus gland before the processing period prevents the formation of antibodies in later life. Thus, one can imagine that during processing, lymphocytes arrive at the thymus, some with reactive compounds on their surface and others without. Those with reactive compounds are destroyed (or suppressed in some unknown manner), thus ridding the body of lymphocytes whose surface proteins react with those proteins and substances characterizing the fetus itself. The remaining lymphocytes, those that had not reacted with circulating substances, are "cocked" during processing and, like cocked pistols, are ready to "fire" at any subsequently encountered foreign substance.

SUPPRESSION OF THE IMMUNE REACTION

The chain of events that is set in motion within the immune system by the presence of a foreign substance involves the rapid and sustained division and redivision of those lymphocytes whose surface antigens react with that substance. Cells that divide repeatedly, however, are dangerous to life; under many circumstances, they would be considered cancerous. Consequently, the immune system requires a control, a means for turning itself off. This turning off is also accomplished by lymphocytes. Thus, there are two sets of lymphocytes: those that augment the production of antigens and sensitized cells, and those that suppress the immune reaction. The normal course of events includes, first, production of an antibody (circulating or bound on cell surfaces) and, second, the suppression of these protective procedures.

AIDS victims suffer from "opportunistic" infections or from cancer of a sort that occurs in organ-transplant patients because they lack the normal protection provided by the body's lymphocytes. Blood tests performed on AIDS victims reveal that the cells that suppress the immune reaction outnumber those that enhance the formation of antigens.

The cause of AIDS is not yet known with certainty. Those persons who initially contracted the disease included extremely sexually active homosexual men, hemophiliacs, and Haitian immigrants who arrived in the United States in small, crowded boats and were subsequently detained in prison camps. More recently, it appears that AIDS can be spread by normal, heterosexual contact. Babies have been born with AIDS.

The prime suspect for the cause of AIDS at the moment is a virus. Some persons have suggested that this virus is endemic in a natural reservoir of mammals (green monkeys, perhaps) in Africa and that a mutant strain has only recently entered the human population (in a similar way, many strains of flu virus are thought to arise from viruses that infect pigs in Asia). In human beings, the virus infects and destroys those T cells ("helper" cells) that coordinate the production of antibodies by B cells. Thus, although the overall production of antibodies in AIDS victims may be raised, these antibodies are not aimed at any particular antigen—they are *poly* (= many) *clonal*, rather than *mono* (= one) *clonal* in origin. As a result, protozoa, yeasts, fungi, viruses, and other opportunistic organisms can invade the body and lead to pneumonia, severe intestinal disorders, and brain damage. Cancers and tumors that are caused by viruses also thrive because they are not controlled by normal immune defense mechanisms. Kaposi's sarcoma, you may recall, is such a cancer; it frequently arises in recipients of organ transplants whose immune system has been suppressed in order that the transplanted organ not be rejected by the recipient.

A recent report, however, recalls the initial observation that AIDS was common among homosexual men. It appears that one of the naturally occurring antibodies, one that occurs without being invoked by a foreign substance, does react with sperm. Furthermore, this antibody

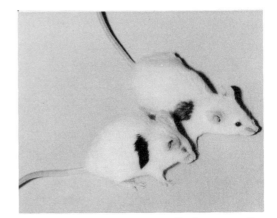

Figure 6.16. Mutant mice that suffer from a genetic disorder called severe combined immunodeficiency are grossly deficient in both T and B lymphocytes. The T-cell deficiency is illustrated here by the *inability* of these mice to reject skin grafts from unrelated black and brown mice. Normally, skin grafts (like organ transplants in man) would be destroyed and sloughed off because of immune reactions involving T cells. For organ transplants to succeed in human beings, the T cells must be suppressed by chemical treatment. *(Photo courtesy of Melvin Bosma, M.D., Fox Chase Cancer Center)*

is strongly depleted both in homosexual men and in persons suffering from AIDS. What sort of regulatory role this antibody might play is unknown; the final word on AIDS has not yet been written.

NOW, IT'S YOUR TURN (6-6)

James Thurber, one of America's greatest humorists, lost an eye as a young boy; it was pierced by a toy arrow. The remaining eye, although unharmed, gradually lost its function so that, eventually, Thurber became blind. Account for the loss of sight in Thurber's unharmed eye. (Hint: Lens and corneal proteins are not among those that occur in the fetal bloodstream.)

INFORMATIONAL ESSAY
Voltaire (1694–1778): *On Inoculation*

American students learn in their biology courses that Edward Jenner (1749–1823) introduced the use of cowpox in immunizing persons against smallpox, a virulent disease that is more often fatal than not. Jenner, of course, did not function in a vacuum; folk wisdom had already identified milkmaids as persons who could, with impunity, administer aid to, comfort, and nurse those who were afflicted with smallpox. The following paragraphs, based on an essay by François Voltaire that was written at least fifty years before Jenner's experiments, reveal that the beneficial effects of inoculation had been common knowledge for centuries. As with many attempts to "date" a scientific advance, close scrutiny reveals that the scientist singled out for acclaim merely stood on the shoulders of his predecessors and contemporaries. Indeed, that is the role of teachers; truly, it is a poor student who cannot see farther than his professor.

On Inoculation

Christian peoples of Europe erroneously claim that the English are fools and madmen. Fools because they give their children smallpox so that they won't catch it; madmen because they deliberately transmit this terrible disease to their children in order to avoid an uncertain future. On the other hand, the English claim that continental Europeans are cowardly and odd. Cowardly, because they refuse to expose their children to a slight discomfort; odd, because in exchange they expose their children to death of smallpox.

Circassian women for ages have given smallpox to their children before six months of age by placing a pustule taken carefully from another child into an incision in the baby's arm. The fate of the pustule in the arm resembles that of yeast in bread dough: it ferments and diffuses through the baby's body.

Pustules from the infected baby are used to communicate the same disease to still other children. When smallpox pustules are unavailable, the inhabitants of Circassia are in as great trouble as other nations when their harvest fails.

Why do the Circassian women deliberately infect their babies with small-pox? Because they are poor and their daughters beautiful. It is in their daughters that they chiefly trade. They furnish the young women for the harems of Turkey, Persia, and other nations whose officials are sufficiently wealthy to buy and maintain such valuable merchandise. The girls are instructed to fondle and caress men, are taught effeminate dances, and are taught how to heighten the pleasures of their disdainful masters. Young girls can be heard repeating their lessons to their mothers much as other children recite their catechism without understanding a single word of what they say.

Frequently, despite taking the utmost care, parents have had their hopes dashed in an instant. For example, having been exposed to smallpox, one family lost a daughter, had a second go blind, and a third disfigured; the parents were ruined. In past times when smallpox reached epidemic proportions, trade was suspended for several years, thus depleting the harems of Persia and Turkey.

A nation that trades minds its interests and exploits every discovery that enhances its commerce. The Circassians were no exception. They observed that scarcely one person in a thousand was attacked by violent smallpox. Some persons may have the disease several times but never fatally the second time. They also noted that the milder smallpox produced small pustules covered by a delicate layer of skin; these never scarred one's face. From these observations, they concluded that should an infant six months or one year of age have the mild smallpox, that child would not die of it, be marked by it, nor be afflicted again.

Therefore, to preserve the life and beauty of their children, it was only necessary to give them smallpox as infants. They did this by inoculating each child with a pustule taken from a second who was infected with the mildest sort of smallpox that could be found.

The experiment could not possibly fail. The Turks, being sensible people, also adopted this custom; at the present time everyone in Constantinople gives smallpox to his children of both sexes as soon as they are weaned.

• • •

I am told that the Chinese have practiced inoculation for one hundred years or more; this fact strongly supports the practice in as much as the Chinese are considered to be the wisest and best governed people in the world. The Chinese, however, do not give the disease by inoculation but by sniffing, as we take snuff. This is a more civilized method, and it produces the same effect. It also proves that had inoculation been practiced in France, thousands of lives would have been saved.

NOW, IT'S YOUR TURN (6-7)

Two of Voltaire's comments are of special interest. One reveals a willingness of the Circassians to accept risks (for their children): "They observed that scarcely one person in a thousand was attacked by violent smallpox." The second (to be found in the second paragraph) reveals how the risk was lowered to one in a thousand: "Pustules from the infected baby are used to communicate the same disease to still other children."

Couching your remarks in terms of natural selection, explain why the Circassians ran only a one per thousand risk of transmitting extremely virulent smallpox to their young children by inoculation.

INFORMATIONAL ESSAY

Some Nonbacterial, Nonviral Diseases of the Third World

Within North America and Europe, bacterial and viral diseases were formerly among the leading causes of death: pneumonia, flu, tuberculosis, diarrhea, kidney infections, diphtheria, and meningitis were among the first ten causes. With the advent of sanitation, sulfa drugs, and vaccinations with antibodies, together with improvements in nutrition and general health care, deaths from infectious diseases (especially those caused by bacteria) have waned. Degenerative diseases, cancer, heart attacks, and accidents are now greater killers than are microorganisms.

Therefore, those who live in developed nations might think that with the control of bacteria and viruses, diseases are nearly a thing of the past. They would be quite wrong; in much of the world, bacteria have never been the overwhelming cause of disease. In the tropics, protozoans and small worms have always constituted a serious threat to health. An ameba (*Entamoeba histolytica*) infects nearly 400 million persons in this world, of whom 100 million exhibit serious symptoms of amebiasis (diarrhea, dehydration, and lesions of the lung, liver, and brain)—the remaining 300 million persons are carriers.

Malaria is another disease caused by a protozoan, not a bacterium. There were one million cases of malaria in the United States in 1937; today, however, more than 250 million persons are afflicted worldwide. Mortality is high among children younger than five. Recall that this is the disease against which sickling hemoglobin (as well as other blood protein mutations) offers some protection: malaria has exerted a powerful selective influence on human populations.

According to one estimate, one-quarter of Africa is barred from agricultural development because of *sleeping sickness* (trypanosomiasis), a disease caused by trypanosomes (protozoans) that are spread from person to person or animals to persons by the tsetse fly. These protozoans eventually attack the central nervous system, causing drowsiness (hence, "sleeping" sickness), convulsions, and a terminal coma.

A small worm (schistosome, the blood fluke) with a complex life history that involves both man and snails affects more than 100 million persons in the world: fourteen million in Egypt alone, and an additional ten or twelve million in China. *Schistosomiasis* is contracted by swimming, wading, or merely dangling an arm in water where the infective form (cercaria) of the fluke occurs. Rumor says that a planned invasion of Taiwan (Formosa) in 1950 was canceled when 50,000 mainland Chinese troops contracted schistosomiasis while practicing amphibious landings in schistosome-infected freshwater lakes. Many fear that the Russian-built High Dam on the Nile River, which is intended to control the river's annual flooding as well as to produce electrical power, will cause a tremendous increase of schistosomiasis in Egypt.

Among the most disfiguring of tropical diseases is *elephantiasis*, a disease caused by a parasitic roundworm (the filiarial worm, a nematode). The small worms become so numerous in the body that they clog many of the lymphatic vessels; as a result, lymphatic fluid accumulates in the legs and, in males, in the scrotum. Huge, misshapen limbs are the result. This disease is particularly hard to treat with drugs because the death and dissolution of so many worms in the bloodstream can cause severe allergic reactions.

Thus, while persons living in developed countries see new leading causes of death take the place of older, now-conquered diseases and wonder what new malady awaits them in years to come, much of the world suffers from diseases known all too well to ancient Egyptians and even to Hippocrates, "the father of modern medicine."

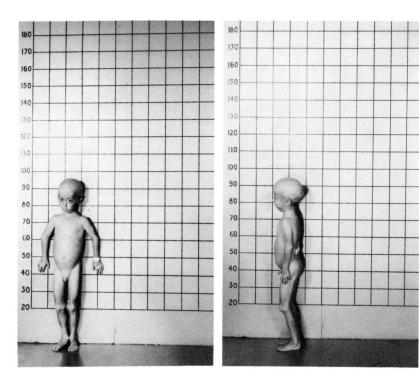

Figure 6.17. *Progeria*, a hereditary disease characterized by growth abnormalities and premature aging. In Part II of this text, now terminating, two views of life were emphasized—its trajectory and its moment-by-moment functioning (its "snapshot"). Those suffering from progeria travel along the trajectory of life at an inordinate speed; after several months of normal development, they reach the physiological equivalent of advanced age (65–70 years) by the age of nine. This and other illustrations of genetic diseases found in this text are reminders that the medical goals of molecular biologists (genetic engineers) are not frivolous ones concerned with cosmetic problems. *(Photo courtesy of V. A. McKusick, M.D., The Johns Hopkins Hospital)*

When Things Go Wrong   199

III
COMMUNITY BIOLOGY

In the next four chapters, aggregations (communities) of individuals, rather than individuals themselves, are the focus of attention. Within any one species of higher organisms, an aggregation of individuals is generally a population—that is, a collection of interbreeding individuals occupying a certain geographical region and persisting through time. Not all aggregations are populations in this sense. Canada geese and many other birds of the North Temperate Zone breed at the northern edge of their geographic range but overwinter without breeding in the south. The large flocks of these birds that gather in southern Texas or Louisiana, for example, are not populations in the sense the term population is used here. Similarly, two groups of human beings who occupy the same geographic area (Anglo-Saxons and Pakistani immigrants in London, for example) may not freely interbreed; such groups represent two rather separate populations according to our definition. Intermarriages that do occur, however, will slowly erode the distinction between the two populations. Modern Brazilians represent an exotic mixture of Spanish, black, and local Indian populations.

Populations pose special problems for biologists. Although they are composed of individuals, populations cannot be understood by a study of one, or even a few, individuals. In this way biology differs from the other natural sciences. The chemist, in order to understand a certain substance (table salt, NaCl, for example) can succeed by analyzing whatever tiny sample he might obtain: the properties of the substance are determined (or defined) by the properties of its individual molecules. The individuals of biological populations are not identical, however—not even nearly so. Consider diploid organisms, such as people, mice, flies, dandelions, fir trees, and lice (that is, most

Figure III.1. Canada geese migrate from Northern Mexico and the Southern United States to the Arctic Circle each spring; nesting, breeding, and the rearing of young take place in the far North. In the fall, these birds—parents and their young—migrate south once more in enormous V-shaped flocks. People fortunate enough to live on one of the flyways used by Canada geese know the thrill of hearing their distant honking and watching their overflight twice each year: north in the spring, south in the fall. *(Photo courtesy of the Tennessee Valley Authority)*

higher plants and animals). Each individual carries two copies of each gene (that is what the word *diploid* means); tens of thousands of genes make up the genetic information needed for normal growth and development and for survival. Recent estimates suggest that, on the average, from one-quarter to one-half of the thousands of genes in any one individual are represented by two *different* forms (*alleles*). The number of combinations of genes that *might* have arisen (but could not, because earthly populations are too small to include them) after the formation of parental gametes (eggs and sperms) and fertilization far exceeds the total number of electrons in the entire universe.

In each generation, then, a population consists of a collection of genetically unique individuals, unlike each other, unlike their parents and more remote ancestors, and unlike their children and all future descendants. Sexual reproduction, the sharing within each individual of genes formerly carried by two different ones (the parents), supplies the cohesion of populations. Sexual reproduction makes the population, rather than the individual, the subject of study in population biology.

Populations evolve; individuals do not. Evolution, in its simplest form, is a change in the frequencies with which different alleles occur in populations. The *frequency* of an allele is a property of the population; individuals possess either two, one, or no copies of a given allele. (In the same way, the proportions of Dodges, Fords, Plymouths, and Pontiacs in a city might be a characteristic of that city; nevertheless, each one-car family has only one or the other of these four types of automobiles.) Changes in gene frequency occur because the individual carriers of certain genes and gene combinations survive and reproduce; others do not. The extent and precise nature of these changes, however, are constrained by a need for local genetic harmony. In each generation old combinations of genes are broken up, reshuffled, and dealt anew to an entirely fresh collection of individuals.

A progressive genetic change might easily accompany a progressive sequence of environmental alterations (the onset of a prolonged drought, for example), thus maintaining the population's ability to inhabit its "chosen" locality. Eventually, in fact, the harmonious interactions of the genes present in one population might come to be disrupted or destroyed if an influx of genes from another, nearby population were to occur. The genetic changes arising in response to this event are those that are expressed in the mating behavior of individuals: further evolution gives rise to reproductively isolated *species*. Thus, the evolution of populations consists of genetic change *within* populations and of the origins of reproductive barriers *between* populations: *phyletic change* and *speciation*.

Generally, populations of several species occupy any one locality. A suburban lot, for example, may be occupied by house wrens, robins, cardinals, squirrels, a host of insect and plant species, and microscopic forms of life too numerous to mention. The coexisting members of each species extract their livelihood from their surroundings: plants from the air and soil, animals and fungi from their plant and animal neighbors—dead or alive. A population of any one species persists by garnering a portion of the available resources for its own use; garnering these particular resources, in fact, more efficiently than local competitors do. Careful observation reveals that coexisting populations (of wrens and squirrels, for example) do not utilize precisely the same local resources. Populations of two species that seek identical resources do not coexist for long; eventually, one displaces the other. If the second species cannot discover an alternative resource, it becomes (locally, at least) extinct.

Competition for resources takes one of two basic forms: *competition* (and subsequent allotment) or *scramble*. The individual members of many species compete for available resources. Winners claim enough (usually in the form of private territories) to see them through all but the very worst of times. Returning once more to the suburban lot: it is unlikely that more than one or two pairs of house wrens would take up residence there; others would be ruthlessly excluded. The brilliant red male cardinal might dash from one light pole or treetop to another, laying claim to not one, but several, lots and lawns. Male robins, especially in early Spring before territorial boundaries are well defined, can be seen fighting furiously in defense of their chosen

Competitive Exclusion

One of the basic principles of ecology states that two species of organisms whose requirements for food and shelter (that is, for *resources*) are identical cannot coexist indefinitely—one species will eventually displace the other. The principle can be proven mathematically and observations tend to support it. Furthermore, careful studies of coexisting species—that is, *closely related* species—as a rule disclose that such species do *not* use resources in precisely the same manner. To the chagrin of those who seek absolute answers, there remain those rare pairs of coexisting species, however, that seem to make identical demands on the environment. Have investigators overlooked some subtle but important difference? Are these seemingly coexisting individuals sustained in number by immigrants from localities where the two species do not coexist? Has enough time elapsed for displacement to have occurred? Fortunately, these puzzling situations are sufficiently rare to pose exciting problems for field ecologists.

In order to lead to competitive exclusion, the shared resources must be in short supply. For example, all higher forms of life rely on the oxygen that is present in the earth's atmosphere, but oxygen is not normally in short supply. Consequently, aerobic organisms do not tend to displace one another through competition for oxygen. However, food particles suspended in pond water may be limited in quantity. Insects, such as water boatmen, may compete for these particles. When two species of water boatmen coexist in the same pond, it frequently happens that members of one species are about twenty percent larger than those of the other. Differences of a similar magnitude are seen in the sizes of many closely related but coexisting species of animals. These differences in size are thought to reflect the outcome of natural selection leading to lessened competition: large individuals seek one source of food, while small ones seek another source. These differences presumably represent the first steps on the paths leading to the *adaptive radiation* of related organisms.

N

100 YDS.

Figure III.2. Two species of flycatchers that nest in close proximity. Their nesting territories touch but do not overlap. Close inspection reveals that one species prefers cleared areas (unshaded) while the other prefers forested areas. *(Copyright, Columbia University Press; reproduced with permission)*

worm-hunting and nesting grounds. Unsuccessful competitors usually become homeless wanderers, unable to obtain mates and to reproduce, often falling prey to predators. Successful competitors, in contrast, because they establish familiar home territories, usually rear their young successfully (not always, however—the neighbor's cat is the *bête noire* of many nesting birds). The total size of the local population, because it is limited by the number of territories, is fairly constant through time.

Scramblers do not acquire or defend territories. Each individual, like a child on an Easter egg hunt, grabs all the resources that it can manage. Common examples of scrambling competitors are larval flies, maggots. A female fly will often lay hundreds of eggs on even a small amount of food—rotting fruit in the case of fruit flies, rotting manure in the case of barnyard flies. Each maggot begins eating the moment it hatches from its egg, and it continues to eat furiously until it is grown. If the amount of food is too small to support the number of larvae growing on it, however, none may get enough and they all may die.

Because they are opportunists, populations of scramblers fluctuate wildly in size—tremendous numbers one season, virtually none the next. Because local populations often become extinct, organisms that

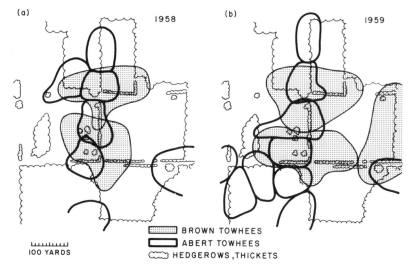

(a) 1958   (b) 1959

100 YARDS

BROWN TOWHEES
ABERT TOWHEES
HEDGEROWS, THICKETS

Figure III.3. Two species of towhees that nest in close proximity. Unlike those of the flycatchers in the preceding figure, the nesting territories of the different species of towhees overlap completely haphazardly. Note, however, that within each species, the nesting territories touch but do *not* overlap. The overlapping of territories established by individuals of the different species suggests that individuals of the two species utilize different resources; this difference is also suggested by the consistently larger territories of the one species. *(Copyright, Columbia University Press; reproduced with permission)*

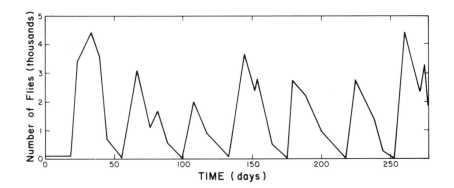

Figure III.4. Wild fluctuations in the size of a laboratory population of the sheep blowfly, a serious pest of Australian sheep. This population was fed rotted liver and was provided with a sugar solution to prevent dehydration. When the number of adults is large, larvae are so overcrowded on the small piece of liver that none survives. Only an occasional surviving larva or a long-lived adult female prevents the population from becoming extinct at each "crash." The one (or perhaps two) surviving female, however, lays the eggs needed to start the entire "boom and crash" cycle once more.

compete by scrambling tend to be excellent dispersers (dandelions are scramblers within the plant kingdom), and are highly fecund (each female is capable of leaving dozens, hundreds, or even thousands of offspring). In contrast, organisms that establish territories usually have many fewer young per mated pair.

As was the case in Part II, the human species has been chosen to illustrate population biology in the chapters in this part. Our species, although it has developed its peculiarities, has had an evolutionary past. Like other species, it has been and continues to be constrained by its environment. Human populations have had to cope with their physical environment in obtaining water, warmth, and shelter, and with their biological environment in getting food, while deterring competitors and predators—both large and small.

NOW, IT'S YOUR TURN (III-1)

How would you classify the behavior of human beings? Do we, like wrens and raccoons, establish territories? Or are we like maggots, scrambling for resources? Does it seem that some persons behave one way and others the other? Does it seem that with respect to some resources, people behave one way, while with respect to others, they behave the other? The answers to these questions are not simple: the word "people" includes groups, such as American Indians, Africans, and Europeans, but it also includes farmers, laborers, and city dwellers. Resources extend from food in the home that is to be shared among family members, to building lots in suburbs, to earth's store of fossil fuels.

7
POPULATIONS

Throughout the previous section, we peered *within* the person, within the single organism. Now, we will study the biology of aggregations of individuals and the interplay between these aggregations and the environment within which they live. In this and the following chapters, we begin a study of *population biology.*

Population biologists ask two basic questions about any population: *What kinds* of individuals compose the population? *How many* of these individuals are there? Thus, their concern has two foci: kinds and numbers.[1] In the past, population geneticists asked about *kinds,* and paid scarcely any attention to number. Ecologists, on the other hand, cared intensely about *number* and the factors that control number (that is, the *size* of the population); they assumed, however, that all individuals were of the same kind—differing, perhaps, only in age. More recently, persons have come to realize that *kind* and *number* are not independent; each one has a tremendous influence on the other.

The Mendelian Population

The word "population" has several meanings. A population is sometimes a collection or sample of items—this is a *statistical* population. Suppose we (i.e., the authors) are interested in the mortality of Japanese beetles that have been exposed to one of the insecticide powders on sale at hardware and garden supply stores. We cannot possibly test *every* Japanese beetle. *If we could, these beetles would be extinct.* Because we cannot test *all* of them, we must be satisfied by testing a *sample* of Japanese beetles. That sample is a *statistical population.*

---

[1] In considering numbers, it is necessary to be *quantitative.* Although the mathematics of population biology can be formidable, in this book it does not exceed high-school algebra.

Figure 7.1. A cage used for studying laboratory populations of the vinegar fly. Parental (adult) flies are placed in the cage. The holes in the floor are plugged either with empty cups (as shown on the left) or with cups containing *Drosophila* medium. Fresh cups of food can be provided in an orderly sequence and at suitable intervals (five cups per week, for example) so that each cup remains in the cage for three weeks. During that time, the eggs that were laid on the food surface on the first day will hatch, larvae will grow and pupate, and adult flies will emerge from the pupae and join others within the cage. The now-exhausted food cup can be removed and replaced by a fresh one. The semicircular shape of these cages allowed two of them to encircle a radiation source, with each cup at the same distance from the source. The purpose was to study the effect of continuous radiation exposure on populations.

Suppose we collect several hundred beetles in a garden in Virginia. By exposing *sub*-samples of 25 beetles to the insecticide for 5, 10, 15, 20, 25, and 30 minutes, we may find that few or none die after a 5-minute exposure, all die after a 30-minute exposure, and various proportions die with the intermediate exposures, and that the proportions of dead beetles increase with the length of exposure. From such data we can estimate the exposure that will kill 50% of the beetles. We can also calculate the distribution of mortality with respect to duration of exposure.

Having made our study, we would like to tell others about this insecticide and its effects on Japanese beetles; however, we must be careful. We really know only what happened to the several hundred beetles that we captured and tested; we cannot be sure that the ones that escaped in the garden would be affected the same way. We cannot be sure that our findings on the beetles in Virginia would agree with similar tests run in New York or Georgia; the beetles may differ, or uncontrolled experimental conditions involving temperature, the beetles' previous diet, and many other subtle sources of variation may complicate matters. Despite these uncertainties, our results and those obtained by other persons in other places at other times can be combined and analyzed so that we can speak with some assurance about the effect of our particular insecticide on Japanese beetles. This knowledge has been gained through analyses of *statistical populations*.

NOW, IT'S YOUR TURN (7-1)

You have probably not studied statistics; nevertheless, can you give an account in your journal of the flaw in the following experiment?

A gynecologist who operated a fertility clinic was not sure which of two techniques, *A* and *B*, was the better for predicting a woman's most fertile period. Each technique was based on several physiological measurements, including body temperature, signs of ovulation, and urine composition. Each technique predicted the supposedly "best" day for sexual intercourse with the highest probability of becoming pregnant. The doctor decided to run a test.

In the first month of the test, he used technique *A* for each of his patients. If a patient became pregnant that month, she was scored a success (+), for *A*. If a patient did not become pregnant, the doctor then tried technique *B* the following month. If the patient became pregnant that month, she was scored a success (+), for *B*.

After treating 300 women, the doctor found that the success rate for *A* was 60%. The remaining 120 women, of course, had by then been subjected to technique *B*; of these 120 women, only 25%, or 30, became pregnant. On the basis of these data the doctor reported that *A* was the better technique, since its success rate was over twice as large (60% vs. 25%) as that of the alternative technique.

The word "population" is used by census takers as well as by statisticians. The population of the United States exceeds 220,000,000 persons; this number is arrived at by simply counting all persons living within the fifty states. It is arrived at by taking a *census*. The population of the United States has grown steadily; the total number of persons counted during each ten-year census has always exceeded all previous numbers. People move, of course—from farms to the cities, from cities to suburbs, from East to West, and from North to South—and this information (and more) can be gleaned from census data. Using an earlier analogy, the population data obtained by the census is a snapshot of the population of the United States.

Turning from the snapshot of a population to its trajectory, we can project the population's existence through time. This is the concern of the population geneticist and of the population biologist, and this population with a time dimension is referred to as a *Mendelian population*. The term Mendelian refers to the name of Gregor Mendel, an Austrian monk. Mendel did the first thorough analysis of the patterns of inheritance of individual traits (seed color, seed shape, height, and others) in a higher organism (the garden pea). His studies established the mathematical rules that describe the passage of genes from parents to offspring. These rules are discussed in a later chapter that deals with genetic counseling.

To emphasize the time element in the notion of the Mendelian population, consider the following example. A small plastic and screen cage (say, 12″ × 18″ × 7″) was designed so that small cups of fresh fly food could be placed in it at daily intervals, and old cups (in which larvae have grown and have pupated, and from which the adult flies have now flown) could be removed. At the University of Edinburgh, some flies (*Drosophila melanogaster*, the vinegar fly) from Kaduna, Africa, were introduced into such a cage in the early 1940s. During the late 1960s, some twenty-five years after the fly population was started, one could still obtain "Kaduna" flies from the geneticist at Edinburgh. To the best of our knowledge, the population still exists—now nearly fifty years old. The flies in a population cage live between twenty and thirty days. In such a cage, kept at 25°C, a fruit fly generation is estimated to span two weeks, perhaps a bit longer. Thus, over 1200 generations of flies have lived in the Kaduna population cage (that many generations of humans, for whom a generation equals twenty-five years, would span about 30,000 years). Still, since no migrants have

Figure 7.2. A male (*left*) and a female (*right*) of the vinegar fly, *Drosophila melanogaster*. Before the widespread use of bacteria and viruses in genetic studies, these flies were the favorite research organism for geneticists. Today, because of the amount of information known about this fly and because hundreds of specially constructed mutant stocks exist, developmental geneticists have returned to it once more. The species' Latin name indicates that the fly loves dew and that it has a yellow abdomen; the subgenus name—*Sophophora*—proclaims that it is a bearer of knowledge.

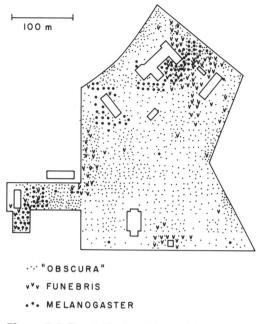

Figure 7.3. The distribution of three wild species of *Drosophila* around various buildings in a German farmyard. The *obscura* flies are scattered more or less uniformly, but the other two species tend to cluster. Clearly, however, the *funebris* and *melanogaster* flies do not cluster in the same areas or near the same buildings. The tendency of living organisms to subdivide and apportion the areas within which they occur is overlooked by many persons or, if not overlooked, ignored as if it were of no consequence. *(Copyright, Columbia University Press; reproduced with permission)*

come into the cage, one can refer to the "Kaduna" population. Surely, it is one of the oldest laboratory Mendelian populations in existence.

NOW, IT'S YOUR TURN (7-2)

1. We have described diploidy (the possession of pairs of genes by most higher organisms), and the apportionment of chromosomes during meiosis, and the reduction of chromosome number in gametes (which contain only one-half the number that is characteristic of body cells), and we have pointed out that each individual obtains half his genes from one parent and half from the other. Using this information, take a moment to try to anticipate Mendel's rules. For example, using the gene A, where the father carries A_1 and A_2 and the mother carries A_3 and A_4, what kinds of combinations of alleles at this locus can the children carry? And in what proportion would you expect the different combinations to occur? Now try other parental genotypes, such as an A_1A_1 father and an A_2A_2 mother. Now try A_1A_1 and A_1A_2, and A_1A_2 and A_1A_2. (Don't worry if you make a few messy pages in your journal working these out; Mendel's notebook wasn't all that neat, either.)

2. The concept of a Mendelian population may seem to resemble that of an old family axe, one which in thirty years has had two new heads and five new handles. Despite these replaced parts, it is called the "old" axe. There is a sentence in the account of the Kaduna population, however, that rules out any real similarity between the concepts of a Mendelian population and that of the family axe. Can you identify the sentence, and then contrast the two concepts?

How Many?

How many individuals are to be found in a Mendelian population? In the case of the Kaduna population of *D. melanogaster*, we would guess several thousand adult flies at any moment. We can be that certain only because the population is confined to a laboratory cage. For populations living "outdoors," it is much more difficult to assign a definite size to a Mendelian population. Indeed, Mendelian populations of free-living organisms range in size from small local aggregates (say, of snails inhabiting the grassy area at the center of a busy traffic circle, or of Portuguese living in a remote valley, high in the mountains of Madeira), through less isolated and therefore interconnected aggregates inhabiting a limited geographical area, to the entire species. By definition, the members of one species do not interbreed (freely, at any rate) with members of other species; this *reproductive iso-*

lation provides the cage that encloses this largest Mendelian population—plastic and screening are not necessary.

Even if the members of a species are spread more or less uniformly over a wide geographical area (as human beings, woodchucks, cabbage butterflies, and many plant species are), mating between individuals does not occur at random (that is, as a chance event) over that entire area. Leaving aside college students who just happen to come together from widely separated areas at precisely the age when marriage—or living with someone—seems important, most persons choose their spouses from persons living in their town or school district, or belonging to their church. Daily travel to a place of employment and an opportunity to meet fellow employees during working hours has enlarged the size of human Mendelian populations, especially in the United States, since the 1940s.

The size of the local, or Mendelian, population of human beings can be estimated by using the proportion of cousin marriages among all marriages. Now, cousins are a strange lot. First of all, and for good reason, there is a taboo against cousin marriages. In many countries, cousins must receive special permission in order to marry. On the other hand, cousins are often of approximately the same age, and they can easily fraternize, needing neither formal introductions nor family acceptance. Consequently, cousins do meet, fall in love, and marry, despite the taboo.

How many cousins does one have? Naturally, that depends entirely upon the number of aunts and uncles one has and on the number of children each of these aunts and uncles has. We can work here only with average numbers of children. Suppose the average number of children per family is two (2). Your father, then, has one sib (brother or sister) and your mother also has one; that makes two uncles, two aunts, or one aunt and one uncle. Each of these two has two children; that makes four cousins, on the average. If our average is incorrect (two children per family would mean a nongrowing population), the resultant error could be substantial; if the average number of children per family were three (3), you would have four aunts and uncles and twelve cousins. Four children per family would result in twenty-four cousins. Obviously, the calculated number of cousins is greatly influenced by the average number of children used in making the calculations. By resorting to a formula [b = the number of children per family; the number of cousins equals, then, $2(b-1)b$], we can use 2.3 as the average number of children (the population of the United States has been growing consistently); the number of cousins, then, is calculated to be $2 \times 1.3 \times 2.3$ or approximately six (6).

Imagine a young woman going out into the world with the intention of choosing a husband. In the world, she finds many young men; after a great deal of courtship, trial and error, and emotional pain, she fi-

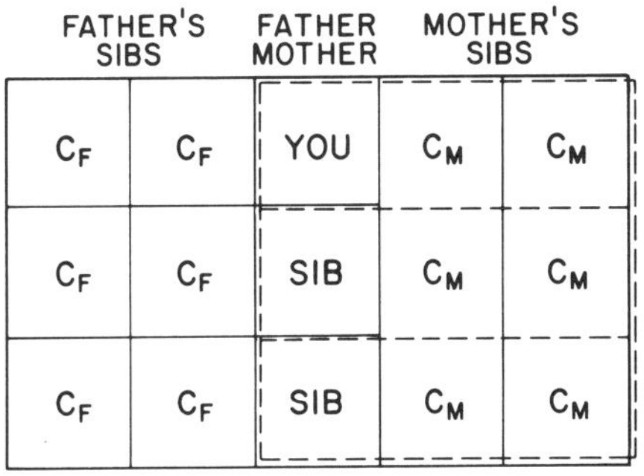

FATHER'S SIBS		FATHER MOTHER	MOTHER'S SIBS	
C_F	C_F	YOU	C_M	C_M
C_F	C_F	SIB	C_M	C_M
C_F	C_F	SIB	C_M	C_M

Figure 7.4. Diagram illustrating how to calculate the number of cousins a person has, provided that the average number of children per family is known. Note that on both the father's and the mother's side of the diagram is an area representing cousins that equals $b(b-1)$ or, in the illustration, 3×2. Consequently, the number of one's cousins equals $2b(b-1)$ as given in the text. C_F, cousins on the father's side; C_M, cousins on the mother's side.

nally chooses one for marriage. What is the probability that this choice will be a cousin? The probability (c) equals the frequency of male cousins among all young men encountered. The young men, collectively, are one-half of the total population (n). And so, the frequency of male cousins among all men equals $[2(b-1)b]/2$ divided by $n/2$ or simply $2b(b-1)/n$; this also equals the probability of cousin marriage, c. Consequently, n, the size of the population, can be estimated by dividing $2b(b-1)$ by the frequency of cousin marriages.

We have already decided to use six (6) as an estimate of $2b(b-1)$. We shall, in fact, assume that six applies to many populations. In rural Sweden, the frequency of cousin marriages has been as high as 7% recently; the population, then, consisted of only 85 or so adults of marriageable age. Stated differently, each boy or girl had 43 persons from which to choose a mate. In an Alpine community in Switzerland, the number of possible marriage partners for any young man or woman seeking a mate was 25 ($c = 12\%$, $n = 50$). The choice is larger in the United States, as one might expect. Here the frequency of cousin marriages is about 0.06%; therefore, $n = 10,000$, and the number of potential partners from which each person can choose a mate equals 5000. In rural America, however, this number drops considerably.

NOW, IT'S YOUR TURN (7-3)

1. You may have noticed how free and easy the authors of this book have been in making assumptions about average numbers of children and, therefore, the numbers of cousins each of us have. Use your journal to tabulate the numbers of cousins you have, and if you have close friends whose families are known to you, the numbers of cousins they have, as well. The data gathered by the entire class can then be averaged. Similarly, the data can be used to reveal the variation to be found among different families: the lowest possible number of cousins is zero; on the other hand, forty and sixty cousins are not impossible numbers.

2. The genetic consequences of cousin marriages are discussed under genetic counseling in a subsequent chapter. You might, however, see whether your state or church has anything to say on the matter. If you have ties to foreign countries, either by descent, tourism, or through penpals, you might try to learn whether the civil government or the predominant church in any of these countries restricts marriages of cousins.

The population of the world has grown at an ever-increasing rate since ancient (Biblical or, even, Classic Greek) times. The population of the United States has grown steadily since the first national census in 1790. In contrast, the Kaduna fruit fly population was established from fewer than twenty flies, but the population grew to its present several thousand quickly—in two generations at most—and has remained approximately that size. Had the Kaduna population started with a single gravid female, the growth of the population would have followed nearly the same path as it did when started by twenty. Similarly, a single bacterium placed in a culture tube will have produced a billion daughter cells within ten hours; shortly thereafter, the number of cells in the culture tube will remain constant.

For virtually all organisms, population growth follows an *exponential* pattern; that is, the number of individuals in a population *tends* to increase by simple multiples—whatever its size today, tomorrow the population tends to be two times, ten times, or one hundred times larger. Exponential growth cannot continue for long, however. Organisms need oxygen, food, water, and space; when an essential resource is exhausted, population growth will cease, or may even decline.

Exponential growth is a phenomenon whose significance most persons find difficult, or impossible, to grasp. Growth of this sort raises problems that can seemingly be put off until tomorrow—and then tomorrow turns out to be two days from yesterday! One illustrative analogy is the man whose fish pond was contaminated with duckweed. Every little weed in the pond doubled each day, but the man put off cleaning the pool day after day, although he knew that when the surface was completely covered, his fish would die. Finally, half the surface of the pond was covered, and half still open. How many days did the man have in order to save his fish? One day!

A second example, one that bears on our own food supply, involves wheat rust, a fungus that infects wheat plants and drastically reduces the yield obtained by the wheat farmer. These fungi make tremendous numbers of spores, many of which are wasted; let's assume, however, that each infected plant can infect twenty others in turn. An unwary farmer might not be concerned when 5% of his wheat is infected, but he should be. The next cycle, in which twenty times as many plants will be affected, will virtually destroy his crop.

Many wheat (and other grain) specialists realize the tremendous effect the last generation of rust can have on the total yield of grain obtained. By planting mixtures of seed obtained from plants that are each resistant to a number of common rusts, total yield can be increased by 10% to 20%. This is accomplished by slowing down the spread of the rust disease; each infected plant infects fewer plants in the mixed stand because the susceptible ones are scattered here and

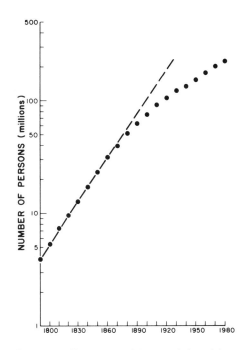

Figure 7.5. The growth of the population of the United States from 1790 to 1980. Notice that for the first eighty to ninety years the growth of the population was *exponential*; that is, the points representing the numbers of persons counted in the first eight or nine censuses fall on a nearly straight line when plotted on a logarithmic (rather than an arithmetic) scale. This fact was not overlooked by a ruler of China who, during the later 1800s, warned his subjects that the number of persons in the United States would soon outnumber those in China, where the number of persons had been constant for a considerable time.

Birth Control

Living organisms, if one can simplify somewhat, have two extreme strategies of competition for resources in perpetuating their own kind. Some species, as we have seen, rely on scrambling: each female produces as many young as possible, and the young fight for food and water. Maggots swarming over a dead mouse or yesterday's dog feces are scrambling in this sense.

Other organisms establish territories; most birds that nest around our homes or in the meadows and forests belong to the territorial group. The territory staked out by each pair of parents can provide food and other resources for the brood they are raising. Competition in this case comes when the young birds (or other animals) attempt to establish territories of their own. To do so, they must dislodge older (and perhaps weaker) birds and, in doing so, compete with their peers. Individuals that fail to establish territories are condemned to wander (a way of life that exposes them to predation) or to aid parents or siblings in raising young that are not their own.

Unlike most species of plants and animals that, with the exception of occasional flushes and crashes, remain nearly constant in number, human beings have steadily increased in number since ancient times. Major cities like Paris and London had fewer than 100,000 inhabitants as late as the 1400s; now many major cities number their dwellers in the millions.

Because human beings understand both how children are conceived and the consequences (both personal and community-wide) of having them, they tend to practice some sort of birth control. Women, that is, reject the idea that they should be continually pregnant. Some important birth control techniques are:

• Abstinence, a complete cessation of sexual activity.
• Adherence to a "rhythm" that avoids intercourse during times when a viable egg is present in the female's reproductive tract.
• The use of mechanical or chemical devices that prevent sperm from entering the uterus; examples are condoms worn by males, diaphragms worn by females, or spermicidal jellies that are injected into the vagina.

(continued on next page)

there at greater distances from one another than they would be in a pure stand. By reducing the number of plants infected by each already diseased plant, the spread of the rust is slowed; consequently, more wheat will be uninfected when it matures and is harvested.

A population may start to grow at an exponential rate, but it cannot grow that way for long. Very quickly, the amount of food, water, nesting space, or other vital resource becomes exhausted, and most offspring then find survival impossible. At that time, competition among individuals of the same population (same species) becomes intense. Eventually, the number of surviving offspring, despite the number of fertilized eggs, seed, or propagules each parent may produce, is reduced to two, equaling the number of parents. When the number of surviving children equals that of their parents, the population no longer grows; it has stabilized.

The conflict between the potential exponential growth of populations (even in populations of constant size, parents produce more than two young on the average; only two survive, however) and the constant (or linearly increasing) supply of food and other essentials was noted in a gloomy essay by the Reverend Thomas Robert Malthus in 1798: As the population increases in series $1:2:4:8:16:32:64:128:256\ldots$, argued Malthus, food can at best increase $1:2:3:4:5:6:7:8:9\ldots$ Whereas the ratio of population to food was at one time 1 to 1, after 9 doublings the ratio is nearly 30 to 1; furthermore, matters become constantly worse. Unlike his optimistic contemporaries, who saw in the French Revolution the start of a Utopian existence for mankind, Malthus could foresee only famine and poverty.

By coincidence, both Charles Darwin and Alfred Russel Wallace read Malthus's essay. They recognized that his account of an exponential increase in the number of organisms in an environment in which vital resources are essentially constant provided an explanation for the culling that is necessary for evolutionary change to occur in the living world.

NOW, IT'S YOUR TURN (7-4)

1. An essential feature of Malthus's argument is the constancy of vital resources. (We can ignore the linear increase he mentioned because even that is a temporary affair.) If the world were infinite in size, exponential growth could continue generations without end. To illustrate the importance of a finite world, imagine that you have deposited $1,000 in a bank at 6% interest compounded annually. Imagine, too, that a second person has deposited $1,000 in the same bank at 8%, compounded annually. Having heard of the better deal made by the second person, you may be envious, but you can still watch your

own account grow. Enter on one page of your journal three columns: (1) my money; (2) his money; and (3) ratio mine/his. If the column covers twenty-five years, you will find that your account has grown to $4291.87; your colleague's account has grown to $6848.48. Now, you learn the bitter lesson concerning finite resources: the bank fails. It claims that it can return only $2000 on those accounts—$773.01 to you and $1226.99 to the other account. You have lost money, while your colleague has gained.

2. Garrett Hardin, noting that the ratios of the corresponding terms of two exponential series themselves form an exponential series, suggested that the implications of this fact are so profound and disturbing that people cannot bring themselves to think of them. In your journal, first prove numerically that Hardin is correct, then list some of the implications of his point. For example, consider two segments of a human population, one of which has a 15-year generation period and the other a 30-year generation.

Malthus's essay has not been without its critics. These persons point out that the world today has many more persons than it did in 1800 and, although we are not all wealthy and well fed, great numbers of us are well-to-do and enjoying good health. To some, today's world disproves Malthus; people, they say, provide the subsistence for other people. In the long run, these persons are going to be proven wrong; Malthus was basically correct. The population boom that accompanied the Industrial Revolution was brought about and has been sustained by tapping and using the world's stores of nonrenewable fossil fuels—coal, oil, and gas. Food has been produced in huge quantities by using land only recently discovered (the Midwestern states were frontier states in 1800!) and, later, by investing more energy in farming (in the way of fuel and fertilizer) than is recovered in the harvested crop.

In an earlier chapter we saw that urine has a higher concentration of salts than does the blood serum from which it is made. In fact, it would be impossible to form urine without the investment of energy: work is required to increase the concentration of dissolved salts. We have now learned the same lesson at the population level: energy is required in huge amounts to "disprove" Malthus.

Still other persons claim that since matter cannot be destroyed, a finite earth is of no consequence: materials will be recycled over and over, just as the water in a river is used over and over by citizens of successive downstream cities. The extreme claim of this sort was made by an economist who said that only economic factors prevent us from collecting the combustion products of gasoline and synthesizing new gasoline from them. These persons overlook the law of thermo-

Birth Control (*continued*)

• The unbalance of the normal female reproductive cycle by use of hormones or hormonelike chemicals by women (the "pill"); chemicals that interfere with the production of viable sperm are also known.

• Tubal ligation, which stops the passage of eggs into the uterus; or vasectomy, which prevents the entry of sperm into the male's seminal fluid.

• Post-conception elimination of the developing embryo can be achieved in a number of ways, one of which involves the removal of the uterine lining (a D&C, *dilation and curettage*). Drugs and hormones can also cause uterine contractions and the abortion of a fetus.

The success of any birth control procedure depends upon human behavior. Few human populations have succeeded in lowering birth rates to match lowered death rates; consequently, human populations continue (over all, at any rate) to grow, thus exacerbating the conflict between the *quality* and the *quantity* of life.

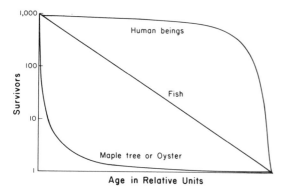

Figure 7.6. The contrasting patterns of survivorship that are characteristic of different organisms. Human beings living in developed nations tend to have little early mortality (neonatal deaths excepted), with nearly every person living until an advanced age. For example, half of those persons born in the United States live to age seventy. In contrast, trees and organisms like the oyster, which produce millions of offspring, have extremely high juvenile mortality rates, but with the few survivors enjoying a long life (in relative units) thereafter. Between these extremes are organisms, like fish, that appear to have a constant probability of dying, regardless of age. This constant probability gives rise to a straight-line survivorship curve when plotted on a logarithmic scale as in this figure.

dynamics that states that useful energy is always lost. In the long run, Malthus will be proven to be right; the seeming contradictions between his predictions and subsequent events arose only because we have used energy at a rate unheard of in 1798.

What Kinds?

A population is described not only by the numbers of individuals it contains but also by their kinds. Because we are discussing populations of a single species, the "kinds" of individuals do not include members of a different species (wrens and squirrels, or apples and oranges). Under the term 'kinds of individuals' we might include those of different ages, those of different physical characteristics stemming from environmental variation (seeds that fall on stony soil produce smaller plants than those that fall on more fertile soils), and genetic differences.

Before we discuss genetic differences between the individual members of populations, some words should be said about age distribution. Different organisms have different age-distribution patterns. A maple tree in a forest, for example, sheds thousands of seeds ("fruits" is a more precise term) every year. Because of the overhead shade and the absence of any space for growing, the seedlings that sprout from these seeds quickly die. Every year, year after year, the seedlings die. However, once the old maple tree itself perishes, there is space for seedlings to grow—sunshine and space. This could lead to the eventual survival of *one* of the thousands of that year's crop of seedlings, but not necessarily: the odds are about 7:1 that the successful replacement of the old tree (for maples, oaks, and beeches, but not for evergreens) will be a shoot developing from the old tree's roots and drawing upon the nourishment that has been stored in this large root system.

In Western societies human beings have an age distribution and a pattern of survival that is nearly the opposite of that illustrated by the maple tree. Nearly all babies that are born survive; mortality during the first month of life is about fourteen deaths per 1000 live births, and about five additional deaths occur during the next eleven months. The mortality then falls quickly, becoming low enough that, for example, half of the persons born in the United States live to age seventy.

A third pattern of survival is one in which the probability of living through each year seems to be a constant; a fifteen-year-old fish, having successfully lived so long, has the same chance of living until age sixteen as a one-year-old fish has of reaching age two, or a five-year-old of reaching age six. If only eighty percent of the fish of each age bracket survive through the following year, only three or four fish of every hundred will live to age fifteen. These oldsters seemingly have neither gained an advantage because of accumulated experience, nor suffered a disadvantage because of their age. Their risk of dying remains constant.

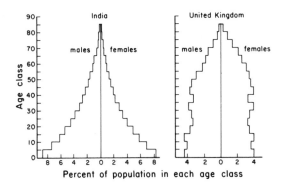

Figure 7.7. The age composition of the population of India (a developing country) and the United Kingdom (a developed country). Note the relative proportions of young to old persons in each country. Almanacs and similar books of knowledge give life expectancy tables for the United States; using their data, construct a diagram for the U.S. similar to those above.

NOW, IT'S YOUR TURN (7-5)

1. In his beautiful book *A Sand County Almanac,* Aldo Leopold describes the results of a study involving chickadees: "of the 97 chickadees banded during the decade, 65290 was the only one contriving to survive for five winters. Three reached four years, 7 reached 3 years, 19 reached 2 years, and 67 disappeared after their first winter." He continues, "I can only speculate on why 65290 survived his fellows. Was he more clever at dodging enemies?" Enter Leopold's data into your journal and see whether you can say something about the long life of 65290.

2. Using references provided by your instructor, insert in your journal diagrams illustrating the age distributions of the populations of India and of the United States, and comment on (1) possible causes for the obvious differences in the shapes of these diagrams, and (2) the consequences of the different age structures. The latter comments can touch on national defense, social security, marketing of manufactured goods, or any other relevant topic that interests you.

Genetic differences among the members of a population are of special interest to the population biologist who is interested in the adaptation of organisms to their environment and, largely as a result of this continual process of adaptation, the gradual evolution of organisms from one recognized structure or morphology to another. In this chapter, we discuss existing variation and its origin.

Individuals differ. If this fact has not come to your attention before, take time now to consider your classmates, two at a time, and note the differences between them. You probably have a number of friends whom you recognize and address by name when you see them. How is this recognition possible? You acknowledge your friends by name because each one differs physically from the others in a number of respects; otherwise, people would form one confusing mass of humanity, looking as much alike as identical twins.

The less familiar persons are with lower organisms, the more these organisms look alike. The extreme view can be expressed, as it was about redwood trees, as: "You see one [*organism*], and you have seen them all." Cats, to those who loathe them, are cats; to a cat lover, even individuals of the same general classification—tabby, tiger, Maltese,

Manx, or Siamese—have strikingly different appearances. Not many persons can or would care to be able to recognize individual flies; *Drosophila* geneticists can and do. One worker, when handed pairs of flies from two laboratory populations, correctly identified the populations from which they were taken in ninety percent of all trials; the differences he used in identification existed even though the populations had been initiated some seventy generations earlier with the same founder flies.

Observation reveals that individuals—both between and within populations—differ from one another. Some differences are discontinuous so the individuals can be separated into distinct classes, such as red, cross, and silver foxes; black and agouti (the standard, wild-type coat color of rats and rabbits and other rodents) hamsters; cyanide-producing and non-cyanide-producing wild clover; paired-fruited and single-fruited silver maples; red and black ladybird beetles; and a seemingly endless list of similar *polymorphisms* (two distinct forms existing in the same population). In addition to variation that causes individuals to fall into distinct classes, there is continuous variation; individuals differ from one another but do so over a continuous range of values. The height of human beings is one example; their weight is another. Tall parents tend to have taller than average children; short parents tend to have shorter than average children. Consequently, height has a genetic basis. Because cultural traditions, such as diet, pass from one generation to another in human pedigrees much as genes themselves do, the genetic basis for height in the human population could be disputed. Comparable observations, however, have been made on experimental organisms—egg size, wing length, and numbers of small hairs in the fruit fly, for example—that possess no culture. In many of these organisms, genetic contributions to the *metric* trait (one that varies in a measurable way) can be assigned to individual chromosomes by means of special breeding tests.

One can safely claim that every trait that has been examined in any

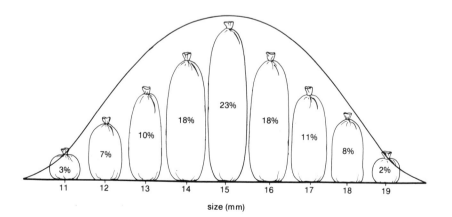

Figure 7.8. A diagram purporting to illustrate the bell-shaped (or *normal*) distribution of sizes of kidney beans. The horizontal scale gives the length of individual beans in millimeters. The heights of the bags are proportional to the numbers (or percentage) of beans in each size class. As so often is the case with observations of this sort, most individuals fall near the mean or average value; individuals differing from the mean decrease in frequency as the deviation increases. What error has been neglected in the preparation of this diagram? (Hint: The volume of a bean 19 mm long would be approximately five times that of a bean 11 mm long.)

size (mm)

organism where a genetic analysis was possible has proven to have genetically caused variation. This claim extends from the heat-, cold-, ether-, and radiation-resistance of fruit flies to the adrenaline levels in lemmings and voles. The point that these numerous studies make is that even in local populations, individuals differ genetically.

The observations and studies described so far tell us that genetic variation exists and that it can be demonstrated for every conceivable aspect of the phenotype. Nevertheless, these studies have been unable to give an answer to one simple question: what proportion of gene loci, on the average, are occupied by two different alleles, and what proportion by two seemingly identical alleles? The answer to this question requires that either the gene itself or, at least, the immediate gene product—a protein, preferably an enzyme—be available for study. The ability to make such a study came in the early 1960s through the development of two experimental techniques: the separation of proteins carrying different electrical charges within a high-voltage electrical field (*electrophoresis*) and histochemical methods that revealed specific enzymes in what was otherwise a hopeless mixture of various proteins.

By using electrophoresis for separating the proteins of any organism—serum from mice and voles, tissues from horseshoe crabs or pine trees, or whole squashed fruit flies—and histochemical staining of specific enzymes, geneticists have discovered (1) that one-half or more of all proteins exist within local populations in two or more forms, (2) that from one-quarter to one-half of all gene loci within local *populations* exist in two or more forms, and (3) that from one-quarter to one-half of all gene loci within *individuals* are represented by alleles that can be shown to differ. These observations totally disprove the view that each enzyme can have one, and only one, functional structure, and that individuals of a species are precise copies of one another in the same manner that molecules of the same chemical are exact replicas of one another. Not only do individuals "differ" in a superficial morphological sense, but electrophoresis has also revealed that they differ in their molecular structure: *the observable variation involves the bulk of all gene loci of most populations of most organisms.*

Individuals of populations differ genetically; that is a fact whose demonstration consists of showing differences in enzymes that are specified by genes at scores of gene loci. Sickle cell hemoglobin, you may recall, differs from normal hemoglobin by the substitution of one amino acid for another in a polypeptide chain nearly 150 amino acids in length, or in a hemoglobin molecule (containing both α and β amino acid chains) of nearly 300 amino acids. Sickle cell hemoglobin does not function properly; it forms needle-shaped molecules when oxygen tension is low (in capillaries, for example, or in the spleen). The all-prevailing variation in protein structure that has been revealed

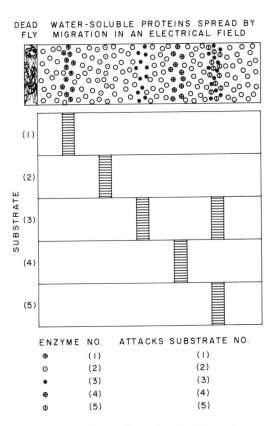

Figure 7.9. A diagram illustrating (*top*) the hodge-podge of soluble proteins that migrate from a squashed fly (seen at the *left*) under the influence of a high-voltage electrical field (*electrophoresis*). Exposure of the gel containing these electrophoresed proteins to various substances (substrates 1 through 5) reveals with considerable accuracy the positions of enzymes capable of attacking these substrates. Such studies have revealed that many enzymes do not exist in a single form (one band) but in two or more forms that migrate on the gel at different speeds, just as normal and sickle-cell hemoglobins do. (*Copyright, Columbia University Press; reproduced with permission*)

Table 7.1. Variation in gene-enzyme systems among fruit flies (*Drosophila melanogaster*) inhabiting the Eastern United States

Gene-enzyme system[a]	Genotype	Fruit fly population			
		New York	Ohio	Kentucky	South Carolina
Pgd	A/A	10	6	5	0
	A/B	0	2	2	2
	B/B	0	5	5	7
Est-6	A/A	8	5	3	6
	A/B	0	9	4	0
	B/B	0	2	2	0
Aph	A/A	9	13	3	0
	A/B	0	8	9	0
	B/B	0	1	6	10

[a]Pgd = 6-phosphogluconate dehydrogenase; Est-6 = esterase; Aph = larval alkaline phosphatase.

by electrophoresis does not involve grossly malfunctioning enzyme molecules; at most, these common, variant molecules have somewhat different reactions to temperature.

Many geneticists believe that the actions of the different enzyme molecules are so nearly alike that individuals are as healthy and as well-off with one as with another. That is, variation can be detected by electrophoresis but it is of no consequence for the health or survival of the individual organism.

NOW, IT'S YOUR TURN (7-6)

Electrophoretic techniques permit us to survey the proteins specified by individual gene loci and to ask for what proportion of all loci an individual possesses different forms of a particular protein, rather than a single form. Earlier observations on existing variation had been almost anecdotal. The distinction between the two types of observations is important. Show that you understand the distinction, and the limitations of the earlier observations, by commenting on the following observation: Black and women professors are to be found among virtually all university and college faculties. What does this fact say about the equality of opportunity for black and women professors? What additional data are needed before one can speak of proportions of black professors or women professors more precisely?

Can the level of genetic variation in a population remain constant, or is it always changing? Is it always decreasing, for example? In discussing the origin of species, Darwin assumed the presence of heritable (the word "genetic" had not yet been coined in Darwin's time) variation in populations. His views were challenged by a young engineer who claimed that, even if variation were to arise, it would quickly be lost. Suppose, to illustrate the engineer's argument, that in a population of 100 individuals whose appearance can be scored numerically, all individuals but one have a score of 1000; the variant has a score of 0. Thus, variation exists in the population. To reproduce, the variant must mate with a normal individual; together they produce two offspring with a score of 500 each (the average of 0 and 1000). The population now consists of 98 1000s and 2 500s. The next generation the population will consist of 96 1000s and 4 750s. The following one, 92 1000s and 8 875s. The variation is obviously disappearing; all individuals are becoming more and more alike. Eventually, the population will become uniform under the scheme suggested by the engineer; 100 individuals all having a score of 990. [Note that 99×1000 (the start) = 100×990 (the end).]

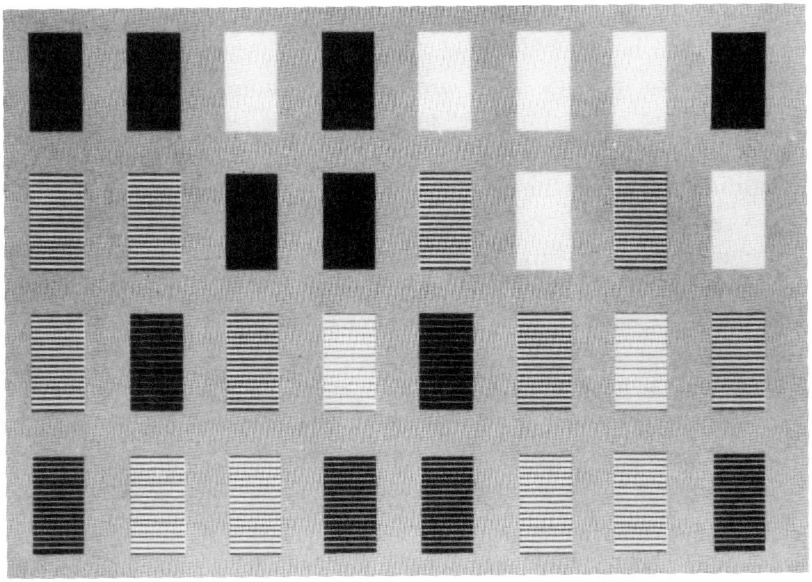

Figure 7.10. A diagram illustrating the loss of heritable variation that would occur if children were merely averages of their parents, i.e., under blending inheritance. (Darwin realized that blending inheritance would be fatal to his theory of evolution by natural selection, but he could not refute it because the patterns of inheritance were not yet understood.) The top row shows eight beakers, four containing undiluted ink and four containing water. The contents of these beakers were mixed according to patterns determined by flipping a coin. The second row shows the beakers after the first flip (there are now four beakers with grey liquid), the third row after the second flip, and the bottom row after the third flip. Clearly, the contents of the beakers are rapidly becoming uniform; that is, variation between beakers is vanishing.

Of course, the young engineer could have looked around and noted the variety of appearances exhibited by his friends, and even by his own relatives. This evidence, which was as available to him as it is to each of us today, did not deter him from his calculations "proving" that variation is quickly lost from populations. If heritable variation is lost rapidly from populations but these populations nevertheless remain variable, then there must be a correspondingly large source of variation that renews the amount of variation each generation. Darwin sought the source of this variation but failed to find it.

What we have learned of genetics so far is sufficient to prove that the young engineer who criticized Darwin was wrong: variation is *not* lost rapidly from a population in the way that his calculations suggest. Darwin, therefore, did not need to seek a source for new variation, one that flooded the population with enough new variation every generation to counterbalance the seemingly large loss.

Suppose that an enzyme—alcohol dehydrogenase, for example—exists in two forms that can be detected electrophoretically. The two forms, named for the rate at which they move in a high-voltage electric field, are called "fast" and "slow." Similarly, the two alleles that exist at the alcohol dehydrogenase locus can be referred to as the "fast" and "slow" alleles, not because they move, but for the proteins whose synthesis they control.

Imagine a population of 100 individuals each of whom possesses two alleles at the alcohol dehydrogenase locus; the population contains, therefore, 200 alleles at this locus. Suppose, now, that 120 of these are fast alleles and 80 are slow (there must be some number of each of the two alleles, and the sum of these numbers must equal 200). In proportions, 0.60 of all alleles are fast, 0.40 are slow.

Now, men and women do not conduct electrophoretic examinations of their beloved's enzymes when they contemplate matrimony. Marriages (or matings, if we are discussing animals or plants other than man) are entered into at random with respect to the alcohol dehydrogenase locus. (The same, of course, is not true in the case of height or skin color; the heights and skin colors of marriage partners are correlated.)

If marriages are contracted at random, the sperm and egg cells that give rise to the children of the next generation also meet at random. The proportions of fast and slow alleles among each type of germ cell (remember, gametes carry only one copy of each gene) are 0.60 fast and 0.40 slow. Therefore, 0.36 (0.60 × 0.60) of the children will carry two fast alleles, 0.48 (0.60 × 0.40 *plus* 0.60 × 0.40) will carry one fast and one slow allele, and 0.16 (0.40 × 0.40) will carry two slow alleles. If the population remains at 100 individuals, 36 will be "fast/fast," 48 will be "fast/slow," and 16 will be "slow/slow."

We can now count up the numbers of fast and slow alleles just as we did in starting our calculations. The number of fast alleles equals 2 × 36 *plus* 48, or 120; the number of slow alleles equals 48 *plus* 2 × 16, or 80. These are precisely the numbers we obtained for the first generation; they persist from generation to generation unchanged. And, we might note, the types of individuals in the population will tend to reappear in the same proportions (36FF : 48FS : 16SS), generation after generation.

NOW, IT'S YOUR TURN (7-7)

You may have noticed a similarity between the calculations performed above and one that is encountered in first-year algebra: $(a + b)^2 = a^2 + 2ab + b^2$. In our example, $a = 0.60$ and $b = 0.40$. Repeat the calculations on gene frequencies and zygotic frequencies (that is, frequencies of different types of individuals) using values 0.50 : 0.50, 0.80 : 0.20, and 0.90 : 0.10. Notice in the last case, although there would be 20 slow alleles in the population of 100 individuals, only 1 of those 100 persons would ordinarily carry 2 slow alleles: the square of a small fraction is a correspondingly smaller fraction.

The Interaction of Numbers and Kinds

Before they joined forces and became population biologists, population geneticists and ecologists both studied populations, but each with little concern for what the other was doing. As a population geneticist, one can speak of a population in which the frequencies of two alleles are 0.6 and 0.4 and, therefore, the three types of zygotes have expected frequencies of 0.36, 0.48, and 0.16, but without specifying the number of individuals. In fact, within a population of 5 individuals, one might encounter 6 alleles of one sort and 4 alleles of another (a total of 10 alleles). One cannot have zygotic proportions of 0.36, 0.48, and 0.16 among 5 individuals, however. Calculations in percentages or frequencies are purely abstract.

On the other hand, ecologists assumed that all individuals were identical, that these individuals were merely units to be counted and tabulated for whatever purpose the ecologist had in mind.

The migratory locust can serve to illustrate the interaction of numbers and kinds. (This locust was responsible for the plague of locusts that struck Egypt in Biblical times.) However, before we discuss the locust in detail, a point about the genetic control of insect development must be appreciated first. A variety of insect phenotypes can be produced by a single genotype under different developmental stimuli. Butterfly eggs, for example, hatch into caterpillars that have no obvious resemblance to their parents; the production of a hormone brings about a change in development. The caterpillar passes through a pupal stage, and then emerges as an adult butterfly. The genomes of flies and beetles have corresponding abilities: they program the development of maggots and grubs, but in response to the proper signal (a growth hormone), they also program the adult insect.

The locust genome responds to hormone signals released as the result of crowding. Under uncrowded conditions, locusts hop and fly around in local areas, where they are generally known as grasshoppers. Females lay tremendous numbers of eggs, and local populations tend to swell. The grasshoppers become so numerous that they often jostle each other physically. Under the stimulus of crowding and jostling, the grasshoppers produce hormones that lead to subsequent generations of grasshoppers that are narrower in build and possess longer than normal wings; these individuals are the migratory locusts. Rather than having solitary habits, locusts are gregarious. They collect in large numbers, flying together as groups. Eventually, on some signal, hordes of locusts will take to the air and fly for miles, seeking new fields. Containing millions of insects, the swarms destroy any field of grain or other crop on which they land. Their migratory flight may take them hundreds of miles. The *number* of individuals in this example results in the switching of individual genomes from one developmental pattern (the solitary, sedentary grasshopper) to another (the gregarious, migratory locust).

The reverse interaction, in which kind affects number, is illustrated by a mutant gene that was discovered in certain laboratory populations of one species of fruit fly, *Drosophila willistoni*. The mutant allele, a dominant one, causes its larval carriers to leave the food cups just

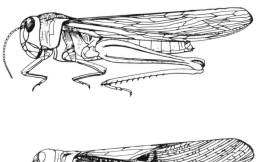

Figure 7.11. Two phenotypes of the migratory locust of Africa: the upper illustration shows the solitary phase; the lower one shows the migratory phase. The different phenotypes reflect the degree of crowding to which the insects are exposed; the migratory phase develops in response to hormones induced by the more or less constant jostling that occurs within a growing locust population. The obvious differences pertain to flight. The solitary phase has large rear legs, suitable for jumping; the migratory phase has smaller legs but is more streamlined for flight. Other differences not apparent in the illustration should be mentioned: locusts in the migratory phase lay eggs more frequently, are shorter-lived, develop faster, and collect in large groups (are gregarious). Migratory groups will move on foot as much as one mile per day; they can fly tens or even hundreds of miles in search of food.

Figure 7.12. An example of the extensive variation that occurs in the shell-color patterns of some species of snails. This diversity may play any one of several roles: the different patterns may serve to conceal the animals under different circumstances, they may serve to disrupt a predator's search pattern, or they may serve in thermal regulation, by the differential absorption of the infrared (heat) rays in solar radiation.

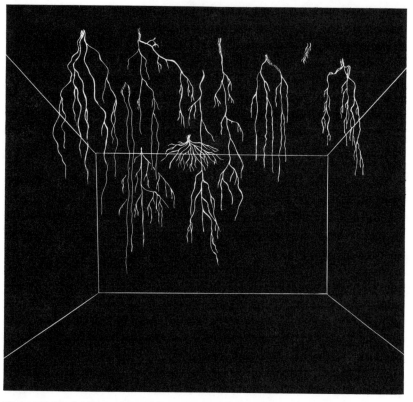

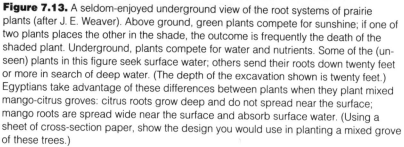

Figure 7.13. A seldom-enjoyed underground view of the root systems of prairie plants (after J. E. Weaver). Above ground, green plants compete for sunshine; if one of two plants places the other in the shade, the outcome is frequently the death of the shaded plant. Underground, plants compete for water and nutrients. Some of the (unseen) plants in this figure seek surface water; others send their roots down twenty feet or more in search of deep water. (The depth of the excavation shown is twenty feet.) Egyptians take advantage of these differences between plants when they plant mixed mango-citrus groves: citrus roots grow deep and do not spread near the surface; mango roots are spread wide near the surface and absorb surface water. (Using a sheet of cross-section paper, show the design you would use in planting a mixed grove of these trees.)

before pupation, and to pupate on the floor of the cage. Populations that lack this mutant characteristically contain 4000–4500 adult flies; similar populations containing the mutant allele contain more than 7000 flies. By pupating outside the food cups, on the floor of the population cage, the larval carriers of this dominant gene avoid the danger of drowning in the culture medium during their pupal stage, since active larvae may dislodge pupae and cause them to sink into the moist food. By pupating on the floor of the cage, the individuals carrying the dominant gene also avoid being removed from the cage and discarded when old, exhausted food cups are exchanged for fresh ones.

The polymorphism of sand dollars serves as a second example of the influence of kind on number. The sand dollar is an *echinoderm*, a relative of the starfish. The disc-shaped skeletons of sand dollars are a familiar sight on flat, sandy beaches. Living sand dollars are preyed upon by sea birds. Because the sand dollars have many color patterns, it is believed that their variety of forms causes difficulty for the feeding birds. The birds are unable to construct a simple image that will serve in seeking out sand dollars. If a bird forms such an image, it will overlook many individuals in its search; on the other hand, without such an image, the bird must proceed slowly, carefully examining each object on the beach that might possibly be an edible sand dollar.

Two Sorts of Genetic Variation

Many reasons can be advanced for thinking that a population of dissimilar individuals has advantages over a population of more or less identical ones. Plants, for example, can take better advantage of the soil if their root systems differ—some remaining at the surface of the soil, others going deeper. Flies having dissimilar genotypes but competing for limited food can be shown to survive in greater numbers than do nearly identical flies under similar circumstances; presumably, slight differences in metabolic requirements allow a mixed population to better utilize scarce nutritional resources.

A population of dissimilar individuals can arise as the result of many genotypes, or it can result from many fewer genotypes, each of which allows its carrier to be flexible in his or her development. Examples of plastic responses have been cited for other purposes: genotypes that produce maggots, caterpillars, and beetle grubs also produce the mature flies, butterflies, moths, and beetles. The geno-

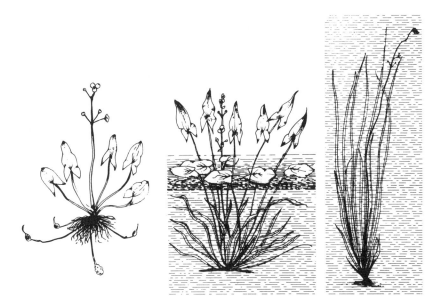

Figure 7.14. An example of the individual phenotypic diversity that can be brought about by environmental differences. The arrow leaf (*Sagittaria sagittifolia*) has one morphology (rigid leaves, upright stems) when growing on land (*left*), a second (limp and flexible leaves and stems) when growing under water (*right*), and a mixture of the two when an underwater plant breaks through the surface of the water and grows in air (*center*). Each form meets challenges posed by its environment: rigid leaves and stems would be broken by water currents or would cause the plant to be uprooted; limp leaves and stems would never serve to gather sunlight for a land plant being crowded by neighbors.

Figure 7.15. The varying hare in its winter and summer pelage. The color of the fur responds to the number of daylight hours: the change to short days leads to white fur; the change to long days leads to brown fur (as well as to mating and rearing of young). This ability to respond to changes in the lengths of days is under genetic control. *(Copyright, The Biological Sciences Curriculum Study; reproduced with permission)*

type of the solitary grasshopper is the same as that of its migratory (plague locust) form. Plants that grow both on land and in the water can have strikingly different phenotypes—even combinations that are part one and part the other—in and out of water. The various worker and soldier castes of ant and termite colonies reflect the reaction of the same genotype to chemical substances that supplemented each individual's diet during its development.

The social castes of India, which originated 4000 years ago, represent a system based on the assumption that diversity results from genetic variation. Originally organized on the basis of color and physical appearance, and later perpetuated by birth, the caste system represented an attempt, in effect if not in purpose, to create genetically uniform populations, each differing from the other and, because of their unique genotypes, each specialized for carrying out certain tasks and for filling certain occupations. The futility of this lengthy experiment has been revealed during the past few decades, following the abolition of the caste system in India; old prejudices linger, of course, but one finds persons of formerly low castes now occupying positions in all walks of life.

Life in the United States and European nations reflects an attitude contrary to that upon which the Indian caste system was based. For example, at least in the last century, most countries have relied on the citizen-soldier in times of war: clerks, factory workers, and farm hands are converted into infantrymen, tank commanders, and pilots. On another level, the recent explosion of "How-to" and "Do-it-yourself" books illustrates at least the faith of authors and publishers that each human genotype produces an individual who is capable of performing many tasks in an acceptable manner.

Doubtless, there are genetic endowments that allow certain persons to excel in certain fields. Just as few of us are built to be Olympic runners, champion prize-fighters, or professional football or basketball players, few of us can compose music, create our inner visions on canvas, or visualize forms within blocks of stone and release them by sculpting. Attempts are often made to subdivide each human ability, and to assign a certain fraction to the environment and the remainder to genetics (one hears, for example, that ten percent of human behavior is probably genetically controlled); a safer statement would be to say that among those who excel in any undertaking, genotype and environment were both favorable, both pulled together.

NOW, IT'S YOUR TURN (7-8)

1. Variation occurs in all populations, including human populations. Some variation results from exposure to diverse environments,

and some results from the different genotypes that individuals possess. To quote a former President: "The world is not fair."

Some persons claim that variation of any sort provides a basis for placing individuals on a linear scale ranked from best to worst. Certainly, runners can be ranked according to the time required to run, for example, 200 meters; the ranking extends from the fastest to the slowest.

Consider the ways in which your friends and fellow students differ. Can you ascribe "better" or "worse" to any of the differences of which you are aware? Is your ranking based on personal attitudes, or would the rankings of most persons agree with yours? What is the meaning of individual rankings if variation itself is "best" for the population— that is, if a population consisting of any one type would be inferior in some sense to one possessing many sorts of individuals?

2. As human beings, we tend to rely heavily upon our eyes for understanding what is happening in the world around us. We also rely on our ears, less so on our nose, and perhaps least on taste and touch. For example, we examine objects carefully before we either touch or taste them. The world is different for other mammals. A finicky, intelligent cat can sit by an opened refrigerator door and inventory the entire contents of the refrigerator with a single breath and an extended nod of the head. To offer such a cat food from an opened can of a less-desirable brand of cat food is useless if an opened can of a better brand is in the refrigerator.

To get some sense of how society might have been organized had human beings relied on their sense of taste rather than on their vision, your class may be divided into those who can and cannot taste PTC (a subtance related to urea; some persons cannot taste it, while others say that it is extremely bitter). Once the class is divided into tasters and nontasters, take a look at those who have been included in your category. The variety you see may suggest why the caste system of India failed to produce genetically differentiated subgroups within the Indian population.

INFORMATIONAL ESSAY
Metrical Traits

Metrical traits are of two sorts: those in which the units are unambiguous in number (the number of eggs laid by a hen in one year; the number of hits obtained by a major league ballplayer in a single season) and those that are truly continuous and that can be measured only to within a certain error (height, weight, and speed, for example). Barring carelessness, there is no error when a poultry geneticist reports that a particular hen laid 185 eggs in one year; of course, we are not told if all the eggs were of the same size. When a sports announcer says that a fighter weighed in at 146¼ pounds neither we nor he knows whether the fighter weighs 146 pounds and 3½ ounces or 146 pounds and 4½ ounces; the accuracy of the scale is limited. Similarly, a person may claim to be 6' 2" tall, but is it really 2" or is it 1⅞" or 2⅛"? With continuous variation, only one value is true, to be sure, but our measuring devices allow us only guesses.

NOW, IT'S YOUR TURN (7-9)

Following a mile race, the official time for one of the contestants is given as 4 minutes 21.013 seconds. A mile equals 5280 feet. Given the speed with which the runner ran, calculate the distance involved in measuring elapsed time to 1/100th of a second. How about 1/1000th of a second?

Metric or mensuration data generally (but not always) follow a common pattern: few individuals exhibit the extremes, and more and more individuals exhibit intermediate values, with the majority falling near or at the average (or *mean*). Some distributions, such as the distributions of annual incomes of U.S. citizens, are not symmetrical, but are skewed—one tail of the distribution curve is much longer than the other. In discussing such distributions, it is sometimes useful to talk about a value on either side of which are found one-half of all observations (the *median*), or the value that includes more observations than any other (the *mode*). Here, however, we discuss only the mean and the distribution of observations about the mean.

Suppose a series of observations are made as follows (the first number is the value, the second, the number of times it was observed): 5, 1; 6, 1; 7, 2; 8, 3; 9, 5; 10, 4; 11, 7; 12, 10; 13, 15; 14, 12; 15, 8; 16, 10; 17, 5; 18, 3; 19, 2; 20, 1. First of all, plot these data in your journal. Note that the values form what is commonly called a bell-shaped (or normal) distribution curve.

If all of these observations could be equalized so that every observation was identical, what would be that *average* (or *mean*) value? This value is obtained as the sum of the products of all values times the number of occurrences, divided by the total number of observations. A pocket calculator helps in making these calculations, but they can also be done by hand. The sum of $(5 \times 1) + (6 \times 1) + (7 \times 2) + \ldots + (19 \times 2) + (20 \times 1)$ equals 1171. The total number of observations equals 89. Therefore, the average equals $1171/89$ or 13.157303. Six decimal places have been given merely to show that 1171 divided by 89 does not yield a simple solution. Because none of the observations has been made in values other than whole units, there is little reason to be more precise than to say that the average value equals 13.16.

Bell-shaped, or normal, distribution curves have two important characteristics: the value of the mean and the extent of the dispersal about the mean. The usual calculation of dispersal is given as the *variance* of the distribution. Of course, one could say that the observed distribution had a mean of 13.16, with a range extending from 5 to 20, but such an account does not convey any sense of how the observations tapered off toward the extreme values. It does, however, convey a sense of symmetry about the mean: $(5 + 20) / 2 = 12.5$, a value not far from the mean.

The variance of a distribution is defined as the average squared deviation from the mean. Let's dissect that definition: the deviations from the mean can be written as $5 - 13.16$, $6 - 13.16$, $7 - 13.16$ ... $18 - 13.16$, $19 - 13.16$, and $20 - 13.16$. Each of these terms gives the deviation of the observed value from the mean. The average squared deviation is obtained more or less in the same way as the average itself. A *squared* deviation from the mean is no more than $(5 - 13.16)^2$, $(6 - 13.16)^2$ ... $(20 - 13.16)^2$. The average of these squared deviations is calculated as the sum of all squared deviations times observations, divided by the total number of observations. Although computational shortcuts exist, most hand calculators can handle the calculations easily. The sum of squared deviations times the number of observations for each equals 821.7984; dividing this number by 89 equals 9.2337. This is the variance of the distribution.

To say that a series of observations has a bell-shaped distribution is merely descriptive (i.e., the curve resembles the silhouette of the Liberty Bell in Philadelphia). To say that it is a *normal* distribution is to say something more precise. Defining the *standard deviation* as the square root of variance (3.04 in our example), the distribution (if it is normal) will have 0.32 of all observations one standard deviation or farther removed from the mean, 0.045 two or more standard deviations removed, and 0.003 three standard deviations removed. In our example, 13.16 ± 3.04 equals $10.12-16.20$; thus 0.32×89, or 28, observations should fall at 10 (or lower) or 17 (or higher), and 27 observations actually fell at these values or beyond. Continuing, 13.16 ± 6.08 equals $7.08-19.24$; thus, 0.045×89, or 4, observations should fall at value 7 (or less) or 20 (or more), and 5 observations actually fell at these values or beyond. Finally, less than one observation should have been seen at value 4 or below or 22 or above; in fact, no such observation was made. Thus, the observed values in our example can be described as *normally distributed*, with a mean of 13.16 and a variance of 9.2337 (which allows the calculation of the standard deviation). Armed with only this information, a person who had not seen the actual observations could construct a hypothetical curve in an attempt to visualize the distribution pattern of the original data.

Data of the sort listed here (which, incidentally, could have represented the number of eggs laid in a month by the various hens in a flock of 89 chickens, or an 89-year record of the inches of rainfall in some arid locale) are frequently used to compare the mean values of two comparable sets of data. The poultry farmer may, for example, want to compare the performance of his flock with that of a neighbor who uses a different brand of chicken feed or owns a different breed of chickens.

The entire testing procedure is not given here, but it is worth describing the expected variation of the mean of the population. To gain insight, think of all the undergraduate classes that you attend. The students of each class differ in height—perhaps from 4' to 6'2". For each class, however, you can imagine an average height. If you now compare these averages mentally you will note that they are much more similar (they vary less) than the actual individuals of any one class. None of your class averages is 4', nor is any 6'2". On the contrary, the averages probably all lie somewhere between 5' and 5'6".

The degree to which the mean or average value can fluctuate from sample to sample (the *variance of the mean*) is related to the variance of the individual observations about the mean in an extremely simple way: the variance of the mean equals the variance of the distribution divided by the number of observations. In our example, the variance of the mean equals 9.2337/89, or 0.1037. The standard deviation of the mean (also known as the standard error) equals $\sqrt{0.1037}$, or 0.32. If another sample of 89 individuals were to be taken from the same population from which the first sample was taken, the calculated mean of the second sample would probably not equal that of the first. Of many repeated samples, 68% of all means should fall between 13.16 ± 0.32 (12.84–13.48), about 95% should fall between 13.16 ± 0.64 (12.52–13.80), and all but one in a thousand should fall between 13.16 ± 0.96 (12.20–14.12).

A statement can be made concerning sample means: Having calculated the mean of a sample, there are two things of which you can be sure: (1) the mean of the sample is *not* equal to the mean of the population from which the sample was drawn, but (2) there is no other value that is more likely to be correct than the calculated mean. The same two items apply to personal experiences or to knowledge gained (in class) from someone who honestly reports his experiences or those of other persons: what you learn is not the *truth*, but no other (unsubstantiated) claim is likely to be more nearly truthful.

Population Growth and Its Limitation by the Environment

Population growth is a topic that mathematically talented biologists enjoy discussing. In our discussion here we use an entirely hypothetical population: all individuals are female, they reproduce without fertilization, and all their offspring are female. Individuals live one generation only.

The total number of starting females is N_0. Each female on the average produces R offspring. The number of individuals in the next generation equals $RN_0 = N_1$. The number in the following generation equals $RN_1 = N_2 = R^2N_0$. And the next: $RN_2 = N_3 = R^3N_0$. The number of individuals in any generation (t) obviously equals N_0 (the original number) times R raised to the t power (R^0, R^1, R^2, R^3, . . . R^t). This pattern of increase is known as exponential increase; microbiologists refer to this stage in bacterial growth within a culture tube as the *logarithmic growth* phase.

The number of generations need not be large before the expected number of individuals becomes impossibly large. Physical resources— space, nutrients, water, or physiologically essential trace elements— become limiting. One can infer this limit to population size and even test it empirically, but how can one describe it mathematically? If the limiting process can be described mathematically in the form of a *model*, then predictions can be made about what will happen if the population is manipulated experimentally.

A commonly used calculation proceeds as follows: At best, the environment can support only K individuals, so each individual must require $1/K$th of the environment in order to survive, and N individuals would require N/Kths of the environmental resources. The unused remainder of the environment, $1 - (N/K)$ or $(K - N)/K$, is the factor that limits growth; when the number of individuals reaches the limit, $N = K$, then $(K - N)/K = 0$, and the population can no longer increase.

The growth of a population at a given time can be expressed as

$$N_{t+1} - N_t = RN_t - N_t = N_t(R - 1) = rN_t$$

where $r = R - 1$. If we include the damping effect of the exhaustion of resources, the expression that describes the growth of a population, ΔN (read "delta-N"), is

$$\Delta N = rN[(K - N)/K].$$

When N is small, the increase is virtually rN; when N approaches K, the increase approaches zero.

The equation as given allows N to become larger than K; that is, it allows the number of individuals to exceed the number that can be

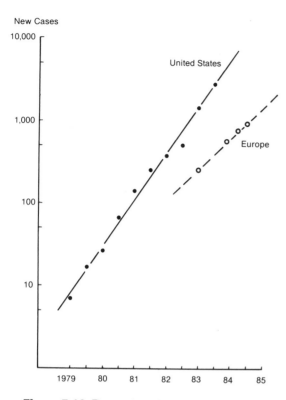

New Cases

Figure 7.16. The number of new cases of AIDS reported on a (generally) semiannual basis in the United States and Europe. The data suggest that during the years covered the number of new cases increased exponentially on both continents. Remember that under exponential increase, the time required for the number of new cases to increase from 1000 to 10,000 is no more than that required for the number to increase from 10 to 100. The rate of increase in Europe seems to be somewhat less than that in the United States; in fact, the dates of origin of the disease on the two continents may have coincided.

supported by (that can survive in) the specific environment. In this case, ΔN, the change in population size, becomes negative, and the population shrinks.

You might take some numerical values for r, N, and K, and (1) calculate the change in population size (ΔN) for each generation, (2) adjust the population size according to your result, and (3) repeat the calculation for the next generation. Do not be surprised if your curves are not smooth; the equation allows for large oscillations. These oscillations are interesting biologically because when N greatly exceeds K, all the individuals of a natural population could die (of overcrowding).

INFORMATIONAL ESSAY
Epidemiology

The population growth of some organisms corresponds to the spread of a disease in others. According to Daniel Defoe's *Journal of the Plague Year*, London's plague epidemic of 1665 began with the deaths of two French sailors who were boarding in a house on Drury Lane. A third death soon occurred in the same house. Then, after a lapse of some weeks, during which the hopes of many were allowed to rise, a fourth person died in a different house. Before the epidemic had run its course, 100,000 or more persons had perished. (At that time, London was a city of only 400,000–500,000 persons, not today's sprawling city of seven million inhabitants.)

The numbers of AIDS victims reported by the Centers for Disease Control (CDC) illustrate the early exponential spread of another disease. The number of new cases reported in the first and second half of each year from 1979 to 1982 are as follows: 1, 6, 17, 26, 66, 141, 249, and (incomplete data) 593. Plotted on semi-log paper, these numbers form a remarkably straight line: the number of new cases more than doubled every six months. (Nearly 5700 new cases were reported from May 1984 to April 1985. Given an exponential increase in disease victims, is this an unexpected number?)

For diseases that leave their victims either dead or immune to subsequent infection, unimpeded exponential spread cannot continue for long: the number of susceptible individuals decreases, thereby reducing the chance that a victim will infect others. Following the technique described in the preceding essay, we can say that during the exponential phase, the expected number of new cases equals the number of ill persons multiplied by the average number each one will infect:

New cases "tomorrow" (N_{t+1}) = Number ill "today" $\times$ I

where I is the average number each ill person infects.

To illustrate a disease running its course, a term that will eventually reduce N_{t+1} to zero must be introduced into the model. This term can be

$$1 - \frac{\text{Persons (alive or dead) who have been ill}}{\text{Total number of persons in the population}}$$

This can be written as $1 - (N_{\text{total}}/P)$, where P = the total number of persons in the population. Thus, the entire course of an epidemic can be written as

$$N_{t+1} = IN_t\,[1 - (N_{\text{total}}/P)]$$

The course of a disease in a city of 100,000 persons, where each person is said (at the outset) to infect two new persons, is shown in Figure 7.17. Under the *mathematical model* that we have used (others are possible, of course), some 15,000 persons never become infected, but the numbers of new cases do grow nearly exponentially for about ten cycles of infection. After another five or six cycles, when the epidemic is rampaging, however, the number of ill persons peaks and then quickly wanes. After another five or six cycles, newly infected persons have become medical rarities: the epidemic has passed. It will return again only when sensitive persons once more constitute the bulk of the population.

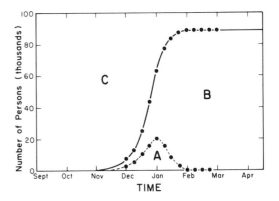

Figure 7.17. The course of an epidemic in a hypothetical city of 100,000 citizens. Each ill person is assumed to infect two others. Thus, when the disease first arises, the number of infected persons increases exponentially: 1, 2, 4, 8, 16. . . . The number of persons who have had the disease (it is not fatal) also increases nearly exponentially at first; when the above numbers are summed (from the left), the totals equal 1, 3, 7, 15, 31. . . . However, when the number of recovered persons becomes large, each ill person is not likely to infect two others, because one of the "infected" ones is often resistant to the disease. At this point, the number of ill persons declines, and the total number of infected persons in the population levels off, leaving about 15,000 persons unaffected by the illness. A = persons who are ill at the specified time; B = persons who have been ill and have recovered; and C = persons who have not yet contracted the disease.

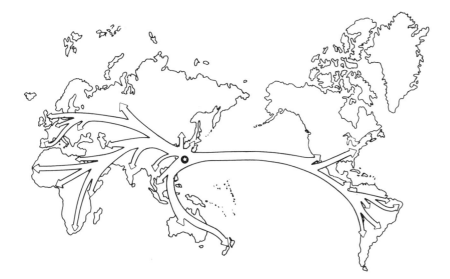

Figure 7.18. The worldwide spread of a modern epidemic, the Hong Kong flu. This strain of influenza was first discovered in Hong Kong(*) during July 1968. Later that year it was detected throughout Southeast Asia, Australia, New Zealand, on both coasts of the United States, and in the major cities of South America. By early 1969 it had swept through the remaining countries of the world. Air travel facilitates the spread of diseases that are not dependent upon sanitary conditions (cholera, for example, spreads in a different pattern). Hong Kong flu entered the United States by way of California and New York, sites of our largest international airports.

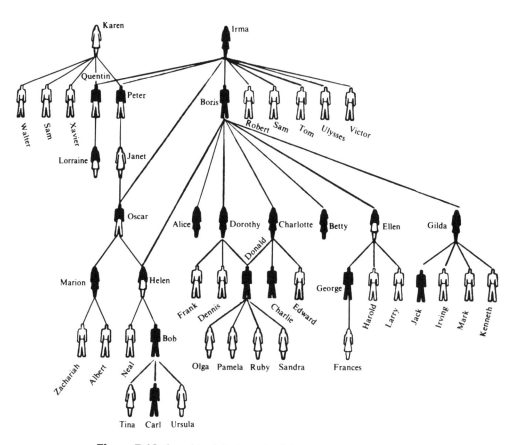

Figure 7.19. An epidemiologic study of the spread of two sexually transmitted diseases, syphilis and gonorrhea (shaded figures have one, or the other, or both) in a small American city. Lines connecting the figures indicate sexual contact. First names are given to emphasize that these are real persons and not mere numbers in a statistical table. Of the approximately fifty sexual contacts that connect this group of persons, one is homosexual (Bob–Carl). The proportion of homosexual contacts is similar in other (and larger) epidemiologic studies, as well.

INFORMATIONAL ESSAY

A Manmade Epidemic among Australian Rabbits

In 1859, a gentleman who was fond of hunting released twelve pairs of English wild rabbits near Geelong, Victoria, in southeastern Australia. From those twenty-four individuals rabbit offspring spread over most of Australia, establishing themselves as the major pest of agricultural and range lands.

Following several unsuccessful attempts to control the spread of the hordes of rabbits that were devastating the landscape, a smallpox-like organism, the myxoma virus, was introduced. This virus is endemic in wild rabbit populations of South America. Scarcely any evidence of its existence is found there, other than an occasional rabbit with a growth on its skin that persists for several weeks. If fluid from such a tumor is injected beneath the skin of a healthy South American rabbit, it will also develop a growth that persists for several weeks. If, however, this fluid is injected beneath the skin of a European wild rabbit, a severe, lethal disease results: the head and the eyes become swollen, and tumorous growths occur throughout the body, not just at the site of the original inoculation.

Mosquitoes are the natural vector of the myxoma virus. The transfer is purely mechanical; the virus does not reproduce within the mosquito's body. If a mosquito inserts its beak into a virus-infected growth on one rabbit and then bites a second rabbit, virus particles clinging to its beak may then be injected into the second animal, thus spreading the disease from the one animal to the next.

In 1950, several wild rabbits were captured in Australia and infected with myxoma virus. During the following six or seven years, the resulting viral disease, *myxomatosis*, spread throughout southeastern Australia, along the southern coastal area, and to the region around Perth, 2000 or more miles to the west.

A particular introduction of viruses was studied near a somewhat isolated lake in October 1951. In this region, some 5000 rabbits could be counted during a walk of about 400 yards—less than a quarter of a mile. After the release of a number of infected rabbits, the number of healthy ones counted during those quarter-mile walks dropped to fifty by Christmas. Instead of healthy rabbits, there were hundreds of sick ones. The number of healthy rabbits increased once more during the Fall of 1952, only to decline sharply as the epidemic recurred.

The myxoma virus and the wild European rabbit now coexist in Australia, with the number of rabbits far below what it was during the years when rabbits overran the countryside. Coexistence has been achieved not merely by the scarcity of rabbits, although scarcity is important. Remember that a mosquito, in order to spread the disease, must bite two animals, presumably within some definite period of time (during the lifespan of the mosquito, if no other factor establishes a shorter interval), and the farther animals are from one another, the less likely a mosquito is to find one animal, let alone two.

The virologists working on the myxoma viruses of Australian rabbits have shown that both the virus and the rabbits changed during the years following the introduction of myxomatosis. During the early 1950s, strains of the virus were extremely virulent and killed infected animals quickly. During the late 1950s, the tested viruses proved to be less virulent, and many failed to kill even English rabbits.

The rabbits also changed. There was a steady decline in the proportion of deaths among nonimmune, wild Australian rabbits tested with a standard (unchanging) laboratory strain of myxoma virus. The mortality rate declined from a high of 90% or more at the start, to 40% to 60% after three or four years, and to 20% after seven years.

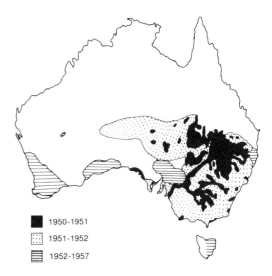

1950-1951

1951-1952

1952-1957

Figure 7.20. The spread of the fatal disease myxomatosis among the rabbits of Australia. The site of the original (1950) introduction of the disease is marked by a star. The area in which infected rabbits had been seen by 1951 is shown in black, by 1952 in stippling, and by 1957 in horizontal shading. No data are available for much of Australia, but we can say with certainty that the spread of the disease has been at least as great as is shown here.

Table 7.2. Changes in the virulence of myxoma virus strains obtained from wild populations of Australian rabbits (1950–1959)[a]

Year of test	Survival after infection (expressed as percent)	
	<28 days (90%+ mortality)	>28 days (60% or less mortality)
1950–51	100%	0%
1951–52	100%	0%
1952–53	91%	9%
1953–54	91%	9%
1954–55	74%	26%
1955–56	58%	42%
1956–59 (average)	65%	35%

[a] The rabbits tested were laboratory-raised animals that had never been exposed to myxoma viruses.

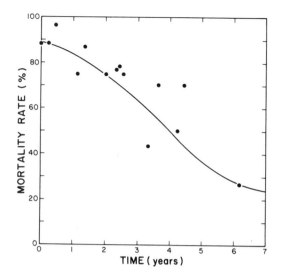

Figure 7.21. Evidence that after the introduction of myxomatosis, the surviving Australian rabbits gradually developed an immunity to the disease. Recalling what was said in an earlier chapter about the generation of immunoglobulins through random rearrangements of certain genes during the formation of lymphocytes, we can use these data to suggest that the overall repertoire of randomly generated immunoglobulins is subject to change under the operation of natural selection.

An evolutionary response of hosts and parasites to each other's existence depends, of course, on the existence of genetic variation. With genetic variation, the common pattern of coexistence is for the host to develop resistance and for the parasite to lose virulence. The passage of smallpox viruses through Circassian children by deliberate inoculation, as described by Voltaire, is an example given in an earlier chapter. Presumably, the pustules chosen to provide inocula were from babies exhibiting mild disease symptoms; no one would choose to inoculate a healthy baby from a pustule on a dying baby.

Pathogens like viruses are subjected to conflicting selection pressures: within an infected individual, the more aggressive (virulent) organisms may out-multiply the others. In fact, if transmission is no problem (as under crowded conditions), the virulence of a disease may increase with time. In contrast, when transmission from one victim to the next becomes a major problem, pathogens that allow infected individuals to survive for a time while building up large populations of mildly pathogenic organisms within the infected individual are at an advantage over those that kill their victims quickly. For the victim, the selectively favored pathogenic traits are experienced as a loss of viral virulence.

8

SELECTION: NATURAL AND ARTIFICIAL

Genetic variation exists within all populations, including those of our own species, *Homo sapiens*. This variation is a demonstrable fact. Once it exists, as we learned in the previous chapter, there is no tendency for it to be lost—contrary to the belief prevalent in Darwin's time, a belief stemming from an ignorance of hereditary mechanisms.

What is the source of this variation? Errors in the replication of DNA. If DNA replicated perfectly during every cell division, in every bacterial division, and in the division of every virus particle, life on Earth—if it existed at all—would be like the first form that ever arose. Mistakes are made, gene and chromosomal mutations do occur, and these mistakes give rise to the variation we see in all organisms. If existing variation has no tendency to disappear, there need not be an enormous supply of new mutations in order to accumulate a great deal of variation. To use an analogy, if a bathtub is securely plugged, the slightest leak of the faucet will eventually fill it to overflowing.

On the other hand, mutant genes causing physical abnormalities are removed from populations; affected individuals may fail to survive or, if they should survive, they may fail to reproduce. An individual who dies because of a genetic mutation removes a copy of the abnormal gene from the population. We have already seen that the total number of alleles in a population can be counted: their number is twice that of the number of individuals in the population. Therefore, when a copy is lost because it affects the health or fertility of its carrier, the remaining frequency of that harmful allele is measurably decreased. A population of 5000 individuals carries 10,000 copies of each gene. If there are 1000 copies of an allele, a_1, at the A locus, and if this allele "kills" one of its carriers, the overall frequency of a_1 has decreased from 1 in 10 to 999 in 9998. The decrease from 0.1000 to 0.0999 is much larger than most mutation rates. At this point, notice we said above, "if this allele 'kills' one of its carriers," but the elimination of a gene by death or sterility need not be absolute. Not all carriers of mutant alleles need die, nor do all need to be sterile. The mutation is removed from the population if it causes only a weakness or a proneness toward dying and sterility.

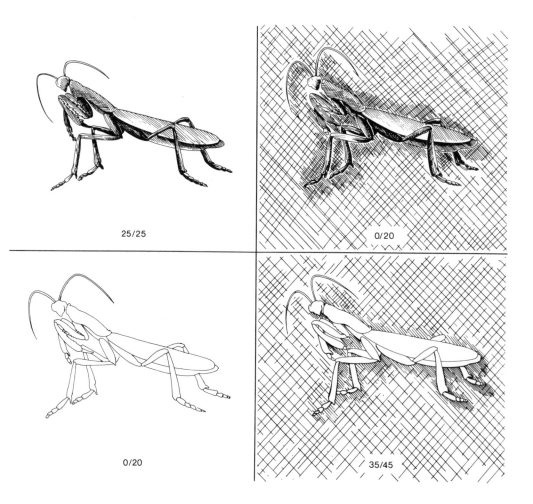

Figure 8.1. Differential predation on praying mantises: a century-old experiment. Praying mantises are either green or tan. An early investigator tethered individuals of both colors in green (fresh) and brown (dead) grass. Over a span of nineteen days, no green insects tethered on green grass had been eaten; in the same period, no brown insects tethered on brown grass had been eaten. In contrast, all twenty-five green mantises that had been tethered on brown grass were eaten by birds within eleven days and thirty-five of forty-five brown insects tethered on green grass had been eaten within nineteen days. The matching colors—green on green and brown on brown—greatly reduce loss to predation. Unmatched colors—green mantises on brown grass or brown ones on green grass—result in heavy predation.

25/25

0/20

0/20

35/45

NOW, IT'S YOUR TURN (8-1)

A biologist once claimed that there was no evidence of natural selection acting upon two forms of a beetle, light and dark, because in the stomachs of toads he had found both color types. Suppose he had made a careful count and had found that, among living beetles, the proportion of light to dark was 70:30, whereas among the remains in toads' stomachs the proportion was 55:45. Would you say that, at least for the toads' diet, selection was occurring? (One must assume, of course, that the number of beetles counted was adequate for forming a judgment. What, for example, would be your conclusion if each pair of proportions—among both living and eaten beetles—was based on twenty beetles?)

The frequency of a mutant allele in a population can be changed if its carriers do not survive as well or reproduce as rapidly as individuals who do not carry this mutant; that is, if mutant carriers are less "fit."

Many years ago, William E. Castle, as a student at Harvard University, showed the consequences with respect to gene frequency of culling from a population all individuals carrying two copies of a particular gene. The consequences are not hard to follow: Assume that the starting frequencies of a_1 and a_2 are 0.50 each; then the frequencies of a_1a_1, a_1a_2, and a_2a_2 individuals will be 0.25, 0.50, and 0.25, as we learned in the last chapter. Castle now culls all a_2a_2 individuals, leaving only a_1a_1 (0.25) and a_1a_2 (0.50) individuals. Because only 75% of all individuals now remain, the frequencies of the two remaining types must be recalculated: a_1a_1 (33⅓%) and a_1a_2 (66⅔%). The frequencies of a_1 and a_2 have now become 66⅔% a_1 (= 33⅓ + ½ × 66⅔) and 33⅓ a_2 (= ½ × 66⅔). From relative proportions of ½:½, the proportions of a_1 and a_2 have become ⅔:⅓.

We can run through one more generation of culling. This time we shall express frequencies as fractions: the frequencies of a_1a_1, a_1a_2, and a_2a_2 are now 4/9 [= (⅔)²], 4/9 = 2 × ⅔ × ⅓), and 1/9 [= (⅓)²]. After culling of the a_2a_2 individuals, only 8/9 of the population remains; consequently, the frequencies of a_1a_1 and a_1a_2 become 50% and 50% (½ and ½). The frequency of the allele a_1 is now 75% [½ + (½ × ½) = ¾] while that of a_2 is 25% (½ × ½ = ¼). A pattern has now emerged; under culling, a gene whose initial frequency is 50% (= ½) drops to 33⅓% (⅓), then to 25% (¼), to 20% (= ⅕), and so on through ⅙, ⅐, ⅛, ⅑. . . . Selection alters gene frequencies. The example used here involved artificial selection, since a_2a_2 individuals were deliberately culled by the experimenter. Figure 8.2 shows that natural selection eliminates a recessive lethal allele from a population in essentially the same manner.

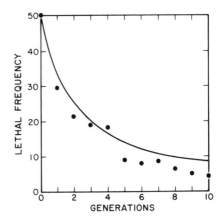

Figure 8.2. The elimination of a recessive lethal gene from a population of vinegar flies. Because the original frequency was 50% (1/2), the subsequent frequencies (in fractions) should have been 1/3, 1/4, 1/5, 1/6, 1/7, 1/8, 1/9, 1/10, 1/11, and 1/12, as explained in the text. The actual frequencies seem to be consistently lower than the expected ones (nine out of ten points on the curve lie below those that were expected). It seems, then, that this lethal gene was harmful to its heterozygous carriers, as well as being fatal to homozygous ones.

NOW, IT'S YOUR TURN (8-2)

The sequence of frequencies of a recessive gene following the culling of all homozygous (a_2a_2) individuals continues through all subsequent frequencies: ½, ⅓, ¼, ⅕ . . . 1/100, 1/101, 1/102 . . . 1/1000, 1/1001,

Table 8.1. The ages of some cultivated plant species

More than 4000 years old

Apple	Fig	Pear
Apricot	Flax	Rice
Banana	Grape	Soybean
Barley	Mango	Tea
Cabbage	Olive	Turnip
Cucumber	Onion	Wheat
Date palm	Peach	

From 2000 to 4000 years old

Alfalfa	Cotton	Radish
Beet	Pea	Rye
Carrot	Lettuce	Sugar cane
Celery	Oats	Yam
Cherry	Plum	

New World species

Avocado	Pineapple	Sweet potato
Bean	Potato	Tomato
Corn (maize)	Pumpkin	
Peanut	Squash	

Table 8.2. Some domesticated animals: the approximate time that they have been domesticated, and an estimate of the number of varieties that have been selected for

Common name	Years domesticated	Number of varieties
Dog	10,000 +	200
Cat	4000	25
Cattle	6000	60
Sheep	7500	50
Goat	7500	20
Pig	6000	35
Horse	5000	60
Rabbit	3000	20
Duck	3000	30
Chicken	5000	125
Canary	500	20
Parakeet	150	5

$\frac{1}{1002}$. . . . This fact has an important bearing upon the effectiveness of certain heroic measures (such as sterilization) that were once advocated by eugenicists for bettering human populations. Suppose individuals carrying two recessive alleles appear once in every 40,000 births; these individuals survive but are severely handicapped. What would be your response to a program under which these individuals would be sterilized? (Hint: What would be the frequency of the responsible gene in the population if homozygous carriers have a frequency of $\frac{1}{40,000}$? If culling occurred, what would be the sequence of frequencies over the first four or five generations during which affected individuals were removed? How many generations would elapse before the starting frequency was reduced by one-half? Finally, if these individuals are severely handicapped, is it likely that they are even now leaving any children?)

Selection in Agriculture

The frequencies of genes (or, allelic forms of genes, if we are to be precise) in populations can be altered by the differential reproduction of individuals of different genotypes, of different genetic endowments. The altered gene frequencies lead in turn to new and altered proportions within the population of individuals with different genotypes. The relationship between the frequencies of two alleles at one locus and those of the three possible genotypes was described previously as an example of the binomial expansion $(a + b)^2 = a^2 + 2ab + b^2$. No matter how many gene loci are involved in producing a particular phenotype, nor how many alleles exist at any of these loci, there exists a binomial expression, complex as it may be, that describes the expected proportion of every possible genotype. If individuals possessing these genotypes have different rates of reproductive success, then gene frequencies in the population probably change.

Fortunately for primitive man, he did not need to understand algebra in order to manipulate recently domesticated or even semi-domesticated plants and animals for his own purposes. The key to the success of early agriculture lay in the understanding that like begets like. Early "experimenters" observed how familial traits are passed down from generation to generation, although the exact patterns of inheritance were too complex for them to comprehend. (The experimental analysis that eventually simplified matters so that inheritance patterns could be understood was done by Gregor Mendel.) Although various physical traits seem to disappear in a family tree from time to time (they "skip" generations), primitive peoples appeared to have understood that if a particular trait was desirable, one should continue breeding the family or line in which it was last seen. Our early ancestors were so successful at agriculture that all major foodcrops

Figure 8.3. The progress pigeon breeders have made in altering the wild rock dove into a number of fancy varieties of domesticated pigeons. The figure that has been positioned between the rock dove and each modern example (Pouter, Jacobin, and Fantail) shows how each variety appeared a century or more ago.

Pouter

Jacobin

Fantail

Wild Rock

and all major groups of domesticated animals were developed before historic times; even most major breeds of animals predate the modern science of genetics. Knowledge of genetics has permitted advances under selection to be made faster than they once were, and to be made more economically than they would be without genetic knowledge, but the material on which modern animal and plant breeders operate—cattle, hogs, poultry, small grains, and vegetables—were initially developed by illiterate prehistoric human beings.

Although one cannot reconstruct the exact course of events that occurred when the first plants were domesticated, one can observe the events that occur when agronomists decide to domesticate a wild plant species today. The pace of change is probably much faster today because modern plant scientists can specify more precisely what traits they are seeking. Primitive farmers would have had no such preconceived ideals; when they began planting and harvesting their own gardens, they probably had their personal convenience in mind. Having plants near home may have pleased not only the lazy members of the family group but also baby-toting mothers and small children, who were most likely to fall prey to predators while out gathering wild seeds and fruits.

Artificial and Natural Selection
I. M. Lerner, 1958

It must be understood that *natural selection* is really not an *a priori* cause of any phenomenon observed in nature, in the laboratory, or on the farm. Natural selection is a term serving to say that some genotypes leave more offspring than others. Natural selection has no purpose. It can be deduced to have existed and its intensity can be measured only *ex post facto*. For any given generation, natural selection is a consequence of differences between individuals with respect to their capacity to produce progeny. The individuals who have more offspring are fitter in the Darwinian sense. To speak of natural selection as causing one array of individuals . . . to have offspring, and another . . . not to, is a tautology.* This fact, however, should not prevent us from attributing a major part of evolutionary change, viewed in retrospect, to natural selection.

Artificial selection, in contrast, is a purposeful process. It has a goal that can be visualized. It may, indeed, be the immediate cause of changes in the genetic composition of populations. The breeder or the experimenter can prevent at will certain individuals in the population under his control from reproducing themselves, thereby decreasing the frequency of their particular kind in the next generation. The fitness of the group of individuals selected, that is, the number of offspring it will leave, is predetermined by the selector to be greater than that of the group rejected.

Natural selection, even under laboratory or farm conditions, can proceed without artificial selection. The latter, generally speaking, does not exist in pure form. Man can decide which individuals are to be discarded or prevented from reproducing themselves (i.e., which individuals will have a fitness of zero). Yet, among the individuals he wishes to be represented by offspring in the next generation, there may still be natural selection. Some of those chosen to be parents may be found to be sterile and thus are destined to join their culled contemporaries in leaving no progeny; among the others there may be a wide variation in fitness, attributable to differences between their genotypes. The fact that natural selection usually accompanies artificial may . . . play a role of considerable importance in attempts to improve domestic animals.

*Tautology: Redundancy consisting of needless repetition of meaning in other words; for example, "audible to the ear" and "my father is male."

Figure 8.4. A survey of the variation that exists among the different breeds of dogs. Considering size alone, the variation exhibited by these breeds is much larger, say, than that separating wolves, coyotes, and foxes. Why, then, are all the varieties of dogs said to belong to a single species, *Canis familiaris*? *(Copyright, The Biological Sciences Curriculum Study; reproduced with permission)*

Figure 8.5. The variety of garden vegetables created by artificial selection from a single species of plant, *Brassica oleracea*. The diversity of the varieties easily matches that of the dogs shown in the previous figure. In the case of these garden vegetables, the chief variant is the site of food storage within the different plants: (a) terminal buds in cabbage, (b) flowers in cauliflower, (c) stem in kohlrabi, (d) stem and flowers in broccoli, (e) leaves in kale, and (f) lateral buds in Brussels sprouts. *(After J. R. Harlan)*

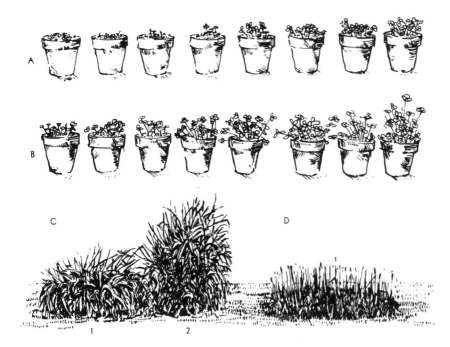

Figure 8.6. The effect of natural selection on the growth habits of clover plants grown for several years (A) in a pasture under grazing by cows and (B) in an adjacent hayfield that was not grazed. The differences illustrated here are inherited; they are manifested by plants grown from seed planted in an experimental garden. Under grazing, the surviving plants were those so low in habit as not to be readily eaten; in the hayfield, the surviving clovers were those whose stems were tall enough to compete successfully with grasses and other hayfield plants. Grasses also responded to the pressures of natural selection in the hayfield and pasture. The two grasses from the pasture (C1 and D) are contrasted with the tall grass that inhabits the hayfield (C2). At seed-bearing time, plant C1 will abandon its low habit and become erect like C2; plant D is a dwarf that never will become tall. As in the case of the clovers, low-growing grasses best avoided grazing cows in the pasture.

Israeli plant breeders recently collected seeds from a wild *Trifolium* species—a species closely related to peas. The original seeds were sown, and new seeds were collected from the dense stand of plants that resulted. Each year the seeds of all plants were combined, mixed, and sown, and, when it was ready, the next year's crop was harvested. The amount of seeds harvested each year was allowed to increase so that larger and larger harvests were obtained. After some ten years of domestication, the characteristics of the domesticated seeds and plants were compared with a new sample of wild seeds obtained from the same wild population.

The characteristics of the wild seeds and plants had scarcely changed in the intervening years: most seeds were dark, weighing between 1.1 and 1.2 mg. When the integument of the seed was not scarred manually, only three to seven percent of the seeds germinated. Wild plants tended to be nearly prostrate; the height of the plants was only one-third the length that they extended along the ground. Different wild populations did differ considerably in seed color; although the sampled population had mostly dark seeds, other populations as little as 10 kilometers away had somewhat smaller, yellow-orange seeds.

What were some of the changes that were brought about *without intent* by the plant breeders who raised the peas for ten years in their experimental fields? Seed size nearly doubled, and yellow-orange seeds increased from ten to seventy percent of all seeds. The original seeds were hard, irrespective of their color; the domesticated population had primarily soft seeds, which unscarred showed fifty-five per-

Tailoring the Crop to the Machine

Because of the extensive use of machines in modern agriculture, artificial selection is aimed largely at adapting the crop to the demands of mechanization. Uniformity is the goal: uniformity in time of development, uniformity in the size of seeds that are to be planted, uniformity in the height or position of the part that is to be harvested, uniformity in harvest time (time of ripening), and, if the crop is to be canned or packaged, uniformity in the size of the head, ear, or fruit. The use of mechanical pickers requires that the crop be resistant to bruising and subsequent spoilage, as well.

cent germination, rather than a germination frequency that was less than seven percent. Plants became more erect. They developed fewer branches but became heavier, nevertheless; the weight of individual branches nearly doubled. Their growth rate changed: the domesticated plants were about three times heavier than the wild ones after forty days' growth. Finally, many wild plants would not self-fertilize (pollen from one plant or flower would not fertilize eggs from that same plant or flower); several of the domesticated plants were fully self-fertile.

These changes have been labeled *the syndrome of domestication*. The traits most involved were seed softening, increase in seedling vigor, change from prostrate to erect growth, and self-compatibility. Whereas wild stands are sparse (perhaps because of the low germination of unscarred seeds), the domestic stands were dense. The denseness itself put a priority on rapid, erect seedling growth if the plant were to produce seed for harvest. Wild stands contain plants of various ages; the domesticated plants were of one age only. The seeds planted under domestication were six months old or younger; only soft seeds can germinate at that age. The change in seed color is explained by assuming that dark seed pigment is a fungicide, and that fungi soften seeds, thus allowing them to germinate. After their studies, the Israeli plant breeders appreciated for the first time the intensity of the selective forces that they had unintentionally turned loose on these *Trifolium* plants.

An earlier statement suggested that an appreciation of how "like begets like" played a strong role in domestication. The experience described above points out another source of selection: unintentional

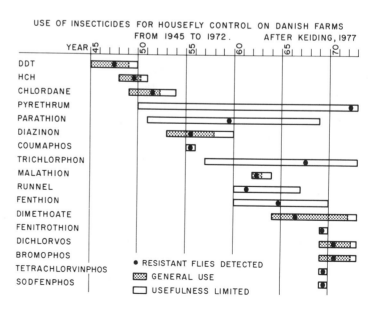

Figure 8.7. The race between the development of insecticides by chemists and the development of insecticide resistance by flies in Denmark. The occurrence of a solid circle (●) in the horizontal bar opposite each insecticide shows when flies with a genetically based physiological resistance to that insecticide were first detected. Sometimes the flies developed resistance quickly; other times they required five, ten, or even twenty years. Generally, the usefulness of an insecticide vanishes shortly after the flies' resistance to it is detected. *(Copyright, Columbia University Press; reproduced with permission)*

selection that exists simply because the procedures followed by domesticators (and this would include the most primitive prehistoric farmers) are not identical to those followed by wild plants left to themselves. The wild plants that did not adapt to domestication disappeared; those whose characteristics coincided with the demands (intentional or otherwise) of the farmers became predominant.

NOW, IT'S YOUR TURN (8-3)

Modern breeders seem to overlook at times the importance of offspring in determining a parent's role in a population: a population takes on the characteristics of its breeding members. Breeders of many animals decree that certain body conformations or stances are desirable, and award championships to animals most nearly resembling these idealized forms. Can you make a guess as to why, in many popular show animals, genes with recessive lethal effects have sometimes appeared in the breed, and then spread through the breed with startling speed? Examples that come to mind are bulldog calves in shorthorn cattle, the *nude* gene in mink, and hip dysplasia in show dogs. (Hint: The carrier of one recessive lethal allele survives; although we use the term "recessive lethal," the gene may have discernible effects of one sort or another on its heterozygous—single-dose—carrier.)

Figure 8.8. An aerial photograph revealing how modern agriculture demands vast, uninterrupted stretches of farmland. This is the type of agriculture that relies heavily upon energy subsidies provided by oil. It is also the type of agriculture that leads to monocultures—that is, thousands of acres planted to one crop—and to the infestation by insect pests and fungal diseases that such "culture conditions" make possible. *(Photo courtesy of the Dow Chemical Company)*

Modern agriculture virtually demands that immense, uninterrupted stretches of land be planted to a single crop; mechanical harvesters work more efficiently under such uniform circumstances. (One of the scientific journals reported several years ago that Midwestern archeologists studying remnants of the American Indians of the region were hurrying to beat the arrival of the bulldozer. It seems as if the slight mounds that mark ancient Indian burial grounds were somewhat too uneven for newly developed harvesting machines, and so the ground was to be made more nearly level for their operation.)

Large areas planted to a single crop are, of course, an invitation to a feast for any insect pest or disease organism. The enormous expanse of one-crop farmland is to a single gravid female insect pest (imagine a white cabbage butterfly and thousands of acres of cabbage!) what a sterile bacterial culture tube is to a single bacterium. In fact, wheat rust is a microscopic fungus, and its culture medium each spring is the expanse of wheat fields that extend from the southern borders of Texas, across the Midwestern and Plains states, and into Canada.

Much of the effort that is invested in wheat breeding is spent devel-

Table 8.3. Changes in the frequencies (given to the nearest 5%) of wheat rust varieties found in the United States over a period of fifteen years. These changes represent responses by the rust to the planting of rust-resistant varieties of wheat

Year	Frequency of rust variety						
	A	B	C	D	E	F	G
1930	5	0	0	35	30	20	0
1932	5	0	0	10	45	25	0
1934	0	0	20	20	5	0	35
1936	10	5	5	5	20	0	45
1938	0	5	0	0	15	0	65
1940	5	35	0	0	10	0	45
1942	0	25	0	0	25	5	30
1944	0	20	0	0	25	0	45

oping plants that are resistant to various strains of rust. The mutants are found already existing in the many strains of wheat—wild and domesticated—that exist in the world, or they arise by mutation at rare intervals in wheat strains already at hand

Wheat rust and wheat breeders are locked in a deadly serious battle. If wheat is attacked by rust *alpha* (Greek letters are used for rust types; Latin letters for corresponding resistant types of wheat), the wheat breeder must quickly come up with an otherwise commercially acceptable wheat of type *A*, a wheat that is resistant to *alpha*. Then, however, rust type *beta* arises, and there is much activity among wheat breeders to come up with wheat type *A* and *B*, resistant to both *alpha* and *beta* rust. Failure to do so means disaster in the nation's wheat-producing states.

In actual practice, the procedure is somewhat different from that described above. Using a variety of rusts known to differ from one another, and working *outside* the United States because some of the rusts are too dangerous to bring into this country, plant geneticists attempt to develop good commercial wheats, each strain of which is resistant to some combination of rust types.

Wheat breeders are aided by one lucky fact: rusts that are able to attack a certain type of wheat cannot reproduce as rapidly as those that cannot attack that type—*if the susceptible wheat is not available*. Thus, rusts *alpha* and *beta* (which can attack wheat types *A* and *B*) will compete with one another if grown on type *B* wheat alone: *beta* rust cannot grow and is eliminated, while *alpha* plus *beta* rust grows more slowly than *alpha* rust alone, and therefore is also gradually eliminated from the expanding rust population.

The differential growth of wheat rusts has been a boon for Canadian wheat growers because whereas the rust that invades the southern portion of the wheat-growing sector of North America in the spring is often capable of attacking Canadian wheat, by the time the advancing rust population has reached Canada the infectious types have been eliminated because they are useless in infecting American wheat types. Despite this "handicap" under which wheat rust operates, there are many breeders who claim that the eventual outcome of the rust–wheat breeder battle will be won by the side that comes up with the last mutant gene. Will the breeder come up with a type that rust will be unable to match? Will wheat rust match each and every mutant variety of wheat the breeders produce? Or, will the rusts come up with an infective type against which the breeders will find no genetic protection?

Genetic variation is essential if a population is to respond to selective forces, whether they are natural or artificial. Sometimes even the professional breeders overlook that simple fact. Figure 8.9 shows the increase in the bushels of oats per acre on U.S. farms from about 1910.

The early increase resulted from a continuous search carried out worldwide in an effort to find the highest-yielding variety of oats possible. After the world's supply of high-yielding oat varieties was exhausted, a program of artificial selection was begun. The best inbred lines of oats were identified, and each year the very best individuals of those lines were chosen to provide seed grain for the farmers. For fifteen to twenty years the number of bushels harvested per acre remained constant. The best individuals of an inbred line (which has no genetic variation) do not differ genetically from the average ones, or even the worst ones. The best individuals are those that happen to encounter a much better than average environment: the best soil, the best soil drainage, the best access to sunlight, and other optimal conditions. No gain could be expected by selecting from genetically uniform material.

The steady gain in oat production since 1932 has resulted from the adoption of a new breeding and selection technique: high-yielding inbred lines of oats are crossed (hybridized), and a further generation is raised. Because two inbred strains are not likely to be genetically identical, the second generation consists of individuals possessing a variety of recombinant genotypes. Selection of the best of *these* individuals, because being "best" now has a genetic as well as an environmental basis, leads to a permanent improvement in yield. This improvement has continued steadily since 1932.

At this point, a statement in praise of modern agricultural genetics might be made. Our prehistoric, historic, and pre-Mendelian ancestors accomplished near-miracles using no more than the principle of "like begets like." Nevertheless, the application of modern genetic knowledge has led to more rapid progress than the earlier techniques, and perhaps even to progress that would otherwise never have been made.

One modern technique of plant (or animal) breeding that may have been unavailable to prehistoric or even historic (but pre-Mendelian) breeders is hybridization and utilization of the exceptional qualities hybrids often exhibit. That modern agronomists capitalize on hybrid vigor does not mean that they understand its molecular or physiological basis. However, today's plant and animal breeders know how to arrange a series of crosses to exploit a phenomenon recognized even centuries ago: crosses between unrelated (not too unrelated, and not too closely related) organisms often yield progeny possessing many superior, commercially desirable properties. For instance, the mule, a hybrid between the horse and the ass, was known in Ancient Greece. Nevertheless, an organized breeding program of the sort needed to produce "double-cross" hybrid corn, the backbone of today's corn production in the United States, simply could not have been sustained under conditions of primitive agriculture. In contrast, how-

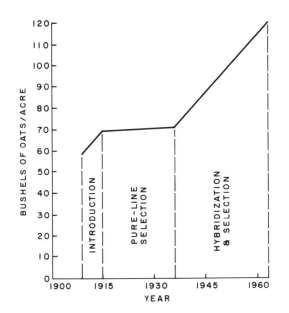

Figure 8.9. The yield (bushels/acre) of oats in the United States between 1910 and 1965. For a period of twenty years (1915–1935), there was no increase. During that time selection for improvement was carried out on "pure" lines—that is, on strains of oats that possessed little or no genetic variation. Yield did not increase again until selection was practiced on the genetically variable offspring obtained by intercrossing (or hybridizing) these pure lines. A response to selection, either artificial or natural, requires genetic variation. *(Copyright, Columbia University Press; reproduced with permission)*

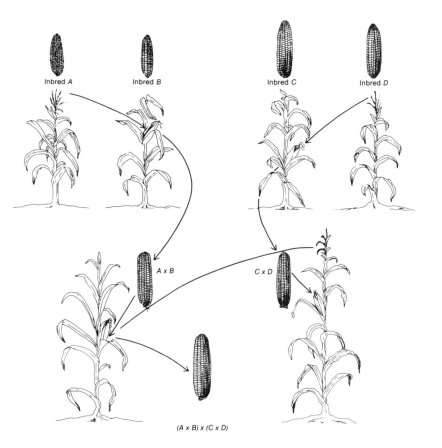

Figure 8.10. The systematic crosses that are needed to produce double-cross hybrid corn seed for use by farmers. Four inferior inbred corn varieties are crossed: *A* with *B* and *C* with *D*. These single-cross hybrid seeds are planted, and the pollen from one (*C* × *D,* in the diagram) is used to fertilize the silks of the other (*A* × *B*). The resulting seeds [(*A* × *B*) × (*C* × *D*)] are sold to the farmer for planting; they produce uniformly vigorous plants. The single-cross plants are also vigorous, but the number of such seeds available is dependent upon the inbred female plant (*B* or *C* in the diagram). Frequently, because the inbred strains are rather frail, the number of seeds in the first generation is small.

Inbred *A*   Inbred *B*   Inbred *C*   Inbred *D*

A x *B*   *C* x *D*

(A x B) x (C x D)

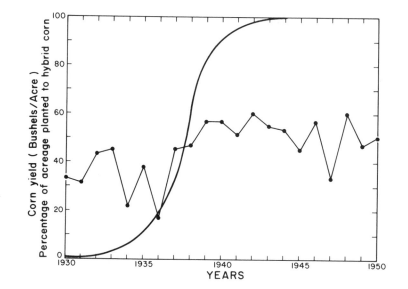

Figure 8.11. The rapid increase in the proportion of all corn acreage that was planted with hybrid corn during the years from 1930 to 1950. The corn yield in bushels per acre is also shown. Before 1938, the yield fluctuated between twenty and forty-five bushels per acre; after 1938 it fluctuated between thirty and sixty. The introduction of hybrid corn immediately raised the total production of corn by about fifty percent.

ever, the consequences of an occasional hybridization of wild and semi-domesticated plants were collected and perpetuated by early farmers, who were experts at recognizing and appreciating grains possessing new and desirable traits.

NOW, IT'S YOUR TURN (8-4)

The science of genetics has advanced so rapidly in the past quarter century—since the publication of the Watson-Crick model of DNA structure and the development of molecular techniques suitable for exploring all ramifications of this model—that it is essentially co-extensive with biology: every discipline within biology now pays homage to genetics and recognizes the influence genetics has within its domain. On the other hand, texts on animal breeding, while citing genetics on the one hand, often refer to "blood" as if it were genetic material. The horse breeder's adage, "Breed speed with speed," is scarcely better than the ancient "like begets like." Perhaps you can do better. Pretend that you are an early New England farmer, raising sheep, and cursing because low stone walls and fences separating fields (the stones that were cleared from the fields *had* to be placed somewhere) are not high enough to keep your animals from escaping. Then, a lamb (male) is born whose legs are exceedingly short, although its body is normal. "Here," you say to yourself, "is an animal that will not jump my fences." Outline a breeding program that you would follow in an attempt to develop an entire herd of short-legged sheep.

Character Displacement

We now return to natural populations of both plants and animals and to the matter of utmost concern to these organisms: successful reproduction. To anthropomorphize may be dangerous; nevertheless, nearly every species of plants and animals on the earth faces extinction, just as human beings do, unless male and female germ cells can meet, unite, and start the new, replacement, generation. Regardless of prevailing emotional, moralistic, spiritual, or empirical attitudes about sex, human beings would not exist if they did not practice it. Neither would spiders, praying mantises, sowbugs, or petunias. Somehow, the mature individuals who possess a supply of sperm or eggs must recognize one another, be attracted to one another, and be equipped to bring their gametes together. Plants that rely on intermediaries, such as wind or insects, need to make a few slight changes in the needs listed above: they must attract the insect pollinator and

Figure 8.12. Some of the bill sizes and shapes that have evolved among the Hawaiian honey creepers. These birds, like Darwin's finches of the Galapagos Islands, have descended from a few early island colonizers. As the shapes of the bills suggest, they are used in different ways (pecking, probing, and crushing) to obtain different kinds of food (crawling and burrowing insects and hard-covered seeds).

ensure that the pollinator's next stop is a flower of the right species. As one might expect, wind pollinators rely on sheer numbers: enormous numbers of pollen grains are released into the wind, with the (statistical) expectation that at least one of them will fall upon each receptive style.

Organisms, including human beings, in which males and females must meet and form individual mating pairs need some sort of communication signals to bring about this personal contact. When male high-school and college students offer to carry textbooks for their female classmates, they are carrying on a tradition as old as higher life itself. Among insects, males may offer morsels of food to females in an effort to become the father of offspring. Roosters (often after eating quite a bit themselves) cluck and call nearby hens to share newly found food. Male bower birds, in a charming likeness to human behavior, build elaborate structures of sticks and straw, decorate them with flowers, stones, and trinkets of carefully selected colors, and entice females to enter. The bower, like a young man's apartment or van, has nothing to do with "housekeeping"; it is a courtship device corresponding to an invitation to dine out or to go to the movies.

Closely related species of organisms that occupy the same locality encounter particularly difficult problems. Almost by definition, matings between males and females of different species are barren; either hybrid offspring are not produced or, if they are, their own offspring are inviable. The cost of producing inviable offspring is enormous. In terms of natural selection, an individual may as well have been born dead, or sterile, as to have his or her gametes unite with those produced by an individual of a different species.

The outcome of selection favoring *intra*specific, rather than *inter*specific, matings is a sequence of signals or of behaviors that ensure that mating pairs are of the same species. Within single localities, even closely related species have widely divergent mating calls, plumages, or other mate-recognition signals. Under the pressure of natural selection, closely related birds inhabiting the same area may develop very different plumages—so different that ornithologists may first assign them to different genera or families. And yet, a related bird on another continent may have retained the original plumage and thus resemble one of the coexisting forms. However, the two similar forms may have been separated for such a long period of time that their DNAs are by now demonstrably different. Thus, similarity of plumage does not mean no change in DNA, the genetic material; on the other hand, marked differences in plumage can occur rapidly, in too short a time for DNA divergence to occur.

That closely related species inhabiting the same area (*sympatric* species, as opposed to *allopatric* species, which are species inhabiting different localities) exaggerate their mating signals (calls, dances, plumages, or whatever) or mating behaviors (seasons during or places at which mating occurs) is a phenomenon known as *character displacement*. Thus, frog species living apart (allopatrically) may have essentially the same call; where they occur sympatrically, however, the calls of different species diverge. Plants of different species living apart may have similar floral structures; the same plants living sympatrically develop different floral anatomies and, consequently, attract very different pollinators. Fruit flies perform mating dances, and some even have spectacular (to a fly) markings on the head. If two related species coexist, though, both the head markings and the mating dances diverge.

NOW, IT'S YOUR TURN (8-5)

Devise an explanation for an old naturalist's claim: If you want to hybridize two closely related species, collect the males and females that you intend to mate from different localities, not from the same place. That is, individuals of the two sexes should be allopatric in origin, not sympatric.

By the time they are observed by human beings, most biological phenomena have accumulated a tremendous history. As a consequence, cause and effect become all but impossible to disentangle. When two related species of birds with very different plumages coexist sympatrically, was it because of the coexistence that plumages diverged? Or were the birds able to coexist because the plumages differed? In some instances definite answers may never be obtained, although one can imagine heroic experiments in which all of one (of a pair of) species is eliminated from an area in an effort to see what changes, if any, the remaining species now undergoes. A group of closely related birds—Darwin's finches—on the Galapagos Islands also provide a clue. Two species of this group of birds live alone on two islands: A on one, B on the other. The beak depths (beak depth is a trait that is closely related to the types of seeds and fruit that birds can eat) of the two species on these two islands are nearly identical. On two other islands, both species are found. On both of these islands, the beak depths of the birds differ, as would be anticipated. However, A's beaks are deeper than B's on one of these two islands, but B's are deeper than A's on the other. Thus, it seems as if a *difference* is selected for in sympatry; which species is large and which small seems to be of lesser importance, and is probably dependent upon the characteristics of the original founder immigrants.

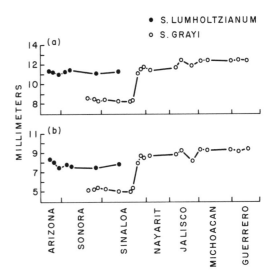

Figure 8.13. An illustration of the abrupt change in reproductive characteristics that often occurs in areas where two closely related species coexist. In this case, the organism is a flower: one species, *Solanum lumholtzianum*, exists in Arizona and in northern Mexico, where it overlaps *S. grayi*, which extends much farther south into Mexico. Where the two exist alone, their flower sizes (a = style length; b = anther length) are nearly identical. Where *S. grayi* coexists with *S. lumholtzianum*, however, its flowers are considerably smaller. In the area of overlap, the two species are visited by different sorts of insects, thus precluding accidental hybridization. *(Copyright, Columbia University Press; reproduced with permission)*

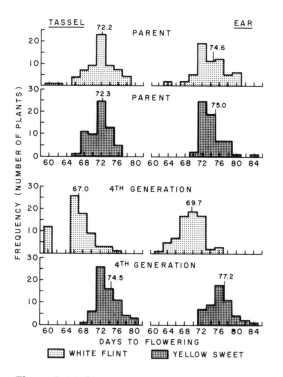

TASSEL   EAR

PARENT

PARENT

4TH GENERATION

4TH GENERATION

DAYS TO FLOWERING

⬛ WHITE FLINT   ⬛ YELLOW SWEET

Figure 8.14. Observations demonstrating that the reproductive behavior of a plant (time of flowering in corn, *Zea mays*) can be changed by natural selection. "Days to flowering" are the number of days that elapse between the time seeds are planted and the appearance of tassels and silk on the plant. The two parental varieties were nearly identical—72 days until tasseling, 75 days until silk appeared. After four generations, during which the seeds were planted as a mixture and ears showing signs of hybrid crossing were discarded, the White Flint variety forms tassels and silk one week before the Yellow Sweet; each variety has shifted its time of flowering in bringing about this sexual isolation. *(Copyright, Columbia University Press; reproduced with permission)*

Experiments have been done that show that mating and other behaviors can be modified by natural selection. These experiments show, almost incidentally, that behavior, like any other aspect of the phenotype, is subject to genetic control. For example, an experiment with corn has clearly demonstrated that the timing of pollen and silk formation can shift markedly during even four generations of selection. Two genetically distinct strains of corn (whose hybrid offspring could also be recognized), but which formed tassels and silk at the same time after sowing, were sown together in a field. The harvested ears were examined, and those possessing the most inter-strain hybrid seeds were discarded. The remaining seed was harvested and planted the following year, and again the ears bearing many hybrid seeds were discarded. In the fourth year of selection, tassel and silk formation in the two strains differed by nearly a week, and few hybrid seeds were formed. Character displacement, then, *can* arise as the consequence of natural selection, although this experiment does not prove that it always does.

NOW, IT'S YOUR TURN (8-6)

A biologist has argued that no one can demonstrate by means of an experiment what *natural* selection can accomplish. This argument places natural selection beyond the reach of experimentation. The term "natural selection" was used above in discussing the results of the corn experiment. Can you sort out what natural selection is, and what is artificial selection in a laboratory or field study? There were biologists in the early 1930s who, being impressed by the seeming uniformity of wild-caught fruit flies, denied that spontaneous mutations of the sort detected in laboratory tests occurred in wild populations. Notice that this claim also leads to a hopeless, Catch-22, situation: to detect the origin of *new* mutations, one must carry out controlled matings in the laboratory.

The males and females of a given species that live together in a particular location find one another, recognize one another, and succeed in mating with one another because of similar habitat preferences and through a series of reproductive or mate-recognition signals. Organisms (flies, flowers, and even bacteria and yeast) communicate with one another by means of auditory signals (peepers, songbirds, and katydids, for example), visible cues (zebras, birds, and flies), and chemical signals (aquatic organisms, gypsy moths, and dogs, for example). The development of new mate-recognition signals and a gradual improvement in their effectiveness can be demonstrated in populations consisting of individuals that carry genes derived from each of two similar, but normally reproductively isolated, species. In one study of this kind, three of four hybrid fly populations tested could be shown to possess mate-recognition procedures that differed not only from those of the two parental species, but also from each other. The fourth hybrid population tested in the experiments could not be differentiated from one of the parental species with respect to mate choice. In this same study, an improvement in the ability of the flies of each of the three hybrid populations to recognize and accept one another as mates while rejecting flies from the other populations was demonstrated. The novel mate-recognition systems underwent improvement through time. This is the essence of evolutionary change.

Because a great deal of communication transpires in the search for and successful courtship of a potential mate, the existence of eavesdropping in the natural world may not be surprising. To be sure, cats, dogs, coyotes, and all other hunters eavesdrop; that is, they use their keen sense of hearing to locate their prey. In one informal demonstration of this point, a pet cat confronted with recorded bird calls played on a stereo system crept stealthily toward a somewhat distant collection of hanging plants, as if that were the most reasonable place to find a bird.

An elaborate system of visual signals has arisen among fireflies. The flashes you see these insects making on warm evenings in late spring and early summer serve to bring males (those that are flying) and females (usually sitting quietly, giving the proper response to the male's signal) together. A whole complex of codes has developed that permits many species of fireflies to exist in the same area, but with little interference among the different species. Females give species-specific responses, and only to the characteristic flashes emitted by males of their own species.

Here, however, is where eavesdropping also occurs. Females of one particular species respond normally to males of their own species as long as they are unmated (virgin). Having mated, however, they change their behavior. Upon seeing the flashes emitted by males of any one of three different species, the now-mated females respond with a signal that is appropriate for the female of the flashing male. This response lures the flying male, and when he alights in preparation for mating, the eavesdropping female catches and eats him—she is carnivorous.

The signaling and false signaling used by fireflies emphasize the complex set of neural responses that can be elaborated through information accumulated in an insect's DNA. First there are the mating signals themselves: the males emit their specific patterns, and the females respond at the proper moment with their counter signal. The

Figure 8.15. The use of flashes for communication in fireflies. The male fireflies shown here use time between flashes, numbers of flashes emitted at short intervals, patterns of flight while flashing, and—of course—color of flash as species identification signals while seeking mates. *(After J. Lloyd)*

eavesdropping carnivorous females have built into their nervous system four responses: the proper one for the species, and the three different responses that correspond to responses that are appropriate for females of different species. What is the trigger that determines which of the three false responses a non-virgin female will give? The signal given by the male of that other species.

NOW, IT'S YOUR TURN (8-7)

Too often, an account like the one given immediately above about the predatory female takes on an "ain't Nature grand" aspect, and is stored away as trivia to be brought up during embarrassing pauses in party talk. There really should be more to these examples than that. They should be seen as descriptions of sets of amazing events that demand explanation: How do mating signals arise and become perfected within a species? How might the act of mating cause a female to change from a mate-seeking to a prey-seeking animal? How might eavesdropping and luring originate and become perfected? How can a repertoire suitable for three different prey species be assembled?

Here is an example you can think about and respond to in your journal: Certain female wasps hollow out long, blind tunnels in plant stems. Starting at the rear of the tunnel, a female will gather together a store of paralyzed insects as food for a single offspring, lay an egg on these insects, and then seal in the egg and its larder with a partition. (The overall length of the cell containing the egg may be about one inch.) The female then assembles a new store of paralyzed insects, lays another single egg, and seals off the second cell with a partition. She continues to do this, cell after cell, until she has filled up the en-

Figure 8.16. A sketch representing a tunnel constructed within a block of wood by one type of solitary wasp (female) and filled with eight chambers, each containing food (mostly paralyzed caterpillars) and an egg. When the grub that has emerged from the egg develops into a mature wasp, it can leave only by the tunnel's entrance. Furthermore, if a mature wasp were to pass through a chamber containing a grub or pupa, the immature individual might die. How can a female schedule the hatching of her progeny so that they mature in the correct order? How can she tell her progeny which is the way out?

tire tunnel. When she has finished there may be ten or twelve cells in the tunnel, each containing a single egg plus its provisions.

Now, the progeny wasps that finally emerge can leave the tunnel only by its original entrance; that is, the newly emerged wasps cannot dig their way out by going through the side of a cell or out the blind end of the tunnel. What sorts of problems can you see with respect to the times at which the young wasps hatch, and their knowledge about which way is out? What kind of information can the mother transmit to her children at the time she lays each egg?

Hints: The developmental times for the two sexes differ; furthermore, a female wasp can decide whether she will produce a daughter (she, herself, allows the sperm to meet the egg) or a son (sperm are excluded, and the egg develops unfertilized, thus producing a son with only half the normal chromosome complement). Second, developmental time for either sex depends upon the amount of food provided for growth; let's say the more food, the longer larval development will last. Third, the partition that is built to seal off each cell need not be flat—it can curve. In addition, it has two surfaces, whose texture is controlled by the mother wasp as she builds it.

Using these hints, construct a scheme that will allow larvae to hatch and emerge from the tunnel in a way that makes it unnecessary for any newly hatched wasp to intrude into cells containing immature progeny wasps, and that ensures that each newly hatched wasp will leave by the original entrance.

A scientist who studied these wasps proposed as a title for a paper, "Communication between two generations of wasps." The title was rejected by the editor of the journal to which the paper was submitted; biological communication was unknown, or at least unappreciated, at that time.

9

EVOLUTION

Unlike chemistry and physics, biology deals with units that have had, and have recorded, a history. Living organisms contain within their DNA a record both of the past and of the heritable adjustments that have been made in response to problems of the past. The materials of living cells are, of course, the materials of organic and inorganic chemistry; the behavior of these substances conforms to laws that apply to the physical sciences. The organization of these substances within the cells and the organization of cells, tissues, and organ systems within individuals are governed by instructions contained in DNA.

The instructions contained in the DNA of any of today's organisms—sunflowers, chipmunks, sowbugs, or human beings—was not arranged *de novo* either yesterday or even last week. Those instructions have evolved from the dawn of life on earth. The histories of all organisms living today have identical total durations; they all trace back to the origin of life. What differs among these histories are the lengths of time during which today's organisms shared common pasts. Sunflowers and human beings separated and struck off on independent historical pathways much earlier than did chipmunks and human beings; the latter went their separate ways long before the great apes (including man) departed from one another, striking out on their separate lineages.

Dobzhansky, one of the great modern evolutionists, wrote not long before his death that "nothing in biology makes sense except in the light of evolution." Dobzhansky's statement emphasizes the contrast between evolution and the religious belief that all organisms were created in their current form by an intelligent and willful act of a Supernatural Being. Many of the "senseless" creations cited by Dobzhansky involve organisms that are adapted for life in rare, exotic, or harsh environments: flies that live in puddles of crude oil, bacteria and algae that eke out a living in the near-boiling water of hot springs, or—one of the most exotic—several species of fruit flies that have become associated with land crabs. These flies lay their eggs near the eyes of the crab, and their larvae feed on the bacteria-laden urinelike wastes that ooze from pores beneath the crab's eyes. The crab ingests the waste once more, together with the bacteria that have broken down the otherwise poisonous substances. Presumably, the crab eats an occasional fly larva that ventures too close to his mouth. There is an interesting footnote to the fruit fly–crab relationship: an entirely different species of fruit fly was forced experimentally to adapt to a life on an "artificial crab"—a plastic and astroturf contraption over which dribbled dilute human urine and which was seeded with soil bacteria. Within one year, this population of fruit flies had made the adjustment to the new mode of life, an adjustment that seemingly prevented them from surviving on what had been their standard *Drosophila* culture medium for many previous years.

A second aspect of Dobzhansky's statement concerns the different meanings that similarity conveys in biology as opposed to chemistry. If two chemists working on opposite sides of the world synthesize a new organic compound having a specific chemical formula—let's say, lysergic acid, LSD—the physical properties of the two batches of substance will be identical: melting point, heat of vaporization, vapor pressure, solubility, whatever is measured will prove to be the same. The properties of the chemical substance are derived from the properties of the atoms of which it consists. The identical properties of the two batches of substance exist even though no atom in either batch of LSD may ever have existed in that particular chemical compound before. The constant properties of chemical compounds are well known, and for that reason chemists are able to compile and profitably use manuals and other handbooks.

Contrast the above with a seemingly comparable biological example. In the United States, one of the most common of all birds is the robin (*Turdus migratorius*). In the northeastern states, the arrival of these birds, more than any other sign, is taken as a harbinger of Spring. There is also a robin in Great Britain: the Robin Redbreast of "Who Killed Cock Robin?" There is also a blackbird. Despite their common names, an observant person who is acquainted with the American robin would never claim that it and the British robin are related; however, he would, despite the difference in coloration and the different names, recognize the British blackbird as an American robin in disguise. The contours of the body, the carriage, the characteristic movements made and poses struck while hunting worms and grubs, and other behaviors are such that our observer would claim a *relation-*

ship by descent between these non-identical birds living on opposite sides of the Atlantic Ocean. If a British friend were to suggest that his blackbird might instead be related to the American blackbird, he would immediately be smothered with protestations. Identical common names carry no weight in establishing biological relationships. Nor is identity in appearance necessary; the British blackbird and the American robin are recognized as relatives (and, fittingly, the scientific name of the British blackbird is *Turdus merula*) despite differences that set them apart.

Notice the difference in expectations for chemical compounds and living beings. For the former, we expect identical compounds to have identical properties, even if they have been synthesized independently and for the first time in the history of the universe. In the case of organisms, we do not expect identity in those suspected of having been descended from a common ancestor (or even in offspring of the same parents). Furthermore, although differences may exist between organisms, if a series of resemblances are detected, the biologist claims a relationship by descent, even though the specimens that he examined came from widely separated areas of the world.

Skeptics—not just Creationists, but also those who continue to demand that all claims of common descent be justified or verified—will not be impressed with what would be the reaction of nearly all amateur and professional ornithologists in claiming a relationship by descent for the American robin and the British blackbird. Those who want the claim of common descent to be proven each time it arises would insist on returning to first principles: What are the facts that compel one to believe in evolution? Actually, there are some; they have been listed by R. C. Lewontin in a recent, uncompromising editorial:

It is a *fact* that the earth, with liquid water, is more than 3.6 billion years old. It is a *fact* that cellular life has been around for at least half of that period and that organized multicellular life is at least 800 million years old. It is a *fact* that major life forms on earth were not all represented in the past. There were no birds or mammals 250 million years ago. It is a *fact* that major forms of life are no longer living. There used to be dinosaurs and Pithecanthropus, and there are none now. It is a *fact* that all living forms come from previous living forms. Therefore, all present forms of life arose from ancestral forms that were different. Birds arose from nonbirds and humans from nonhumans. No person who pretends to any understanding of the natural world can deny these facts any more than she or he can deny that the earth is round, rotates on its axis, and revolves around the sun.

To Lewontin's roster of facts, I might add one more: It is a *fact* that if one deals a number of playing cards onto a table, the first card dealt will be at the bottom of the pile, and the last one on its top. This relationship between temporal order and relative physical position applies to geological layers of the earth's surface just as it does to playing cards on a tabletop.

Predicting the Course of Evolution: *In Vitro* Studies

Purists who hope to remain impartial to the claims of evolutionists can lean back in their chairs (rattan or teak from Southeast Asia), gaze out their windows (upon a scene of maple, oak, or locust trees of the northeastern United States), and ask, "Why, if it is so important, does the theory of evolution not have predictive power?" The reason why the predictive power of evolution is weak is basically the same reason why the skeptics' parents would not predict that their children would be in the northeastern United States, or, if they were, that they would be occupying rattan or teak chairs. The possibilities available both for evolution and for mobile middle-class persons are simply too great for specific predictions. General predictions, yes. If one has a large pegboard with a single opening at the top, and many collecting tubes at the bottom, one can predict that a steel ball dropped in at the top will come to rest in one of the collecting tubes at the bottom. What is up, must come down!

But the skeptic asks, "In which bottom tube will the ball come to rest?" That cannot be predicted in advance. Let a thousand tests be made using a polished ballbearing, and one will be prepared only to offer odds as to which of the collecting tubes the dropped ball will enter. Unless there are horrible flaws in the board's construction, one would expect that a thousand steel ballbearings dropped in at the top would form a bell-shaped distribution among the various collecting tubes: most balls dropped should tend to bounce back and forth and end up under or nearly under the entrance point. Some, however, will bounce farther away, either to the left or to the right, and will come to rest in more remote collecting tubes.

If a person cannot predict the course of a steel ball that has been dropped into the top of a pegboard containing one hundred or more protruding pins and fifteen or more collecting vials, how can he be expected to predict the course of evolution in a world already containing two million or more species of plants and animals? A number of pins can shunt a falling steel marble into unpredicted paths; two million species and their innumerable interactions render the fate of any novel event uncertain.

In Vitro Evolution

For fifteen or more years, the late Dr. Sol Spiegelman and his colleagues have studied the self-replication of RNA molecules *in vitro*. The "solution" within which the RNA molecules self-replicate is complex, to be sure. It consists largely of the ruptured contents of bacterial cells, which, although no longer containing whole bacteria, contain proteinaceous and enzymatic cellular contents. To this rich culture medium can be added the nitrogenous bases needed for the

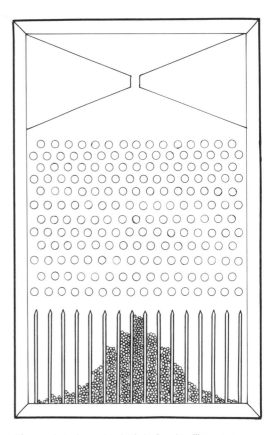

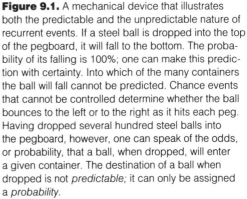

Figure 9.1. A mechanical device that illustrates both the predictable and the unpredictable nature of recurrent events. If a steel ball is dropped into the top of the pegboard, it will fall to the bottom. The probability of its falling is 100%; one can make this prediction with certainty. Into which of the many containers the ball will fall cannot be predicted. Chance events that cannot be controlled determine whether the ball bounces to the left or to the right as it hits each peg. Having dropped several hundred steel balls into the pegboard, however, one can speak of the odds, or probability, that a ball, when dropped, will enter a given container. The destination of a ball when dropped is not *predictable*; it can only be assigned a *probability*.

construction of new RNA molecules. An energy source must be provided because the synthesis of new RNA strands (using an old strand as a model) requires energy. As was just suggested, the system also requires a model suitable for copying, a molecule to get things started. That model is the single-stranded RNA molecule that represents the RNA content of a RNA-bacterial phage known as $Q\beta$.

NOW, IT'S YOUR TURN (9-1)

Predictability has become extremely important in some biological circles, so much so that, rather than describe the results of Spiegelman's experiments, it seems worthwhile to lead you through them, while you make periodic entries in your journal.

First, the RNA of the $Q\beta$ bacteriophage is treated by the cellular machinery as if it were a messenger RNA (mRNA); bacterial ribosomes and associated enzymes use information contained within the RNA strand to synthesize a nonbacterial protein. This new protein then combines with a number of the *bacterial* enzyme molecules to make an enzyme (a synthetase) that, using the original $Q\beta$ molecule as a model, synthesizes a complementary copy. Complementary models are copied from the original, and duplicates of the original model are copied from the complementary strands. Eventually, in processes governed by well-understood feedback mechanisms, still another part of the $Q\beta$ phage strand (and, by now, its many, many duplicates) is translated to form the protein coat that is part of the $Q\beta$ phage and that is responsible for the phage's ability to infect new bacteria once the original infected cell ruptures and releases hundreds of new, mature phage particles.

In Spiegelman's *in vitro* experiments, the naked RNA molecule ($Q\beta$ phage RNA without its coat) was introduced into a small vial containing the complex "nutrient" mixture consisting largely of ruptured bacterial cells. After fifteen minutes, a small portion of the first tube was transferred to a second. After another fifteen minutes, a small portion of the second tube was transferred into a fresh, third tube. This process was continued for many transfers.

From the beginning, the quantity of RNA in each tube was monitored by two methods: One method measured the actual amount (*weight*) of RNA present, and the second measured the *number* of phage particles capable of infecting bacteria. The results showed that the rate of RNA synthesis increased from transfer to transfer, but that the rate at which the infective particles were formed decreased. Your first task is to account for these observations. (Hint: Lest you have forgotten, these culture tubes do not contain live bacteria; therefore, coat proteins for the $Q\beta$ RNA in the transfer tubes are unnecessary; they *are* needed, however, for a particle to be scored as "infective" when tested on live bacteria.)

Having repeatedly found that Qβ RNA reproduces more and more rapidly (and having arrived at the solutions to the questions posed above), Spiegelman and his coworkers tried a series of starvation experiments, in which they reduced the concentration of one or another of the nitrogenous bases needed for RNA replication, or they poisoned the solutions with an inhibitor of RNA replication. In the meantime, they had become adept at analyzing the precise sequence of bases that made up the replicating molecules.

First, the number of bases of the replicating RNA was found to be only about six percent of the number in the original Qβ molecule. Did you anticipate this finding? If not, go back to the earlier questions and think about them some more. Second, under starvation and with the addition of an inhibitor, replication of RNA slowed down for some time, but then speeded up once more. What do you think these periods of speeding up might represent? How might the analysis of base sequences be used to check your answer? (Recall that the composition of RNA, like that of DNA, is precise: at each site there can be only one—but *any* one—of four bases, *U, A, G,* or *C*. This precision makes it unnecessary for you to suggest complex, untestable answers to the above question.) When the replicating RNA had "adapted" to the presence of a chemical inhibitor of replication and was reproducing itself nearly as fast as the original molecule without the inhibitor, the following test was set up: the inhibitor-resistant and "wild-type" molecules were tested in the presence of the inhibitor and in its absence. In the presence of the inhibitor, the resistant molecules replicated most rapidly. What about in the absence of the inhibitor? How do you account for the answer you have given?

The final experimental observation to be considered is the following: If the complex "culture" medium is not seeded with Qβ RNA or any other RNA molecule, it sits quietly for some time, after which RNA molecules appear and replicate. It seems as if the replicating enzyme can spontaneously put together combinations of bases to form a brand new RNA molecule, and that occasionally one of these new strands possesses the properties needed to act as a template for further replication. The composition of the self-assembled RNA in any tube can be analyzed in detail—to the exact sequence of bases if need be. Several cultures containing near-starvation levels of the bases needed for RNA replication and an additional number of cultures in which the concentration of needed bases was high were studied. Events transpiring in any one of these cultures were of course independent of those taking place in any other; there was no cross-contamination. How would you explain the following results? The RNA molecules that succeeded in replicating within the "starvation" vials differed considerably in composition from vial to vial; the RNA molecules that succeeded in replicating within the vials with ample concentrations of bases were very similar in composition from vial to vial. Relate your answer to the claim that natural selection is opportunistic. [Note: The presence of a self-replicating molecule in a vial depletes the store of resources (purine and pyrimidine) in that vial, thus effectively preventing the subsequent origin of a competing self-replicating molecule.] The depletion of available resources was the only reason the original, full-sized $Q\beta$ molecule could replicate for any time in the initial tests: inefficient as it was for an *in vitro* environment, it was vastly superior to any random sequence of purine and pyrimidine bases that the enzyme molecules might have assembled by accident.

The experiments carried out by Spiegelman illustrate many of the aspects of evolution that one would easily have predicted in advance. There is a logic to evolutionary change. Not in all details, perhaps, because an outsider does not know what base substitution will increase the affinity of an enzyme for its RNA substrate, or what substitution will offset the presence of a replication inhibitor. Aside from such predictions, which would require information on the physical chemisty of both enzyme and RNA molecules, there is ample room to make rather specific predictions. If all worked out well above, you succeeded in making at least some of them correctly.

What Is Life?

The ability of a chemical molecule to specify its own reproduction (as DNA does) or the reproduction of its complement, which can be copied in turn to produce more molecules like the original (as $Q\beta$ RNA does), is no small matter. When this replicative ability continues faithfully despite base substitutions, the replicating molecule acquires the ability to respond to environmental challenges (recall the replicating molecule under starvation conditions and in the presence of an inhibitor). The molecule can now be considered living. This point has been made emphatically by H. J. Muller:

I think the most fundamental property distinguishing a living thing. . . . is its ability to form copies of itself. We call this "reproduction"; but such copies must include innovations—mutations—that distinguish a given living thing from its parents. It is this property of not merely reproducing itself but also reproducing its mutant types that inevitably led the first multiplying objects through the three-, four-, or five-billion year course of evolution by which all present-day living things, including ourselves, have gradually taken shape under the directing influences of natural selection.

Confronted with a fellow panel-member who insisted upon including such things as dynamic equilibria, flow of matter and energy, specific synthesis, and architectural complexity as properties of living things, Muller responded to the moderator: "In my opinion all the properties he mentioned are results of the evolution of living matter by the mechanism Darwin called 'natural selection' . . . adaptation comes as a result of evolution." Later, in explaining once more where his definition of "living" lies, Muller repeated: "I should draw the line where the Darwinian process of natural selection begins to come in, and that is at the appearance of replication of a self-copying kind—that is, the replication of mutations." That, in essence, is what Spiegelman's work on the $Q\beta$ phage RNA and its many molecular descendants has been all about. Most persons would prefer to define life in terms of growth, movement, irritability, and reproduction; the point made repeatedly by Muller, however, is that these characteristics arose through evolution (and, presumably would arise again at another time or on another planet) following the origin of self-copying molecules—molecules that also replicate mutational changes.

Darwin's Evidence for Evolution

A book entitled *On the Origin of Species by Means of Natural Selection, or the Preservation of Favoured Races in the Struggle for Life* was published in Great Britain in 1859. Its author was Charles Darwin. The effect of this book on the beliefs and intellectual concepts of human beings cannot be overestimated. In it, in some 500 pages, were mustered the arguments that would convince most persons of the Western world that the profusion of living forms that inhabit the earth did not arise, species by species, through the creative acts of an omnipotent God; rather, these forms of life, *all* forms of life, are related by descent. Any two forms of life, according to Darwin's logic, share a common ancestor at some more or less remote period in the past. If any two forms share a common ancestor, then we all do: life as we know it has arisen but once.

Two aspects of Darwin's overall argument must be separated at this point. The first encompasses the reason for believing (1) that natural rates of increase will cause all kinds of organisms to exceed the number that the environment can support, (2) that the resultant stress will lead to competition between members of individual species, and in the death or sterility of many of them, (3) that the competing individuals vary with respect to their ability to compete and survive, (4) that at least part of this variation is inherited (Darwin, of course, knew nothing of genes and chromosomes; he was born too soon), and, therefore, (5) that the genetic composition of populations changes gradually with time, leading eventually, as the title of his book sug-

The Galapagos Finches

Great ideas have often arisen under strange and unexpected circumstances. Archimedes, a scholar of Ancient Greece who was asked by his king to test the purity of a gold crown without marring it in any way, saw while bathing the means by which the test could be made. He is said to have run naked in the street, shouting "Eureka!" Newton was twice struck: once, literally, it is said, by a falling apple; a second time in church while watching the pendulum-like swaying of a hanging ornament. These moments apparently yielded ideas concerning gravity and the motion of planets. Sir Arthur Fleming, the discoverer of penicillin, got his inspiration from a common observation of microbiologists: fungal colonies growing as contaminants in bacterial cultures are frequently surrounded by a clear zone within which bacterial cells fail to grow. At some point Fleming realized that the fungus *inhibited* bacterial growth; that realization led him to the isolation of penicillin.

Charles Darwin accepted a position as naturalist and captain's companion aboard the *Beagle* on a global voyage of exploration and map-making that would last nearly five years, from December 1831 to October 1836. During that voyage, Darwin's chief occupation was to collect fossils, plants, insects, fish, reptiles, birds, and mammals at every opportunity. Such opportunities arose in the Falkland Islands, in South America, on small islands in the Pacific, in Australia, and in New Zealand. The material Darwin obtained during the voyage of the *Beagle* still constitutes one of the major collections of all time.

Upon leaving England in 1831, Darwin held the prevailing view of living creatures and their origin: each species had arisen as a special creation of a Divine Creator.

The apple figuratively struck Darwin as he observed the flora and fauna on the Galapagos Islands, barren oceanic rocks that straddle the Equator about five hundred miles west of South America. The bulk of all plants consisted of cactuses. The larger forms of animal life inluded giant tortoises, marine lizards a yard or more in length, land lizards, and birds. The latter were not numerous in kind, but were, in Darwin's word, "curious."

(*continued on next page*)

gests, to the origin of (new) species. These arguments have been discussed throughout the last three chapters; they need not be repeated again.

The second aspect of Darwin's overall argument was his search for evidence that would support the conclusions he had already reached on an abstract or intellectual level. Of course, no one's thoughts are so well organized and ideas so clear that one can say "Here is what I must prove. Where should I now find my evidence?" Darwin was the naturalist aboard the *Beagle* during its five-year voyage from Great Britain to South America, thence around the southern portion of the world, returning to South America, and then to Great Britain once more. The observations he made during those years—in the Falkland Islands, in South America, and especially in the Galapagos Islands— shaped his ideas concerning the variation of species and sent him searching for a cause that would explain the variation he continuously encountered. Thus, the patterns of variation that he saw made him receptive to Rev. Malthus's essay on overpopulation. That essay suggested the basis for a natural selection among wild organisms that might in turn lead to new species. With that idea firmly in mind, Darwin saw the need for additional supporting evidence.

Geographical Distribution

To many (but by no means to all) persons today, it may seem quaint to pose as an alternative to evolution the notion that all living things were created separately by a Supreme Being. In the mid–1800s, this possible alternative was not considered quaint. On the contrary, the Biblical account of Creation was the account accepted by nearly every thoughtful person. To suggest an alternative possibility was to assume the burden of proof: the *status quo* always possesses enormous power simply by being the *status quo*.

Among Darwin's observations were those concerning the species of birds, reptiles, and mammals inhabiting the Falkland Islands (islands lying off the east coast of southern South America), South America itself, and the Galapagos Islands (islands off the west coast of South America, more or less on the Equator). The fauna of each island system differed from the corresponding fauna of South America. Especially striking, however, was the similarity that could be detected, despite obvious differences, between the Falkland Islands inhabitants and those of eastern South America, and the corresponding similarities (again, despite obvious differences) between Galapagos birds, for example, and the birds of western South America. Darwin reasoned that a creator who had created fauna for bleak oceanic islands might have made special forms that could live successfully on both the Falklands *and* the Galapagos; that is, the fauna on the two island groups would have resembled one another. Instead, however, each island group had animals most nearly resembling those of the portions of South America from which migrants might have arrived. In that case, however, one had to admit that once the settlers had arrived on the islands, they diverged in appearance from the continental forms, even to the point of being different species. On a smaller scale, Darwin noted that individuals representing corresponding species but inhabiting the East and West Falkland Islands, or inhabiting any of the ten principal islands of the Galapagos Islands, also differed from one another. Again, if these organisms were created, the Creator must have worked on a fine scale, indeed. But, if they had arrived from South America (east or west coast, depending upon the island group), and had accumulated differences since their arrival, then differences between the inhabitants of different islands could be seen as part of that same process of gradual divergence.

In studying fossils at various locations along the *Beagle*'s voyage, Darwin made another observation: the fossils of any locality resemble the present-day organisms of that same locality more closely than they do the present-day forms of other localities. This, in Darwin's view, suggested that the present-day forms of each locality are descendants of the earlier fossils; if so, however, they had changed in appearance during the intervening millennia.

Classification

Darwin seized upon naturalists' ability to classify animals in a hierarchical system as evidence for evolution. Early classifiers had assumed that they were cataloging God's handiwork; they did not seek any sign of common descent as they set about their work. After Darwin (and Lamarck, as well) started a search for supporting evidence, he realized that a system of classification based on several similarities, rather than one or very few (a classification of plants based on flower color alone would have proven useless to Darwin), should reveal patterns of descent. Gaps in the classification represented extinctions of possibly intermediate forms. In making this argument, Darwin referred to the corresponding one used in describing the relationships between languages (and of course no one has ever contended that languages represent special creations of an omnipotent Being).

Morphology

The aspects of morphology and anatomy that Darwin accepted as evidence of descent with modification were the diverse shapes that are assumed in different groups of organisms by what is, in fact, the

The Galapagos Finches (*continued*)

Having entered considerable information into his journal on October 8, 1835, Darwin then wrote as follows: "The remaining land-birds form a most singular group of finches, related to each other in the structure of their beaks, short tails, form of body, and plumage. . . . Seeing this gradation and diversity of structure in one small, intimately related group of birds, one might really fancy that from an original paucity of birds in this archipelago, one species had been taken and modified for different ends."

An idea that has been expressed is much like a genie that has escaped from a bottle—neither disappears again. The idea that a single species had undergone modifications leading to the variety of finches Darwin observed on the Galapagos Islands developed with time; by 1859 it had developed into a theory of evolution resulting from natural selection— a theory expounded in *The Origin of Species*.

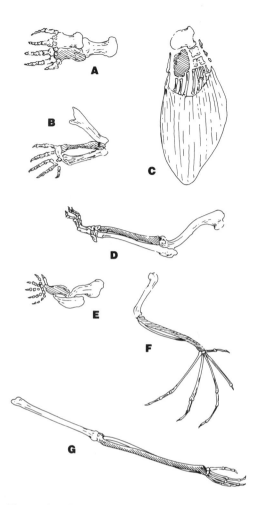

Figure 9.2. Drawings of the forelimbs of various vertebrates, emphasizing the three bones—humerus, ulna, and radius—that are common to all forms. The functions served by these forelimbs range from swimming and flying to burrowing underground. Conceivably, the various functions might better have been met by separately designed limbs; instead, they are met by modifications of one basic structure. The use of variations on a common theme, Darwin reasoned, was evidence supporting his theory of evolution. The radius has been shaded in each example; the humerus is the upper, single bone. The figure shows the forelimbs of (A) a whale, (B) an alligator, (C) a fossil fish, (D) a dog, (E) a mole, (F) a bat, and (G) a gibbon.

same structure. There are three main bones in the human arm: the one of the upper arm (*humerus*) and the two of the lower arm (*radius* and *ulna*). Beyond the radius and ulna lie the wrist and finger bones. That same pattern (but with bones of various shapes; "distortions" in our anthropocentric point of view) can be found in the flippers of whales, porpoises, and seals, in the forelegs of dogs, cats, and bears, and in the wings of bats. Their corresponding parts can also be found in the forelimbs of birds (where they form wings), lizards, and amphibia.

The forelimb is not the only example that might have been cited; one anatomical feature after another could be brought out for examination. Those who preceded Darwin looked upon these features as demonstrating the use of a common design by the Creator of the universe, but in doing so they made a tacit, untenable assumption: that the Creator was a being of limited imagination. Swimming, flying, digging, walking, and grasping organs, serving diverse functions as they do, might better have been designed with their specific tasks in mind. The three major bones of the human arm can by no stretch of the imagination be construed as the optimal mode for fulfilling the many functions that the forelimbs of different organisms serve.

NOW, IT'S YOUR TURN (9-2)

In recent years, a favorite question for use during students' oral examinations has been discovered by correspondents for the journal *Science*. Reference to this question has recurred as if it demands a substantial modification in modern evolutionary thought: Why haven't animals evolved wheels? Take time to enter into your journal some of the features that must accompany wheels in order that they be useful for locomotion, the problems such features would pose for living tissue, and the problems they pose concerning the utility they must exhibit at every stage of their evolutionary development.

Embryology

The adult forms of various organisms differ considerably. It is the adult, of course, that must succeed in eking out a living and, while it is doing that, it and its mate must also proceed with reproduction. Either eggs must be laid or young must be cared for by their parents.

Dissimilar as adult forms might be, embryos often tend to resemble one another. This is true not only for vertebrates—mammals, especially—where the embryo passes through a series of superficially comparable premature stages, but also for marine invertebrates, where

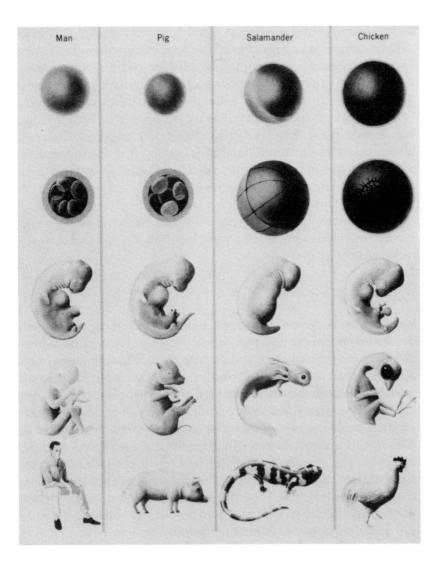

Figure 9.3. The embryologic development of human beings, pigs, salamanders, and chickens. All four organisms begin life as a fertilized egg, which starts the individual's development with a series of cell divisions. Because of the large amount of yolk the chicken egg contains (the yolk of a chicken's egg is a single cell—the egg cell), the entire cell does not divide; instead, it develops cleavage furrows. As development proceeds, all embryos begin to look alike; the indentations that can be seen behind and below the head in every case correspond to the gill arches of fish embryos at a corresponding stage. Finally, as development nears completion, each embryo takes on an appearance similar to that of its corresponding adult individual; these late embryos can be readily identified as to type of organism. The similar courses taken during the early development of many organisms, invertebrates as well as vertebrates, suggested to Darwin that these various organisms had descended from common ancestors. *(Copyright, The Biological Sciences Curriculum Study; reproduced with permission)*

larval forms provide a motile stage of life, and where adults are commonly sessile.

The similarity between early embryonic forms gave rise to the idea that individual development reveals a brief resumé of the history of the species in past times: *ontogeny recapitulates phylogeny*. In brief, this "law" suggests that although the adults of higher organisms diverge greatly, their embryos do so to a much lesser extent. Furthermore, the characteristics of the highest organisms have been added on to those of lower ones, so that, for example, the mammalian embryo still passes through a stage during which it possesses gill arches, much as the embryos of fish do and did in the past. Taken literally, of course, this law is nonsense: there is no "fish" stage in the development of a mammal. Taken less literally, the idea has considerable rationale. Each

Social Darwinism

Six years before the Darwin-Wallace papers were published (see page 280), Herbert Spencer, an English philosopher, claimed that the pressure of subsistence benefits the human race. Having read Malthus (as had both Darwin and Wallace), Spencer argued that the struggle for existence, by placing a premium on skill, intelligence, and self-control, had stimulated progress by selecting the best individuals during each generation. "Survival of the fittest" is a Spencerian phrase.

In sharp contrast to views expressed in Chapter 1, Spencer wrote, "My ultimate purpose . . . has been that of finding for the principles of right and wrong in conduct at large, *a scientific basis* [italics added]." A lack of adaptation to prevailing conditions, he claimed, is evil. Because organisms, including human beings, adapt, evil tends to disappear. Evolution, which mostly reflects adaptive changes resulting from natural selection, is progress. Because he equated change (evolutionary change, in particular) with progress (and, hence, good), Spencer was led to oppose state aid to the poor: "clear the world of them, and make room for better." If a person lives, it is good that he should live; if he dies, it is better he should die. In addition to state help to the poor, Spencer opposed state-supported education, sanitary regulations, housing regulations, and the regulation of medical practice.

Spencerian philosophy entered the American scene in the late 1800s as *Social Darwinism*. At that time, the growth of railroad companies and other large businesses was regarded merely as the survival of the fittest. Andrew Carnegie, having read Darwin and Spencer, saw, as a flash of light, the truth of evolution: "All is well since all grows better."

The fundamental weakness of Spencer's approach to a study of society was his assumption that science can provide a basis for identifying what is good and what is evil. Because he chose biological evolution as his source of ideas and analogies, Spencer believed that whatever was "natural" was by definition good. If starvation and a lack of nesting territories control the number of red-winged blackbirds, then starvation and lack of housing are beneficial for the human population.

(*continued on next page*)

higher organism develops from a fertilized egg, a seemingly nondescript, microscopic bag of biochemicals, membranes, and enzymes. From that beginning, a complex, multibillion-celled organism must develop. A Creator, as a Master Engineer, might have planned independent pathways from egg to adult for each organism, but evolution, which virtually by definition builds on what has been accomplished before, lacks the ability to construct theoretical plans in advance; evolution resembles a tinkerer. It modifies existing structures. Amphibia are descendants of fishes; reptiles are descendants of amphibia; birds and mammals are descendants of reptiles. One would expect, then, to see traces of this common ancestry appear during each organism's complex trajectory of life. Darwin, with good reason, claimed the observed "recapitulation" as supporting evidence for his belief in a common ancestry of all organisms.

Rudimentary Organs

Around each person's ears are a series of small muscles; properly developed and innervated, they would allow us to twitch and maneuver our ears as horses, pigs, and dogs do. Within the perimeter of our outer ear shell is a thickened, in-turned portion; everted, it would provide a point to our ear similar to that of many primates. Between our small and large intestines is a small *appendix*, an organ best known for the trouble it gives us, *appendicitis*, and the remedy provided by the surgeon, an *appendectomy*. [In the rabbit, the corresponding organ (the cecum) is large, and serves as a pouch within which plant material is digested by bacteria.] In the inside corner of each human eye, there is a small patch of tissue that swells with some allergies; in cats, the corresponding tissue forms a third "eyelid," one that can cover the eye from the side rather than from the top or bottom.

All of these items are examples of rudimentary organs. They are found throughout the animal kingdom. Darwin saw in them additional evidence for a common descent of all organisms. Confronting the standard alternative view of his day (origin by independent creation), he could ask: "If the different animals have had independent origins, why do we find that what is a functional organ in one species is a rudiment or an aborted vestige in another?" Why should a useless, unnecessary vestige be included in the plan of one organism *unless* the ancestral individuals of the two organisms were the same species, and the form now possessing the vestige simply had no need for that organ?

The theory of evolution is often criticized for a lack of predictive power. In one sense, the criticism is true. There are too many variables in the world, and events move in unexpected directions—especially if we consider long-range predictions. In a second sense, the statement is unfair. Had Darwin been an ordinary man, he would have suggested (in several papers or a small book) that, in his opinion, all organisms have descended from a common ancestor. His supporting data would have been taken largely from notes compiled during the voyage of the *Beagle*. Following that, and amid a background of argument and controversy, there would have appeared in support of Darwin's views, taxonomists, morphologists, anatomists, embryologists, and others, each of whom would have introduced his published findings with words such as these: "Darwin (1859), using the many interesting observations that he made while occupying the position of naturalist aboard the *Beagle* during its circumnavigation of the globe, has suggested that all organisms alive today share a common ancestor, and that today's diverse forms of plants and animals have gotten to their present state through a process of natural selection, a favoring of well-adapted individuals and a culling of the weak and infertile ones. If Darwin's thesis is correct, it should follow that. . . ." Here, of course, the writer would have described the supporting data to be found in morphology, embryology, or elsewhere. Looking back at a collection of such papers, we would say that Darwin's ideas *had* predictive value. Unfortunately, perhaps, Darwin was not an ordinary man; he brought *all* the available information together in his book, *On the Origin of Species*. Evidence that would normally have been predicted by Darwin's views was actually presented as part of the initial argument. Following that presentation there was little for anyone to do until entirely new types of data could be collected.

Modern Evidence of Evolution

Modern evidence of evolution comes from a variety of fields that either did not exist or were not fully developed during Darwin's lifetime. One of these sciences is cytology, which had its origins in the 1600s, when the first microscopes appeared. The development of biological dyes and of the apochromatic objective for the compound microscope occurred, however, near the time of Darwin's death. The second science, or cluster of sciences, is molecular biology. This science had its beginnings in the 1950s, when the colinearity of DNA and proteins was demonstrated, and when modern techniques were developed for the study of proteins, DNA, RNA, and other large molecules. Colinearity means that successively arranged base-pair triplets on DNA correspond, in the same sequence, to successively arranged amino acids in a protein molecule; letters as they appear in a line on a typed page, for example, are *not* colinear with the arc of type on a standard typewriter.

Social Darwinism (*continued*)

Events that occur in nature are not, on that account, necessarily good, or for the best. This is not to say that the human population is not part of the natural world, or that the ultimate survival of human beings is not dependent, for example, upon green plants. We are, indeed, dependent upon that thin film of life that covers our Earth, the *biosphere*. For problems like overcrowding, a reduction in the birth rate offers a solution that is preferable to an increase in the death rate. Our brains may be unable to make hepatic decisions, but there are problems whose solutions are best arrived at by rational thought. Such problems need not be solved "in Nature's way"—a way that relies largely upon the modification of the species' genetic endowment.

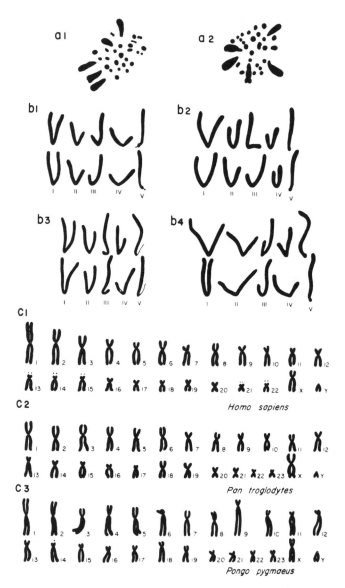

Figure 9.4. Similarities in the overall chromosomal morphologies of related organisms. In modern terminology, descent from a common ancestor requires that the genetic material carried by that ancestor be transmitted in an unbroken "line" ("interlocking cable" would be a more appropriate phrase for sexually reproducing organisms; see Figure 11.2) to today's descendant. The shorter the period of separation, the less change one would expect to see in even gross chromosomal morphology. Yucca and agave chromosomes are represented at the top; four species of lilies are illustrated next; at the bottom are karyotypes of man and two other great apes.

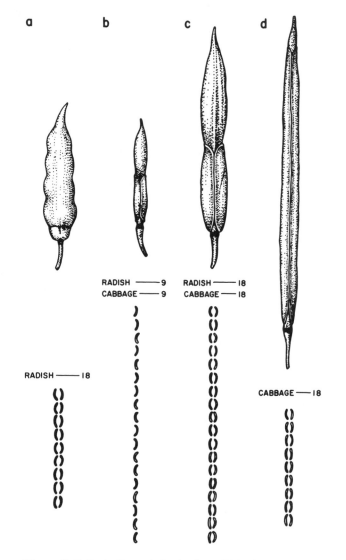

Figure 9.5. "Instant" speciation in plants. Biological species are defined as populations that are reproductively isolated from other, similar populations. The need for each individual to possess an entire complement of chromosomes causes sterility in many hybrids, because, during egg and sperm (pollen) formation, the mismatched chromosomes will not segregate in an orderly manner. The hybrid between a cabbage and a radish is one such sterile hybrid; the nine chromosomes obtained from each parent allow it to live, but they will not pair properly during meiosis. Either by accident or after chemical treatment, cells may arise carrying duplicated chromosomes. If these cells give rise to a flowering branch, egg and pollen will contain nine radish and nine cabbage chromosomes; the union of these pollen and egg cells produces individuals with the doubled chromosome number. These *polyploid* individuals are fertile when crossed among themselves, but they produce sterile offspring when crossed with either parental (cabbage or radish) species. Convince yourself that these sterile hybrids would carry one complement of paired chromosomes plus one set (either radish or cabbage, depending upon the nature of the cross) consisting of single chromosomes only.

Cytology

The first of the many contributions that cytology has made to evolutionary studies concerned the general morphology of the chromosomes of various groups of plants and animals. As one might suspect, these morphologies do not vary haphazardly. On the contrary, organisms that are identified as being closely related (and, by that, we mean closely related by descent) possess chromosomes whose morphologies are more nearly similar than are those of chromosomes from unrelated organisms.

A second tremendous advance was the discovery of truly giant chromosomes in flies, mosquitoes, and gnats (dipteran, or two-winged, insects). These enormous chromosomes (whose total length in a single nucleus may exceed 1 mm), when magnified a thousand times by a high-quality microscope lens, permit the cytologist to make exceedingly detailed maps of the wide, narrow, solid, and granular bands that appear along their lengths. As one might expect, morphologically similar species prove to have chromosomal banding patterns that are largely the same. In those instances where viable hybrid larvae can be obtained through interspecific crosses, the similar regions are found not only to *look* alike, but also to pair with one another, thus proving that they are the same regions by descent.

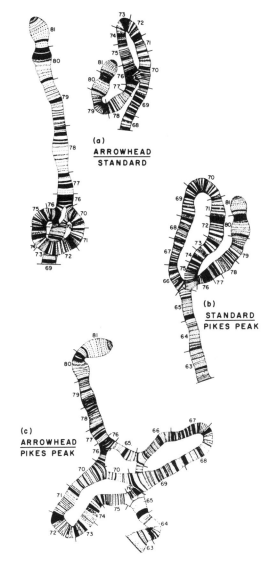

Figure 9.7. Drawings of paired giant chromosomes observed in salivary gland preparations of larvae produced by wild-caught females of *Drosophila pseudoobscura*, a relative of the vinegar fly. The chromosome structure of any species varies from one individual to another; physical rearrangements are extremely common. The rearrangements shown here are *inversions;* each illustration is of a pair of dissimilar chromosomes that have been forced into a loop because of the band-by-band attraction in these chromosomes. If the sections in the Standard arrangement are numbered sequentially, Arrowhead has the region 70–76 inverted and Pikes Peak, the region 66–75. Confirm the two inverted sequences in the complex pairing of Arrowhead and Pikes Peak chromosomes.

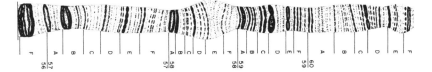

Figure 9.6. A short section of one of the giant chromosomes from the larval salivary gland cells of the vinegar fly (*Drosophila melanogaster*). This illustration shows the exquisite detail that can be discerned in these chromosomes using no more than high-quality light microscopes. People whose lives have been spent studying these chromosomes can easily detect the transposition of bands from their normal position to a new one, for example, after X-ray exposure. They can detect inversions (the rotation of a section of chromosome) as small as two-banded ones. Close inspection of the figure reveals that two chromosomes (one maternal, one paternal) are closely entwined to give what appears to be a single structure; the pairing of the two chromosomes results from a band-by-band attraction.

The giant chromosomes also allow us to detect a large number of naturally occurring structural alterations in chromosomes, alterations that were undetected in preparations of the small chromosomes of most cells. The majority of these structural rearrangements involve the rotation of a segment of the chromosome (sometimes a short segment, at other times a long one) through 180°—*inversions.* For reasons explained in a subsequent essay, most of these inversions lie totally within one of the chromosome arms and do not involve the cen-

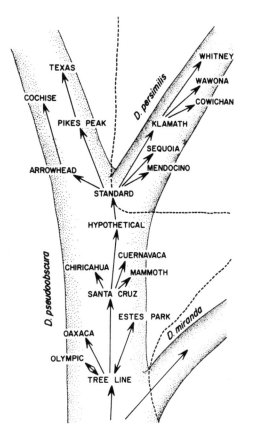

Figure 9.8. An arrangement of inversions that reconstructs a sequence of events involving three species of *Drosophila*. An analysis of chromosomal inversions can lead to a reconstruction of the temporal sequence in which the individual inversions actually occurred. The inversions themselves do not identify ancient gene arrangements as distinct from recent ones, but when a sequence spans several species, as this one does, the degree of relatedness of the different species provides additional information regarding order. In the case illustrated, *D. pseudoobscura* and *D. persimilis* are closely related and share one gene arrangement (Standard); *D. miranda* is more distantly related and shares no gene arrangements with the other two species. Because Tree Line is more nearly like the gene arrangement commonly found in *D. miranda* than are the others, the phylogeny can be built with Tree Line of *D. pseudoobscura* near its base.

tromere. These rearrangements themselves provide evidence about the origin and relatedness of species. In order to have a segment rotated, a chromosome must be "broken" in at least two places; because these breaks are presumably rare events, however, the majority of inversions will be the result of the minimum number of required breaks—two. Thus, if a cytologist is confronted with the three gene arrangements:

I. A B C D E F G H
II. A E D C B F G H
III. A E D G F B C H

he or she will note that I and II differ by an inversion involving two breaks (the involved segments are BCDE/EDCB), and that II and III differ by an inversion involving two breaks (the involved segments are CBFG/GFBC), but that I and III differ in a manner that requires *four* breaks—one each between A and B, C and D, E and F, and G and H. Therefore, the cytologist would postulate a *phylogenetic* sequence in which double-headed arrows indicate uncertainty as to direction:

I ↔ II ↔ III

Arrangement I probably did not give rise to III, however, nor did III give rise to I. Figure 9.8 illustrates a lineage of inversions that has been traced through three *Drosophila* species.

With improved staining techniques, it has been possible to reveal small, differentially staining regions of standard, metaphase chromosomes. As a consequence, what were formerly indistinguishable chromosomes (indistinguishable because they were of nearly identical size) can now be separated on the basis of staining patterns. A spectacular use to which these new techniques have been put has been to compare the chromosomes of the orangutan, the gorilla, the chimpanzee, and man. The homology of the light and dark bands occurring within the chromosomes of these four species is seemingly absolute.

Molecular Biology

The contributions made by molecular biologists to evolutionary studies have, for the most part, involved the comparative sequences of amino acids in homologous proteins of different species, and, more recently, the overall similarity of the DNA of different species.

The outcome of both protein and DNA studies has been clear: the more closely related two groups have been judged to be on the basis of morphological and other "classic" taxonomic traits, the more similar (as an overwhelming rule) their proteins have proven to be, and the more nearly identical the base-pair sequences of their DNA have proven to be. An occasional exception has proven the value of these

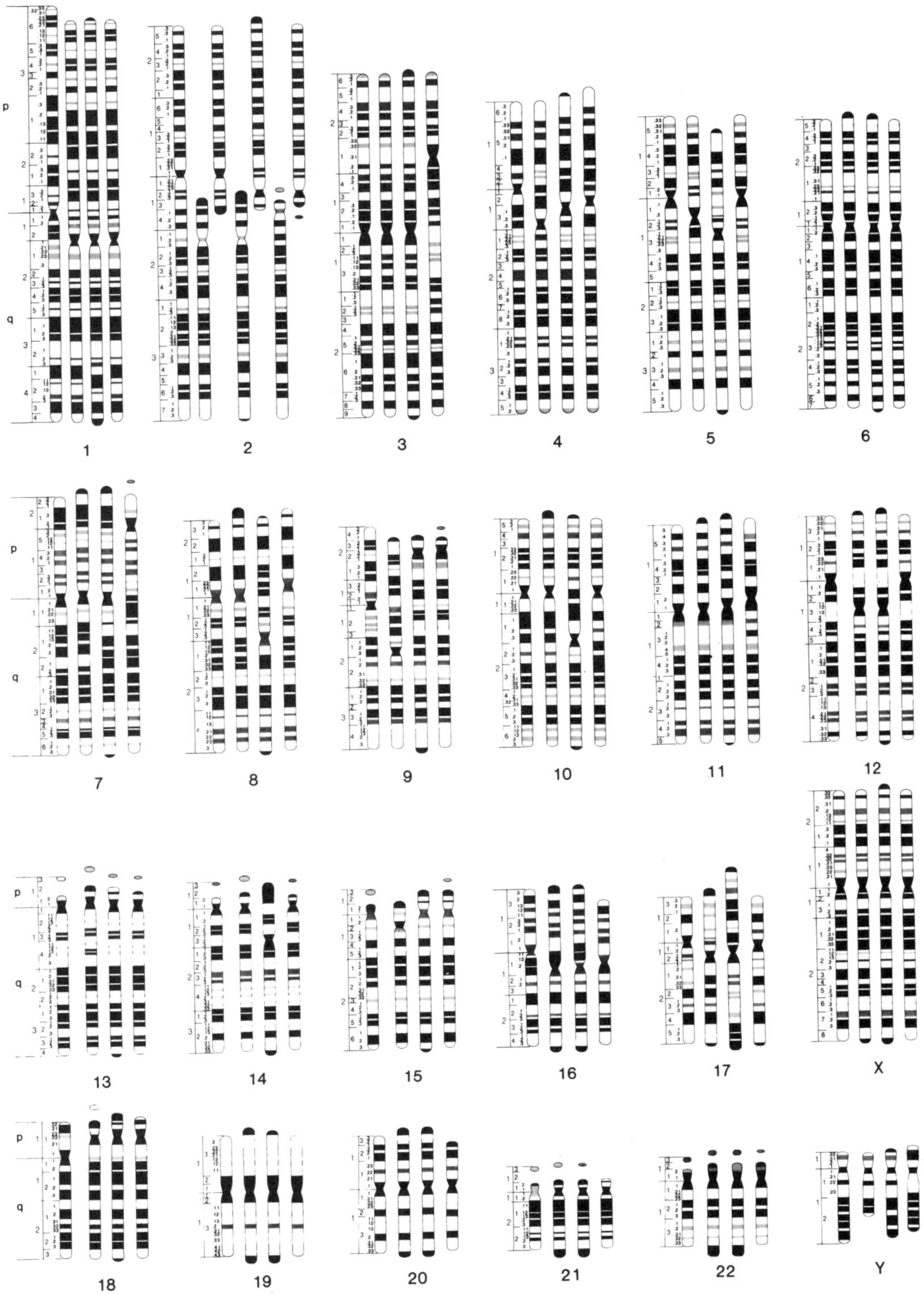

Figure 9.9. The banding patterns of chromosomes from man, chimpanzee, gorilla, and the orangutan (left to right, respectively). The chromosomes have been arranged to emphasize the extensive homology in chromosomal banding patterns among these four primates. These homologies, which characterize the genetic material itself, lend substantial support to the theory of evolution—support of a kind that Darwin could not have imagined. *(Photo courtesy of J. L. Yunis, University of Minnesota Medical School; © 1982 by the American Association for the Advancement of Science)*

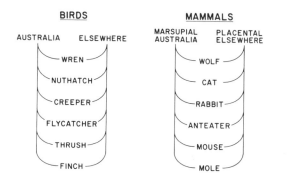

BIRDS		MAMMALS	
AUSTRALIA	ELSEWHERE	MARSUPIAL AUSTRALIA	PLACENTAL ELSEWHERE
WREN		WOLF	
NUTHATCH		CAT	
CREEPER		RABBIT	
FLYCATCHER		ANTEATER	
THRUSH		MOUSE	
FINCH		MOLE	

Figure 9.10. The convergent evolution of distantly related organisms that results from an adaptation to corresponding ways of life—that is, to similar ecological niches. For tens of millions of years, Australia has been a physically isolated, island continent. Although isolated from the rest of the world, the birds and mammals of Australia developed a variety of morphologies whose counterparts are to be found in Europe, Asia, and the Americas. The corresponding morphotypes are so close among birds that, until DNA–DNA hybridization proved otherwise, relationships were thought to be within the morphologies; for example, an Australian wren was thought to be more closely related to a wren from elsewhere than to an Australian nuthatch. This view proved to be wrong. A similar error was not made for mammals because Australian mammals are marsupials, whereas their counterparts in the rest of the world are placental.

molecular procedures. For example, in a survey of crabs (Crustacea), one investigator found great similarity in protein structure in all crabs, as expected, except for one type—the *horseshoe* or *king* crab! It is well known among biologists that the horseshoe crab is not a close relative of any other living organism—it resembles spiders as closely as it does crabs—and many persons would place it in a class by itself. At any rate, despite being fooled by the common designation, "crab," the molecular biologist did confirm what taxonomists had known for many years.

Comparisons of the DNA sequences of organisms belonging to different groups have been especially rewarding when these groups include individuals of bizarre morphological appearances. Birds of paradise, for example, have such exotic appearances and differ so much from one another that some taxonomists thought them to be an artificially created group of only distantly related birds. Studies of DNA homologies reveal, to the contrary, that they are closely related. Under the need for different species of these birds to achieve reproductive isolation, visual recognition signals involving color and shapes of feathers evolved rapidly (these correspond to the character displacements discussed in the preceding chapter). Taxonomists, like these birds themselves, use visual clues in drawing their (in this case, erroneous) conclusions. DNA–DNA hybridization, the laboratory procedure by which the homology of genetic material is revealed, is a more objective technique.

DNA hybridization has revealed startling information about the birds of Australia. These birds, although different from those of the other major continents, were nevertheless recognizable as to "kind": wren, nuthatch, warbler, and other familiar, informal groupings. For nearly a century, it had been assumed that because they can fly, the Australian birds were the descendants of birds that had invaded Australia and had evolved there, geographically isolated from their original homes. However, DNA studies have proven this interpretation to be wrong. The different birds of Australia are closely related! Australian wrens, for example, are *not* closely related to wrens elsewhere; on the contrary, Australian wrens are closely related to other Australian birds that are not wrenlike.

The study of Australian birds emphasizes the extent to which living organisms adapt physically to meet the demands of their chosen way of life. The necessity for large, rapidly swimming marine vertebrates to possess streamlined bodies is a commonly cited example; as a result, sharks, porpoises, and the extinct ichthyosaurs have similar body shapes. Still another example of *convergent evolution* that is frequently cited compares Australian marsupials and their counterparts among the placental mammals of the rest of the world. Within the Australian marsupial group were included "wolves," "rabbits," "squirrels," and even a marsupial "mole." In these cases, confusion was impossible: a mammal possessing a marsupial pouch differs considerably from one that nourishes its unborn fetuses by means of a placenta. The Australian birds gave no corresponding clue regarding their separate origin. In showing that theirs *is* a separate origin, however, the molecular analysis of DNA homologies has demonstrated, to an extent not previously appreciated, how the interactions between organisms—plants, herbivores, and carnivores, and competitors and cooperators—lead to clearly defined ways of life, to clear-cut physical needs, and, consequently, to striking morphological similarities between even distantly related organisms whose ways of life happen to coincide.

The contributions of modern microscopic and molecular studies support Darwin's contention that all organisms have descended from a common ancestor, that at some point in the past any two species of organisms have been represented by a single species from which both modern ones have subsequently descended. The common ancestor of lizards and human beings obviously lived many millions of years ago, that of chimpanzees and human beings lived much more recently.

NOW, IT'S YOUR TURN (9-3)

In the discussion of the contributions of molecular biology to evolution, one important concept to emerge from molecular studies was omitted because it is a *concept*, not a *thing*: the genetic code. The genetic code consists of triplets of base pairs on messenger RNA that literally spell out the amino acids that are to occupy successive sites in the protein molecule.

Just as Samuel F. B. Morse could have chosen any combination of dots, dashes, and spaces to correspond to the letters of the alphabet, so species of truly independent origins could have come to possess their own genetic codes.

Plant pathologists have learned not only that some plant viruses (viruses that reproduce in plant cells) are transmitted from plant to plant by plant lice, but that the viruses actually reproduce in their insect hosts. Whether the virus reproduces within the plant or within the insect, the protein coat that is specified by the viral DNA turns out to have the same amino acid constitution. What bearing does this observation have on the question of the universality (or at least, near-universality) of the genetic code?

Lysenko and Lysenkoism: The Replacement of Science by Doctrine

The Lenin Academy of Agricultural Sciences of the U.S.S.R. held a momentous meeting from July 31 to August 7, 1948. The tenth sitting, on the morning of August 7th, was addressed by Trofim Denisovich Lysenko, the Academy's president.

"What is the attitude of the Central Committee of the Party to my report? I answer: The Central Committee of the Party examined my report and approved it." The stenographic report adds parenthetically: Stormy applause. Ovation. All rise.

Lysenko's report on that day continued:

"Thus experiments in vegetative hybridization provide unmistakable proof that any particle of a living body, even the plastic substances, even the sap exchanged between scion and stock, possesses hereditary qualities.

"Does this detract from the role of the chromosomes? Not in the least. Is heredity transmitted through the chromosomes in the sexual process? Of course it is.

"We recognize the chromosomes. We do not deny their existence. But we do not recognize the chromosome *theory* of heredity. We do not recognize Mendelism–Morganism."

At the close of Lysenko's report, a succession of geneticists took the floor to recant their previous beliefs and to pledge support of Lysenko's views. The official resolution passed by the members of the Academy reads: "This session notes that to this day scientific research in a number of biological institutes and the teaching of genetics, plant breeding, seed cultivation, general biology and Darwinism in universities and colleges, is based on syllabuses and plans that are permeated with the ideas of Mendelism–Morganism, which is gravely prejudicial to the ideology training of our cadres. In view of this, this general meeting is of the opinion that scientific research in the field of biology must be radically reorganized and that the biological sections of the syllabuses of educational institutions must be revised."

Revised they were. Genetics was removed from all textbooks in the Soviet Union and virtually every Communist nation of Europe and Asia. Because of the stature of the geneticist Hans Stubbe within Germany, East Germany was an exception. Stubbe attempted to repeat Lysenko's experiments on hybrid grafting in tomatoes (the "vegetative hybridization" that provided "unmistakable proof" that heredity resides in every particle of the living body) and failed. Stubbe's skepticism eventually convinced Chairman Mao Tse Tung, who allowed genetic research and teaching to be resumed in mainland China during an era initiated by Mao's words, "Let a hundred flowers bloom."

(*continued on next page*)

Creationism and Creationists

From the day they were first made public, Darwin's views have aroused strong and emotional opposition in some quarters. In an era when the truth of the Bible was taken for granted, and when its interpreter, the village pastor, was likely to be one of the two most highly educated persons in the community (the village doctor was the other), to deny the Biblical account of Creation, as Darwin did, was outrageous. The immediate result was a torrent of ridicule and abuse, not only from religious leaders but also from those scientists who had failed ever to examine their own studies from any other than the all-pervasive, church-oriented perspective. They held that ours is an ordered universe, running its course according to the plans of a Supreme Creator. Scientists are human beings; they are not pleased to see the status of their lives' labors threatened.

Professor Ernst Mayr has made a similar point regarding the nineteenth-century philosophers of science: Darwin had presented his case without any reference to Plato. How could his work be taken seriously if he failed to include a single reference to the man who argued that each living thing is but an imperfect copy of a corresponding *ideal* thing? (Platonists held that, for example, horses vary; therefore, each horse is an imperfect copy of an ideal horse. Chairs vary; therefore, each chair is an imperfect copy of an ideal chair.) Darwin had dismissed Platonism without comment (the horses that we see are the only real horses, and they vary; the chairs in our house are the only real chairs, and they vary). Plato's constant ideal could not evolve; it was reconstituted generation after generation. Darwin's variable populations could evolve; the organisms living at any specified moment were (collectively) the species. The species could evolve continually.

Creationism as it exists today has little to do with the philosophy that prevailed prior to Darwin's publications. Current adherents to Creationism represent a residue of those who originally, having had no better theory, assumed that each species stemmed from an independent act of creation. The majority of people, having become aware of the internal contradictions that afflict the theory of Special Creation, having encountered evidence not generally available until it was organized by Darwin, and having seen the masses of confirmatory evidence provided by cytological and molecular techniques, now accept Darwinian evolutionary theory as the most reasonable, least complex explanation of the vast array of organisms now inhabiting the earth.

Human Culture

At several times during the evolution of the universe, events have occurred for which the outcomes are said to have transcended anything that might have been predicted in advance. Under the influence of solar energy and local electrical discharges, the methane, water, and carbon dioxide of the primitive atmosphere gave rise to an enormous number of what are normally referred to as organic molecules: amino acids, purines, and pyrimidines, among others. The first molecule that could make copies of itself, however, brought about changes on earth that would have been difficult to foresee, because the consequences of molecular *self-replication* are fantastically complex. The replicating molecules would have increased in number (however slowly), 1, 2, 4, 8, 16, 32, 64, 128, . . . In the meantime, small molecules that might have formed an alternative kind of replicating molecule were utilized in the replication of the first one. (This, you may remember, was one of the outcomes of the Spiegelman experiments on self-replicating RNA.)

Replicating molecules—the start of life itself in Muller's terminology—generated biological evolution. The secret to evolution lay in a chemical that could reproduce itself and, furthermore, could reproduce its own alterations—mutations—as accurately as it did its original, starting configuration.

After millions of years of biological evolution, human beings arose. Here, again, past events were transcended. Whereas DNA molecules accumulate information gradually, by having first one variant and then another subjected to the outcome of natural selection, human beings accumulate information in their brains. Whereas genetic information is passed from parent to offspring, human knowledge is passed from teacher to pupil, from generation to generation, and from group to group, regardless of familial relationships. Whereas lower animals accumulate information from birth to death, human beings first through oral and then through written history (not to mention today's electronic revolution) have accumulated the experience of generations, and they have made (or can make) this experience available to every living individual.

The result of human cultural development is a tremendous acceleration in the processes of change: culture evolves. Feedback mechanisms that successfully modulated the interactions of individuals within and between species under biological evolution are grossly inadequate for modern culture. By the time an error in judgment has been detected, and someone starts to say, "I don't think . . . ," the first problem has changed, and a new one has already taken its place. Power—raw physical power—that was unimaginable only a century ago has been concentrated in fewer than a dozen hands, each of which has a red button near its index finger. Additional raw physical power has become available to collections of ruthless individuals in the form of explosives. Human culture has indeed transcended biological evolution; it remains to be seen, however, whether it has the

Lysenko and Lysenkoism (*continued*)

In retrospect, it appears that many of Lysenko's claims were based on scientific naiveté (ignoring contamination as an explanation for the occurrence of grains of rye within a harvest of experimentally treated wheat) and even fraudulent experiments. In the dozens of claimed transformations of one species into a second, no new species were reported; only previously known forms arose—exactly as one would expect of contamination.

More than genetics was destroyed in the Soviet Union; many Russian geneticists were destroyed or imprisoned, as well. N. I. Vavilov died in prison; I. Agal was executed; S. G. Levit died in prison. Most of the geneticists who survived imprisonment were "rehabilitated" after the downfall of Lysenkoism and the revival of genetics during 1965–1966. A delegation of Soviet geneticists traveled to Brno, Czechoslovakia, in 1965 and placed a wreath at Mendel's monument to commemorate the 100th anniversary of Mendel's publication on inheritance in peas.

Our account of this sorry chapter in the history of biology can be closed with a quotation from the Russian martyr N. I. Vavilov. Let it be the last time words such as these must be spoken in defense of scientific integrity:

We shall go to the pyre,
we shall burn, but we shall
not renounce our convictions.

Cultural Evolution

To an extent far, far greater than any other animal—including their near relatives, the great apes—human beings have developed an alternative to physical heredity: cultural heredity. The passing of information from generation to generation is no longer limited to the transmission of DNA from parents to offspring by way of germ cells. Learning is no longer a cyclic phenomenon beginning with the newborn individual and ending, in the sense that all that has been learned vanishes, with the individual's death.

Cultural heredity follows from the transmission of learned experiences from one individual to another. Using gestures, word of mouth, writing, and other means, humans exchange knowledge within or between generations, and between unrelated or distantly related individuals, as well as between parent and offspring. The printed word (and, in modern times, electronic devices) has put at the disposal of any individual the accumulated knowledge of all past generations of human beings.

Just as a gradually changing physical heredity has resulted in a steady evolution of living forms, so has human culture evolved through the accumulation of cultural information. The resulting evolutionary changes are reflected in the tools that human beings have manufactured in different ages. Personal weapons have evolved from stone axes to swords, to crossbows, to rifles, and finally to automatic weapons. Cooking utensils have evolved from crude metal or clay pots to electric fry pans and crock pots. Some items change rapidly; others, like combs, change little. Uncertainty in the direction of change is revealed by the buggy whips on early automobiles and the rocket fins on later ones. Ways of life, as well as the tools needed for living, also change: agriculture, for example, has displaced hunting and gathering nearly everywhere on earth. Today's major religions have existed only for several thousand years, and it is unlikely that they will exist in their present forms several thousand years from now.

Theoreticians have attempted to formalize cultural evolution in the same manner that population geneticists and ecologists have formalized organic evolution. Many proposals have been made, none of which needs be discussed in detail. Each proposal attempts to identify the smallest unit of culture (a unit corresponding to the gene in physical heredity) and then to follow its fate as it spreads within or is eliminated from individual populations, and how it spreads from population to population if it is successful. The smallest unit of cultural heredity seems to be the *idea*, the new way of viewing a problem or the new way of expressing a thought that occurs in one person's mind at one moment. One such *idea* was Darwin's comment on the Galapagos finches written aboard the *Beagle* on October 8, 1835: "One species had been taken and modified for different ends."

stability that is necessary to lead mankind through a sustained, controllable cultural evolution or whether, because of inadequate feedback mechanisms and stabilizing influences, human culture will eventually destroy itself.

INFORMATIONAL ESSAY
Chromosomal Aberrations

At the time of their discovery, no one would seriously have asked whether the chromosomes of a species varied from individual to individual. Life itself seemed so miraculous that only the most perfectly operating genetic mechanisms, it seemed, could possibly sustain it. Certainly, when the question involved enzymes, most biologists thought of species as they thought of chemical compounds and crystals: only one enzyme molecule could possibly be the correct one; all other forms of the molecule (if any existed at all) must be abnormal by definition.

Today, the enormous molecular variation that exists within individual species is recognized. Living beings—lice, ants, mammals—are not analogous to individual crystals. Each differs from all others. Except for identical twins (or, in armadillos, identical quadruplets), which are known to be genetically identical, every organism alive on earth today—every organism of every species of plants or animals—probably differs from *all* organisms of the past and will probably never be duplicated in the future. That is the numerical or combinatorial power of the genetic variation now known to exist.

The development of fine microscope lenses and biological stains opened the way for the detection and study of chromosomal variation. A combination of genetic analyses and cytological observations (the first suggesting that all was not as it should be; the second attempting to see why not) led to an understanding of *translocations*, exchanges of chromosomal material between dissimilar or *non-homologous* chromosomes. Very often, because of the pairing of homologous regions of chromosomes, translocations could be detected as *rings*, rather than *pairs*, of chromosomes during meiosis. Such rings reveal the homologies of, and the physical alterations undergone by, chromosomes carried by many related species—especially plants.

Inversions are chromosomal aberrations arising by the rotation of a segment of DNA. They can be of two sorts: those where the rotated segments include the centromere (*pericentric inversions*) and those where they do not include the centromere (*paracentric inversions*). The consequences of the inclusion or exclusion of the centromere within the rotated segment are tremendous with respect to fertility and, therefore, with respect to the fate of a newly arisen inversion. The centromere, you may recall, controls the movement of its chromosome; of the divided centromere, one copy goes to one daughter cell while the other copy (and its associated chromosome) goes to the other daughter cell.

Crossovers—that is, exchanges between homologous chromosomes—that occur within pericentric inversions lead to gametes that carry duplications of some chromosomal regions and that completely lack other regions. Such chromosomes give rise to genetically unbalanced (inviable) zygotes, thus causing the heterozygous parent to be partially sterile. In *Drosophila*, crossing over is limited to females; thus, females heterozygous for pericentric inversions suffer from partial sterility. Of course, sterility tends to eliminate the pericentric inversion from the population.

A paracentric inversion need not confer partial sterility upon its carrier: the orientation of chromosomal strands following recombination causes the functional egg to receive a non-crossover chromosome. Because each egg receives a non-crossover chromosome, fertility is not reduced. As a consequence, *para*centric inversions are commonly observed in those dipteran species whose giant chromosomes allow their detection; *peri*centric inversions and translocations, on the contrary, are rare today, and were seemingly rare throughout the evolution of any genus of flies. To make the contrast between "common" and "rare" more specific: tens of thousands of *para*centric inversions have shaped the chromosomal evolution in the genus *Drosophila*; a dozen or fewer *peri*centric inversions have left any trace of their existence, either in the present or in the past.

INFORMATIONAL ESSAY
Continental Drift, Land Bridges, and Wallace's Line

Whenever the subject matter of a science slowly changes, that science develops a split personality: it must deal with both *statics* and *dynamics*. By definition, evolution involves change; however, the rate of change is so slow that one can discuss either the dynamics of evolution (change, itself) or the statics of evolution (the status quo of existing organisms). The same split confronts geologists and those biologists (paleobiologists) whose research material consists of fossil organisms. Many persons study or are concerned with the earth as it appears today. Among these would be the ecologists and physiologists who study organisms as they are found at the present time; also among these persons are the map-makers, those persons who construct maps showing that Blacksburg, Virginia, is 540 miles from Ithaca, New York, and—except for a few miles at either end—most of the distance is traversed by an Interstate Highway—I 81.

In contrast to map-makers, however, are the geologists, who point out that the distance from North America to Europe—that is, the

Alfred Russel Wallace

A letter dated June 20, 1858, written to the Secretary of the Linnean Society by Charles Lyell and Jos. D. Hooker, two of the Society's distinguished members, read in parts as follows:

"My dear Sir—The accompanying papers, which we have the honour of communicating to the Linnean Society, and which all relate to the same subject, viz. the Laws which affect the Production of Varieties, Races, and Species, contain the results of the investigations of two indefatigable naturalists, Mr. Charles Darwin and Mr. Alfred Wallace.

"The gentlemen having, independently and unknown to one another, conceived the same very ingenious theory to account for the appearance and perpetuation of varieties and of specific forms on our planet, may both fairly claim the merit of being original thinkers in this important line of inquiry; but neither of them having published his views, though Mr. Darwin has for many years past been repeatedly urged by us to do so, and both authors having now unreservedly placed their papers in our hands, we think it would best promote the interests of science that a selection from them should be laid before the Linnean Society."

The essay by Alfred Russel Wallace was entitled, "On the Tendency of Varieties to Depart Indefinitely from the Original Type." Wallace wrote that, unlike domestic varieties, which, if left to themselves, return to the normal form of the species, wild animals present a very difficult case: one cannot decide which is the *variety* and which the original *species*. Indeed, many so-called varieties outlive their parental species and give rise in turn to varieties of their own.

"The life of wild animals is a struggle for existence." This claim by Wallace was buttressed by the observation that the total number of individuals of any species remains roughly constant; therefore, if offspring are twice as numerous as their parents, twice as many individuals must die annually as make up the species. Numbers of individuals in populations, he pointed out, are determined not by numbers of offspring produced, but by the amount of food that sustains the population. Despite the size of their litters, wild cats are never numerous—their food supply is precarious.

Wallace discounted Lamarck's suggestion that progressive changes in species stem from attempts of animals to increase the development of their own organs—by volition. On the contrary, the talons of falcons and cats arose through the differential survival of groups that "had the greatest facilities for seizing their prey."

(*continued on next page*)

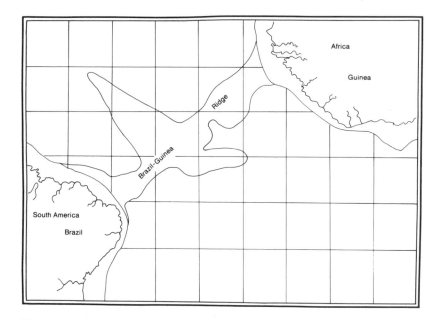

Figure 9.11. An outline of the land bridge that supposedly connected eastern South America (Brazil) with Western Africa (Guinea). Land bridges of this sort were proposed to account for the geographical distributions of some modern-day plants and animals. Visualizing a mass of land extending across the deepest part of the Atlantic Ocean was difficult; however, it was even more difficult for some persons to believe that several types of organisms would by accident have nearly identical geographic distributions spanning both sides of the Atlantic Ocean. It is now known that continents move, and that South America once abutted Africa. Neither chance nor land bridges need be evoked to explain many of the once-puzzling geographical distributions of living organisms.

width of the Atlantic Ocean—increases about as rapidly as one's fingernails grow. In the course of one's lifetime, the width of the Atlantic Ocean increases approximately one's height—not an important factor for airline pilots, perhaps, but a systematic change that has great importance over geologic time.

For decades, botanists and zoologists have recognized that similar plants and animals have strange, but consistent, distribution patterns involving widely separated continents. How were such distributions to be accounted for? Chance, of course, was always a possibility: a seed from an Ithaca, New York, silver maple tree may even today be miraculously carried to Shanghai, China, by high winds, the flight of some deranged bird, or by a junketing academic. Chance can account comfortably for a single unexpected observation, but it is stretched beyond belief when a series of species show similarly strange, disjunct distributions.

Biologists who despaired of relying on chance as a suitable explanation found themselves postulating an alternative, but equally debatable one: land bridges. One such bridge was said to connect Africa and Brazil, although it would have transversed the deepest part of the Atlantic Ocean. How does one choose between two highly improbable explanations?

A related and equally perplexing problem arose off the coasts of Borneo and Sumatra. Alfred Russel Wallace, an expert on the flora and fauna of Southeast Asia, claimed that, along a line that separated the island of Bali from an equally small island, Lombok, only fifteen miles away, the flora and fauna changed drastically, as if the regions had been oceans apart. The validity of Wallace's line has been discussed and argued about for a century, as are many statements that seem to have a kernel of truth but for which persons can find many exceptions.

What, in fact, is the truth behind Wallace's line? The space between Australia and New Guinea (to the south and east) and Sumatra, Java, Borneo, and the Philippines (to the north and west) was in the past an ocean, not a strait only a few miles wide. The land mass that includes Australia and New Guinea has been moving slowly but steadily to the north, and it has only now made contact with the southeastern part of Asia.

The map in Figure 9.12 shows that India was also separated from Asia, and has only "recently" made contact, thus becoming Asia's subcontinent. Why are the flora and fauna of India not distinct, as are those of Borneo and Australia? Probably because the contact was made along the entire base of an inverted triangle; no water barriers separate India from Asia as they separate Borneo and Australia from the mainland and the large islands of southeast Asia. The "top" of the subcontinent of India is too long to permit the peninsula to remain in isolation.

What evidence exists suggesting that continents move? Early evidence was suggested by those who noted that the eastern bulge of South America seems to coincide with the western indentation of Af-

Alfred Russel Wallace (continued)

The idea that living forms evolve apparently came to Wallace during a bout with a fever he contracted while collecting specimens in Malaysia. The checks—war, disease, famine, and the like—that act on man must, he thought, act on animals as well. He then recalled the even greater multiplication of animals, and realized the checks on them must exceed those on man. "While pondering vaguely on this fact, there suddenly flashed upon me the idea of the survival of the fittest—that the individuals removed by these checks must be on the whole inferior to those that survived."

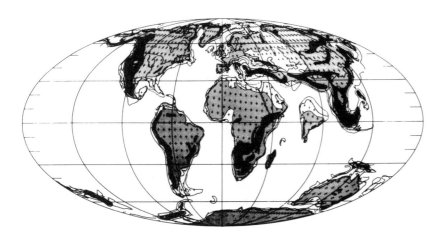

Figure 9.12. A map of the earth as it appeared forty-seven million years ago, during the Eocene epoch. Of particular importance are the positions of Australia, Malaysia, and India. Wallace's line, as explained in the text, separates the geographic distributions of two contrasting faunas that seem to have developed oceans apart, although the islands on which they are found may be only fifteen miles apart at the tip of Malaysia. In the Eocene period, Wallace's line would have fallen within the wide expanse of ocean separating Australia and New Guinea from Malaysia and the Philippines. During the intervening years, Australia and New Guinea have nearly collided with Malaysia, but the once isolated fauna are still largely intact. India, in the meantime, has collided with the land mass of Asia; the collision has produced the Himalayan mountains, which include Mount Everest. *(After A. M. Ziegler, C. R. Scotese, and S. F. Barrett, 1982)*

rica. Two more bits of evidence have now been added to this first piece. One involves droplets of glasslike material that are found scattered worldwide. The origins of these droplets would make a good deal of sense if the continents on which they are found were at one time contiguous. The second involves the traces of the earth's magnetic field that are embedded ("frozen" in place) within the earth's crust. This embedding occurs only within a narrow temperature range: lower, for example, than the temperature at which material oozes from the earth's interior in the mid-Atlantic but higher than the temperature of this material after centuries of contact with the waters of the ocean. Thus, as the magnetic poles of the earth have switched positions during the history of the earth, the changes in the position of, say, the north magnetic pole have left their mark in the ocean floor. The ocean floor is zebra-striped with "north" and "south" bands; furthermore, these bands are symmetrical on either side of the place of upwelling—the mid-Atlantic ridge.

Consequently, the thoughtful proposals concerning land bridges were on the right track: the "bridges" have proven to be land masses that were, in fact, in contact. Wallace's line, which seems to separate flora and fauna that have developed oceans apart, does exactly that: some thirty million years ago, Australia and New Guinea were an ocean removed from Sumatra, Java, Borneo, and the (present-day) Philippines.

INFORMATIONAL ESSAY
More on Ecological Convergence

The complexity of biological phenomena is illustrated by a seeming paradox: diverse organisms *converge* in form as the result of evolution, but related forms undergo *adaptive radiation* as they adopt dissimilar habits and morphologies. The concept of an ecological niche has revealed that these seemingly contradictory events are really two faces of a single coin. Because an ecological niche is composed of many elements, convergent evolution reflects changes in organisms that are undergone in conforming to the elements that are *common* to many niches, and adaptive radiation reflects changes that are undergone in conforming to those elements that *differentiate* one niche from another.

In every region of the globe, the physical characteristics (soil, rainfall, incident sunshine, temperature, and the like) of the area determine the types of plants (flora) that can grow there and the local fauna that can be supported. Historical factors cause otherwise similar areas to have different local flora and fauna, although ease of dispersal can, of course, erase what would otherwise reflect historical differentiation.

Within any one area, and (in a sense) superimposed upon all areas, are guild relations. A *guild* is a collection of organisms of different species whose means of maintaining and perpetuating themselves are similar. In the desert, ants and small rodents compete for seeds that are produced by the same plants; they are otherwise so different, however, that one would not place them in the same guild. Similarly, hordes of migratory locusts may deplete crop plants and cause starvation among domestic cattle and their owners, but no guild relation is involved here. On the other hand, the grazing mammals of the world—the bison of the Western prairie and the wildebeests and zebras of Africa—can be viewed as representing a guild: they are the massive animals that reap the harvest of the plains. Flycatchers of all sorts also represent a guild. Guild members, in short, use similar methods of making a living.

Because the organisms in a guild face similar challenges, no matter what species of animal is involved, they often exhibit similar characteristics. Small burrowing mammals possess huge front paws, sturdy shoulder girdles, greatly reduced ears and eyes, and pointed sensitive snouts, features that are found in the true mole (an insectivore), the marsupial mole, and other burrowing rodents. Anteaters also exhibit common features, such as an elongated snout, a sticky tongue, heavy claws, and tough skin.

The number of ways by which animals can make a living is not large: herbivores eat plants, and carnivores eat herbivores or other carnivores. This point is emphasized by the parallels that can be drawn between the marsupial fauna of Australia and the placental fauna of any other major continent. The names given to the Australian marsupials by settlers and others reveal parallels: the marsupial *wolf*, the marsupial *cat*, marsupial *mice*, the marsupial *mole*, the *rabbit* bandicoot, the flying phalanger (also known as the flying *squirrel*—but it is a marsupial, not a placental, mammal), and the wombats, which are similar to badgers.

The adaptation of living organisms to the environment, however, is not governed by a set of precise rules nor is it achieved with the use of calipers and micrometers. Historical accidents have determined what plants and animals are found in various parts of the earth. Marsupials, for example, are believed to have arisen in South America, and to have spread across what was later to become Antarctica into the land mass now known as Australia. At the time this spread occurred, the three continents were joined. Once their homeland had separated from other land masses, the Australian marsupials underwent a tremendous radiation, adopting all the common ways by which mammals support themselves. However, there are no marsupial bats.

The recent studies (mentioned earlier in this text) on the degree of homology between DNAs of various species have revealed the extent to which convergence can cause a resemblance between unrelated birds. For many years, the world's leading ornithologists studied the birds of Australia and interpreted their observations as meaning that Australia had been repeatedly invaded by birds from other continents. DNA–DNA hybridization analyses have revealed that, on the contrary, the Australian birds that so closely resemble their counterparts on other continents are *not* related to these other birds: they are related to other, morphologically dissimilar Australian birds. Natural selection acting upon members of a guild leads to greater similarities (convergent evolution) than had ever been suspected previously.

INFORMATIONAL ESSAY
Taxonomy and Systematics

The naming of living organisms constitutes the field of study known as taxonomy. The system currently used by all taxonomists was developed by Carolus Linnaeus (1707–1778). The Linnean system is also known as the *binomial system of nomenclature*. Each species is given a two-word name, known as the genus and the specific epithet. The first name (the genus) is a noun and the second (the specific epithet) is an adjective describing the first. For example, the scientific name of the grizzly bear is *Ursus horribilis,* "horrible bear." The names of organisms are written in Latin because Latin is a dead language and will not itself evolve. Furthermore, Latin is the parent language of many present-day languages and its use can transcend many language barriers. The genus and specific epithet identify the species. Thus, *Homo sapiens* refers to the species of "wise men."

Species are defined as populations of organisms that are *reproductively isolated* from other, closely related populations. The key to understanding the (biological) species concept is the phrase "reproductively isolated." Reproductive isolation may occur in many ways. It may precede mating: organisms may occupy different geographic localities, may have different mating times, or may exhibit different behavioral or courtship patterns. Of course, isolating mechanisms may break down and allow individuals from two previously isolated populations to interbreed. The offspring may die, or if they develop to maturity, they may be sterile. Mules are sterile hybrids. On the other hand, the offspring may be fertile; if such intergroup matings are common, reproductive barriers may break down in what is known as a "hybrid zone."

Scientific names serve as a scientific tool. Acceptable names have purpose or utility, they are not arbitrary, and they reflect relationships to other, similar organisms. The science of classifying organisms on the basis of their evolutionary relatedness is called systematics. Hence, all similar species should belong to the same *genus*, all similar genera to the same *family*, all similar families to the same *order*, all similar orders to the same *class*, all similar classes to the same *phylum*, and all similar phyla to the same *kingdom*. For example, the systematic classification of the grizzly bear is:

Kingdom:	Animalia
Phylum:	Chordata
Class:	Mammalia
Order:	Carnivora
Family:	Ursidae
Genus:	*Ursus*
Species:	*horribilis*

While taxonomists might be viewed by some as pursuing abstract—even useless—knowledge, taxonomy actually has important applications: for example, the misidentification of an insect can be as costly to a farmer as the misdiagnosis of a disease and the erroneous administration of a medicine can be to a patient and his or her family.

10
COMMUNITIES OF SPECIES

Recall the gardens, parks, forests, or even vacant lots that you have seen. Was there any place among them where you encountered one, and only one, species of living things? Do not let names fool you. Just because a park is called the Redwood Forest or Cyprus Gardens, it does not mean that only redwood or cyprus trees grow there. Unless you have visited some extraordinarily exotic place, you have probably never seen a locality that was inhabited by one species, and one species only.

If such a place were to exist, the organism would most likely be a plant, because plants can utilize water, carbon dioxide, and solar energy to make more plant material. Animals, however, need either plants to eat or, in the case of carnivores, animals that have already fed on plants. Even plants die, however, and must decompose—a process requiring microorganisms. The closest approach to a single-species region is found in the vast expanses of modern farms that are planted with a single crop (monocultures). In economic terms, these farms are supposed to be the most efficient and the most profitable; this judgment, if correct, may apply in the short term only. A great deal of fertilizer must be applied to these fields, because the nutritional needs of the individual plants do not vary; every single plant makes the same nutritional demand as the others. Energy is needed, of course, in the manufacture of fertilizers. Tremendous amounts of energy are also required in the manufacture of the pesticides that are needed to prevent or suppress what would otherwise be a rapid, even exponential, growth of insect and other pest populations. Over all, as we have seen earlier, most crops yield only approximately the same number of calories as were invested in planting, caring for, reaping, and distributing the harvest. The sources of this energy are the fossil fuels—oil, coal, and natural gas—that exist on earth in finite quantities only. The analyses of agricultural economists take the availability of fuel as given; in the future, this assumption will be wrong.

NOW, IT'S YOUR TURN (10-1)

1. During the 1930s, a scientific paper was published that described a species of small spider that was the only form of life occurring on glaciers in Switzerland. In discussing how these spiders might survive, the author suggested that parents might eat their own young, and that the larger offspring might eat their smaller sibs; in this way, the survivors would obtain their nutrients. Reply to this paper in your journal.

2. Using the same logic as the author of the spider paper described in the preceding paragraph, economists and government officials have recently stressed the possibility that crops be fermented to produce alcohol that can then be used as a fuel. Assume that the fermentation of 100 calories of grain yields 50 calories of fuel, and that 100 calories of fuel (oil, gas, and coal) are required to produce the 100 calories of grain. Suggest a means by which one allotment of fuel might easily be doubled.

If most places on earth are occupied by not one, but many, species of organisms (both plants and animals), additional questions arise: Are the communities—that is, the mixtures of species that are found in different localities—stable in composition? Or, at least reasonably so? If so, how did they come into existence? Two extreme possibilities come to mind: First, that species, like atoms, have certain properties that result in the formation of communities, just as the properties of atoms result in their combining to form molecules. Thus, the complex of plant life that collectively forms a meadow may be preordained to do so (given the opportunity), just as carbon, oxygen, and hydrogen are preordained to produce glucose, if given an opportunity to do so.

The second possibility is, of course, that species that live together for any length of time not only adjust to one another's presence but also come to rely upon that presence—to adjust to the other's presence in such a manner that, if one species were to be removed, members of the others would actually suffer.

The second possibility can be verified in any number of instances involving extreme adaptations. The sloth, for example, spends nearly all its hours hanging upside down under the branches of trees in tropical rainforests, munching slowly on leaves. The three-toed sloth cannot stand up; when it is placed on a flat surface, its four legs simply spread-eagle.

Sloths were not originally built as they are today, and they did not take to an arboreal life because they were anatomically constructed to do so. The current version of the sloth is the result of its species' association with the trees, during which it has evolved physical changes that suit it for its upside-down way of life. Should the trees of the tropical rainforest disappear, so would the sloths that now dwell in them.

Squirrels are also arboreal animals that have adapted their habits for a life in and near trees. As buildings have displaced trees in many urban locales, the squirrels have vanished. In another example, the tails of woodpeckers contain modified feathers that are used like a telephone lineman's climbing spikes: the bird clings to the bark with its claws, leans back on its tail, which gives it secure support from beneath, and then pounds away at the bark with its beak. Like the squirrels and sloths, woodpeckers have adapted to life on trees.

Experimental Evidence on Interspecific Interactions

The information presented in the preceding two or three paragraphs is no more than a recitation of anecdotal evidence. Among the several million species of living organisms, there can be billions or even trillions of possible interactions. Consequently, a strong case for the mutual adjustment of two coexisting species to each other's presence cannot be based on a recitation of individual cases, even though the cited list might contain several hundred instances. For example, one cannot prove that professors are generally dishonest by citing a half-dozen or even a dozen instances where professors have failed to pay their debts, or by naming one or two who have been convicted of shoplifting.

What sorts of things are observed when different species of the same genus are forced to develop together in a confined space? To the extent that they demand identical resources, of course, they interfere with each other's survival—provided that the total amount of the resource is limited, or that the individuals do not encounter their needed resource purely by accident. If an essential resource is not limiting (that is, enough of it exists for the survival of all individuals), then slight differences in the overall requirements of the two species may lead to a greater total production (in either number of individuals or total quantity—*biomass*—of all individuals combined) when two different species are grown together than when they are grown separately in monocultures.

Information about the interaction of two plant species is sometimes obtained by planting seeds together in a series of garden plots or containers using a design known as a *replacement series*. Imagine, for example, six medium-sized flowerpots. In one are planted twenty seeds of round, red radishes. In the second are planted only sixteen seeds of round, red radishes plus four of a second variety, a white, icicle (elongated) radish. In the third, the numbers are twelve red plus eight white; in the fourth, eight red plus twelve white; in the fifth, four red plus sixteen white; and in the sixth, only twenty seeds of the white icicle variety are planted. Notice that there are twenty seeds in every flowerpot; the only difference between the pots is the relative proportions of the two types of seed: one type decreases, while the other increases.

The radishes are allowed to grow for two months, after which they are harvested, washed free of dirt, and the inedible leaves and fibrous roots are removed. Only the edible portions of the two types of radish remain. Suppose our total yield of radishes per pot is as follows (pot 1 had only red radishes, pot 6 only white ones, and pots 2–5 had the proportions described above; numbers are grams):

Pot	1	2	3	4	5	6
red	300	240	180	120	60	0
white	0	80	160	240	320	400
total	300	320	340	360	380	400

(Plot these figures on a sheet of graph paper for inclusion in your journal.)

What do these numbers (and your graph) indicate? First, the total grams of red radishes decreased as a straight line (linearly) as the number of red seeds per pot decreased, and it seems that red radishes tend to weigh about 15 grams. Second, the total grams of white radishes increase linearly with seed number; thus, it seems as if an average white radish weighs 20 grams. Furthermore, the sum of the weights of the two types of radish increases linearly from pot 1 to pot 6. Given the conditions as they have been described, the total yield *should* increase linearly. For example, starting with the 300 grams in pot 1 as the base, pot 2 had four white radishes, each of which was 5 grams heavier than the reds they replaced $[300 + (4 \times 5)] = 320$. Pot 3 had eight white radishes, each of which was 5 grams heavier than the reds they replaced $[300 + (8 \times 5)] = 340$. The yields of the remaining pots can be calculated in the same way.

At this point, we might conclude that because all radishes grew and because the average weight of each type, red or white, was constant, these growing radishes did not interfere with one another. That conclusion would be hasty, however. We now plant only *one* seed in each flowerpot. Now our harvest might show that the average weight of red radishes is 30 grams and that of white radishes is 40 grams. Therefore we know that twenty radishes growing in a pot *do* interfere with one another; the crowded radishes are only half the size of those that were grown singly. The straight lines of the graph, then, tell us something different: radishes interfere with one another's growth when crowded, but each radish's (red or white) effect on the growth of another (again, red or white) is the same. In terms of crowding, twenty radishes share the available but limited resources, which are sufficient to permit each radish to grow to only one-half its possible size.

Before citing a more practical example, we shall pretend to raise radishes once more, this time obtaining a different set of data. The number of pots and the numbers of red and white seeds are just as they were in the first example above: 20 red:0 white; 16:4; 12:8; 8:12; 4:16; 0 red:20 white. Having harvested the radishes, washed them, and removed tops and fibrous roots, we now collect the following data, where the listed numbers are weights in grams:

Pot	1	2	3	4	5	6
red	300	288	252	192	108	0
white	0	144	256	336	384	400
total	300	432	508	528	492	400

(Plot these figures on a sheet of graph paper for inclusion in your journal.)

What do these numbers indicate? First, of course, the data show that the greatest total harvest of radishes came from one of the intermediate pots, one that contained eight red and twelve white radish seeds. Second, the yield of neither red nor white radishes, when examined by itself, is linear; each increases most rapidly when the number of that type is low, and each tends to flatten out as that type becomes the predominant type. Each type of radish (red or white) tended to be large when it was rare, despite the presence of sixteen or twelve of the other sort. Crowding effects are more extreme between members of the same type of radish than between members of the two different types. This, then, is the key to the position of maximum yield: loss of yield through crowding is primarily an *intra*varietal phenomenon.

Darwin's words about competition have been scarcely improved upon since he wrote them (which reveals what an extremely thoughtful and astute man he was):

The dependency of one organic being on another, as of a parasite on its prey, lies generally between beings remote in the scale of nature. This is often the case with those which may strictly be said to struggle with each other for existence, as in the case of locusts and grass-feeding quadrupeds. But the struggle almost invariably will be most severe between the individuals of the same species, for they frequent the same districts, require the same food, and are exposed to the same dangers. In the case of varieties of the same species, the struggle will generally be almost equally severe, and we sometimes see the contest soon decided: for instance, if several varieties of wheat be sown together, and the mixed seed resown, some of the varieties which best suit the soil or climate, or are naturally the most fertile, will beat the others and so yield more seed, and will consequently in a few years quite supplant the other varieties.

And, somewhat later he said, "As species of the same genus have usually, though by no means invariably, some similarity in habits and constitution, and always in structure, the struggle will generally be more severe between species of the same genus, when they come into competition with each other, than between species of distinct genera."

NOW, IT'S YOUR TURN (10-2)

1. Mixtures of two particular species of oats yield much more than does either species alone if the mixtures are planted in deep soil. The greater yield nearly disappears if the mixture is planted in shallow soil. Explain how this could happen. Keep in mind that the roots of one species tend to spread near the surface of the soil, whereas those of the other species tend to penetrate deeper into the soil. (Different root structures—round and elongated—were also emphasized in our hypothetical radish experiment.)

Mixtures of different grasses and various forage crops do not always yield more than single varieties or species grown alone. Nevertheless, one tabulation shows that although 39 mixtures yielded less than the poorer-yielding species, 73 mixtures yielded more than the higher-yielding one. Similarly, 85 mixtures yielded less than the average of the two tested species (but as much as, or more than, the poorer one), whereas 117 yielded more than the average (but no more than the better-yielding species). Refer to the informational essay on testing for goodness-of-fit, and decide whether the numbers (39 versus 73 and 85 versus 117) of each of these pairs differ significantly. (Hint: If they do not differ, one would expect them to be equal: 56 versus 56 in the one case and 101 versus 101 in the other.)

2. Did you read carefully what Darwin said? Did you understand it? Explain, then, why he said that species of the same genus *usually* have some similarity in habits but *always* in structure. Why is it that habits may or may not be similar, but structures always are?

Before we leave the experiments on interspecific interactions, still another possibility should be mentioned: species can interfere with one another, so that mixtures of the two yield fewer total survivors than one would expect. To visualize this, imagine what would have happened if the Christians who were thrown to the lions in Rome's Colosseum had been armed with knives and spears. The outcome would have been that the Christian–lion mixture would have yielded fewer Christians, *and* fewer lions. Most experiments in which the vinegar fly, *Drosophila melanogaster,* and its close (virtually indistinguishable) relative *D. simulans* have been raised together have demonstrated that the two species interfere with each other. The total number of flies obtained from mixtures is smaller than one would expect from results obtained by studying each species separately.

Why Do Stable Communities Exist?

Experimental studies have shown that species in close proximity may seem to ignore one another, may enhance each other's survival, or may interfere with each other's normal development and survival. Thus, the species' characteristics must influence the outcome of encounters between them. Organisms do have their biological characteristics, and it is reasonable to expect that an already existing characteristic might influence the possible incorporation of an additional species in an established community.

Introduced or alien plants and animals represent enormous experiments on community structure. For example, starlings were introduced into the United States in 1890; they are now part of the natural community throughout this country. The house (or English) sparrow is also an immigrant, having been introduced in New York City in 1850; it has displaced a number of formerly common birds and has spread throughout the United States, Mexico, and Canada. The sparrows have also differentiated geographically, so that several subspecies (races) now exist in the United States. An example of a temporarily *un*successful (accidental) importation is provided by the fire ant. For many years after their introduction into the United States, these ants were to be found only near Near Orleans, Louisiana. The widespread fire ant infestation that causes so much unhappiness in the southeastern states of the United States today apparently stems from a second introduction of fire ants from South America. Today's fire ant arose through the hybridization of ants from the two separate importations: they represent a natural experiment in recombinant DNA. A similar event may have taken place in Australia with respect to the fruit fly that is an orchard pest there: a sudden increase in the geographical range of these flies has been attributed to a hybridization between closely related species. Actually, Europeans introduced many plant and animal species into Australia and New Zealand. Nearly all of the introductions have been disastrous from the viewpoint of the local flora and fauna: placental mammals have displaced many of their marsupial counterparts, rabbits virtually overran Australia until the deliberate introduction of a lethal virus disease reduced their numbers drastically, and a cactus overran much of that continent until a cactus-eating moth was introduced as well. Today, the offending cactus exists as sparse, isolated plants, while the moth ekes out its existence in Australia searching for and feeding on this now scarce commodity.

The intrinsic properties of an organism that have developed during earlier times and at other places play a role in determining whether that organism has any place in a novel community setting. Once the decision to reject or to incorporate has been made in favor of incorporation within the community, a train of evolutionary changes are set in motion; these changes affect both the newcomer and the older members of the community. These changes are needed for (and, having been made, tend to promote) community stability.

Character Displacement

Character displacement has been mentioned earlier in this text. As we discussed previously, organisms cannot afford to choose members of a different species for mates, because interspecific hybrids are often sterile or produce high proportions of inviable offspring. An individual who indulges in an "erroneous" mating wastes his or her time and gametes. The outcome of selection favoring intraspecific (correct) matings is a diversity of mating signals, plumage colors and patterns, flowering times, sizes of flowers, mating sites and times, and other factors involved in mate selection and mating. The divergence arises within local areas (and local communities); this conclusion is based on the frequent similarity of these same characteristics that is observed when comparing species living at some distance apart.

Ecological Specialization

Biologists are familiar with two seemingly contradictory evolutionary patterns: *adaptive radiation* and *convergent evolution*. The first refers to the variety of lifestyles adopted by the descendants of what was originally a small group (perhaps a single gravid female) of progenitors; the second refers to the obvious similarity between diverse, unrelated forms that occupy the same habitat and that have had to respond to similar environmental challenges.

The paradox posed by these two seemingly opposite responses of organisms to their environments disappeared once scientists had gained an understanding of the *ecological niche*. Like many biological concepts, this one has caused considerable controversy: Is an ecological niche something actually occupied by some organism? Or is it something that can exist and be described even though it is not occupied? We hold the latter view: the ecological niche can be viewed as a collection of points, each of which represents one state of each of many aspects of the environment. Viewed in this way, the niche is a cluster of points in many dimensions that correspond to points on a standard graph where each point is represented as $P_{(x,y)}$ or, more simply, (x,y). If temperature, t, humidity, h, altitude, a, and sunlight,

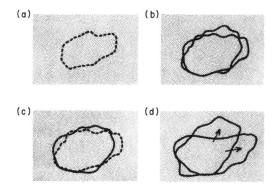

Figure 10.1. A schematic representation of (a) an ecological niche and, in those cases (b, c, and d) where two species occupy extremely similar niches, the tendency either for one to become extinct or for the two to undergo diversifying evolutionary changes. The individual dots in the background represent the theoretical points discussed in the text, each of which represents a combination of individual states for many environmental factors, such as light, moisture, pH, and temperature. The dashed line in a outlines a theoretical, but unoccupied, ecological niche that encompasses many points. Two species occupying two slightly dissimilar niches are shown in b. If the niches shown in b are too similar, one species will tend to displace the other, which may become extinct, as shown in c. Alternatively, the two species may evolve (d) so that they come to occupy niches differing from the original ones. Such changes (which lead to *adaptive radiation*) may permit the two species to coexist. On the other hand, adaptation to the conditions specified by the shared points of the two niches leads to *convergent evolution*.

s, are the environmental factors that determine a niche, each point would have given values of *t, h, a,* and *s.*

Two ecological niches are identical if all points of one correspond to points included within the second, and vice versa. Two niches are *not* identical if either contains points *not* contained within the other. A difficulty arises because many, if not most, environmental factors have continuous distributions, and these must be divided into discrete values. A second difficulty arises because man's mind and vocabulary can imagine and describe factors in the environment that many organisms cannot distinguish. These are not fatal problems, merely problems to be aware of.

All else being equal, organisms that occupy otherwise uninhabited portions of ecological niches will leave more offspring than will those that share (unwillingly, of course) portions of their niches with other species. Differential reproduction is the basis of natural selection and adaptation. As a consequence, the niches actually occupied by species tend to become different as the nature of the individuals belonging to each species changes: the divergence of occupied niches is perceived as adaptive radiation.

To differ, two niches need not differ in *all* points; many points can be held in common. Air and its fluid dynamics present a common challenge to flying organisms, water and its physical characteristics present a common challenge to swimming animals, and soil presents similar challenges to many burrowing animals. Meeting these common challenges gives rise to convergent evolution.

Coevolution

Two coexisting forms can undergo long-term evolutionary changes in concert, with each change in one evoking a response from the other. An example that illustrates the ability of one organism to track changes that occur in another is provided by one of the mustards (*Camelina sativa,* subspecies *linicola*) and domesticated flax.

Flax is one of the oldest of domesticated plants, and fossil mustard seeds have been found in caves that were inhabited by prehistoric people (caves 10,000 or more years old). Flax, which was grown for its fiber, was continually selected for tall, erect growth in dense strands. The mustard, *Camelina sativa* (which may have been raised originally for mustard oil), seems to have been a contaminant of flax since the earliest days of agriculture. Today, *Camelina sativa linicola* grows in flax fields and mimics the flax plants. Its seeds are harvested with those of flax and are sown with the flax seed the following season. As a flax mimic, the mustard has undergone a number of changes: it is taller, more slender, and less branched than the non-mimic members of the species. Its seeds are retained in their pods until the flax is harvested.

After flax is harvested, it is threshed, a process that yields the seeds for next year's crop. Chaff and other debris (including seeds of unwanted weeds) are removed from flax seed by winnowing: the threshed seeds are dropped across a horizontal air current, which carries off light material, deflects flax seed a certain amount, and lets heavy material simply drop. Over past centuries, the flax mimic—the mustard plant—has developed seeds that are blown the same distance as, and are collected with, the flax seed. Thus, the flax farmer plants both flax and mustard seeds when he sows next year's flax crop. In this instance, coevolution consists of the evolution of flax under millennia of domestication, and the evolution of the mustard (the flax mimic) in a way that permits it to continue to grow and thrive as a weed in fields of flax. What is judged a weed today seems to have been a useful crop plant originally. Consequently, weeds are not necessarily plants that spread from woods, hedgerows, and wild meadows and invade cultivated fields. Weeds are simply plants—even "domesticated" ones—that grow where they are not wanted.

The coevolution of insect and plant species can be reconstructed from relationships that are too specialized and too intimate to make sense under alternative explanations. Orchids, for example, both mimic the appearance of female wasps and exude an odor that stimulates mating behavior in male wasps. The male wasp's attempts to mate with the orchid's fake female wasp parts pollinate the orchid flower.

Ants have special relationships with several tropical plants. These plants furnish (1) sugar solution, (2) protein-rich bodies on their leaves, and (3) suitable ant living quarters, either in thorny appendages or within the hollow stems. In return, the ants patrol the plant and destroy all potential insect pests. They clear the ground around the base of the plant, thereby reducing competition with other plants. At times, the ants destroy tendrils of climbing plants that might otherwise harm their host. By harboring ant guests, the host plants reduce the numbers of their competitors and eliminate insect herbivores. Both plant and ant profit from this relationship.

If you visit an aquarium, you should be able to observe the special relationships between large tropical fish, such as the groupers, and smaller "cleaner" fish. Most aquaria place the small cleaner fish in larger fish's tanks because the smaller fish are important to the larger fish's health. The small fish remove parasites from the skin of the larger ones; they will even swim into a larger fish's mouth, cleaning around the teeth and inside the mouth. They will also swim into and clean the larger fish's gill chamber when it raises its gill cover. In coral reefs, the cleaner fish have cleaning stations, much like carwashes or beauty shops, to which the larger fish come as customers. One biologist removed the cleaner fish from a coral reef and found that the "customers" disappeared; those that remained after two or more weeks were covered with sores.

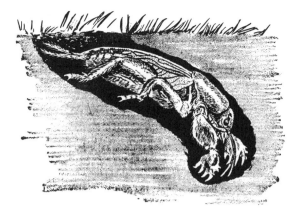

Figure 10.2. Mammals that burrow underground, whether they are marsupials, rodents, or insectivores, come to possess remarkably similar morphologies: paddle-like forelimbs, huge shoulder girdles, fur that can be brushed in either direction, reduced eyes, and a sensitive snout. The mole cricket shown here, which is a burrowing cricket, occupies an ecological niche that poses problems for it that are in many respects similar to those posed for burrowing mammals by their niches. Hence, the cricket's forelegs resemble moles' forelegs because they have made similar adaptations to digging in soil. They are an example of convergent evolution. *(Sketch from* The Animal Kingdom*)*

When describing the courtship of fireflies, we mentioned that the female of one species, once it had attracted a male of its own kind and had been fertilized, used its attractive flashes to capture males of three other species as prey. Having lured a male of one of these other species, the carnivorous female would eat him. The same sort of trickery has also evolved in fish that imitate cleaner fish. For example, small blennies imitate the guild markings of cleaners and thereby gain access to a larger fish, ostensibly for doing a cleaning job. Once near enough to the large fish, the blenny will take a bite of flesh from the larger fish and retreat.

NOW, IT'S YOUR TURN (10-3)

In your journal, give an account of the behavior of the carnivorous female firefly: Having been fertilized, why doesn't she lure males of her own species to their deaths? Why only males of other species? (These questions are not as simple as they appear to be; you may want to experiment with logical—maybe mathematical—answers on scratch paper before turning to your journal.)

On the Size of Ecological Communities

Groups of diverse organisms live together in communities. These communities are not haphazard collections of individuals, nor are they rigid or static collections that remain forever unchanged. Ecological communities are collections of organisms—species of organisms—whose interrelationships tend toward stability and whose adaptations or evolutionary adjustments to the existing conditions tend to stabilize those very conditions.

First, there are the producers, the green plants. These organisms range from algae through grasses, bushes, and vines to trees, but they all utilize sunlight, moisture, and carbon dioxide to produce the enzymes and the thousands of cellular substances that make life—theirs and others'—possible.

Second, there are the herbivores. These range from aphids, leaf miners, and katydids to voles, rabbits, deer, and bison. Herbivores are unable to synthesize protoplasmic substances from water and carbon dioxide, so they do the next best thing: they eat the producers. Of course, the "hervibores" also include mushrooms, toadstools, and fungi—non-green plants that also utilize ready-made organic matter.

Third, to complete a small but essential cycle, there are the decomposers—bacteria and a variety of microorganisms. In fact, 200–500

CAMBRIAN FAUNA

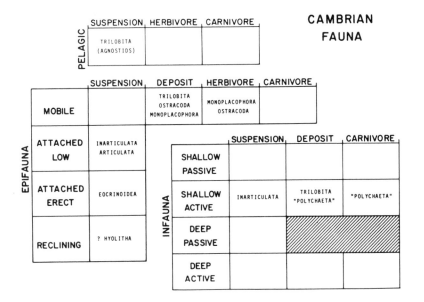

PELAGIC

SUSPENSION	HERBIVORE	CARNIVORE
TRILOBITA (AGNOSTIDS)		

EPIFAUNA

	SUSPENSION	DEPOSIT	HERBIVORE	CARNIVORE
MOBILE		TRILOBITA OSTRACODA MONOPLACOPHORA	MONOPLACOPHORA OSTRACODA	
ATTACHED LOW	INARTICULATA ARTICULATA			
ATTACHED ERECT	EOCRINOIDEA			
RECLINING	? HYOLITHA			

INFAUNA

	SUSPENSION	DEPOSIT	CARNIVORE
SHALLOW PASSIVE			
SHALLOW ACTIVE	INARTICULATA	TRILOBITA "POLYCHAETA"	"POLYCHAETA"
DEEP PASSIVE		(hatched)	(hatched)
DEEP ACTIVE			

Figure 10.3. Organisms' use of a marine environment during different geologic eras. To a large extent, evolution has led to an ever more efficient use of the environment. Organisms have evolved that can swim (pelagic organisms), lie on the surface of the ocean floor (the epifauna), or burrow into the floor itself (the infauna). Furthermore, these organisms can hunt (the carnivores), graze (the herbivores), suck up debris (deposit), or live on suspended matter (suspension). The three diagrams show that, whereas during the Cambrian Era (500–570 million years ago) many possible modes of life were not utilized, these modes did evolve during the Paleozoic Age (225–400 million years ago) and especially during the Mesozoic–Cenozoic Age (25–200 million years ago). *(Diagrams courtesy of R. K. Bambach, Virginia Tech)*

MIDDLE & UPPER PALEOZOIC FAUNA

PELAGIC

SUSPENSION	HERBIVORE	CARNIVORE
CONODONTOPHORIDA GRAPTOLITHINA ? CRICOCONARIDA		CEPHALOPODA PLACODERMI MEROSTOMATA CHONDRICHTHYES

EPIFAUNA

	SUSPENSION	DEPOSIT	HERBIVORE	CARNIVORE
MOBILE	BIVALVIA	AGNATHA MONOPLACOPHORA GASTROPODA OSTRACODA	ECHINOIDEA GASTROPODA OSTRACODA MALACOSTRACA MONOPLACOPHORA	CEPHALOPODA MALACOSTRACA STELLEROIDEA MEROSTOMATA
ATTACHED LOW	ARTICULATA EDRIOASTEROIDA BIVALVIA INARTICULATA ANTHOZOA STENOLAEMATA SCLEROSPONGIA			
ATTACHED ERECT	CRINOIDEA ANTHOZOA STENOLAEMATA DEMOSPONGIA BLASTOIDEA CYSTOIDEA HEXACTINELLIDA			
RECLINING	ARTICULATA HYOLITHA ANTHOZOA STELLEROIDEA CRICOCONARIDA			

INFAUNA

	SUSPENSION	DEPOSIT	CARNIVORE
SHALLOW PASSIVE	BIVALVIA ROSTROCONCHIA		
SHALLOW ACTIVE	BIVALVIA INARTICULATA	TRILOBITA CONODONTOPHORIDA BIVALVIA POLYCHAETA	MEROSTOMATA POLYCHAETA
DEEP PASSIVE		(hatched)	(hatched)
DEEP ACTIVE		BIVALVIA	

MESOZOIC – CENOZOIC FAUNA

PELAGIC

SUSPENSION	HERBIVORE	CARNIVORE
MALACOSTRACA GASTROPODA MAMMALIA	OSTEICHTHYES MAMMALIA	OSTEICHTHYES CHONDRICHTHYES MAMMALIA REPTILIA CEPHALOPODA

EPIFAUNA

	SUSPENSION	DEPOSIT	HERBIVORE	CARNIVORE
MOBILE	BIVALVIA CRINOIDEA	GASTROPODA MALACOSTRACA	GASTROPODA POLYPLACOPHORA MALACOSTRACA OSTRACODA ECHINOIDEA	GASTROPODA MALACOSTRACA ECHINOIDEA STELLEROIDEA CEPHALOPODA
ATTACHED LOW	BIVALVIA ARTICULATA ANTHOZOA CIRRIPEDIA GYMNOLAEMATA STENOLAEMATA POLYCHAETA			
ATTACHED ERECT	GYMNOLAEMATA STENOLAEMATA ANTHOZOA HEXACTINELLIDA DEMOSPONGIA CALCAREA			
RECLINING	GASTROPODA BIVALVIA STELLEROIDEA ANTHOZOA			

INFAUNA

	SUSPENSION	DEPOSIT	CARNIVORE
SHALLOW PASSIVE	BIVALVIA ECHINOIDEA GASTROPODA	BIVALVIA	BIVALVIA
SHALLOW ACTIVE	BIVALVIA POLYCHAETA ECHINOIDEA	BIVALVIA ECHINOIDEA HOLOTHUROIDEA POLYCHAETA	GASTROPODA MALACOSTRACA POLYCHAETA
DEEP PASSIVE	BIVALVIA	(hatched)	(hatched)
DEEP ACTIVE	BIVALVIA POLYCHAETA MALACOSTRACA	BIVALVIA POLYCHAETA	POLYCHAETA

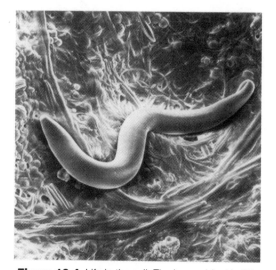

Figure 10.4. Life in the soil. The large object in this scanning electron micrograph is a nematode, a simple, unsegmented worm. Nematodes are among the most numerous of "higher" organisms. Some, like the one in the photograph, are free living; they eat the bacteria and algae that can also be seen in the picture. Others are parasitic in plants, and still others are parasites of vertebrate and invertebrate animals. Nematodes are so numerous that, it has been claimed, a ghostly outline of the earth and all its plants and animals would remain if all matter other than these worms were to vanish. *[Copied with permission from Scanning Nature, by D. Claugher; © Trustees of the British Museum (Natural History); Cambridge University Press, Publishers]*

pounds of microbes are to be found in every acre of rich soil. These microbes, by decomposing both plants and animal carcasses, complete the breakdown of organic compounds until only water and carbon dioxide remain. Again, this same carbon dioxide and water is recycled by green plants into organic compounds.

Fourth is a group that was left out in the short cycle completed above, the carnivores. The carnivores are the insects, fish, amphibia, reptiles, birds, and mammals that live on other animals, not on plants. Examples are the tiger beetles, the dragonflies, the shrews, the weasels, the wolves, and large and small cats. Because energy is lost at every trophic level in the biological or ecological community, the number of carnivores a community can support is severely restricted.

Finally, there are the *top* carnivores, the carnivores whose diet consists substantially of other carnivores. These individuals must be among the rarest of all organisms; a tremendous investment of energy has gone into the pyramid of life leading up to what top carnivores consider appropriate food.

What has all this to do with the area that is occupied by a community? Only this: if the top carnivores are recognized as integral members of the biological community, then the area over which the community extends must be large enough to encompass the home territories of several breeding pairs of the top predators. Inbreeding, as we discussed earlier and will discuss again under genetic counseling, frequently leads to serious genetic disability. Therefore, a reasonably stable community must allow for some fifty pairs of adult breeding pairs of top carnivores.

As we have outlined it here, the area occupied by an ecological community can be considerable. A pair of ravens (oversized crows, in a sense), for example, may occupy a home territory of four square miles. Fifty pairs would then require two hundred square miles—an area ten miles wide and twenty miles long. A pair of falcons may require as much as ten square miles; fifty pairs, then, would need five hundred square miles—a square between twenty and twenty-five miles on a side. The sparrow hawk requires about one-half of a square mile. (Sparrow hawks are the small hawks often seen hovering over the median of interstate highways, searching for small birds and mammals.) If its territory included only the median and nothing on the sides of the interstate, one might expect to encounter a sparrow hawk about every half-hour. With all the uncertainties of these calculations (not every sparrow hawk will be hovering as one drives by, and one does at times need to watch other cars, not birds), the estimated half-hour interval is not bad.

Large, carnivorous mammals, as one might expect, need larger territories than those of hawks, falcons, and owls. A pack of eight timber wolves requires 540 square miles of territory. Twenty-five packs (to

Figure 10.5. An experimental study of species survival in patches of trees that are left uncut or unburned during the clearing of Brazil's massive rainforests. As part of an economic development program, Brazil is destroying much of the rainforest that occupies the enormous Amazon River drainage basin. Biologists, making the best of what may eventually prove to have been an enormous error, have persuaded the authorities to leave patches of trees as islands (of 1–20,000 acres) whose species—both plant and animal—can be subjected to long-term study. If the account given in the text is approximately correct, even a 20,000-acre "island" (30 square miles) will be inadequate to support even one pair of jaguars, Brazil's largest cat. *(Drawing based on photo by R. O. Bierregaard)*

provide reasonably unrelated breeding animals) need 13,500 square miles—a square nearly 120 miles on a side. A cougar requires about twenty-five square miles; 100 adult cougars—fifty mating pairs—would require 2500 square miles—a square fifty miles on each side.

The numbers that have been given here are important to the establishment of public parks and wildlife preserves. If, for example, a proposed wildlife preserve of 10,000 square miles is modified into a "string of pearls" (that is, into ten small parks of 1000 square miles each) for ease of public access, the timber wolves in that territory are eased out of existence. A thousand square miles will not support even two packs of eight individuals each. The mountain lions and bears of the area are also placed in jeopardy. In Brazil, scientists have been conducting experiments on the appropriate sizes of natural reserves to be left untouched in the Amazon jungle. If they underestimate the needed sizes of these reserves, they will have failed to account for inbreeding—the effects of which will not appear for several generations. One must allow for about fifty breeding pairs in order to avoid the harmful effects of inbreeding.

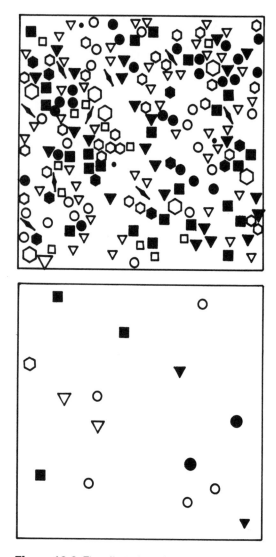

Figure 10.6. The effect of grazing by rabbits on the species richness of a British meadow. The top chart indicates the position and numbers of plants of various species in a small area of an open meadow; the bottom chart shows the few species left within a similar area from which rabbits had been *excluded*. The symbols represent yarrow, vernal grass, quaking grass, bluebells of Scotland, dogtails, bumblebee weed, wild thyme, and other meadow "weeds." In Britain, rabbits themselves are controlled by natural predation; in Australia, where introduced rabbits had no natural enemies, they multiplied tremendously and seriously overgrazed the land (see Figure 7.20).

Top carnivores also play a special role in ecological communities that has not yet been mentioned: to a great extent, these carnivores preserve the existing variety among their prey. This effect follows from the search patterns used by predators in seeking their prey. Suppose, for example, that groundhogs are common; predators that are able to utilize groundhogs as prey will achieve their greatest success—they will hunt with greatest efficiency—if they prey primarily on these common rodents. The consequence of this pattern of hunting is that the number of common prey organisms is reduced, while other, rarer prey species increase in frequency. Herbivores play a similar role in maintaining the variety of plant species on which they graze. On islands off the coast of Britain, there were meadows grazed upon by rabbits that contained nearly 100 plant species, but when the rabbit population was destroyed, the number of plants dropped to about five species of grass.

Thus the chain governing the size of ecological communities extends from the top predators through the lower ones to the herbivores and then to plants: the variety of life that exists at each level depends greatly upon the organisms found at the level above. At the apex (perhaps in the form of an exotic hawk) is the top predator, which has a powerful role in the maintenance of the variety of living things—even of the grasses and herbs—of the entire community.

Of course, extinction is an enemy of biological diversity. Whatever the number of species present in an ecological community, if the number of individuals representing any given species drops to zero, that loss permanently decreases the variety and increases the sameness of the community. Communities inhabiting small, isolated areas can become depleted of biological diversity rather rapidly. Not only is the stabilizing influence of carnivores removed, but also chance events become extremely important if many species are represented by only a few reproducing individuals each.

Greenbelts, hiking trails, and rustic bicycling paths that link otherwise modest or small community parks help maintain the diversity of life found in these parks, because animals can migrate between parks if they use the rather wild and often unused connecting pathways. A park that has produced a bumper crop of chipmunks, for example, can provide the migrants that may restock another community's park that has lost its last chipmunk inhabitant. On a larger scale, deer and other large mammals generally found in Northern Europe are occasionally sighted in Austria and northern Italy; they appear to move south through the hunting preserves that are common in Europe, and along the extensive hedgerows that run between European farms connecting forested areas.

NOW, IT'S YOUR TURN (10-4)

A colloid suspension consists of one substance (the dispersant) dispersed within a second substance (the matrix). Cream is a colloid suspension of fat droplets dispersed in water (see Figure 10.7). When agitated, cream turns to butter—a colloid suspension of water droplets dispersed in a matrix of fat. Keeping in mind that the properties of these two colloid suspensions, cream and butter, are quite different, we can regard farmland and wooded areas as two components of a colloid suspension: woodlands can be dispersed in a matrix of farmland (as in much of America's Midwest) or farmland can be dispersed in a matrix of woodland (as is frequently the case even in Western Europe).

Note the different consequences of these two patterns of land use. What is the role of nature trails, greenbelts, and biking paths in determining the nature of our farmland–woodland "colloid suspension"?

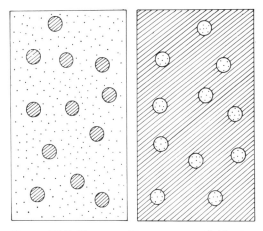

Figure 10.7. Diagrams of two common colloid suspensions: heavy cream and butter. The difference between the two substances depends on which is the dispersant and which is the (nondispersed) matrix. In cream, fat is dispersed in water. However, when cream is sufficiently agitated, the fat becomes the matrix within which droplets of water are dispersed. The outcome is butter, a substance quite different from cream.

Figure 10.8. The utilization of land on the Great Plains. Using the colloid suspension analogy of the preceding figure, the central part of the United States can be seen as a matrix of fields of corn and grain. An aerial view of this land shows roads running north to south and east to west at one-mile intervals. Within each of the square miles marked off by these roads, one can see outlines of the original 160 acres (one-quarter square mile each) given to early homesteaders. The pattern has been eroded in places, of course, but it is still common for the four original homesteaders' houses to be found clumped at the center of the square mile their holdings occupied. In this case, the dispersant is the combination of houses and shade trees dispersed in an otherwise uniform matrix of farmland. *(Photo reproduced by permission from This Sculptured Earth, © 1959, Columbia University Press)*

Figure 10.9. A countryside in which grain fields are embedded within a matrix of forested areas and hedgerows. Unlike wildlife in areas where small clumps of trees are scattered in vast expanses of cultivated fields, the natural wildlife of the area shown in this figure migrate relatively freely along hedgerows from one wooded area to the next. In terms of "island biogeography"—the study of the number of species that can be supported by isolated islands—the landscape illustrated here represents one large island for wildlife because easy access routes are provided in the form of hedgerows between cultivated fields. *(USDA aerial photograph)*

Endangered Species

About 300 BC, Ptolemy I established a library at Alexandria, Egypt. His library was to become one of the truly great libraries of the world. At its peak, it contained nearly a half-million scrolls (which is equivalent to between 100,000 and 200,000 volumes by modern standards). Every ship arriving in the port at Alexandria was searched; all books were removed, taken to the library, and copied, and the copies were returned to the vessel. The library at Alexandria kept the originals. The Alexandrians also had an arrangement with the Athenians. Books from the library at Athens were taken to Alexandria for copying, and a deposit equal to nearly $100,000 in modern currency was left in Athens as a guarantee of the books' return. Again, the copies, not the originals, were returned to Athens from Alexandria, but the Alexandrians did tell the Athenians that they could keep the deposit.

Some six or seven hundred years after the library was founded, it was destroyed. The destruction was not a single act; rather, it was spread over many years, during many invasions (starting with Julius Caesar), sieges, and lootings.

Today, we know of the library only through contemporary accounts. In what sense have we today suffered as a consequence of the destruction of this library, one of the greatest libraries of all time? Do you think the library contained things we might have known, but that we do not know? These are impossible questions to answer: the information contained in those ancient scrolls vanished when the library vanished.

The rise and fall of the library at Alexandria has been described here in some detail because its destruction is in many ways comparable to the loss of a species, plant or animal, through extinction. The DNA of any species is comparable to a library; it contains all the information accumulated by that species from the beginning of life on earth, information that through all years except the last kept representatives of that species in existence. The DNA of all existing species has histories of identical lengths; that of human beings is no longer, and no shorter, than that of roses of Sharon or sowbugs. During its history, however, the DNA of each species accumulated information that permitted its carriers to cope successfully, if not perfectly, with the particular problems they faced. We cannot yet read DNA, but that is a historical accident: we have been born too soon. We *are* able to determine the base sequences of rather long fragments of DNA. By recognizing common, overlapping sequences, we can arrange these fragments linearly and reconstruct the base sequences of sizable lengths of DNA. However, the meaning of these sequences (except for those specifying the amino acid composition of proteins) is not yet known. Therefore, the loss of any species corresponds to the loss of a library, a library known to contain many volumes, but whose contents must forever remain a mystery.

The loss of a library forms only one analogy for the loss of information contained in the DNA of a species lost through extinction. Another analogy is based upon the interrelations that exist between the member species of complex communities. Who, for example, would suspect that cutting down a forest would endanger a species of salmon? It did in one case: the loss of the trees increased the water temperature in the forest's river, rendering the stream unsuitable for the salmon's spawning. Similarly, seashore development and the destruction of estuaries threaten commercial marine fishing because young fish must grow up near the shore, where they find food among sea grasses.

The loss of first one species, then a second, and then a third is a matter of some concern. While it is true that the loss of the mud darter, a small fish that lives in certain eastern streams, would not bring civilization to an end, one cannot continue to make that claim as the number of lost species continues to increase. A fellow passenger who forcibly extracts a rivet from the side of an airplane while it is in flight may not endanger you and the rest of the passengers, but neither he nor the rest of you can be sure of that. If this passenger were to persist in searching for and extracting additional rivets—each of which has an unknown function—a wise fellow passenger would request that the rivet-collector be restrained. That analogy roughly describes the status of knowledge surrounding species and their role in communities. We do know that the rare top predators are often keystones that hold the community together; the role of other member species is too often completely unknown.

INFORMATIONAL ESSAY
Testing Goodness-of-Fit (Chi-square)

In the course of their studies, biologists often make decisions on the basis of fairly small numbers. Physicists and chemists, as a rule, deal with extremely large numbers. For example, chemists know that methane, CH_4, has a molecular weight of 16; 16 grams of this gas, therefore, contain 6.02×10^{23} molecules. At 0°C and one atmosphere pressure, 16 grams of methane will occupy 22.360 liters. How different a science is when it has such numbers, and such precision, at its disposal! Consider on the other hand the poultry farmer who must get rid of one of two hens; one has laid 75 eggs in the past 120 days, but the other only 63. In many respects, the second hen is the better, but is she better where it counts: the number of eggs she will produce?

Statisticians have devised arithmetic procedures that permit one to decide whether observations differ from one another or from an expectation based on a prior belief. Suppose, for example, that you match coins with a stranger during your lunch hour. Later, after dinner, you wonder whether it was reasonable that he won sixty tosses and you won only forty out of the total 100 tosses. Was this fellow a con artist who has actually stolen your money by cheating?

If he were an honest man, one can say that his expected number of wins would have been fifty and yours fifty; coin tossing is a game of chance in which both parties have equal chances of winning. What, though, is the expectation if the stranger cheated? Unfortunately, there is none. If he were a cheat, did he cheat on all throws, or only on some? If he cheated ten times, on which ten tosses of the hundred were they? Only one assumption, one expectation, one *hypothesis*, makes a definite statement: the stranger did *not* cheat, therefore, each of you should have won fifty times out of the hundred tosses.

The next step is to test this hypothesis (the *null* hypothesis). The test takes the form of a table:

	You won	You lost	Total
Expected (E)	50	50	100
Actual (observed=O)	40	60	100
difference (O−E)	−10	+10	0
(difference)^2	100	100	
(difference)^2/E	2	2	
Sum of (difference)^2/E = 4 = Chi-square			

Chi-square (χ^2), as you can see, is the name given to the number one obtains by adding all the (difference)^2/Es—the more items to be added, the larger the Chi-square. Also, the larger the Chi-square value, the more the observed numbers have deviated from those expected on the basis of the hypothesis. How big a Chi-square would make one suspect the validity of the hypothesis? This depends upon the "degrees of freedom"; the smaller the degrees of freedom, the smaller Chi-square must be to make one hesitant about *accepting* the hypothesis.

When a hypothesis provides a basis for listing expectations as it has here (50:50), the number of degrees of freedom is one less than the number of categories (2 − 1 = 1 degree of freedom). The logic can be seen if you recall that whatever the number of losses, the number of wins is fixed (no freedom) because there were 100 tosses. Consequently, if you were to say, "I won forty-five times," I could reply, with no additional information, that you then lost fifty-five times. There is one degree of freedom to this analysis.

The Chi-square table accompanying this essay (Table 10.1) tells us that *p*, the probability of observing deviations *as large or larger* than those actually observed, is about five percent. If the stranger were an honest coin tosser, there is only one chance in twenty that he would have won sixty or more tosses out of the 100. Nothing is absolutely sure in the world of rather small numbers, but most persons interpret one chance in twenty as grounds for *rejecting* the hypothesis that was being tested. That is, one might reject the hypothesis that the stranger was honest; consequently, one would conclude that he was dishonest. On the other hand, if this decision were to mean that the stranger, if arrested, would be sentenced to jail, one might prefer probabilities as low as one in a hundred, or one in a thousand before robbing him of his liberty. Statistical calculations can always lead to a Chi-square, and through the Chi-square to a *probability* (the probability, remember, of seeing deviations *as large or larger* than those actually observed if the hypothesis were true). The interpretation of that probability and the nature of the decisions that are to be based on it are the responsibilities of the interested party. As a rule of thumb, however, probabilities as low as 0.05, 0.01, or 0.001 are generally called significant—that is, the hypothesis is rejected.

Table 10.1. Chi-square values for use in testing "goodness of fit." Choose the proper degree of freedom in the leftmost column and compare your calculated Chi-square with the values listed in the table. At the top of the column above each value listed in the table is the probability of obtaining a Chi-square value *this large or larger* if, in fact, the hypothesis being tested were true.

Degrees of freedom	Probability (p)			
	0.50	0.05	0.01	0.005
1	0.46	3.84	6.64	7.9
2	1.39	5.99	9.21	10.6
3	2.37	7.82	11.34	12.8
4	3.36	9.49	13.28	14.9
5	4.35	11.07	15.09	16.7
6	5.35	12.59	16.81	18.5
7	6.35	14.07	18.47	20.3
8	7.34	15.51	20.09	22.0
9	8.34	16.92	21.67	23.6
10	9.34	18.31	23.21	25.2

NOW, IT'S YOUR TURN (10-5)

Despite having read this essay, some of your classmates may say that the phrase $p = 0.05$ cited in the example here means that there is only a five percent chance that the stranger was honest. Read the definition once more, and note in your journal the difference between the definition and the statement in the previous sentence. One never errs in attempting to make statements as precise as possible.

INFORMATIONAL ESSAY
Island Biogeography

Offshore islands present biologists with a variety of natural laboratories. An old observation, repeatedly confirmed, is that, other things being equal (distance from the nearest shore, for example), large islands harbor more species of a related sort (members of the same or of related genera, that is, organisms that go by the same familiar name, such as ant, lizard, or fruit fly) than do smaller islands. On the other hand, near-shore islands of a given size harbor more species than do similar sized islands that are farther offshore. For example, as the size of islands in the South Pacific increases from one square mile to one hundred thousand or more square miles, the number of ant species found on these islands increases from five or less to one hundred or more. The number of lizard species on the islands increases from two

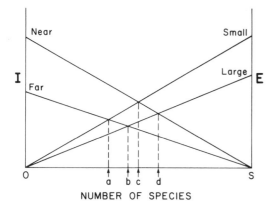

Figure 10.10. The interrelation between rates of immigration (*I*) and extinction (*E*) in establishing a (dynamic) equilibrium number of species (*a*, *b*, *c*, or *d*) on an oceanic island when the total number of species available is *S*. The rate at which new species (immigrants) can arrive at a *near* island is greater than that for a *far* one. The rate at which species can be lost by chance (or otherwise) is greater for a small island than for a large one. Intersections of immigration and extinction lines denote points of equilibrium where the rates of immigration equal those of extinction. Note that near/small islands are expected to have fewer species than near/large ones, and that far/small islands have fewer species than far/large ones.

or three to about one hundred as the islands on which they live increase in size from one square mile to fifty thousand square miles.

Suppose that one were to measure off one square mile of a large island and ask, "How many species do I find in this one square mile area?" The numbers of either ant or lizard species would be considerably higher in the measured area than on islands of equal size.

The explanation for these facts lies in the occasional, accidental loss of a species from a small island, and the long time required before gravid females or mating pairs of the same species arrive from elsewhere—from the nearest portion of the mainland, for instance.

Within a large island or on a continent, the production of more progeny than the local environment can support is commonplace. Animals have mobility, and when they are overcrowded, they tend to move. Earlier we described the migratory locust. Another well-known example is the lemming. During years of gross overpopulation, large numbers of these small mammals reportedly dash over cliffs and fall into the sea, where they drown. Standing inside a marked-off square mile set within a much larger area, an observer records the presence of many displaced wanderers desperately seeking home territories of their own, and each species recorded within the measured mile makes the list of species longer. There are no such migrants on small, offshore islands.

The numbers of species of any common type of organism found on islands can be accounted for, even predicted, by a simple diagram developed by E. O. Wilson and Robert McArthur (see Figure 10.10). The horizontal axis extends from zero species on an island to *S* species, the maximum possible (probably equal to the number of related species found in nearby coastal areas on the continent).

When no species exist on the island, any new immigrant represents the addition of a new species. Let the rate at which new species arrive (such as arrivals in one year) be *I*; islands near the mainland will have a higher immigration rate I_n (for *near*) than will more distant ones, I_f (for *far*). If all species were to be found on an island, immigration could bring in no new ones; therefore, both I_n and I_f tend toward zero as the number of species on the island approaches *S*, the maximum number.

If all *S* species were found on an island, some would be lost by chance within our one-year time limit. The rate of extinction (*E*) would be greater on small islands (E_s) than on larger ones (E_l). On the other hand, if no species were to be found on an island, none could become extinct; hence, the extinction lines lead to zero in the lower left portion of the diagram.

The important points on the diagram are those formed by the intersections of the immigration (*I*) and extinction (*E*) lines; each intersection represents a number of species ($\hat{S}$) at which immigration counterbalances extinction—a point of stable (but dynamic) equilibrium. Notice that, in the diagram, the equilibrium number of species for far islands is smaller than that for near ones; it is a matter of the rate at which new immigrants arrive. Notice, too, that for a given immigration rate, the small island (with its high extinction rate, E_s) harbors fewer species than do the larger islands, with their smaller extinction rates (E_l). The diagram shows the relationship between immigration, extinction, and equilibrium numbers of species very well. Not as clearly shown, but implicit in what we have said, is the impermanence of the actual species. An island may have an equilibrium number of ten species out of a total of fifty (= S), but the names of the ten species recorded during one year may differ from the names of those to be found five years later. The equilibrium *number* is a dynamic, not a static, equilibrium; the actual species may change from year to year—only their number remains relatively constant.

In island biogeography, islands are not defined strictly as bodies of land surrounded by water. For living organisms, any patch of inhabitable territory surrounded by hostile territory is in effect an island. For alpine plants that thrive above the timberline in mountain ranges, each mountain peak that projects above the pine forests represents an island. Water holes and oases represent islands for desert-dwelling animals. Pine forests on the isolated mountain ranges in the Death Valley region of California are islands. The clumps of trees that grow in isolated groves in southern Brazil are also islands. The number of species found in these isolated patches of habitable territory varies according to island size, just as it does on offshore islands.

IV
BIOLOGY AND SOCIETY

Human beings form societies, and these societies tend to acquire properties and characteristics of their own. Therefore, a society cannot be understood simply by examining its member citizens one by one, because a society is more than a collection of individuals. The Woodstock concert of 1969 attracted some 400,000 persons, but in no sense did the collection of young men and women at the concert compare with Denver, Colorado—a city of about the same size. The one was a gathering of persons, while the other is an organized society. The undertakings of a society, once started, develop an inertia that carries them forward, sometimes in the face of individual opposition. Social undertakings become *institutionalized*, gaining a "life" of their own. War, for example, is not the summation of many altercations. On the contrary (as the late war in Vietnam illustrates), wars can be undertaken with the merest public sanction (the Tonkin Gulf resolution, for example), grow in magnitude despite simultaneous growing opposition, and upon terminating, leave scars that require decades to heal. The current Soviet experience in Afghanistan promises to serve as a second example of this kind of institutional inertia.

The chapters in this part of the text deal with human populations at the societal level, a level of complexity that is uniquely human. Paradoxically, the *biology* of societal problems is no more difficult or complex than the biology of wound healing, or of predator–prey relationships. Consequently, it may seem that a discussion of certain biological problems has been needlessly delayed. Because they are not difficult, why have these topics not been covered earlier? In our opinion, it is a matter of placing or assigning emphasis properly. Two examples will illustrate both the question and the response.

No two human beings are exactly alike, and no two populations of human beings are identical. These facts are direct consequences of the mechanisms of heredity and sexual reproduction. They have been appreciated by most geneticists for a half-century or more. They form the biological foundation for an understanding of the origin and differentiation of human races. Despite this background knowledge, however, William C. Boyd could still complain in 1950: "As a result of doctrines embodying the concept of race, but having no scientific foundation, millions of persons have been tortured or killed. Yet nearly everyone, even the *victims* of this delusion, seemed to assume that distinct, physical races do exist and that each possesses well-defined differences in mental abilities and attitudes." Even now, as the year 2000 approaches, there are still many persons who oppose the integration of schools, denounce equal opportunity in housing and employment, and thoughtlessly recite the very doctrines Boyd condemned so long ago.

An understanding of the biological or genetic basis of race and race formation has not led to an easing of racial tensions; thus, we learn that knowledge and behavior are not necessarily correlated. Therefore, the biology of race, in our opinion, should be taught within its truly important context: the institutionalized racism of modern societies. The biology remains the same, but the context is societal, not personal or community, biology.

Persons belonging to different groups—national, cultural (including language), religious, or racial—often distrust and dislike one another. Indeed, outsiders are frequently loathed. Otherwise, the tragic events that have taken place in the Middle East, Northern Ireland, Quebec, Watts (California), Wounded Knee (North Dakota), and Miami (Florida), have no logical explanation. This loathing, however, is not a simple, personal matter; rather, it is an institutionalized emotion.

None of the pressing problems relating to population, food, energy, or other resources can be solved until the matter of racism is resolved. The reason is obvious: to improve its lot, any oppressed group needs power, and power (either in physical combat or at the polls) requires bodies. Because the striving for equal status calls for larger minority populations, these minorities must postpone indefinitely what they perceive as secondary matters: conservation and the preservation of resources. It is in this gloomy societal context that the biology of human variation is presented in this book.

Since the rediscovery in 1900 of Mendel's work on inheritance in the garden pea, generations of students have memorized the ratios with which genetically different individuals appear among the offspring of various parental crosses. Of themselves, these ratios are not genetics; they are the trappings of genetics. Mendel knew this. He stressed these mathematical ratios in his original paper because they proved that *something* behaved according to predictable laws; that something (not the numerical laws by which it behaved) was the heart of heredity.

Today, we know what that something is: it is DNA. Genetics, then, is a science dealing with (1) the chemical nature of DNA, (2) those aspects of the structure of DNA that enable it to store and retrieve biologically useful information, (3) the means by which DNA is accurately replicated, and (4) the mechanisms by which replicated DNA is transmitted from cell to cell, and from generation to generation. Mendel's laws, important as they were to early geneticists, are characteristic of one (admittedly large) group of organisms: sexually reproducing, diploid plants and animals.

Mendelian ratios do not aid in an understanding of cellular biochemistry and energetics or of the wisdom of the body. Nor do these ratios aid in an understanding of ecological communities or of the flow of materials and energy through the biosphere. They are necessary, however, in understanding the odds that are cited by a genetic counselor meeting with prospective parents. That is, Mendelian ratios apply to matters of public health, which is an institutionalized feature of modern society.

The treatment of societal matters in a biology textbook cannot avoid raising some of the issues that have made *sociobiology* controversial. Sociobiology, as defined by E. O. Wilson of Harvard University, is a science that attempts to account for the *behavior* of all organisms (including man) in terms of natural selection and evolution. A simplistic application of biological knowledge to human affairs, however, can lead to what the opponents of sociobiology call biological determinism. Heredity is to the biological determinist what predestination is to the Calvinist: it is a fate that is imposed on the individual at birth by forces beyond his control, that determines the course of his life, and that he had best learn to accept.

The controversy surrounding sociobiology lies beyond the intended scope of this book. Nevertheless, we might recall earlier discussion of the wisdom of the body (wisdom accumulated over millennia by DNA), as exemplified by the liver or by wound healing, and the wisdom of the brain and central nervous system. Obviously, the solutions to certain problems are best recorded in DNA: these are largely chemical and managerial problems arising during the operation of cellular machinery, or in the maintenance, repair, and defense of the physical body. The more complex problems involving the relationship of the individual to its (including sometimes hostile) neighbors are best solved by a complex nervous system, acting through motor organs in second-by-second response to information constantly supplied by sensory organs.

A clear example illustrating the difference between heredity and culture in human behavior is the reaction to pain: all human beings scream with extreme pain; how they scream, and the words they scream, differ from culture to culture. Beyond the nature of screams, however, the biological determinists and their opponents differ in their explanations of aggression and poverty, and in their views concerning the fate of the human species.

11

HUMAN VARIATION

Variation Within Populations: Nongenetic

Individuals differ from one another in many ways. Each person can address dozens of relatives, friends, and acquaintances by name on sight. This fact alone suggests both the extent of visible variation and the complexity of the classification scheme and filing system of the human brain. Human beings can apparently differentiate among as many as 10,000 faces.

Each person undergoes systematic changes in appearance as he or she grows older. Therefore, some of the variation that can be observed between persons in a community or population reflects variation in the ages of these persons. An old, well-established community, just because its citizens represent a wider range of ages, may exhibit a wider range in physical variation than a suburban housing development in which homeowners are drawn from a narrow age group.

Differing ages contribute to the variation observed in human populations. So does sex. Both age and sex, in a sense, are accidents of birth. One's age depends upon when one was born; one's sex depends upon which of two types of sperm fertilized the awaiting egg. That early accident (considered here to be *non*genetic) is responsible for many of the differing physical, emotional, and behavioral characteristics of boys and girls, men and women.

Physical accidents also contribute to differences between individuals. An improper bite that is caused by poor dental care can alter facial features. Growth patterns depend upon inheritance, diet, and exercise. A broken leg that heals improperly can result in labored walking. More subtle, but no less important, are the effects of low-level environmental poisons. Lead is found in the peeling paint of many old tenement houses and in the exhaust gases of automobiles, trucks, and buses. Mercury occurs in much of the nation's water supply and, through accidental contamination, in meat and other food. Carbon monoxide is an extremely poisonous gas, and, like lead, it is present in automobile exhaust gases. All three of these poisonous substances can damage nervous tissue, thus altering the mental activity of affected persons. Long-term exposure to these poisons tends to occur among those who live near the heavily traveled avenues and highways of large urban areas. Exposures to these poisons reflect inequities of opportunity, but because each generation of an exposed family grows up handicapped, such environmentally caused traits may appear to "run in families."

NOW, IT'S YOUR TURN (11-1)

Use your journal to demonstrate your ability to carry out a bit of personal research: Pick a pollutant and, using the *Reader's Guide* or past issues of newspapers and magazines, report on its status as a public concern, its source, its biological effects, and what you consider to be a likely (or desirable) solution to the societal problems it poses. As you prepare your report, consider carefully whether you are acting as an advocate for one point of view, or whether you have attempted to see the problem through the eyes of various interested parties.

Variation Within Populations: Genetic

Age, accidents, exposure to poisonous environmental contaminants, sex (which we have considered to be an accident of conception), and still other factors differentiate individuals. Beyond these sources of variation, however, lies genetic variation, variation that is caused by differences in genetic material acquired by each individual at conception.

Genetic information, as we have learned earlier, is carried by a special chemical, deoxyribonucleic acid, called DNA for short. This chemical, in the human species as in other species, contains information that specifies how thousands of protein molecules are to be constructed from smaller chemical building blocks. In addition to the specifications about *how* to build these large molecules, DNA contains information about *when* to build them: when in each cell cycle, when during development, and in which tissues.

Until recently, geneticists believed that proteins were so precisely adapted to their function that variation in their structure could not be tolerated. While it is true that mistakes in the structure or composition of certain proteins do cause serious inherited diseases, not all genetic variation is harmful. Modern biochemical techniques, primarily electrophoresis, have revealed that one-half or more of all proteins in human beings (and other organisms, as well) exist not in a *single* form, but in two, three, four, or even more different forms. Because each of these forms must be built according to directions carried by DNA, this protein variation reveals in turn that a series of unlike forms of genes (*alleles*) may exist at a majority of all gene loci. At a given locus, a person can carry only two alleles, which may be identical (homozygous) or different (heterozygous); another individual may carry the same or still other alleles of this same gene at that locus. These different genetic make-ups are the basis of genetic variation; in a practical sense, however, genetic variation is ordinarily observed only if the different alleles affect the development, appearance, or health of their carriers.

For convenience, two kinds of genetic variation can be recognized. The first includes the numerous mutant genes and chromosomal errors that result in grossly misshapen embryos and defective newborns, babies who scarcely survive (or fail to survive) because of nutritional or other biochemical deficiencies. It also includes those whose nervous system fails to function properly. Such persons are mentally handicapped; they may be physically handicapped as well. Disorders of this sort were discussed under health and disease in Chapter 6.

The second kind of genetic variation affects physical traits, such as height, weight, hair and eye color, bodily proportions, bone structure, and other aspects of the human body. This kind of variation is accepted as "normal." Individuals who exhibit extreme characteristics, on the other hand, are still recognized as being unusual: the world's tallest, smallest, or heaviest man as well as the bearded lady is still likely to be found touring with a circus. Not all records in Guinness's book are desirable ones to hold.

Many characteristics that are affected by the genes that individuals carry, especially height and weight, are also affected by the food they eat, the fluids they drink, the exercise they get at work or play, and other aspects of their environments. Many of the different physical aspects exhibited by different persons are unimportant with respect to their health.

An important aspect of genetic variation that might otherwise be ignored lies not in the variation itself, but in emotional reaction to variation. Many persons simply cannot cope with variation. The imaginary world within these persons' minds is filled with uniform, stereotyped images. These are the persons who believe that, when you have seen one redwood, you have seen them all.

For example, individuals differ in their ability to smell flowers. One person may be unable to smell flowers of a certain type; those who do

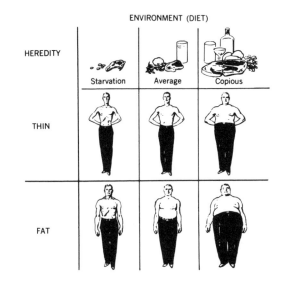

ENVIRONMENT (DIET)

Figure 11.1. The genetic and environmental factors that, together, determine a person's weight. Diets vary: some persons eat very little, while others consume average amounts of food and still others love to gorge themselves. Heredities differ, as well: some persons are predisposed by their genetic endowment to be thin, others to be fat. The actual weight of any one individual is determined by the combination of genotype and environment. A "genetically thin" person who habitually overeats may be considerably heavier than a "genetically fat" person who conscientiously adheres to a restricted diet. Note that, had actual weights, in pounds, been inserted below each of the six men in the illustration, their *mean* weights and *variances* could have been computed, including (1) genetic variance, (2) environmental variance, and (3) using all six figures, total variance (which would equal the sum of the other two). (Compare this figure with Figure 4.16.)

Reactions to a Flower-smelling Experiment

The following are comments made by spectators and test subjects overheard and recorded by the geneticist during his tests of flower-smelling ability:

I do not want to help science; science has never helped me.

Everybody knows that freesias have a lovely odor and to ask if freesias are fragrant is a silly question.

There is something wrong with these flowers, they don't smell at all! You have done something to them.

A perfectly heavenly odor.

These stink. They stink like hell.

What are you trying to prove?

You can't smell these flowers so long after they are cut; but the trouble is the public doesn't know it. They just fool themselves.

Other persons just imagine they can smell flowers and they lie, because freesias don't have any odor.

Too many cigarettes are why so many men can't smell flowers and the women are getting just as bad now.

The flowers have been smelled out.

After I smell, will you tell me the right answer?

Did I vote right?

But, Professor, she voted different. We both can't be right.

Finally, an exchange:

No, I don't like the pink ones; never have liked that color.

But the odor, do you like it?

No, I don't like pink.

smell them, however, may differ widely in their descriptions of the odor they perceive. Both the ability to smell and the characteristic of the perceived odor differs from one type of flower to the next. Thus, someone who is unable to smell freesias may be able to smell petunias perfectly well. Someone else, of course, may question whether the odor of petunias is "heavenly" or "terrible."

About fifty years ago, a semi-retired geneticist described the reluctance of persons to either acknowledge or accept variation. Having left his research laboratory, this scientist remained scientifically active by studying family groups for the ability (or inability) of their individual members to smell flowers. To collect his data, he would set up floral displays at country fairs, where he would then persuade the members of families (fathers, mothers, and children as well as aunts, uncles, grandparents, or cousins) to smell various flowers and fill out a small questionnaire.

Long before he had uncovered the inheritance patterns of flower-smelling ability, this scientist had discovered that many people were simply afraid of being "wrong." Those who had completed the test would loiter nearby in an effort to learn whether their answers were correct or incorrect. To some, the idea that there is no correct way to smell a flower was unthinkable. "In that case," they would ask the geneticist, "how does it smell to you?" Scientists, it seems, know (or should know) all the answers!

Even objects—those things we call by common names, such as "cat," "desk," "towel," and "chair"—differ from each other. In fact, department stores have special sales at which they sell imperfect merchandise ("seconds") at substantially reduced prices. But an imperfect sweater, despite its flaws, is still a *sweater*.

From ancient times until today, persons have had difficulty reconciling the use of the same name for objects that are obviously dissimilar. Plato, a philosopher of Ancient Greece, made a suggestion about this problem that has endured for more than 2000 years. According to Plato, for every object there exists an ideal object of which the real thing is an imperfect example. All objects are imperfect when compared to their Platonic ideals; the little variations in them are accidental deviations from the unattainable ideal.

Plato's views dominated much of human thought well into the twentieth century. For example, a biological species, under Plato's views, is an ideal type—a mental construct—possessing certain characteristics. The individual plants or animals that are assigned to that species are not identical, nor are they expected to be, because they are imperfect representations of an idealized type. This is known as typological thinking: somewhere in the mind is an image for everything that can be named. The image reflects how the thing *should* be; in contrast, real things are only imperfect copies of the ideal.

Plato's ideas have been displaced in many sciences, but they are by no means dead. They still govern the thoughts of many persons. However, Darwin's views have replaced Plato's in modern biology. The world in which we live, in the Darwinian view, contains many kinds of living things, such as plants, insects, birds, and people, the individual members of which differ from each other. The individuals are reality: *there is no ideal*. Therefore, in naming organisms, the biologist must allow for these differences. A population of snails on a small island, for example, does not exist as something separate from the individual snails of which it consists. If these snails differ from all other known snails, then the description of the population must in some manner include a measure not only of the variation among the island snails but also of the difference between them and other known snails. The ideas that we carry in our heads must be able to cope with the reality of variation. Ideas that indicate how things *should* be, and that treat variation as if it were a mistake, cannot deal successfully with the real world.

If we admit to the reality of the objects that we see and touch (and to the nonexistence of imaginary ideals), the notion that there is a correct way to smell a petunia should seem absurd. Petunias emit certain chemical substances. Some persons cannot detect these substances by smell. Others do, but their sensory interpretations may differ. The human population includes all these different persons. There is no right or wrong to biological variation.

Thus the contrast between the Platonic and Darwinian views can be summarized as follows: variation, to Plato, represented errors in the construction of real objects that were intended to be copies of a single, ideal model; it was the model to which the name of the actual objects applied. Darwin denied the existence of the imaginary model; therefore, in his view differences between the individual members of a population are accepted as an important feature of their reality.

NOW, IT'S YOUR TURN (11-2)

Among the flowers tested by the geneticist mentioned in the text were freesias, petunias, and marigolds. If one or two plants of these species can be obtained, bring them to class and test one another's perception of their odors. For each type of plant, list your classmates according to their ability or inability to smell it, and, for those who can smell it, note their account of its odor. As you make these lists, ask yourself if the lists for different plants are the same, or have different plants given different results? Can you see any physical resemblance among those persons who gave similar responses to a particular plant?

Table 11.1. Two lists of human races prepared by one author, in 1955 and in 1962

1. Murrayian	1. Northwest European
2. Ainu	2. Northeast European
3. Alpine	3. Alpine
4. Northwest European	4. Mediterranean
5. Northeast European	5. Hindu
6. Lapp	6. Turkic
7. Forest Negro	7. Tibetan
8. Melanesian	8. North Chinese
9. Negrito	9. Classic Mongoloid
10. Bushman	10. Eskimo
11. Bantu	11. Southeast Asiatic
12. Sudanese	12. Ainu
13. Carpentarian	13. Lapp
14. Dravidian	14. North American Indian
15. Hamite	15. Central American Indian
16. Hindu	16. South American Indian
17. Mediterranean	17. Fuegian
18. Nordic	18. East African
19. North American Colored	19. Sudanese
20. South African Colored	20. Forest Negro
21. Classic Mongoloid	21. Bantu
22. North Chinese	22. Bushman and Hottentot
23. Southeast Asiatic	23. African Pygmy
24. Tibeto-Indonesian Mongoloid	24. Dravidian
25. Turkic	25. Negrito
26. American Indian, Marginal	26. Melanesian-Papuan
27. American Indian, Central	27. Murrayian
28. Ladino	28. Carpentarian
29. Polynesian	29. Micronesian
30. Neo-Hawaiian	30. Polynesian
	31. Neo-Hawaiian
	32. Ladino
	33. North American Colored
	34. South African Colored

Variation Between Populations

The Platonic and Darwinian views of variation within populations differ in ways that become especially important when *different* biological populations are compared. To be useful, the Platonic model must remain constant through time. Under this view, the errors that account for individual variation are repeated generation after generation. For example, in a Platonic world the cabbage butterflies that live in a vacant lot should closely resemble each other from summer to summer. Some persons will see this year's little white butterflies as being the same as last year's; these persons forget that this year's butterflies are the offspring of the earlier ones. Because the butterflies do not appear to change, it seems reasonable to Platonists to believe that each year's crop is another batch of imperfect samples of what should be the unchanging, perfect (or ideal) cabbage butterfly. Such an idea is encouraged by Plato's views of the world.

Because he denied the existence of an ideal model, Darwin's ideas are better suited for a changing world. Each year's crop of cabbage butterflies, to continue our example, may be, but is probably not, identical to the earlier one. There is no fixed model, and so populations, generation by generation, are free to change. The ability to change is essential, of course, to Darwinian evolutionary theory.

If populations can change with time, they will also differ in space. A simple example will illustrate what this sentence means. Suppose one has a laboratory culture of fruit flies and two cages built of lucite and fine screen. One hundred flies are removed from the culture bottle, separated into two lots of fifty each, and placed in separate cages. The flies in one cage are provided with food to which a small amount of table salt (salt, in high enough concentration, is a poison for flies) has been added; over time, the amount of salt is gradually increased. Small, sublethal amounts of copper sulfate (another poisonous salt) are added to the food that is provided in the second cage; the amount of this poison is also increased every few weeks.

Within a year, the populations of flies that live in the two cages (these are the fifteenth to twentieth generation progeny of the original fifty flies) will differ genetically. In the first cage, only flies that are resistant to the toxic effect of table salt are able to reproduce; here, the frequency of genes conferring this resistance to their carriers increases. Thus, in the one cage the flies will be resistant to rather high concentrations of table salt; in the other, to copper sulfate. Put flies of one cage into the other, and they will die. The difference exists because both populations changed with *time*. Because they changed with time and because the directions in which they changed differed, the resulting populations are no longer the same. Local geographic populations undergo genetic change in response to the demands of local conditions. Different demands in different localities cause once similar populations to diverge in various ways. These are the simplest of all evolutionary changes.

Each of us is more familiar with other people than with fruit flies or cabbage butterflies. We know dozens of persons by name or on sight. We are also able to perceive systematic differences between groups of persons. Europeans, Africans, and Asians form groups that can be recognized readily. The words "whites," "blacks," and "Orientals" also reveal that these populations of human beings differ. The division can be made finer, however; Northern (Scandinavian) and Southern (Mediterranean) Europeans are recognizably different. So are the Chinese, Japanese, and Indians. Still finer divisions can be made. Persons who are interested in these matters have succeeded in naming more than one hundred groups of human populations. This type of splitting and resplitting of human populations can continue as long as someone has the patience to recognize ever smaller differences. Or as long as it serves someone a useful purpose.

Groups of human populations whose members resemble one another genetically, and which are sufficiently important for someone to name, are known as *races*. Races named in this manner are *categories of convenience*. The two lists of races shown in Table 11-1 were prepared within six years of one another by the same author. Of the thirty-nine names that appear in these two lists, only twenty-five appear on both. Lists of races change because the judgment of those who compile them changes: what was convenient at one time need not be convenient at the next.

Unlike species, races are not permanent. A species, once its individual members no longer breed with those of other species, is something that will persist for a long time, unless, of course, it is wiped out by a climatic change or other catastrophe. (It may also "disappear" because it becomes so altered through evolution that scientists decide that its members need a new name.) The situation with races differs completely: Members of different races, upon meeting, interbreed more or less freely. Consequently, as conditions change and distribution patterns of populations change, races lose the characteristics for which they were originally named. However, out of the new mixture will appear other clusters of populations that will in their turn come to be known as races.

Viewed through time, a species resembles a well-defined watershed seen from the mouth of the main stream; it is independent of other watersheds. Races, also viewed through time, more nearly resemble the interconnected network of tidal pools and eddies that form, disappear, and then re-form in new patterns on a broad, flat beach.

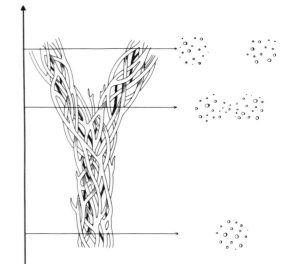

Figure 11.2. An illustration of (1) the meshwork of populations that, during the passage of time, fuse, break apart, and re-fuse *within* a species as well as (2) the splitting of two groups of populations into reproductively isolated species. Until the speciation split occurs, any two or more populations (if brought together physically) could interbreed and produce a fertile, hybrid population. Following the split, reproductive isolation is assumed. According to this diagram, a species does not arise from an Adam and an Eve; from the onset, a species consists of a number of local populations. Consequently, there is no reason to believe that a newly arisen species must ever be genetically homogeneous.

NOW, IT'S YOUR TURN (11-3)

Speciation is an evolutionary process during which once actually or potentially interbreeding forms become segregated into two or more arrays that are incapable of interbreeding. A *species,* the outcome of speciation, is, then, a group or population of actually, or potentially, interbreeding organisms that is reproductively isolated from all other such groups. These definitions suggest that a species is a group of organisms that, in the eyes of most biologists, deserves to be called a species. Contrast this general agreement on a rather precise definition to the arbitrary designations for both lower categories (subspecies and races) and higher ones (genera, families, and even higher categories). In what sense can one claim that categories both above and below the species level are works of art, not of science?

An Analysis of Racial Differences

The individual members of a single population differ from one another. Many of these differences are inherited. Races also differ from one another (otherwise, no one would have named them) and racial differences are also largely genetic. Sometimes culturally imposed characteristics differ among races (tattoos, decorative scars, artificially molded features, and clothing), but they are not, as a rule, used by modern scholars in identifying races.

Is there any means by which the differences between races can be compared to those between individuals within races? Or within local populations? Is it possible to measure the total genetic variation within a species, for example, and then determine where most of it lies? Does it lie between races? Or, alternatively, does it lie within races?

Africans are black and Europeans are white. Doesn't that answer the question we have just raised? Doesn't all genetic variation lie between races? After all, there are only slight differences between the shades of whiteness or blackness of individuals within these two races.

The answer to the above question is No. The contrast between black and white skin color does not answer the question because that is how the races were originally identified. A simple example can illustrate the problem: Suppose all of the rubber erasers in a large school are collected and sorted by color. Many are red, but there are also green ones and blue ones. Some are blue or green with a white layer sandwiched in between. Entirely white erasers are also common. Finally, tan gum erasers are not uncommon.

When the sorting is over, there will be very little variation in color among the erasers in each pile. Even the red erasers may have been sorted into separate piles of deep red and light pink ones. Let each pile be considered a "race" of erasers. Color must be ignored in answering the question, "Is the greater part of the variation among erasers found within or between races?" There is no question that the piles differ in color (unless the sorters were either careless or color blind) and that the erasers in each pile are of the same color. Instead of color, we must now ask about size (weight in grams for example), shape (round, square, or tapered), texture (soft or hard, smooth or rough), composition (with or without sand or emory dust, type of rubber), and any other measurable aspect of erasers.

If we find that blue erasers come in different sizes and that these correspond to the different sizes of red erasers, we conclude that nearly all of the variation in size among erasers is found within "races" and that little or none is found between them. On the other hand, if most of the little round erasers taken from the ends of pencils are red, then there would be variation between eraser races with respect to shape. The important points in answering the question on the distribution of variation are (1) the traits on which the racial classification was made must be ignored and (2) as many other traits as possible must be found and measured.

Modern laboratory techniques have provided geneticists with precise answers for questions that earlier workers could answer only by guessing. These techniques are the electrophoretic techniques for studying protein variation. They reveal variation in the electrical charge of protein molecules, variation caused by differences in amino acid composition. These differences, in turn, reveal corresponding differences in the composition of segments of DNA (genes) that are responsible for building these proteins.

Human races, whether there are five or fifty of them, have not been identified on the basis of protein variation. Like the erasers of the preceding paragraphs, humans have been classified largely on the basis of color. Facial, skull, and other bodily characters have also been used in refining the classification and increasing the recognized number of races. To a great extent, however, the classification of races took place before the techniques for studying the molecular structure of proteins were invented. Consequently, these formerly unseen and unsuspected gene differences can be used to measure and compare the variation that occurs *within* and *among* races.

Genetic variation is measured as the chance, or likelihood, that an individual will carry genes (one from his mother, the other from his father) that specify two different forms of a particular protein. Several examples will make this clear. Suppose that only one form of a particular gene exists in a population. Then, the chance that a person

Table 11.2. Allelic frequencies at seven polymorphic loci in Europeans and black Africans

Locus	Europeans Allele 1	Allele 2	Allele 3	Africans Allele 1	Allele 2	Allele 3
Red cell acid phosphatase	.36	.60	.04	.17	.83	—
Phosphoglucomutase 1	.77	.23	—	.79	.21	—
Phosphoglucomutase 3	.74	.26	—	.37	.63	—
Adenylate kinase	.95	.05	—	1.00	—	—
Peptidase A	.76	—	.24	.90	.10	—
Peptidase D	.99	.01	—	.95	.03	.02
Adenosine deaminase	.94	.06	—	.97	.03	—
Average heterozygosity per individual	.068 ± .028			.052 ± .023		

After Lewontin.

Table 11.3. Examples of the similarities and differences in blood group allele frequencies in three racial groups

Gene	Alleles	Racial group Caucasoid	Negroid	Mongoloid
Duffy	Fy	.0300	.9393	.0985
	Fy^a	.4208	.0607	.9015
	Fy^b	.5492	0	0
Rhesus	R$_0$	.0186	.7395	.0409
	R$_1$	.4036	.0256	.7591
	R$_2$	.1670	.0427	.1951
	r	.3820	.1184	.0049
	r'	.0049	.0707	0
	others	.0239	.0021	0
P	P$_1$	.5161	.8911	.1677
	P$_2$	.4839	.1089	.8323
Auberger	Au^a	.6213	.6419	—
	Au	.3787	.3581	—
Xg	Xg^a	.67	.55	.54
	Xg	.33	.45	.46
Secretor	Se	.5233	.5727	—
	se	.4767	.4273	—

After Lewontin.

might carry two *different* forms is *zero*. Suppose that one form of a gene vastly outnumbers all others, of which there may be several. Again, the chance that a person will carry two *different* genes is small (not zero, however); consequently, the genetic variation in the population is small. In contrast, suppose there are several different forms of the gene, all of which occur with equal frequency. The probability that a person will carry two different sorts is high. The genetic variation in this population is large.

Various proteins for which human beings are known to differ are listed in the accompanying tables. Neither the names of these proteins nor their functions concern us at this moment. The tables show that groups often differ from one another either in the *common* form of the gene or in the relative *frequencies* of the different forms. These differences merely confirm what we already know: races differ from one another.

How much do races differ genetically, compared to differences between individuals of the same race (or of the same local population)? The answer to this question is given in Table 11.4. Data are available for seventeen gene loci. The table shows the chance (or likelihood), H_0, that a person carries two different forms of each of these seventeen genes. It then shows what this chance would be if all populations within a race were combined into one large population (H_1). Finally, it shows what the chance would be if all populations and races were combined into one enormous population that included the entire human species (H_2). Surprisingly, perhaps, H_1 and H_2 are only slightly larger than H_0. Indeed, because of deficiencies in the molecular techniques, the differences calculated from these data are too large; even more variation than that calculated here lies within local populations. Nevertheless, the results are clear: more than ninety percent of all genetic variation within the human species lies *within* individual races; less than ten percent (much less, in fact) results from genetic differences *between* races.

NOW, IT'S YOUR TURN (11-4)

Earlier in this chapter, it was suggested that tests on the odors of flowers be conducted (taster–non-taster tests using PTC would work as well, as would blood-type classifications—A, B, AB, and O) and that class members be separated on the basis of the test results. Following that, we asked, "Can you see any physical resemblance among those persons who gave similar responses to a particular plant?" Explain in your journal how this question relates to the one discussed in the preceding paragraphs in which we asked about genetic variation within, as compared to between, groups.

Racial Differences with Respect to Intelligence

According to many persons, differences between human beings can be used in arranging individuals on a scale reflecting *superiority* and *inferiority*. Some persons run faster than others; the faster runners are considered superior. Some sing better than others; the better singers are superior. Some are more intelligent than others; the more intelligent are superior.

From this argument (which assigns relative *values* to individuals who differ), it follows that races that differ systematically from one another deserve the labels *superior* and *inferior*. For most physical abilities or skills, it is difficult to argue convincingly that superiority or inferiority is important. According to a popular song, gentlemen prefer blondes; it is not easy, however, to convince everyone that blondes are superior to brunettes, or that blue eyes are superior to hazel or brown ones.

Intelligence is another matter, however. After all, human beings differ from all other animals in the extent to which their brains have developed. The human brain is remarkable for its capacity to run through and evaluate within seconds the outcomes of numerous possible actions, and to decide which course of action promises to be the best (usually for the individual concerned; rarely for all human beings). Of course, some personal decisions, such as changing one's employment or buying a new home, may be reached only after days of thought, and perhaps only after the thoughts of two or more brains have been combined (the wife's and husband's, the children's, and the banker's, perhaps). Nevertheless, the human ability to visualize possible future events, and either to avoid or to prepare for them, seems to be unique within the animal kingdom. This ability grows with experience and training. The level of one's mental skills, together with one's ability to profit from experience and training, constitutes one's intelligence.

Persons differ in intelligence, often for genetic reasons. For example, mental retardation is a common effect of many genetic disorders. To understand how this can be, we must understand the brain, which is a strange organ. It makes up only two percent of the body's weight but receives twenty percent of the blood supply, thus suggesting that the brain's demand for oxygen is ten to fifteen times greater, gram for gram, than the body's average. Furthermore, a microscopic examination of the nuclei of brain cells reveals that very little of their genetic material is inactive. When chromosomal material is not used by any cell, it is packed and stored against the nuclear wall, where it can be clearly seen under the microscope. Lastly, the cells of the brain do not divide after birth. Each of us is born with all the brain cells we shall ever have; after twenty years of age, in fact, we begin losing large numbers of them daily.

Table 11.4. Measure of human diversity within populations (H_0), within races (H_1), and for the entire species (H_2) based on seventeen gene loci

| Locus | H_0 | H_1 | H_2 | Amount of total variation | | |
				Within populations	Within races	Between races
Haptoglobin	.888	.938	.994	.893	.050	.057
Lipoprotein Ag	.829	—	.994	.834	—	—
Lipoprotein Lp	.600	—	.639	.939	—	—
Xm	.866	—	.869	.997	—	—
Red cell acid phosphatase	.917	.977	.989	.927	.061	.012
6-Phosphogluconate dehydrogenase	.286	.305	.327	.875	.058	.067
Phosphoglucomutase	.714	.739	.758	.942	.033	.025
Adenylate kinase	.156	.160	.184	.848	.022	.130
Kidd	.724	.930	.977	.741	.211	.048
Duffy	.597	.695	.938	.636	.105	.259
Lewis	.960	.993	.994	.966	.033	.001
Kell	.170	.184	.189	.899	.074	.027
Lutheran	.106	.139	.153	.693	.215	.092
P	.949	.978	1.000	.949	.029	.022
MNS	1.591	1.663	1.746	.911	.041	.048
Rh	1.281	1.420	1.900	.674	.073	.253
ABO	1.126	1.204	1.241	.907	.063	.030
Average				.849	.075	.075

Within populations = H_0/H_2
Between populations, within races = $(H_1 - H_0)/H_2$
Between races = $(H_2 - H_1)/H_2$

After Lewontin.

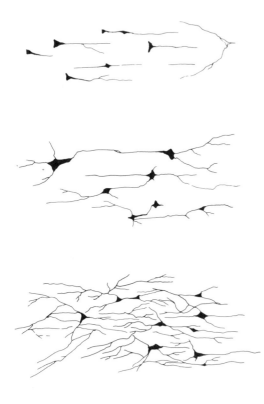

Figure 11.3. The growth and development of neurons within the cerebral cortex over the first two years of postnatal life. The number of neurons does not change; it is determined by prenatal factors, including heredity, maternal diet, and the presence or absence of toxic substances in the expectant mother's blood. The latter, like carbon monoxide or lead, may be an inescapable part of the maternal environment or, like nicotine (and the carbon monoxide of tobacco smoke), alcohol, or miscellaneous chemical substances, may reflect the mother's "lifestyle." The growth of the branching axons is dependent upon the baby's diet and mental stimulation as well as upon heredity and chemicals in the postnatal environment.

These peculiarities of the brain may make it especially susceptible to genetic errors. Apparently, an exceptionally large set of human genes is at work in the brain (not *all* genes are functioning, of course; brain cells obviously don't make hemoglobin, muscle, or cartilage proteins). Therefore, *genetic* mistakes are likely to affect the normal operation of brain cells. Second, the proper functioning of cells within the central nervous system depends upon billions of accurate interconnections between them; the delicacy and required precision of the brain's structure makes this organ especially vulnerable to error. Third, with no cell divisions, the brain has no means by which to discard and replace malfunctioning cells. The toxic metabolic wastes of genetic errors simply accumulate in nerve cells and destroy them. The normal destruction that occurs after the age of twenty may reflect the consequence of a normal and unavoidable accumulation of these intracellular wastes.

The idea of genetic differences with respect to intelligence (a notion that is not entirely false) has been extended by many to a claim that human races *systematically* differ in intelligence. The claim is even more specific: In its most extreme form it says (1) that blacks have a lower average intelligence than whites, (2) that this difference is genetic in origin, and *therefore* (3) that blacks, as a race, are genetically inferior to whites. Because many persons, some of whom are extremely vocal, subscribe to this view, it must be discussed here at length (every member of the U.S. Congress was once provided a complimentary copy of a book promoting racial superiority). The concluding pages of this chapter, then, deal with the *measurement of intelligence*, the study of its *inheritance*, and the bearing that variation within and between races has on the claim of *racial superiority* or *inferiority*.

The Measurement of Intelligence

If intelligence were a trait that varied from person to person as height, weight, running speed, strength, and other physical traits do, it would exhibit a bell-shaped frequency distribution. That is, very few persons would have extremely low or extremely high intelligence, more of them would be more nearly average, but the majority of all persons would have average intelligence.

But how can intelligence be measured? So far, only by tests. Because tests are devised by human beings, however, they can be made to reveal any distribution pattern the test-maker wants. This point is important: measuring intelligence is not like measuring height. Suppose three students were to measure the heights of their classmates. One has a foot rule, the other a meter stick, and the third a broken broom handle. Despite their different tools, they will arrive at comparable conclusions. To transform the measurements of one student into those of another requires only a proper conversion factor, such as one inch equals 2.54 centimeters, or the broken broomstick equals 0.64 meters.

Intelligence tests do not have conversion factors. A test can be constructed for which one-half of all those tested score 0 and the other half, 100 percent. In such a test, the first question would be chosen so that about one-half of those being tested would give the wrong answer; the others, the right one. All subsequent questions would be phrased so that those persons who missed the first would miss the remainder as well, but those who were right on the first one would also answer the others correctly.

A test of this sort would be useless, because another test of the same sort might reverse the standings of those tested. Normally, then, a test is built around many questions that have been carefully chosen to result in scores having the familiar, bell-shaped distribution. The outcome is achieved by controlling the correlations between the proportions of correct answers given to different questions. Clearly, a test for measuring intelligence does not compare with a yardstick that is used for measuring height. A "good" intelligence test is one that has been carefully designed to yield the desired results.

Tests consist of questions that are concocted by human beings. Therefore, the questions will exhibit all of the failings and blunders that are the fault of any group of persons. These failings are illustrated in Figure 11.4, which gives a list of nine questions that have been chosen from a recently published sample of twenty-four questions. A brief critique of four of these questions is given below:

Question 2. The correct answer is given as 24. This answer is obtained by noting that $3 + 6 = 9$, $6 + 9 = 15$, and (therefore) $9 + 15 = 24$. However, $24 + 3$ does not equal 6 and so the construction of the figure is wrong for the suggested type of reasoning.

Question 3. Figure 5 is said to be correct because it cannot be divided into symmetrical halves by both vertical and horizontal lines. Anyone who has doodled, however, will see that Figure 4 is the only one of the five that cannot be drawn without lifting one's pen from the paper—an important consideration for both doodlers and mathematicians.

Question 6. The number 7 in the left-hand figure is obtained, according to the answer book, by subtracting 6 $(2 + 4)$ from 13 $(8 + 5)$. It can be arrived at in a more obvious manner by adding 2 (left leg) and 5 (right arm). Adding the left leg (5) and the right arm (2) on the right-hand figure also gives 7. The "correct" answer is given as 3.

Question 8. The correct answer in this case is obtained by subtracting the top figure from the bottom figure in each column and doubling the answer to obtain the middle number. Thus $6 - 2 = 4$ and $4 \times 2 = 8$, $18 - 10 = 8$ and $8 \times 2 = 16$; $7 - 2 = 5$ and $5 \times 2 = 10$ (the "correct" answer). Notice, however, that the greatest difference in column one is 6 $(8 - 2)$, that in column two it is

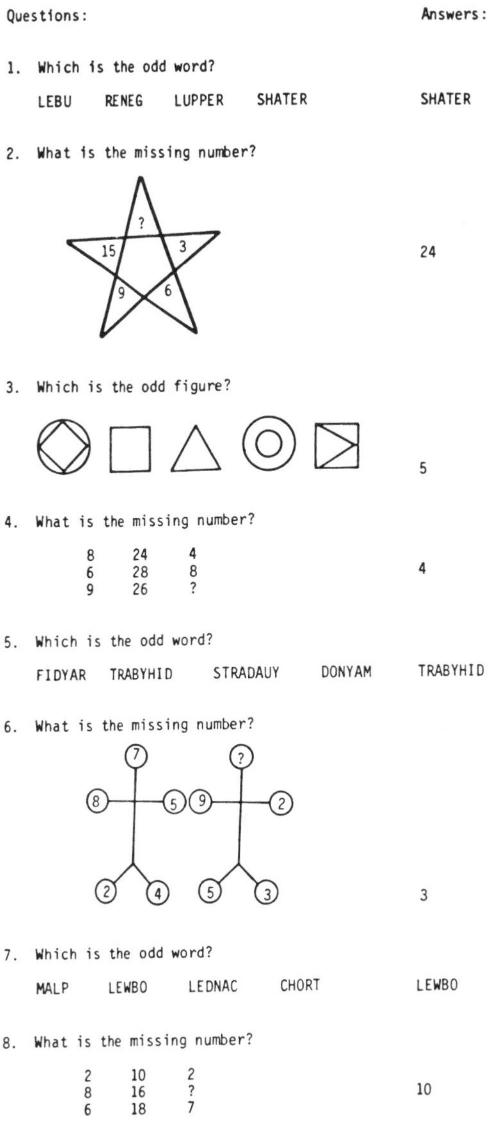

Questions: Answers:

1. Which is the odd word?

LEBU   RENEG   LUPPER   SHATER   SHATER

2. What is the missing number?

24

3. Which is the odd figure?

5

4. What is the missing number?

8	24	4
6	28	8
9	26	?

4

5. Which is the odd word?

FIDYAR   TRABYHID   STRADAUY   DONYAM   TRABYHID

6. What is the missing number?

3

7. Which is the odd word?

MALP   LEWBO   LEDNAC   CHORT   LEWBO

8. What is the missing number?

2	10	2
8	16	?
6	18	7

10

Figure 11.4. Examples of the sorts of questions commonly asked on intelligence tests, together with the "correct" answers. Several of these questions have been chosen for detailed discussion in the text. A further point can be made here: anagrams are commonly used in measuring intelligence. Unless the use of the term "word" in the question is made clear, however, the questions become meaningless. For example, in the first question LEBU is the only word containing the letter B, RENEG has the only double E, LUPPER has the only double P, and SHATER the only A, S, or T. In short, every word in Question 1 is an odd word.

Jacques Barzun on Objective Tests *

What of the questions in the two booklets (*Scholastic Aptitude Test* and *Science*)? I am officially informed that they were selected from actual tests, most of them being copied verbatim and the rest with small editorial changes. Such a process of successive filtering should have eliminated serious blemishes. That the surviving questions are not always clear-cut is implied by the following passage on page 18 of *Scholastic Aptitude Test:*

"As you read through the explanations of the verbal section, you may disagree with what we think to be the correct answer to one or two questions. You may think we are quibbling in making certain distinctions between answer choices. It is true that you will find some close distinctions and just as true that in making close distinctions reasonable people do disagree. Whether or not you disagree on a few questions is not terribly important, however, for the value of the test as a whole is that people who are likely to succeed in college agree in the main on most of the correct answers. It is this that gives the [*Scholastic Aptitude Test*] its predictive power.

"For this very reason, when you find it hard to make or recognize a distinction between answer choices, it is better not to spend much time on that question. It is the whole [*Scholastic Aptitude Test*] rather than any single question in it that makes the test a good indicator of college ability."

The advice in the last paragraph quoted above has serious implications. Consider it in the light of the mutually exclusive propositions: (a) the test contains genuinely difficult questions that are free from ambiguity but call for reflection and cannot be properly analyzed in a short time; and (b) the test is devoid of such questions. If (b) is true, the advice is reasonable, but the test is unworthy of the highly gifted student, since it gives him little if any chance to display his superiority over his merely clever rivals. If (a) is true, the advice defeats the purpose for which the genuinely difficult questions were included, and is tantamount to a plea for superficiality.

* *The House of Intellect* by Jacques Barzun; Harper & Row, New York (1959).

8 (18 − 10), and *therefore*, the greatest difference in column three would be 10. The missing number by this reasoning equals 2 + 10 or 12.

Considerable space has been devoted to these four questions in order to make a point that is not always recognized. "Correct" answers on intelligence tests are often arbitrary. The required answers are not unique. In many instances, the "correct" answer is merely an obvious one: one that was obvious to the test-maker, and the one most likely to be chosen by many persons taking the test. If this is true, however, it raises a question that will shortly be raised again: To which persons among those being tested will the answers be obvious? It also points out an interesting paradox: the highest scorers on intelligence tests may be those persons who are most adept at recognizing the obvious. The obvious answers, presumably, are those that average persons see. Therefore, the high scorers may be the most nearly average, rather than the most intelligent, persons.

The required answer (that is, the "correct" one) to each of many questions on intelligence tests, as we have seen, often proves to be the most obvious one of several equally correct answers. Obvious to whom, however? First, to the test-maker, who overlooks alternative solutions. Second, to test-makers as an intellectually biased collection of specialists. Test-makers converse with one another and discuss test-making problems; out of their conversations and correspondence come ideas for new and better tests. The ideas are limited, however, by the limits of the group itself.

Third, the answers are obvious to those who are "in tune" with test-makers. Here lies the really great difficulty in assessing the scores obtained by blacks on intelligence tests. The questions, the words, and the procedures (many of which, as we have seen, are faulty) used in these tests are devised by college-trained white professionals, who arrive at answers that they regard as correct. The farther removed persons taking the tests are culturally from the test-makers, the less likely they are to hit upon answers that, for arbitrary reasons alone, are said to be the correct ones.

Black educators have repeatedly demonstrated the above point by devising intelligence tests in the language and concepts of blacks. Of course, the test scores of blacks are raised as a result. Indeed, tests can be devised on which blacks consistently outscore whites.

This brings us to the final point in the matter of measuring intelligence. Tests are given for a purpose; they are intended to predict future performance. Specifically, intelligence tests given to school children are given to predict each child's performance in the school years ahead. To a considerable extent, these test scores also predict eventual status in the community, as measured, for example, by annual income.

A predictive test is one that provides a means for forecasting. A hypothetical relationship between test scores and, for example, annual

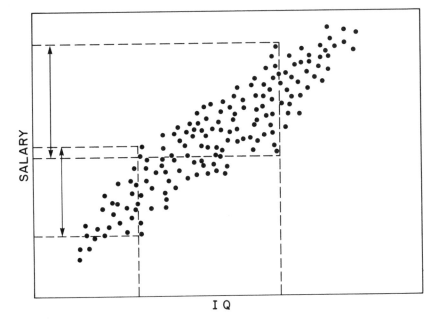

Figure 11.5. A chart illustrating the hypothetical relationship between IQ (as determined during school years) and annual salary attained (let's say) at age 45. The correlation between IQ and salary is not perfect, but persons having the greater IQ score do tend to have the higher salaries. For the two IQ scores chosen as illustrative examples, the salary ranges scarcely overlap. Conversely, persons making a given salary are seen to have obtained a broad range of IQ scores when tested in school. Despite this variation, in this example IQ scores are shown to be of some use in *predicting* eventual salaries.

income is shown in Figure 11.5. The test scores are known, so they are shown along the bottom line of the graph. (This is a convention; the bottom or horizontal axis is used for the *independent* variable.)

The vertical axis represents annual income. This is the unknown (the *dependent* variable) when the test is given, but past experience has shown that the higher the test score, the higher the eventual annual income. If the relationship were perfect, the dots would fall on a straight line. Actually, the relationship is not perfect, and so the dots are scattered in an elongated ellipse.

Unfortunately, the world is not as simple as the make-believe example suggests. Instead, we know that the mean annual incomes of blacks and whites differ considerably. We also know that for many years this difference has been the result of job discrimination, restricted educational opportunities, and the like. Because these facts are known, salaries can be plotted as the independent variable along the horizontal axis, as they are in Figure 11.6. When the predictive test scores are examined, taking into account the difference in annual incomes, a sad but necessary fact is revealed: *The average test scores of*

A Black-Oriented IQ Test

Robert Williams developed the Black Intelligence Test of Cultural Homogeneity (BITCH) because he felt that there should be a test that was as fair to the majority of blacks as the Wechsler Intelligence Scale for Children was to the majority of whites. While the Wechsler measured blacks' knowledge of the white experience, the BITCH measured whites' knowledge of the black experience. The SOB test, son of the original BITCH test, continues the tradition. Listed below are some words, terms, and expressions taken from the black culture. Circle the letters identifying the correct meanings as black people use them. The answers are given at the end.

1. the bump
   a) a condition caused by a forceful blow
   b) a suit
   c) a car
   d) a dance
2. running a game
   a) writing a bad check
   b) looking at something
   c) directing a contest
   d) getting what one wants from another person or thing
3. to get down
   a) to dominate
   b) to travel
   c) to lower a position
   d) to have sexual intercourse
4. cop an attitude
   a) leave
   b) become angry
   c) sit down
   d) protect a neighborhood
5. leg
   a) sexual meaning
   b) a lower limb
   c) a white
   d) food

Answers: 1—d; 2—d; 3—d; 4—b; 5—a.

Figure 11.6. A chart much like the one in Figure 11.5, except that salary (at age 45) is considered here to be the independent variable (and thus it is shown on the horizontal axis). Furthermore, the horizontal scale indicates that two segments of the population differ in the average salaries (*solid triangles*) they attain. In order for IQ scores obtained during school years to be predictive of eventual salaries, they must be scattered in an elongated ellipse of the sort shown in the chart (just as in Figure 11.5). The vertical lines drawn from the mean salaries of the two groups reveal that, in order to be predictive, the mean IQ scores of the one segment *must* be lower than that of the other (see double-headed arrows at left). Tests that do not establish this difference are nonpredictive and, because they do not serve their intended function, are discarded.

blacks must be lower than those of whites if the test is to predict the future successfully.

Other than the general discrimination to which blacks have been subjected for years within American society, the *necessary* difference in the mean test scores of whites and blacks neither reveals nor requires a conspiracy of American educators and businessmen. Ambiguous test questions based on loose reasoning have "correct" answers that are most apparent to those possessing the same background as the test-makers. Test-makers represent the dominant culture in America. Black educators can construct alternative tests on which black children attain high scores, but because such results are nonpredictive in our society, these tests cannot and will not be accepted as valid for use in the nation's schools. At best, they can be used to shock, to educate, or to encourage educational reform. The average intelligence test scores of blacks and whites on standardized tests will become equal only when the distributions of the ultimate status of these two peoples in society have also become equal.

The Inheritance of Intelligence

If intelligence is difficult to measure, understanding its inheritance must, of necessity, be even more difficult. Many persons have tried to study the inheritance of intelligence. Sir Francis Galton, the founder of the eugenics movement in Great Britain, tried his hand at it. In 1869, he wrote a book entitled *Hereditary Genius*. He was unable, however, to separate genetic inheritance from inheritance by "word of mouth." (The latter is cultural inheritance. The main transmission of culture is from parents to children, in that parents are models for their children. Attitudes, values, aspirations, mannerisms, and other traits are easily transmitted from parents to children by speech and example.)

Since 1869, the techniques for analyzing intelligence (measured by intelligence tests and the IQ scores they yield) have become increasingly refined. Nevertheless, the effects of cultural and genetic inheritance have never been successfully disentangled. Attempts at separating these two kinds of inheritance have involved, for example, comparative studies of identical twins who were raised apart and twins who were raised together. Identical twins have identical sets of genes. Hence, according to those carrying out these studies, if identical twins are raised apart, only the environment should differ. If, on the other hand, they are raised together, both genes and culture should be identical. Serious questions concerning these studies (and similar ones made on fraternal twins) arise because the adoption of a child or the placing of a child in a foster home is not a chance (random) arrangement. The future environment of the child is matched carefully to that of its past by the adoption agency. Rechecking some of the early studies has shown that those identical twins who were supposedly raised "apart" actually played together, shared desks at school, lived side by side, and were well aware that they were twins.

The analytical techniques used by those interested in the genetic analysis of intelligence are those used by plant and animal breeders for improving agricultural crops and farm animals. Unfortunately, these techniques, which are highly mathematical, are not applicable to human studies precisely because randomization, the removal of children from their own families and placing them with adoptive ones according to a pattern determined by throwing dice, is impossible.

There is little doubt that intelligence is related to the structure and functioning of the brain, so obviously it is influenced by heredity. Furthermore, brain cells require more oxygen than other cells; this is a measure of their metabolic activity. Four or five times as many different proteins are found in nerve cells as in other cells, suggesting that five times as many genes are active in nerve cells as in normal cells. Consequently, the nervous system, especially the brain, is dependent upon even more genes for proper functioning than are cells in most parts of the body.

Genes affect the structure of the brain; consequently, they influence intelligence. However, this fact provides no support for those who claim that the systematically lower test scores obtained by blacks prove that blacks are genetically inferior to whites in intelligence. As we have seen, the so-called "proof" is no proof at all. Differences between groups need not have the same cause as differences between individuals within groups. For example, height is determined in part by heredity (tall parents tend to have tall children) but this does not prove that the slight stature of the pre–World War II Japanese population was genetically determined. Indeed, the growth of post–World War II Japanese children reveals that the earlier slight physical build was largely cultural in origin; dietary habits in Japan have changed in recent years.

Within- and Between-Race Genetic Variation and the IQ Controversy

When brain cells are examined by appropriate molecular techniques, they prove to contain about 40,000–50,000 different proteins; other types of cells have about 10,000 proteins. This difference is a reflection of the importance genes play in the development of the brain. Not only must nerve cells grow during the first two years of life, but also each one of them must develop thousands of connections with other cells. There is no reason to believe that the formation of these interconnections should be mediated differently than are similar (but simpler) contacts exhibited by other cells. Proteins provide the necessary physical structures; proteins provide the recognition signals; proteins provide the specificity that makes precise contact possible. Of course, genes specify how these proteins are built.

Variation in any of the thousands of genes responsible for the building of nerve cell proteins would introduce variation into the circuitry of the brain. This is the variation that will result in a flawed connection here or an excess of intercellular connections there. Memory for certain details might be impaired in the one case, but an ability to see proportions might be improved in the other.

Genetic variation that influences the specificity of connections between nerve cells in the brain is not to be confused with the mental retardation that accompanies a number of genetic disorders. These disorders involve general metabolic flaws that damage nerve cells through an accumulation of toxic chemicals. They are genetic sledge-hammers that effectively destroy the nerve cells. The thousands of genes that are responsible for the thousands of extra proteins found in nerve cells are the ones that are likely to affect intelligence and other subtle mental attributes.

The striking feature revealed by modern studies of human genetic variation is the vast amount of this variation that is found within local populations. Between-race variation constitutes much less than ten percent of the variation that exists within races. No gene has been found among human races that serves as a diagnostic marker for racial variation. Races differ only in the relative proportions (or frequencies) of various gene forms.

If the genes responsible for the synthesis of proteins in nerve cells behave like those that control the synthesis of enzymes or that determine blood types, nearly all existing variation in nerve cell proteins will be found *within* races; very little will be left over as between-race variation. For this reason, together with the other facts and arguments presented in this chapter, it seems unlikely that races differ greatly with respect to the genetics of intelligence. The observed differences in test scores of blacks and whites are adequately explained by cultural bias, social discrimination, and, to some extent, early brain damage caused by local environmental contaminants.

INFORMATIONAL ESSAY
Genetics and IQ: Fatal Flaws in the Conventional "Wisdom"

Many persons believe that a clear-cut genetic basis for differing levels of intelligence (as measured by IQ tests, and recorded as IQ scores) has been established by educational psychologists and others. These persons have read that IQ possesses a high *heritability*, a phrase that immediately suggests that genes are overwhelmingly involved. Further, they read that whatever causes variation *within* a group or race is the most likely cause of differences *between* groups or races as well. Thus, it would appear that, if genes are responsible for the high heritability within, let's say, the white race, then the consistently low average IQ scores of blacks reflect the inferiority of their genes for intelligence. Finally, although they do not understand the analytical procedures, these persons believe that all that they have learned has been revealed by time-proven genetic analyses of the sort that have led to the improvement of agricultural practices by plant and animal breeders. However, each of the conclusions mentioned in this paragraph is flawed; *the commonly held belief that blacks have been shown to be genetically inferior to whites with respect to intelligence has no foundation whatsoever.*

The first item to be confronted is the second one listed above: whatever causes variation *within* a group is most likely to cause variation *between* groups as well. Imagine that in a study of the vinegar fly (*Drosophila melanogaster*) a genetic basis for the size of individual flies has been demonstrated: large parents produce large progeny, and small parents produce small progeny. Furthermore, a mathematical relationship exists between the sizes of parents and progeny, such that the average outcome of any mating can be predicted with reasonable accuracy.

Two culture bottles of flies are now removed from the stockroom shelf—one from France and the other from Greece. Upon examination, one finds that the flies from France are larger than those from Greece. Does this observation prove that vinegar flies from France are genetically larger than those from Greece? Not at all. The most likely explanation is that there was more culture medium, more yeast, or fewer competing larvae in the culture bottle containing the French flies. The cause of *between*-group variation need not necessarily correspond to that of *within*-group variation. This point has been made a number of times in this text: poverty, poor nutrition, toxic pollution, and substandard housing can lead to a culturally inherited syndrome of ignorance and mental impairment that is transmitted from one generation to the next as consistently as is a family's DNA.

Turning now to the first item mentioned above—IQ has a high *heritability*, we can ask what this statement means. First, however, the term heritability must be defined. Heritability is the ratio obtained by dividing genetic variance by total phenotypic variance. The total variance is, for our purposes, the sum of genetic and environmental variances. The meaning of this definition can be understood if we recall Figure 4.16, which illustrated the growth of plants that were collected at three altitudes, split into three clones each, and then transplanted to the other altitudes. Some variation between the appearances or morphologies of individuals (phenotypic variation) has a genetic basis; that is genetic variation, or, expressed mathematically, *genetic variance*. Some variation in appearance (also phenotypic variation) is caused by the differing environments that individuals encounter. This is environmental variation or, more formally, *environmental variance*. The two types of variation combined equal the total variation among individuals. The computation of variances is such that the total phenotypic variance equals genetic variance plus environmental variance.

Contrary to common belief, heritability does not refer to a particular trait; instead, it is a characteristic of a population with respect to that trait. Heritability, that is, is a *characteristic of the population*—not of a trait. To say that IQ has high heritability does not make sense; the statement that because IQ has high heritability, genes must be largely responsible for intelligence, merely adds to the confusion. In the vinegar fly there is a mutant gene that causes the fly's eyes to be white, instead of their normal brick-red color. The environment has relatively little effect on these eye colors, but we can stipulate that they are either dark or light red, and that the corresponding effect on white eyes is to cause them to be anywhere from milk white to a light cream color. This stipulation merely introduces some environmental variation into our example.

Suppose all flies in a population are homozygous for the normal, red-eyed gene. The population contains no genetic variation; therefore, heritability equals zero. Suppose, instead, that all flies in a population are homozygous for the mutant, white-eyed gene. Again, there is no genetic variation and heritability equals zero. Suppose, now, that some flies in a population are red-eyed and some are white-eyed. Genetic variation now exists, and a certain heritability can be calculated as the ratio of genetic variance over total variance. The genes under discussion have remained the same—red- and white-eyed genes. Under different circumstances heritability either existed or equaled zero. Consequently, heritability cannot refer to the trait; on the contrary, it must refer to a population of individuals. Nor does zero heritability imply that a trait has no genetic basis; the preceding discussion was centered on a mutant gene causing white eyes. Consequently, the conventional or popular usage of the word heritability is wrong.

Our final concern is the means by which the genetics of IQ has been studied. These are the procedures that have been developed by plant and animal breeders and that have been extremely useful in crop and livestock improvement.

In many mathematical procedures, if one inserts numerical values into certain equations, one can make numerical estimates of otherwise unknown quantities. If $y = 2a + b$, one can calculate $y = 4$ if one claims that $a = 1$ and $b = 2$. The expression $y = 4$ makes sense, however, *only* if the assigned values for a and b make sense. Computer specialists have a term, GIGO, that covers these matters: Garbage In, Garbage Out.

The procedures of quantitative genetics depend upon one's ability to randomize environmental effects. In an earlier section, we mentioned the graduate student who put his small supply of fertilizer on his best plant varieties so that it would not be wasted. He did not randomize environmental conditions; therefore, his experimental results are worthless. The same is true of all genetic analyses of intelligence: children raised apart are not randomly distributed in the population. On the contrary, adoption agencies go to great lengths to ensure that foster parents are as nearly similar to the original parents as possible—in race, pigmentation, nationality, religion, and even economic status. The foundation upon which quantitative genetics is built is lacking when its procedures are used by those studying the genetics of intelligence.

A final point involves the resolving powers of different experimental procedures, and their ability to answer particular questions. The resolving power of early microscopes was not sufficient to clearly reveal the composition of sperm cells; consequently, Hartsoeker could detect homunculi because he thought that each sperm *must* contain a little person. One can state the following with absolute confidence: If one set of experimental procedures is not sufficiently refined to provide a reliable answer to a particular question, then a second set of procedures that is cruder than the first cannot provide an answer. If a high-powered light microscope cannot reveal whether a certain structure is double or single, a hand lens will also be unable to resolve the

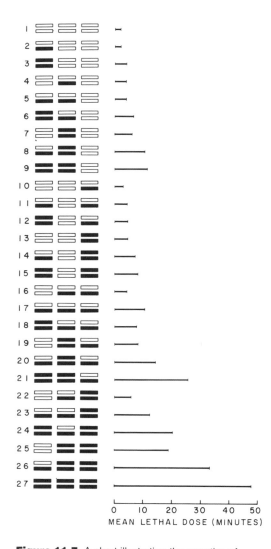

Figure 11.7. A chart illustrating the genetics of DDT-resistance in the vinegar fly (*Drosophila melanogaster*). The rectangles represent chromosomes, of which the vinegar fly has three major pairs. White rectangles represent chromosomes from flies sensitive to DDT; black ones from flies that, through artificial selection, had become resistant to DDT. The bars at the right indicate how much exposure to a certain level of DDT was needed to kill 50% of the exposed flies: the mean lethal dose. In every case but one (17 vs. 18), the addition of a chromosome from the resistant strain increased the mean lethal dose. Because this kind of study follows individual chromosomes, its resolving power exceeds that of studies based on means and variances only. Nevertheless, a comparable study of human beings would fail to resolve the question of racial difference in IQ because of "cultural heredity," nongenetic influences that run in human families because of opportunities that were missed (or denied) in the past. *(Copyright, Columbia University Press; reproduced with permission)*

matter. An electron microscope, with its higher magnification and greater resolution, *might* suffice.

The procedures of quantitative genetics do not rely on the manipulation of individual chromosomes. Instead, they rely on averages. When two parental strains of corn are crossed, each hybrid offspring carries one set of chromosomes from each parent. That is the last precise statement a geneticist can make in this type of mating. When two hybrid individuals are crossed, their progeny, on the average, have equal numbers of chromosomes from each parental strain; however, how these chromosomes are distributed among individual progeny is unknown. Backcrossing the original hybrid individual to members of either parental strain, one expects that twenty-five percent of all chromosomes will come from one strain and seventy-five percent from the other. Again, however, those are expected averages; the chromosomal constitution of individuals is unknown. Contrast the reliance on average distributions of chromosomes from two sources (parental strains) in the previous example to the procedure illustrated in Figure 11.7, where each chromosome can be manipulated individually. The chromosomes in this case come from DDT-sensitive and DDT-resistant strains of flies. The results illustrated here prove that resistance to DDT can be traced to genetic material (the flies in categories 1 and 27 have resistances virtually identical to those of the original strains, even though several generations of crosses elapsed in the synthesis of the illustrated combinations) and reveal the extent to which genes on different chromosomes contribute to DDT resistance.

The procedures that led to the data illustrated in the figure are more refined than those of quantitative genetics: one procedure (that used in studying DDT resistance of fruit flies) can trace the presence or absence of individual chromosomes, while the other (that used by plant and animal breeders) can only follow average proportions of chromosomes in various generations after the initial cross.

Now we can prove that the more refined procedure, although suitable for vinegar flies and perhaps for mice, is not adequate for answering the question, "Does the presence of genes of African origin in an individual lower that individual's IQ?" Take categories 25, 26, and 27 in the figure: Individuals in category 26 differ from those in 25 by the presence of a single black bar in the first position; this bar increases the resistance to DDT of its carriers. Individuals in category 27 differ from those in 26 by carrying two black bars (not just one) in the first position; the second bar increases the level of DDT resistance still more.

What if the black and white bars were chromosomal segments of European and African origin, and the measure was of IQ, rather than resistance to DDT? Would the question of genetics and IQ not be answered? No. The parents and grandparents of those individuals classified as falling into categories 25, 26, and 27 would differ systematically in appearance, with a greater proportion of features and characteristics recognized as black ones occurring among the ancestors of persons in category 26 than in those of 25, and still more among the ancestors of persons in 27 than in those of 26. In human societies, the status of the parents has profound effects on the fate of their children. This applies in all generations: the status of grandparents affected the fate of their children, and their children's eventual status will affect the fate of grandchildren. Even the refined procedures used in the study of DDT resistance in the vinegar fly are unable to eliminate the effect of cultural inheritance in human populations. If the refined procedures are incapable of providing an answer to a question, cruder procedures—in this case, the procedures of quantitative genetics—are also incapable of providing reliable answers.

This essay can be concluded with a cautionary note. Apparently, a study of "racial" IQ has been made using comparisons of the sort discussed above concerning DDT resistance. That study revealed no significant difference between the IQs of persons falling into three categories (such as 25, 26, and 27) where these persons carried 0, 1, or 2 alleles of a sort normally found in Africa. The meaning of "no significant difference" should be kept in mind: a difference as large or larger than that observed would be expected in more than 5% (or 10%, or 25%) of all similar tests. "No significant difference" does not mean "no difference." The argument that has been presented in this essay claims that the test is unsuited for its intended purpose. Furthermore, the argument predicts, should persons persist in repeating what is an unsuitable analysis, they will eventually encounter significant differences, but these will be differences caused by *cultural* heredity. Unfortunately, an inadequate analytical procedure will at that time have been used (as many have in the past) to demonstrate (erroneously, of course) that genes of African origin lower one's intelligence.

On Accepting Evidence

In the first chapter of this text, a question was raised: What can one do if one cannot bring oneself to accept what purports to be scientific evidence? One can disagree, of course, but, in the absence of adequate counter-evidence, one must admit to resorting to faith.

The account given in this chapter about the inadequacy of current genetic analyses for answering questions concerning racial differences in intelligence was once the subject of a university seminar. At the talk's conclusion, the local host commented, "I don't care what you think, I think so-and-so." He then started his analysis by assuming what was really to be proven or tested (namely, that blacks *are* genetically inferior in intelligence). The remainder of his comments consisted of supporting claims that were to account for the assumed inferiority. He simply ignored the carefully reasoned account given in the seminar.

What evidence is accepted and what rejected is an interesting topic in science. Even before Watson and Crick had proposed the now-accepted model of DNA structure (1953), Dr. Rosalind Franklin examined an X-ray photograph of a DNA crystal and noted in her journal that DNA was composed of two (or a higher even number) of anti-parallel strands. Nearly ten years later, biochemists showed, using more familiar techniques, that this was indeed true.

During the 1940s, most of the genetic variation known in human beings was regarded as selectively neutral variation. Among these neutral variants were the alleles that determine the A-B-O blood group system. Later, nearer 1950, certain stomach ulcers were found to be nonrandomly associated with certain blood types. Because "neutral" suggested that blood groups should not be associated with serious ailments or illnesses, many persons claimed that the association with stomach ulcers proved nonneutrality—that is, natural selection was involved in determining the frequencies of the different alleles. Others held back, claiming that selection would not be proven until a relation between ulcers and reproductive success had been demonstrated.

The nature of the evidence that is needed to cause someone to discard a previously held notion varies considerably from person to person. It also varies considerably from science to science. If a science has no generally accepted means for rejecting hypotheses, hypotheses refuse to die. They may become passé, but they're not dead; they only need an ardent spokesman in order to come to life once more.

12

THE ORIGIN AND MAINTENANCE OF HUMAN VARIATION

No two living human beings (excluding identical twins) are genetically identical. No two human beings who ever lived have been genetically identical. Because ten percent of all human beings who ever lived are alive today, the second statement does not add much to the first—but it sounds impressive.

Where does genetic variation come from? What keeps it in human populations? Firm answers to these questions cannot be provided for a great many of the seemingly unimportant variations that make each person unique. In the case of serious genetic defects, two factors are important: *mutation* and *selection*.

Mutation is the change of a gene from one form (or *allele*) to another. Genes are responsible for specifying the structure of protein molecules. Mutant alleles specify altered proteins. Mutations are errors made during gene duplication. Because these errors occur with certain average frequencies, it is often possible to calculate the *mutation rate* for a gene.

Selection is a shorthand term for *natural selection*. If a mutant allele causes a genetic disease that is lethal or results in severe physical disability or mental retardation, a carrier of this allele may not reproduce. Because this individual fails to have children, the responsible gene has, in effect, removed a copy of itself from the population. Natural selection, in this case, is the elimination of alleles from a population through the death or sterility (either complete or partial) of their carriers.

Children who suffer from genetic disease, besides their own torment, cause grief and anguish for their parents and other family members. Anguish is an incalculable price that must be paid for the presence of mutant genes. Another cost is measured in terms of dollars: the costs of hospitalization and extended medical care for those who cannot help themselves. For genetic disease, the second cost is tremendous.

Inheritance Patterns

Each individual carries between five and ten recessive gene mutations capable of causing extremely serious genetic diseases if they were present in double dose. Few of us know that we carry these genes. They were passed to us by our parents, and we shall pass them on to our children without knowing their effect. However, if a man and woman who both carry the same recessive mutant gene marry, the children of this marriage may be unlucky enough to experience the gene's effect. One-fourth of the children of such marriages will receive the mutant gene from each parent, and, as a result, will suffer from gross physical or functional abnormalities, including mental retardation. The children will receive genes they do not want, from parents who do not know they carry them. No one is at fault. Nevertheless, in the United States, about twelve million persons suffer from these accidents of inheritance.

Genetic diseases can be conveniently classified into three categories: dominant, recessive, and sex-linked. The first two, *dominant* and *recessive*, reflect the number of mutant alleles a person must carry in order to show the symptoms of a given disease. A *dominant* gene affects its carrier even though it was transmitted from only one parent. Because every carrier must show the disease symptoms, a dominant trait appears in every generation of the family that possesses it. The inheritance of a dominant trait is illustrated in Figure 12.1. The pedigree chart in Figure 12.2 shows the occurrence of brown teeth in five generations of one family. Note that every individual with brown teeth had a parent who also had brown teeth; this pattern is the hallmark of a dominant gene.

A *recessive* gene has no obvious effect on its carrier unless it is present in a double dose, that is, the affected person has received a mutant form of the gene from *each* parent. Both parents, consequently, must be carriers. Because the gene is recessive, however, the parents need not show abnormal symptoms. Remember that each of us carries from five to ten recessive genes capable of causing gross, even lethal, abnormalities.

Why, then, are we not all genetic defectives, if each of our parents carries five or ten genes for gross abnormalities? The answer is that the defective genes occur among a total of tens of thousands of genes. Thus, it is quite unlikely that any two persons who decide to marry will have the *same* mutant genes. If the genes are represented as

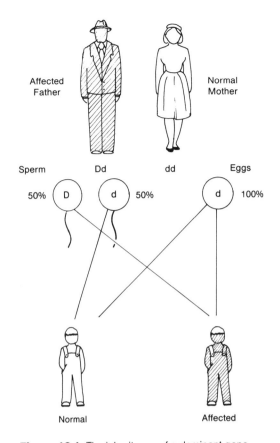

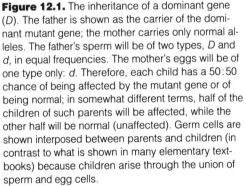

Figure 12.1. The inheritance of a dominant gene (*D*). The father is shown as the carrier of the dominant mutant gene; the mother carries only normal alleles. The father's sperm will be of two types, *D* and *d*, in equal frequencies. The mother's eggs will be of one type only: *d*. Therefore, each child has a 50:50 chance of being affected by the mutant gene or of being normal; in somewhat different terms, half of the children of such parents will be affected, while the other half will be normal (unaffected). Germ cells are shown interposed between parents and children (in contrast to what is shown in many elementary textbooks) because children arise through the union of sperm and egg cells.

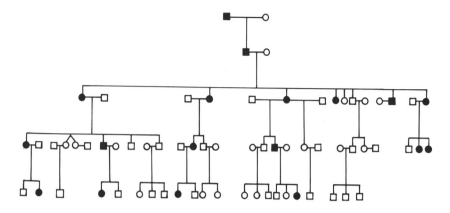

Figure 12.2. A pedigree chart illustrating the inheritance of brown teeth through five generations of one family. Every individual in the chart who has brown teeth has a parent who also had brown teeth. By counting, confirm that about one-half of all children born of couples in which one parent had brown teeth also have brown teeth. [Incidentally, the original wife remarried; her second husband had normal (non-brown) teeth, as did all descendants of the second marriage.]

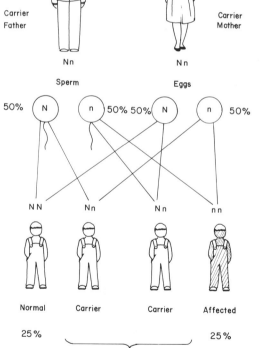

Figure 12.3. The inheritance of a recessive allele (*n*). Both parents are normal in appearance, but both are carriers of *n* (as well as of *N*, the normal allele). Each parent forms two types of gametes; one half carry *N*, the other half carry *n*. These gametes can unite to produce three types of children: *NN*, *Nn*, and *nn*. Heterozygous children, *Nn*, are twice as frequent as the other (homozygous) types (*NN* or *nn*) because they can arise in two different ways. One-quarter of the children born of parents who both carry the same recessive mutant gene will be homozygous for that gene and will exhibit the corresponding genetic abnormality.

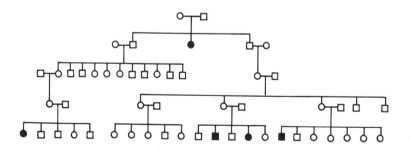

Figure 12.4. A pedigree chart illustrating the inheritance of albinism through five generations. Albinism is a recessive trait caused by a defective enzyme. Notice that, unlike the inheritance pattern of brown teeth shown earlier, the inheritance of albinism represented in this pedigree is not predictable by the parents' appearance—the parents of each albino are normal in appearance.

letters of the alphabet, with capital letters standing for normal genes, a man carrying mutant genes *a* and *k* would have only normal children if his wife carried mutant genes *g* and *y*. However, if she carried *a* and *y*, one-quarter of their children on the average would be *aa* and consequently suffer from a disabling disease.

The inheritance of a recessive gene is shown in Figure 12.3. In Figure 12.4, a pedigree chart extending over five generations (no records exist for the earlier husband and wife, however) shows that persons exhibiting the recessive trait *albinism* most often are born of normal-appearing parents.

The third category of genetic disease, *sex-linked* disorders, is named for the chromosome on which the mutant gene is located. A sex-linked mutation may be either dominant or recessive. Human body cells contain forty-six chromosomes; twenty-three are paternal in origin and twenty-three are maternal. Among the forty-six chromosomes, there are twenty-two pairs. That is, twenty-two of the twenty-three maternal and paternal chromosomes match one another with respect to the function of the genes they carry. Depending upon the person's sex, the twenty-third "pair" may or may not constitute a pair: they are the X and Y (sex) chromosomes. In women, the body cells normally contain a pair of X chromosomes (therefore, women normally transmit one X chromosome to each of their children). However, the body cells of men normally contain two unmatched chromosomes, X and Y. Male germ cells are, therefore, of two types: one-half carry an X chromosome, and one-half carry a Y chromosome. If an egg (which of necessity must contain an X chromosome) is fertilized by an X-bearing sperm cell, the resulting individual is XX, female. If, on the other hand, an egg is fertilized by a Y-bearing sperm, the resulting individual is XY, male.

The Y chromosome, although important in determining the sex of an individual, carries few genes that are involved in genetic disease. Because of the special role of X and Y chromosomes in determining sex, sex-linked genes have a special inheritance pattern. This pattern is represented in Figures 12.5 and 12.6. The pedigree chart in Figure 12.6 shows the inheritance of *hemophilia*, a disorder in which blood fails to clot normally. The illustrated pedigree is largely that of the British Royal Family, beginning with Queen Victoria, who must have carried this recessive, sex-linked gene. Note the large number of hemophiliac sons in this pedigree. Because the Y chromosomes of males carry no normal allele for the mutant gene, the single recessive gene causes males to suffer from the disease. Note, too, that if a hemophiliac male survives and reproduces, all of his daughters are carriers (although they are normal with respect to the ability to form blood clots), but his sons are normal.

Early Recognition of the Effects of Hemophilia

The following paragraph has been copied from the *Code of Jewish Law*, which was compiled by Rabbi Joseph Karo (1488–1575) and published in 1565:

"If a woman lost two sons presumably from the effects of the circumcision, as it was apparent that their constitutions were so weak that the circumcision has caused their exhaustion, her third son should not be circumcised until he had grown up and his constitution became strong. If a woman lost the child because of the circumcision *and the same thing happened to her sister* [emphasis added], then the children of the other sisters should not be circumcised until they have grown up and have a strong constitution."

Note how this exoneration is extended to the sons of sisters and then examine the inheritance of hemophilia among the descendants of Queen Victoria. Can you see why the circumcision of grown men of strong physical constitution would entail no danger?

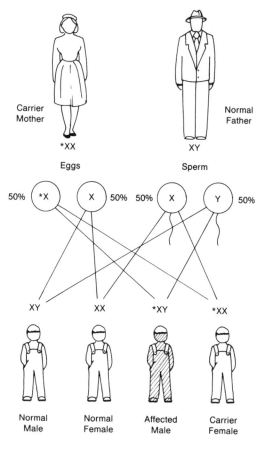

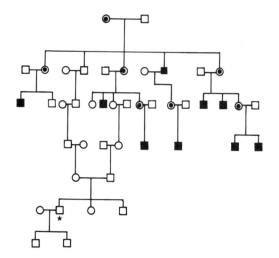

Figure 12.6. A pedigree chart illustrating the inheritance of hemophilia in the royal families of Europe. The initial female carrier (●) is Queen Victoria. The pedigree includes German princes, Russian czars, and Spanish and English kings. The present Prince of Wales is designated by a star (★). Did he or his sons run any risk of inheriting Queen Victoria's mutant gene?

Figure 12.5. The inheritance of a recessive sex-linked trait (*X). The parents are both normal in appearance; however, the mother carries the recessive mutant allele on one of her X chromosomes (*X). Both parents produce two types of gametes in equal frequency: the mother's eggs are one-half X and the other half *X; of the father's sperm, one-half carry an X chromosome, and one-half a Y chromosome. In this case, one-half of the children are females, and one-half are males. One-half of the female children carry the recessive mutation (*X) and, like their mother, are normal in appearance. One-half of the sons receive a normal X chromosome from their mother and are normal. The remaining one-half of the sons receive the recessive mutation (*X) from their mother, and, with no normal allele to "cover" it, show the abnormal phenotype characteristic of the recessive mutation.

NOW, IT'S YOUR TURN (12-1)

1. Perhaps to the chagrin of your professor, genetic counseling has provided the first "official" excuse for discussing Mendelian ratios. Until the late 1930s, Mendelian ratios were the heart of genetics in the minds of many persons; only a few geneticists had enough faith in biochemistry to believe that it would contribute much to an understanding of the gene. In Chapter 7, however, you were asked to calculate the expected proportions with which offspring of different genotypes would occur, given the genotypes of their parents. In effect, you were asked to calculate Mendelian ratios, even though they were not identified as such. Now, to show that you do understand Mendelian genetics, we ask that you enter into your journal the genotypes of the progeny produced by parents of the following genotypes, and the phenotypes of these progeny if the upper-case (capital) letters represent dominant traits (the phenotype corresponds to the capital letter itself; for example, *Mm* individuals would have phenotype *M*).

a) aa × AA

b) Aa × aa

c) Aa × Aa

d) AA × Aa

e) AABB × aabb*

f) AaBB × aaBb

g) aaBb × Aabb

h) AaBa × AaBb

i) AaBbCc × AaBbCc

j) AaBbCc × aabbcc

*For (e) through (j), assume that different genes lie on different chromosomes.

2. In the text, we said that parents pass on mutant genes of which they are unaware to children who do not want them; no one is to blame. This statement is not true for dominant mutations, such as many forms of dwarfism. A dwarf child born of parents one of whom is a dwarf knows precisely which parent *knowingly* risked passing on the mutant gene. What are the factors involved in understanding this situation? Should blame be assessed? What are the rights of married couples regarding bearing children? What are the rights or reasonable expectations that a child might claim? These are not trivial questions: the better reproductive biology is understood, the greater legal claim children have of being free of either genetic or environmentally caused defects.

Mutation

A mutation, as we said earlier, is an error in gene duplication in which one of the two daughter genes represents an erroneous copy of the original DNA molecule. As a result, it carries faulty information regarding the structure of a certain protein. If this protein is an enzyme, the metabolic reaction that depends upon it may not be able to take place, especially if the enzyme entirely fails to function. Waste products accumulate; reactions that should be turned off after running for sufficient time never receive a signal to stop. Cellular machinery is thrown out of kilter, and, through a domino effect, the whole organism can be affected, perhaps at the cost of its life.

Mutation in human beings can be studied more easily with domi-

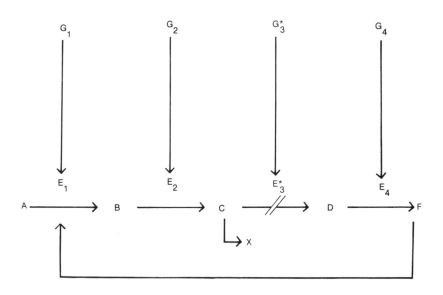

Figure 12.7. The disruption of a normal biochemical reaction by a mutant gene (G_3*). The diagram illustrates a hypothetical biochemical pathway in which compound A is converted to a final product F through the intermediate products B, C, and D. Each transformation (A into B, B into C, etc.) is controlled by a corresponding enzyme (E_1, E_2, E_3, and E_4). The enzymes are the proteins specified by the genes, G_1, G_2, G_3, and G_4, respectively. The mutant form of G_3 results in the production of an inactive enzyme (E_3*). The pathway is therefore unable to proceed from C to D. Compound F, which, once made, normally turns off the pathway by inactivating E_1, is not made; consequently, the transformation of A to B and B to C proceeds uncontrolled. Compound C either accumulates or, after reaching an abnormally high concentration, is degraded by another enzyme to another compound X. The harm done by a mutant gene like G_3* may be caused by the lack of F, by an excessive concentration of C, or by the presence of an unusual compound, X.

nant genes than with recessive ones, since every carrier of a dominant gene can be recognized and, therefore, every one of the mutant dominant genes in a population can be accounted for. Recessive mutations can hide unseen in their phenotypically normal carriers.

A Danish study of *chondrodystrophy*, a common form of dwarfism, can be used to illustrate how the rate at which human genes mutate can be discovered. This form of dwarfism is dominant; therefore, normal parents (who cannot carry the gene) cannot have dwarf children *unless* a mutation occurs during one of the cell divisions preceding germ cell formation. If a mutation should occur, an egg or sperm that would have carried a normal gene at this spot on the chromosome carries instead a newly arisen mutant gene. The mutant gene will be disclosed by the birth of a dwarf baby whose parents are normal.

In the Danish study, about 95,000 births were studied; among these were ten dwarfs of the type illustrated in Figure 12.8. Two of the ten were born to couples in which one partner was a dwarf, so these two cannot be included in the calculations. The number of dwarfs born to normal parents was eight. The conception of 95,000 children requires 95,000 sperm cells and 95,000 eggs, or 190,000 gametes in all. This is important because a mutation to a dominant allele in either gamete, mother's or father's, is sufficient to cause an affected child. Therefore, eight mutations occurred among 190,000 gametes; the mutation rate, then, is approximately 4 in 100,000 (4×10^{-5}).

Mutation rates have been determined for a number of genetic diseases—dominant, recessive, and sex-linked—and the results gener-

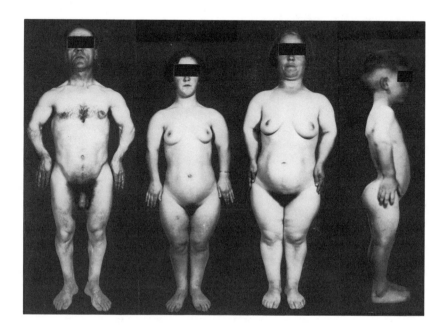

Figure 12.8. Four chondrodystrophic members of one Swedish family. The older man and woman are brother and sister. Their father was a dwarf born of normal parents (presumably by mutation, as described in the text). The younger girl and boy are cousins; she is the man's child, he is the woman's. At one time, an international lightweight weight-lifting champion was a dwarf of this sort. Can you see an advantage he may have gained over his competitors by means of this mutant gene? *(Photo courtesy of S. Nørby, M.D., University of Copenhagen)*

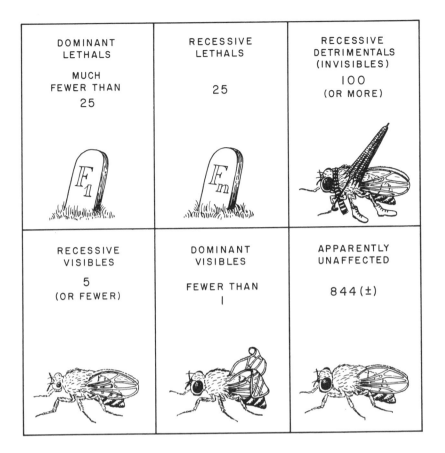

DOMINANT LETHALS	RECESSIVE LETHALS	RECESSIVE DETRIMENTALS (INVISIBLES)
MUCH FEWER THAN 25	25	100 (OR MORE)

RECESSIVE VISIBLES	DOMINANT VISIBLES	APPARENTLY UNAFFECTED
5 (OR FEWER)	FEWER THAN 1	844 (±)

Figure 12.9. A whimsical account of the types and frequencies of gene mutations that might be expected to occur among a thousand spermatozoa of the vinegar fly, *Drosophila melanogaster*, after exposure to 150 r of X radiation. A similar array of mutations might be expected to occur spontaneously, but with a frequency about one-seventh or one-eighth of that shown here. Because the induction of mutations involves changes in DNA, and because human growth and development are controlled by DNA, just as the fly's or any other higher organism's are, exposure to X-rays or other penetrating ("hard") radiation will have corresponding effects on germ cells. These effects will be seen in descendants of exposed persons, possibly for centuries. *(Copyright, Columbia University Press; reproduced with permission)*

ally fall between one new mutation for every 100,000 (10^{-5}) gametes and one for every 1,000,000 (10^{-6}). Although these rates for individual genes are low, it is necessary to recall that there are tens of thousands of genes that are able to mutate. Thus, the *total* mutation rate could be as high as one new mutation in every ten gametes.

NOW, IT'S YOUR TURN (12-2)

In calculating the mutation rate of chondrodystrophic dwarfism, a physical trait that is caused by a dominant gene, we assumed that a dwarf child born of normal (non-dwarf) parents represented a new mutation. Can you imagine any other source for such births?

Mutation vs. Selection

If genetic mistakes arise continually, and if these mistakes have been occurring throughout hundreds of thousands of years, why

Table 12.1. The numbers of children produced by chondrodystrophic dwarfs and by their normal brothers and sisters. Differential fertility of this sort is an example of natural selection.

	Dwarfs (N = 108)	Normal siblings (N = 457)
Children	27	582
Children per parent	0.25	1.27

haven't we all been destroyed? Had there been no corrective device, we might have been. However, mutational errors are corrected in human populations, as they are in populations of other organisms, by natural selection. The affected carriers of mutant genes have fewer children than do other, more normal persons (often they have none at all). Because the total number of offspring produced by normal individuals is sufficient to maintain the total size of the population, the *frequency* of mutant genes among all genes carried by the individuals in a population tends to grow smaller with time.

Perhaps our earlier question should be reversed: Why doesn't natural selection eliminate *all* mutant genes from populations? The answer to this question can be approached through intuition. If there are no mutant genes in a population, none can be eliminated. They can only arise by mutation. Now we let the mutant gene accumulate. Suppose the physical impairment associated with a dominant gene causes its carriers to have only eighty percent as many children as other persons. The total frequency of this gene in the second generation will be 0.80 of that of the first (which equaled the mutation rate), *plus* the entire lot of the newly arisen ones—1.80 times the mutation rate in all. In the following generation the frequency of mutant genes will be $(0.80 \times 1.80) + 1.00$, or 2.44 times the mutation rate. Clearly, the frequency of mutations will increase each generation until as many are eliminated in one generation as arise anew by mutation. The frequency of mutant alleles at this time will be five times the mutation rate because (letting u = mutation rate) $\frac{1}{5} \times 5u = u$. At this time, the input of new mutations equals their loss; this represents a dynamic, stable equilibrium.

Nearly a thousand dominant genetic disorders have been identified in humans by geneticists. If the mutant genes responsible for these diseases arise at a rate of one per million gametes, and if each were to accumulate to five times its mutation rate (five in a million), then the combined frequency of these mutations would be five in a thousand. Because the genetically affected person, in the case of dominant genes, may receive the mutation from either parent, ten in one thousand, or one percent, of all newborn babies might be affected by such genes. This rough estimate of the effect of dominant genes falls comfortably within the five percent of newborn babies who are said to suffer from clear-cut genetic diseases of all sorts—dominant and recessive.

What has been said about the elimination of dominant genes applies as well to recessive ones, but with an additional complication: in the case of (completely) recessive genes, selection eliminates only individuals who have a double dose of the mutant allele. The frequency of such individuals equals the *square* of the frequency of the mutant allele gametes.

Imagine a recessive mutation that causes a disease that is fatal during early childhood. Suppose that one egg or one sperm in every thousand carries this mutant allele. The frequency of affected children among newborn babies would be $\frac{1}{1000} \times \frac{1}{1000}$, or one in a million. The death of these few children would just counterbalance a mutation rate of one in a million from the normal to the mutant gene. The frequency of a *recessive* allele in a human population, consequently, may be a thousand times greater than the rate at which it arises by mutation. For every child that dies of a recessive genetic disease, as many as two thousand persons may carry the same gene without harm. This is consistent with the earlier claim that each person may carry from five to ten recessive mutant genes, any one of which in double dose could kill, maim, or sterilize.

NOW, IT'S YOUR TURN (12-3)

One-third of all X chromosomes in a population are carried by males (XY) and two-thirds are carried by females (XX), provided that the proportions of males and females are nearly equal (50:50). Sex-linked recessive lethals kill their male carriers because the Y chromosome is unable to "conceal" them; these lethals have very little effect on heterozygous females because each lethal gene in one chromosome is concealed by the dominant, non-lethal allele in the other. Explain, either verbally or in mathematical symbols, why the equilibrium frequency for recessive sex-linked lethals in a population equals three times the rate at which these lethals arise by mutation.

Genetic Disorders as a Public Health Problem

Because the mutation rates for individual genes are low, physicians once believed that genetic disease was of little or no importance to public health. These beliefs were destroyed in 1950 by the geneticist H. J. Muller, in an address to the American Society of Human Genetics. The title of his address was "Our Load of Mutations." The speech accomplished two goals. It showed that, contrary to accepted medical beliefs, as much as twenty percent of human illness (including fatal diseases) might have a genetic basis. It also demonstrated that the continued exposure of large numbers of persons to diagnostic X-rays, to industrial radiation, to fallout from nuclear explosions, and to a variety of commonly used mutation-causing chemicals would lead to even greater levels of genetic disease in human populations. Muller doubted whether the human species would survive if continuously exposed to even low levels of radiation.

Figure 12.10. Huntington's disease is a degenerative disorder of the nervous system caused by a dominant mutant gene. Early symptoms include involuntary movements and loss of motor control of muscles. Later, personality changes occur, along with a loss of memory and a decrease in mental ability. The onset of Huntington's disease can occur anywhere between the ages of ten and sixty or more. In most cases the victim is over thirty-five. Consequently, he or she has usually already raised a family; each child has a 50:50 chance of inheriting the disease. Because terminal patients may be helplessly bedridden for five to fifteen years, their care is extremely expensive. *(Photo reproduced with permission of the Pennsylvania Department of Public Welfare)*

A recent booklet published by the National Institutes of Health cites these depressing figures:

• Twelve million Americans suffer from genetic disease.

• More than one-third of all spontaneous abortions are caused by defects in the chromosomes of unborn fetuses.

• At least forty percent of all infant mortality can be blamed on genetic defects.

• About five percent (one in twenty) of all live births are affected by an obvious (although not necessarily fatal) genetic defect.

• About eighty percent of all mentally retarded persons are retarded for genetic reasons.

• About one-third of the admissions to children's hospital wards are made for genetic reasons.

• Each married couple has an average three percent risk of having a seriously defective child.

These figures do not include more recent findings, such as: twenty percent of all heart attacks that occur before age sixty are caused by one or the other of three mutations affecting fat metabolism; about one person in ten is unable, for genetic reasons, to tolerate one or another of many commonly used drugs; and genetic variation may underlie much of the harm suffered by many smokers.

About two thousand genetic disorders are known today. Each year, new discoveries add another hundred or so. About half of these disorders are caused by dominant genes; although each gene may occur in low frequency, family pedigrees for dominant genes are easily understood. Some 150 diseases are caused by sex-linked genes. The X chromosome is only one of twenty-three (amounting to about six percent of the total length of all chromosomes), but the pattern of sex-linked inheritance is so characteristic that these traits are also easily recognized. The remaining eight hundred or so traits are recessive and not sex-linked. These are traits that pop up unexpectedly among the children of normal parents. They are very difficult to study because human families tend to be small. We can guess how many such diseases we *should* know about (instead of 800) by noting the 150 traits assigned to the X chromosome (which we said represents six percent of all chromosomal material). A simple calculation: $.06/.94 = 150/x$ reveals that nearly 2500 (non-sex-linked) recessive diseases should have been recognized by now; that they have not been identified is a measure of the difficulty of such studies.

The Public Health Service once estimated that it costs $20 each day to provide professional care for an institutionalized person; this amounts to about $7500 dollars each year. Inflation, of course, has increased these figures five- or even tenfold. Because many genetic diseases affect persons from early childhood, their care can be extremely expensive. A mentally retarded person who is confined to an institution for forty years requires nearly $1,000,000 in care. The total bill for medical and custodial care provided for those of the twelve million persons who require it is huge, to say the least. In a sense, however, this is the least important cost of genetic disease; the truly important cost is borne by those who suffer from these crippling diseases and by their emotionally shattered parents.

Shedding Our Load of Mutations

Natural selection is the normal cleansing agent that removes mutant genes from any population, including human genes. The victims of mutant genes are often severely handicapped. They may die as newborns or during early childhood or, should they live, they may produce few or no children. Obviously, the death or sterility of afflicted persons controls the frequencies of mutant genes and keeps them at low levels. In a sense, mutant genes commit suicide. Is this the best way to control these genes, however? Must the good health of most individuals be purchased at the price of the death or suffering of a few? Can this misery be avoided?

The answers to these questions are No (natural selection is not the best way to control the frequencies of mutant genes) and Yes (to a large degree, the misery inflicted on human beings by genetic disease can be avoided). One way by which mutant genes may be eliminated and genetic disease avoided is through genetic counseling, accompanied if necessary by therapeutic abortion. In addition, specific treatments are now available for some genetic diseases. These treatments alleviate the symptoms of the disease; they do not alter the mutant gene itself. Finally, molecular biologists are working on techniques that may enable doctors of the future to exchange good genes for bad ones.

Genetic Counseling

Contrary to the old adage, ignorance is *not* bliss! Bliss comes of knowing the truth and, if danger threatens, of knowing how to avoid it. Since Muller made his address in 1950, medical geneticists have made tremendous strides in both detecting and correcting genetic diseases.

The biochemical bases of many genetic diseases are now known. Albinos, for example, occur in all human groups. They are recognized by their pinkish white skin, whitish blond hair, and pink eyes, with red (instead of black) pupils. The basis for these characteristics is a faulty enzyme among those that normally synthesize the brownish-black pigment that is found in the cells of skin, hair, and eyes.

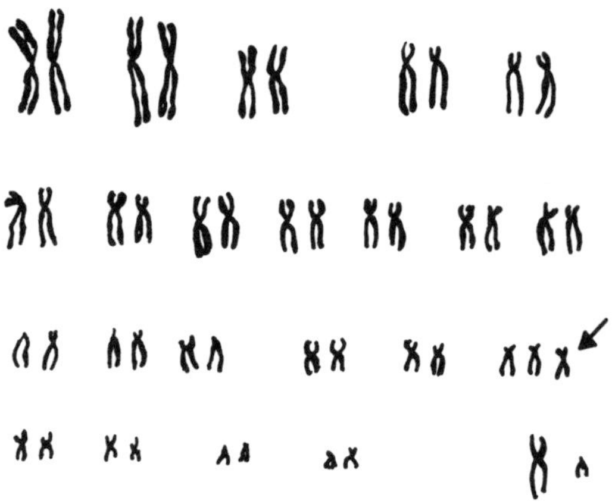

Figure 12.11. The karyotype of a severely malformed newborn who was trisomic for (that is, had three rather than two of) one of the smaller chromosomes (*arrow*). Microscopic examination of embryonic cells obtained by amniocentesis can reveal whether the embryo's chromosomes exhibit abnormalities (extra chromosomes, translocations between chromosomes, or inverted segments within chromosomes). If an observed abnormality warrants it, a therapeutic abortion may be recommended by the attending physician or genetic counselor.

Another genetic disorder, PKU (short for *phenylketonuria*), was first identified in mentally defective patients; the urine of these patients contained a chemical substance called phenylpyruvic acid. Later it was discovered that *all* persons whose urine contained this substance were mentally defective. Today, PKU is known to be caused by an enzymatic defect that blocks the normal metabolic alteration of one substance (phenylalanine) into another (tyrosine).

A third example of the advances in our understanding of the basic causes of genetic disorders is a disease called *agammaglobulinemia* (for which there is no short name). Children with this disease suffer from severe, recurrent infections of all sorts. They are seemingly unable to resist any bacterial or viral disease. The reason has been found by subjecting the blood serum proteins of such children to starch-gel electrophoresis and comparing the results with those obtained by analyzing the serum of normal persons. In the affected children's serum, a class of proteins is missing. It is now known that these proteins, the gamma-globulins, contain the antibodies that protect us from infective organisms (see Chapter 6). Children who lack these proteins are unable to ward off disease; they must be protected by a massive use of antibodies. The genes that would normally lead to the production of gamma-globulin do not function in these persons.

The knowledge that medical geneticists have now accumulated about the biochemical bases of genetic diseases allows them, in many cases, to detect the *carriers* of recessive mutant genes. This is important for prospective parents, especially when both are carriers of the same mutant gene. On the basis of cells obtained from unborn embryos, it is now possible to identify those embryos that, if born, will suffer from severe genetic defects.

Genetic counselors explain the facts of heredity to parents, newlyweds, and genetically afflicted persons who are ignorant of genetics. Very often a man and a woman seeking advice have had one defective child and now wonder about the next one, which may or may not be on its way. Indeed, this is the reason most couples seek advice. The counselor's first task is to explain, as we have done here, that mutant genes are carried by all persons, that no one carries them by choice, and, therefore, that no one is to be blamed for the birth of an abnormal child. The parents should blame neither each other as individuals nor themselves as a couple.

The second task for the counselor is to learn the nature of the defective child's disease and to explain to the parents how this disease is inherited. This information is given as the odds that each subsequent child will be affected. This task is easier if biochemical tests confirm that both parents are indeed carriers of the defective gene. (The detection of many genetic diseases relies on a microscopic examination of human chromosomes; these techniques are advancing nearly as rapidly as the biochemical ones.) Detection and diagnosis aid the genetic counselor and his or her patients tremendously.

Decisions concerning their future reproductive behavior must be made by the husband and wife, themselves; the genetic counselor can only provide information and sympathetic guidance. In the past, the counselor could, in effect, simply quote odds; in many ways he resembled a bookie. Couples had three options: to refrain from having more children, to adopt children, or to take chances with the lives and health of their future children. (One counselor has said, "I tell them the odds are 3 to 1 that the next child will be normal." Russian roulette played with a six-chambered revolver offers odds that are nearly twice as good, 5 to 1.)

Today, troubled couples have still another option: as we mentioned in Chapter 2, an unborn fetus can be tested during the early weeks of pregnancy to determine whether it will suffer from a genetic disease. At present, this test can detect sixty or more genetic defects, and more testable diseases are added each year. If the test reveals that the baby, when born, would be defective, the parents may choose to have a therapeutic abortion. Thus they are able to continue having their own children but with the assurance that no additional child (or no child at all, if they visited the counselor before having their first one) will suffer from the disorder that brought them to the counselor for guidance.

Because of chance events, and possibly because of the differing environments in which different groups of persons lived in the past (such as the crowded ghettos of European cities and the malaria-infested regions of the Mediterranean coast and Africa), certain genetic diseases are often largely confined to specific groups. Sickle-cell anemia is largely a disease of blacks; the mutant gene that is responsible for this disease confers resistance to malaria upon its carriers. This resistance, important as it once was in malarial regions of Africa, is of no benefit to blacks living in North America or Europe. Tay-Sachs disease, a progressive disease of the central nervous system that is invariably fatal, is restricted largely to Jews of European descent. Cooley's anemia (thalassemia) is another blood disorder whose mutant gene confers resistance to malaria. It is frequent among Italians and Greeks whose ancestors lived in the malaria-infested regions bordering on the Mediterranean Sea.

Because "high-risk" populations have been defined for many genetic diseases, genetic *counseling* can be extended into genetic *screening*. The latter, whether voluntary or required by law, consists of screening large numbers of otherwise normal persons to determine whether they *carry* a particular mutant gene. *Genetic screening is an undertaking that must be accompanied by the best possible counseling.* In cases where screening is demanded by law, information is forced upon

The Genetic Counselor's Problem

Nearly a hundred descendants of a Portuguese sailor named Antone Joseph held a family reunion in Oakland, California, one weekend to discuss a still-unnamed disease that has been killing them, their parents, brothers, sisters, aunts, uncles, and cousins continuously since the 1850s. The disease has been categorized as an autosomal dominant trait. That means that each time an affected adult has a child, that child stands a 50:50 chance of developing the disease.

The following comments were made by family members during a genetic counseling session; they illustrate the task facing a genetic counselor:

I was really surprised today to hear about the 50:50 chances. We were sort of told the women in our family had strong genes and that because of this there was probably 90:10 chance of not getting the disease.

She (another family member) wouldn't come and she'd only say she just had her reason. It is a scary thing.

It is time to bring all of this out into the open.

My mother was told that my father had syphilis but I never believed it. The attitude was always there in my childhood that this disease was someone's fault. There was a lot of blame, guilt, and secrecy; our relatives just would not want to discuss it. In some families the children were never told the disease existed. But it's hard to hide when you have relatives dying everywhere before their fortieth birthdays.

My father split from home when he was fifteen because people were dropping dead all around him.

The reason some thought it was dying out is because people are having smaller families. At one time six out of ten kids would get it, but now maybe one out of three gets it and it sounds like a lot less.

My mother had the disease too and once, when I went to visit her, I stumbled a little bit and I saw that she knew I had it. I saw the look on her face. But we made a joke of it. I told her I planned living one year longer than she did. She said, "More power to you."

I definitely have hope they will find a cure for the disease. It's the main reason I won't stop having kids. The people that have so many brains will eventually find a cure. I'm not going to worry about that.

someone who did not ask for it. Ignorance may not be bliss, but neither is uninterpreted information. Without expert and sympathetic counseling, a carrier of a mutant gene can be emotionally crushed by the news that he or she carries a "bad" gene. Perhaps worse, those *not* carrying the mutant gene sometimes feel superior as a consequence. Insurance companies, banks and other lending institutions, and potential employers have been known to discriminate against carriers of mutant genes. These are unwanted and unwarranted outcomes of genetic screening programs because *each* of us (as we have repeatedly emphasized) carries many mutant genes. Those who have been screened know whether they do, or do not, carry one particular mutation of the tens of thousands they *might* have carried. Except for the one particular gene for which the screening was performed, carriers and noncarriers do not differ. As a result of screening, however, potential parents within a high-risk population gain information that should permit them to avoid the devastating experience of having a genetically defective child.

NOW, IT'S YOUR TURN (12-4)

Several industries have undertaken genetic screening of their employees and assignment of types of employment according to an individual's genotype. This practice is not entirely new: During World War II, the U.S. Army Air Force adopted a policy of restricting black soldiers and officers to ground duty. An inherited trait, poorly understood at the time, had resulted in ruptured spleens during flights in nonpressurized cabins, and was suspected of being involved in some fatal crashes of black fliers. Review as thoroughly as you can the various problems raised by genetic screening in industry; these are so numerous that class discussion will be needed to cover even nearly all of them.

Perhaps in responding to the preceding item, you or one of your classmates may have suggested that genetic screening might lessen the expense of industrial safety. For example, it might keep sensitive persons out of polluted areas of a factory. If so, then how might one respond to the suggestion that levels of air pollution be allowed to increase, and the money that otherwise would be spent on pollution-control devices be spent instead on air-conditioning the homes of the elderly and of those suffering from respiratory diseases?

Inbreeding: Cousin Marriages

Many, perhaps most, genetic diseases are caused by recessive mutations: in such cases, each affected person needs to receive mutant alleles from both parents. Because these recessive alleles are carried by large numbers of unsuspecting persons, an occasional mating between "carrier" parents does occur, and an occasional abnormal child results. Indeed, the chance occurrence of a mutation that is lethal to a child or causes a sterile child must, as we have seen, equal the rate at which the normal allele mutates to a mutant one.

The rarer an allele is in a population, the more often its affected carriers prove to be children born to parents who are related. For the most part, they are the result of cousin marriages, but also of marriages that include second cousins, first cousins once removed, and aunts and nephews or uncles and nieces. The reason is that close relatives are more likely to share mutant genes, even though these genes may be rare in the general population. The chance that unrelated carriers of a rare allele will marry decreases as the *square* of the allele's frequency: one per million for a gene whose frequency is $1/1000$, but only one per billion for one whose frequency is $3/100,000$. In this example, the gene frequency has been lowered to one-thirtieth its original value, but the frequency of affected (homozygous) zygotes has been lowered one thousandfold.

The consequences of cousin marriages are illustrated in Figure 12.12. In the figure, the great-grandparents are said to differ completely for the four alleles they carry at the *a* locus: a_1, a_2, a_3, and a_4. Because each great-grandparent carries only two alleles, together they can carry four, at most. For any one of these four alleles, let's say a_1, there is only one chance in two that it will be transmitted to either one of the two children (grandparents). If it has been transmitted to each one, however, there is only one chance in two that either child (grandparent) will transmit the allele to his or her child (a parent). Now, granted that their sexes are appropriate, these individuals are the cousins that might marry. If they marry one another, each one has one chance in two of transmitting the specified allele to their child. The total probability of passage from great-grandparent to the child of a cousin marriage equals $(½ \times ½ \times ½) \times (½ \times ½ \times ½)$, or $1/64$. However, the great-grandparents carried four alleles; because any one of the four could have followed the pattern just outlined, $4 \times 1/64$, or $1/16$, is the probability that any one of the four alleles would become homozygous in the child of a cousin marriage.

The probability that an allele carried by either one of two great-grandparents will find its way *in double dose* to a great-grandchild born of parents who are cousins is 1 in 16. This probability is true for every gene locus; consequently, children of cousin marriages are homo-

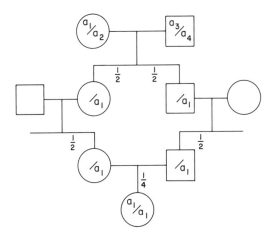

Figure 12.12. A diagram of four generations, illustrating (1) the pedigree of a child born of parents who are cousins and (2) the passage of one allele, a_1, from the great-grandmother through two of her children (grandparents) and, eventually, to the child of the cousins who married. In each generation, there is a fifty percent chance that a child will receive the specified allele, a_1, from its parent; consequently, there is 1 chance in 64 that the child at the bottom of the chart will be homozygous, a_1/a_1. Because a_1 was only one of the four alleles carried by the great-grandparents, the child of a cousin marriage has $4 \times 1/64$, or 1 chance in 16, of being homozygous for one or the other of these four ancestral alleles. *(Copyright, Columbia University Press; reproduced with permission)*

Table 12.2. A summary of early deaths among children of first cousin marriages and of corresponding control marriages[a]

Marriage	Children			Dead (%)
	Dead	Alive	Total	
First cousins	968	5815	6783	14.3
Controls	2326	21,636	23,962	9.7

[a] Twelve large studies are included in this summary. Test these data for significance using a Chi-square analysis.

zygous (that is, carry *identical* alleles) at one-sixteenth, or 6.25%, of all gene loci.

Earlier, we said that although most parents carry five or ten mutant alleles that, if homozygous, would cause serious harm, most children escape because normally the mutant alleles of one parent are not the same ones that are carried by the other. Children of cousins do not have these odds in their favor: what one parent carries, the other might also carry, because they are cousins.

Because of the variation among members of human populations, there is no "normal" human being in a Platonic sense, no *ideal* human being. Consequently, children of cousin marriages, unless they suffer from an identifiable genetic defect, blend in with all other children; they cannot be identified individually. Nevertheless, when large numbers of children are examined, and the average survival of those whose parents are cousins is compared with that of children whose parents are unrelated, children of cousin marriages exhibit a higher mortality rate. A large study of Japanese children has shown that children born of parents who were cousins were slower in learning to walk and to talk, were lighter in weight, shorter, and slenderer, were physically weaker, and lagged behind the other children (on the average, that is) in virtually all school performance grades. The scholastic scores of children of cousin marriages were an average of four to five percent behind those of other students.

A genetic counselor would have to cite these facts if he or she were called upon to advise two cousins who were contemplating marriage. If, however, there were no known abnormalities in the recent ancestry of these cousins, the counselor would be obliged to mention that two siblings in any family often differ by more than five percent in height, weight, and scholastic scores, and that the low-scoring child is not disowned as a consequence. Therefore, after giving appropriate words of caution, the counselor cannot make a tremendous case against cousin marriage. (However, contrary to claims made by some social anthropologists and sociologists, the counselor would issue strong warnings indeed against brother-sister marriages; children of such marriages would carry identical alleles at twenty-five percent of all gene loci. Such children would be lucky to escape the effects of any one of the five to ten major mutant genes carried by each parent.)

Amniocentesis and Abortion

The procedure by which an unborn fetus is tested for the presence or absence of a genetic disease is *amniocentesis*. Because of the sensitivity of modern biochemical tests and microscopic examinations, cells grown from those obtained during amniocentesis are usually sufficient to reveal whether the fetus will suffer from genetic disease.

Because many abnormalities result from the absence of specific enzymes, the biochemical tests applied to the embryo's cells are tests for these enzymes. Similarly, the embryo's cells are tested with respect to the microscopically visible genetic material (chromosomes) they carry.

Amniocentesis is relatively safe in comparison with other surgical procedures. On the other hand, because a needle is inserted through her abdominal wall and into her womb, the mother risks bacterial infection and the baby "risks" mechanical injury (such as a pierced eye). Because of these risks, although they are low, physicians would rather avoid performing an amniocentesis. The physician is willing to assume the risk of amniocentesis only because a genetically defective child could mean even greater human suffering.

At times, a pregnant woman who has undergone amniocentesis and has learned that her child will be defective chooses nevertheless to carry her child to term (birth). *That choice is hers.* Her doctor, her husband, and society at large must accept it. In the minds of many persons, there are no ethically or morally acceptable grounds on which an unborn fetus can be snatched from a woman's body without her consent. On the other hand, most women recognize that the entire testing procedure is predicated on the assumption that genetically defective embryos will be aborted. Otherwise, why should a woman undergo physical risk to both herself and her unborn child in order to obtain knowledge if that knowledge is not to be used?

NOW, IT'S YOUR TURN (12-5)

The sole reason for performing amniocentesis is to gain information concerning the genetic constitution of the developing embryo with respect to possibly severe inherited disorders; that information is then used in deciding whether or not the embryo or fetus is to be aborted. There is, of course, a growing movement in the United States to prohibit *all* abortions. What do you see as the *biological* issues in the abortion controversy? What are the issues involving personal rights? Do you recognize any issue (or issues) as being absolute and, therefore, beyond compromise? In a class discussion, search for inconsistencies and contradictions in your and others' positions.

Chemotherapy

A newborn suffering from PKU, phenylketonuria, is unable to metabolize phenylalanine properly. As a result, phenylalanine accumulates in the baby's body cells. Eventually, alternative metabolic processes come into action and siphon off some of the excess; the chemicals created in these "bypass" reactions are eventually excreted from the body by way of the kidneys.

Untreated, a PKU baby will suffer from severe mental retardation because the excess phenylalanine damages the cells of the nervous system. On a diet that lacks phenylalanine, however, the dangerous excess is avoided. To avoid mental damage, PKU babies are literally starved for phenylalanine—a starvation that would be inexcusable in the case of a normal baby. PKU, however, leaves little choice: mental retardation for life or temporary starvation for an amino acid that is essential for normal growth and development. Fortunately, withholding phenylalanine from the PKU baby's diet for a brief time avoids brain damage and allows the baby an excellent chance of being normal.

Two obvious treatments are available for those who, because of a mutant gene, lack an enzyme controlling one step in an essential metabolic pathway. First, substances that lie beyond the genetic block may be added to their diet, thus ensuring that the desired end products of the metabolic pathway will be synthesized. Second, should an intermediate substance accumulate, thus causing severe toxic damage (as in the case of PKU), the initiating substance (such as phenylalanine) can be withheld from the diet. A good deal can be done for at least a few genetic diseases merely by adding some substances to, and removing others from, the victim's diet.

Galactosemia is another example of a genetic disease for which a simple change in the diet represents an effective treatment. Babies suffering from galactosemia have high levels of galactose (one of many sugars) in their urine, body fluids, and tissues. It happens that galactose and glucose are two simple sugars that, when chemically combined, form the "double-sugar" (disaccharide) *lactose*. Lactose, also known as milk-sugar, is one of the main components of milk. The digestion of lactose proceeds first by splitting it into the two simpler components.

Newborn babies who are galactosemic are able to carry out the first step in the digestion of lactose, because their enzymes split it into galactose and glucose. Furthermore, they can utilize glucose perfectly well. They cannot, however, utilize galactose because of their defective enzyme. As galactose accumulates, it causes enlargement of the liver, cataracts and blindness, and mental retardation. An understatement in one textbook reads: "The baby does not thrive."

Galactosemia is treated by removing galactose from the baby's diet—the sooner the better. In effect, this means taking the baby off its milk diet and substituting an artificial formula containing glucose or sucrose, sugars that the affected baby's enzymes can handle with ease. Under this modified diet, the defective enzyme is not called upon to function.

New Genes by the Ounce?

When tests reveal that a fetus is genetically defective, its parents may choose to have a therapeutic abortion or, in cooperation with their doctor, to continue with the mother's pregnancy while preparing to treat the newborn baby with special diets, supplementary vitamins, antibiotics, or by other methods appropriate for the specific disease.

The dream of molecular geneticists is the substitution of good genes for bad ones. In this way, defective enzymes would be replaced by those that function properly. No longer would it be necessary to alter the diets of those who have genetic diseases; the defective genes themselves would be replaced. This dream is so vivid to some scientists that they speak of the day when DNA for most genes will be available at the pharmacy, just as penicillin and aspirin are available today.

Bottles of genes are not yet available, of course. Furthermore, before one gene can be substituted for another in human beings, a great deal more must be known than is now known. Even so, a "library" of the genes of the small vinegar fly has already been assembled; another has been established with human DNA. In fact, such libraries are probably available for two dozen species. Compiling DNA libraries would have appeared impossible a few years ago. More than that, making such libraries would not have even been *thought of* a few decades ago. The building of these libraries represents the first, giant step toward the artificial manipulation of genes; the technical procedures, consequently, deserve to be described in some detail.

The building of a library of vinegar-fly genes requires both flies (*D. melanogaster*) and bacteria (*E. coli*). The flies provide the DNA that is to be saved; the bacteria serve as library shelves where the DNA can be stored. The secret is to get a small amount of DNA (that representing one gene, if possible) from the fly into the bacterium, where it will be copied each time the bacterium divides or, as we shall see below, even more often than that.

The experimental procedure by which fly DNA is inserted into bacteria is illustrated in Figure 12.13. It happens that *E. coli* have, in addition to their circular DNA chromosomes, smaller circular pieces of DNA that carry genes that confer resistance to antibiotics. These extra little circles—plasmids—divide faster than the bacteria and can be passed from one bacterium to another. Thus, one bacterium with a particular plasmid can "infect" an entire culture that was previously free of that plasmid. This type of infection, involving as it does genes for antibiotic resistance, provides a marvelous natural defense (for bacteria) against penicillin, tetracycline, streptomycin, and other antibiotics.

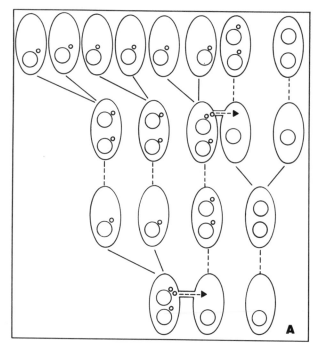

Figure 12.13. A: The reproduction of a bacterium and the spread through the bacterial population of an extrachromosomal plasmid (small circle). At the bottom left is a bacterium that is about to divide but that at the moment is transferring a plasmid into a plasmid-free cell. The first cell then divides, while the other prepares for division by duplicating its (large) circular chromosome. The plasmid in the latter cell also duplicates. Any one of the three cells can now infect another plasmid-free bacterium, eventually causing the infection of the entire bacterial culture.

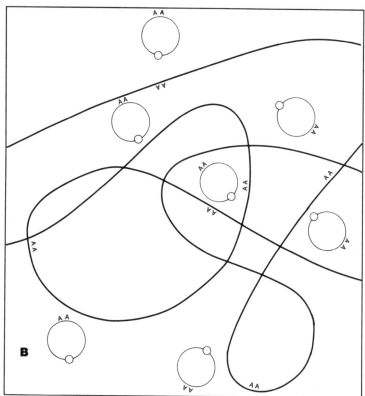

B: A mixture of small, circular DNA plasmids and much longer DNA strands that have been isolated from a fly, a human being, a rat, or any other higher organism. The regions designated AA are those at which a particular enzyme can cut DNA; note that each plasmid has only one AA region, whereas AA regions occur sporadically along the longer DNA strand.

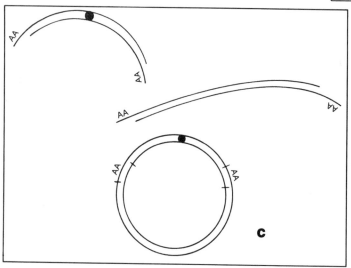

C: At the top are shown a fragment of DNA that has been cut at each of two AA sites (staggered cuts that leave "sticky" ends) and a circular plasmid that has been opened by a cut at its only AA site. The bottom diagram shows the incorporation of the foreign DNA into the plasmid by the union of the matching sticky ends (AA).

Enzymes are known that, when added to a solution of DNA, will cut the DNA molecules only at specific places. If DNA is extracted from flies in long strands, the strands can be clipped into shorter segments by one of these enzymes. It also happens (by chance) that the same enzymes snip plasmids in one place only; they open up the circle. Because the clipped DNA ends are "sticky," the cut pieces tend to rejoin. In a mixture of bacterial plasmids and fly DNA, it frequently happens that a short segment of fly DNA becomes inserted into a plasmid, thus making a DNA circle somewhat larger than the normal plasmid (a "hybrid").

To identify these "hybrid" molecules, the DNA is mixed with antibiotic-*sensitive* bacteria. The plasmid DNA can enter the bacteria without being destroyed; fly DNA alone would be digested if it entered a bacterium. After a brief wait, the bacteria are placed on culture media containing an antibiotic—tetracycline, for example. Only tetracycline-*resistant* bacteria grow in these cultures; so the bacteria that grow must have picked up a plasmid, with its gene for antibiotic resistance (the original bacteria, remember, were sensitive to the antibiotic).

The experimenter now has several bacterial colonies, each of which has acquired a plasmid from the complex fly–plasmid DNA mixture. The plasmids can be isolated and their sizes measured. Those that are larger than expected must carry extra DNA. It is a simple job, then, to find out if the extra DNA is from the fly or whether two bacterial plasmids have fused to give a double-sized circle.

When fly DNA is carried on a bacterial plasmid, there is scarcely any limit to the amount that can be obtained for experimental purposes. If it were necessary, a railroad tank car could be filled with liquid culture medium and inoculated with the bacterial strain carrying the hybrid plasmid. This would eventually yield a pound or more of the hybrid DNA, an amount that probably exceeds the total weight of this one gene in the entire world's population of vinegar flies.

Elegant experimental procedures already exist that promise to usher in an era of artificial genes; even as these words are written, even more techniques are being developed. However, there are serious research problems to be overcome before such genes can be put to practical use. First, many molecular geneticists worry about inserting "naked" human DNA into *E. coli*. It is feared that much of mammalian DNA corresponds to the RNA of cancer-causing viruses. The concerned biologists claim that potentially cancer-producing DNA should not be inserted into an organism that lives so intimately with man (*E. coli*). This concern will delay research, perhaps, by confining it to certain laboratories that are equipped with special safety features, but it will not stop research.

A second and more serious problem involves the insertion of a gene

Figure 12.14. A child with Down syndrome, the least serious of the abnormalities caused by the presence of an extra (very small) chromosome. Persons with Down syndrome are severely handicapped, both mentally and physically. They are not able to care for themselves as functional members of society. The child shown here illustrates the joy these persons can experience; however, if he outlives his parents, he will experience despair over a loss that to him will be incomprehensible. *(Photo courtesy of the National Foundation—March of Dimes)*

into a person's cell in such a manner that it will function at the proper time. For example, no tissue makes hemoglobin except the blood-forming cells of bone marrow. The only cells containing hemoglobin are red blood cells. Defective hemoglobin will not be cured if a normal hemoglobin gene, incorporated into the cells of a child, makes hemoglobin in the cells of the skin, and muscles, and brain cells as well as in the bone marrow. Relatively little is known about the control of gene action in higher organisms, and much more must be learned before gene substitution becomes a plausible treatment for genetic diseases.

Postscript

This chapter deals with abnormal genetic variation, its maintenance in human populations, and its implications for public health. The last are a matter of concern to society. In order not to disrupt the discussion of essentially biological and medical aspects of genetic diseases, we did not make a detailed analysis of the financial impact of genetic disease.

However, these costs can be easily estimated, and they are indeed tremendous. (In order to make them realistic for the 1980s, the dollar figures cited below have been inflated about fivefold.) About 5000 new cases of Down syndrome occur each year. Each victim requires about $1,250,000 of lifetime care. Consequently, the new cases that arise each year in the United States demand a $6 billion commitment for health care. Other diseases make their demands. The 50,000 persons who suffer from sickle-cell anemia, and who require emergency hospital care as often as five times each year on an average, require a further annual commitment of some $125 million. Obviously, the annual medical bill for the care of the twelve million sufferers of genetic disease easily amounts to billions of dollars.

Suppose, however, that all genetic diseases were to vanish tomorrow. Suppose, also, that all persons who now care for the sufferers of genetic diseases were to claim that they possess no other skills, and they went on welfare. The total cost to the nation in dollars would not change much. Dollars cycle within a nation's financial apparatus in the same way that the elements carbon, oxygen, nitrogen, and phosphorus cycle within the biosphere.

Dollar costs aside, the truly important aspects of genetic disease are first the suffering endured by those who are afflicted, and second the anguish of their parents and siblings. The purpose of genetic counseling, of genetic screening, and of medical therapy is the alleviation of human suffering. As for the dollars, they might possibly be spent for more pleasant and worthwhile purposes if the incidence of hereditary disorders were lower. That, however, is about the best for which one can hope.

13
HUMAN COMMUNITIES

Human beings are social animals. Not the only ones, of course: ants, bees, and termites also form societies. Higher primates, such as monkeys, baboons, gorillas, and chimpanzees, also live in well-organized social groups. In sheer magnitude, however, human societies lie far, far beyond their nearest nonhuman counterpart. Ancient Greece, in which many of the moral and practical problems confronting communities of man were discussed at length, seems in retrospect a tiny and simple society. Today, the United States is one of a half-dozen nations with populations of over one hundred million, nations that are desperately attempting to solve serious internal problems. Simultaneously, these nations and dozens of smaller ones strive to make the United Nations a functional forum for the conduct of international affairs.

It would be fruitless to study ants or honeybees in an effort to find solutions for the many problems confronting human societies. On the other hand, it may be possible to discover general principles by which one might

- identify the advantages of being a social animal
- measure the individual's sacrifice in accepting this advantage
- identify the responsibilities a social individual assumes, and
- assess the forces that stabilize social structure.

By examining human societies with these biological principles in mind, we may gain insight into our own lives, although the illustrative examples themselves may appear absurdly trivial.

Simple Aggregations

Many lower animals aggregate in large numbers: schools of fish are one example, migratory locusts are a second. Because some sort of aggregation is a necessary first step in the organization of a society, it is pleasing to learn that aggregations may be useful in themselves. Because the act of aggregating may be useful, a propensity for aggregation can easily evolve through natural selection.

Ten frogs living on the edge of a circular pond can illustrate the advantage of togetherness. First consider ten frogs that space themselves evenly around the edge of the pond. Say also that a water snake lives in the pond. When the snake is hungry, it swims around, unseen, underwater; suddenly, it surfaces near the pond's edge and snatches the nearest frog. For each of the ten frogs, the odds of being eaten are one in ten. (Be sure that you agree with the last sentence.)

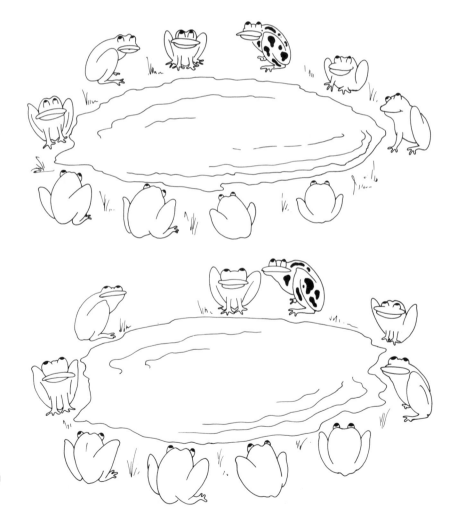

Figure 13.1. The increased safety that accompanies togetherness ("herding behavior"). At the top, each of the ten frogs equally spaced around the circular pond has one chance in ten of being snatched by the snake swimming in the pond. These odds apply to the spotted frog as well as to all others. In the lower diagram, the spotted frog has joined a neighbor. It and its companion form one of nine targets for the snake; furthermore, if they are attacked, only one will be snatched and eaten. Hence, the odds of being eaten have dropped to one in eighteen for the spotted individual—about one-half what they were before.

Now, consider ten frogs around the same pond, inhabited by the same snake, but, in this case, one frog differs from the others: it prefers to sit next to another frog, rather than keeping its distance like the others. As a result, the ten frogs tend to divide the rim of the pond into nine equal sectors, one of which contains two frogs side by side, instead of a single frog. Now, when the snake attacks, the probability of being eaten for each singlet frog is one in nine—slightly higher odds than before. For the two frogs sitting side by side, however, this probability is halved, because the snake can snatch only one, so the probability that the company-seeking frog will be eaten is one in eighteen. The same is true for the frog that tolerated the other's company. Togetherness has nearly halved the probability of being eaten. That advantage is far from negligible. If avoiding the snake were an important problem for these frogs, the obvious outcome would be ten company-seeking frogs in a single heap. Most often, the snake would surface too far away to snatch any one of them. Still, even within the heap, the safest place of all would be in the center; therefore, a good deal of jostling and elbowing would take place as each frog that found itself on the outside tried desperately to get inside the heap once more.

Such seemingly senseless, but intuitively sound, milling about occurs whenever a predator threatens a herd of grazing animals or a school of fish. The prey flee from danger, but they do so as a compact herd or school. The stragglers and other isolated individuals are the ones most likely to be seized. Rioting mobs of humans exhibit the same cluster pattern; the rioters who are listed as casualties in the newspaper are most often those who became isolated from the mob proper (or, worse, isolated persons who by accident happened to be near the riot scene as innocent bystanders).

NOW, IT'S YOUR TURN (13-1)

One hundred and one rioting European students flee from the riot police. As they run across a large square, they split into five equal groups, except for one student who finds himself alone. A policeman with a riot gun makes a decision to fire one shot at random into the fleeing groups (including the group of one). Show that, for the isolated student, the probability of being shot would drop from 1 in 6 to 1 in 105 if he could quickly join a larger group.

Altruistic Behavior

The company-loving frog of the previous section did not leave his isolated spot and settle next to another frog as an intelligent act or as an act of charity. The assumption was that he differed (genetically) from the other frogs. In seeking company, however, he lowered the odds of being eaten for both himself and his neighbor. Consequently, an advantage lay not only with one frog's seeking company but also with the other's being willing to tolerate a close neighbor (this tolerance is also assumed to have a genetic basis).

"Advantage," as used above, refers to selective advantage, an increased chance of living and reproducing. In Chapter 12, we learned that genes that cause death and sterility tend to become rare in affected populations. Conversely, genes that favor survival and reproduction become more common as time goes by. Consequently, if seeking company and tolerating it are to any extent caused by the genes the two pond-dwelling frogs carry, then these genes will become more and more common in serpent-threatened frog populations. Under the circumstances, aggregation in froggy clusters would eventually be a common practice among frogs faced with this simple threat.

Samuel Butler, a nineteenth century English writer, claimed that a hen is an egg's way of producing another egg. This claim has now been refined and modernized by E. O. Wilson of Harvard University, who says that an organism is only DNA's way of producing more DNA. Although these views might be regarded as cynical, they are realistic ones. The many sorts of organisms living today carry within themselves libraries of hereditary information that have been tested through the ages for their ability to construct living, fertile creatures.

The normal means by which DNA successfully reproduces itself is through the good health and high fertility of its individual carriers. Harmful mutant genes, as discussed in the previous chapter, are continually removed from populations by the death or sterility of their defective carriers. Normal genes remain in the population for the converse reason; that is, because of their carriers' good health and high fertility.

Genes need not remain in populations only because their carriers survive; they may remain because some of their carriers sacrifice themselves for others. The famous British biologist J. B. S. Haldane, who has been mentioned in this text before, once commented that his genes (those he himself carried) would be as likely to be passed on to future generations if he saved the lives of two of his brothers or of four of his cousins, even though he might lose his own life in so doing. This comment describes the selective advantage that may reside with unselfish, or *altruistic*, behavior.

Many examples of altruistic behavior can be found among lower animals. The small bird that sounds an alarm at the sight of a hawk, for example, may save the lives of other members of a flock, but it does so at the risk of revealing its own location. By sounding an alarm, the bird risks its life. Many birds and mammals routinely protect not only their own offspring but also those of other members of their group or herd. This behavior often has fatal consequences. Because the members of a group are usually related, however, the DNA carried by the martyred individual still increases among future group members.

An especially clear illustration of altruistic behavior and of a mating behavior that makes it profitable occurs in a small bird, the Tasmanian hen, which is found in Tasmania, an island near Australia. The Tasmanian hens breed, as most birds do, within clearly marked territories. In some instances a territory is occupied by a pair of birds, one male and one female. In others, however, it is occupied by three birds, *two* males and one female. The mated trios produce larger clutches of eggs and raise more chicks to maturity than do the pairs. Now, the two males of the trio are always brothers. Because of this relationship, the DNA that they share (by way of their parents) is multiplied in the batches of sons, daughters, nephews, and nieces that these brothers raise with their shared mate.

The rudiments of altruistic or social behavior can also be seen in the behavior of certain species of small wasps. In some species a female wasp may rear a brood by herself—making a "paper" nest, foraging for insect prey, laying eggs, and protecting the nest from harm. On the other hand, two sisters may cooperate, with one of them playing the role of a "maiden aunt." Eggs are laid by only one of the two sisters, but both defend the household and share its burdens.

How many additional progeny wasps must a maiden aunt and her sister, the mother, raise in order to compensate for the aunt's self-imposed sterility? Not as many as one might guess, because of the peculiar genetic system of wasps, bees, ants, and other Hymenoptera. Among these insects, males have only a single set of chromosomes; females have two sets. Sisters, then, receive identical genes from their father (he has only one set to give them) and, in addition, they have a 50:50 chance of obtaining the same one of the two alleles of their mother's genes. Consequently, sister wasps are, on the average, identical for seventy-five percent of all the genes they carry, rather than fifty percent, as are human sisters and those of most other higher organisms. As a result, the mother-aunt pair does not have to double the number of successfully reared progeny; they need only raise one-third again as many progeny as a single female in order to compensate for the aunt's sterility. Interestingly enough, it is precisely within the Hymenoptera that one finds the most numerous examples of complex social organization. Social systems seem to have

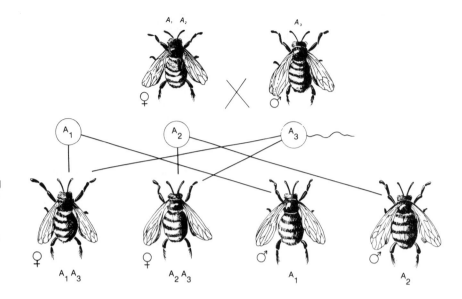

Figure 13.2. The inheritance pattern of bees and other Hymenoptera (wasps, hornets, and ants). Females have two sets of chromosomes (they are diploid, as are most higher animals); males, however, have only one chromosome set (they are haploid). If the two alleles of a female are labeled A_1 and A_2, and the one allele of the male is labeled A_3, their daughters will be of two sorts: A_1A_3 and A_2A_3. Two sisters chosen at random from among these daughters will either be identical (let's say, A_1A_3) or they will share the half of their alleles (A_3) contributed by their father. Because these two possibilities are equally frequent, two sisters share, on the average, seventy-five percent of their genes. Thus, two sisters raising a brood together can easily increase the size of that brood to a number that offsets the virginity of the helper female.

arisen independently but repeatedly throughout this order of insects. When its cost in terms of progeny is low, social organization arises frequently.

NOW, IT'S YOUR TURN (13-2)

1. Using diagrams based on your knowledge of inheritance patterns in higher (diploid) organisms, explain why any inherited tendency for altruism must cause the altruistic individual to aid more and more individuals the less closely they are related to him (or her), if the genes involved are to increase in frequency in a population.

2. Biologists, at times, borrow common terms to describe the behaviors of animals: altruism, rape, cooperation, slave, and aggression, for example. Comment on the advantages and disadvantages that accompany the use of common terms in science. One professor (a historian) claims that his first and most difficult task in facing each year's class of students is to "destroy their vocabulary"; otherwise, at the end of the term he realizes that each time he used a word, the students had a different definition in mind. How does this man's problem relate to the use of common terms in scientific discussions? Which would you find easier: Learning a new term or learning a second meaning for an old term?

Diversity and Social Structure

The most complex individual in which all body cells are identical is a 32-cell colony of simple, water-dwelling cells that fail to separate as they divide. Each cell has one or two flagella; the small globule swims through the combined effort of all 32 cells. Close relatives of these primitive "animals" make smaller aggregates of 16, 8, and 4 cells. The lower limit, of course, consists of individuals that exist as single cells.

Beginning with organisms that are nearly as simple as these 32-cell collections and continuing through the most complex plants and animals, individual organisms are composed of collections of *specialized* cells that carry out special functions, such as digestion, defense, motion, and internal coordination. The cells that are responsible for internal coordination form the nervous system in all but the very simplest animals. That is, they form the communication network that makes complex organization and cell specialization possible. The nervous system, for example, enables a tentacle to transport newly captured prey to the mouth.

Like individuals, communities face complex problems and have a variety of needs. The individuals who enter into a community structure must be prepared to perform the various functions needed for community survival. Ultimately, the reward for the cooperative efforts of the diverse members of a community goes to the DNA of the reproducing individuals. Genetic material that leads to a successful community (or society) is passed on to descendants, who tend to recreate similar communities. Human beings are unique because they have attained the most complex societies without restricting reproduction to specialized breeding individuals. The same cannot be said of the human body itself, nor even of a hive of honeybees; in both of these "societies," special organs (reproductive organs) or special individuals (queen bees and the short-lived drones) are directly responsible for reproduction.

The diverse needs of a successful society can be met by its members in several ways. Permanent castes may be established, or individuals may perform different roles at different ages, or individuals may "volunteer" to fill particular needs according to their skills. For example, termites and ants have worker and soldier castes as well as subcastes of each. The development of each individual is controlled by diet and other chemical stimuli that are in turn closely controlled by the immediate needs of the colony. Among honeybees, individual workers are "assigned" to various jobs as they age. At different ages they act as nurses for the young or as nest builders, they fan their wings in order to ventilate and control the temperature of the hive, and they forage for food. However, individual variation is tolerated. Some workers never cease nursing the young, while others begin foraging for nectar

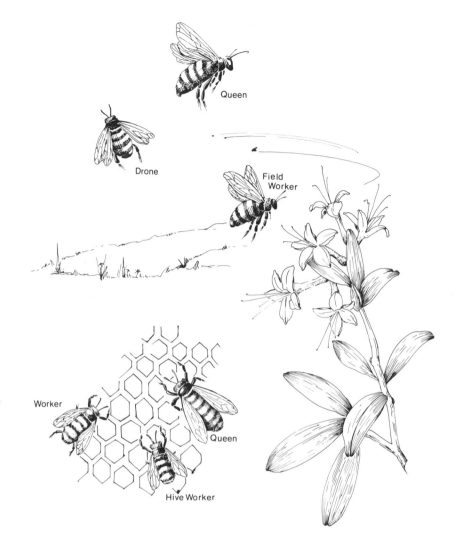

Figure 13.3. The three castes of the honeybee: the queen (fertile female), the drone (fertile male), and workers (sterile females that maintain the colony). The drone's only function is to find and mate with a fertile female. Mating occurs during what is called nuptial flight. Drones are then left to die. The fertilized female (the queen) lays eggs that hatch into sterile females, workers. The workers care for the queen and for the young through all stages of development. Maintenance of the hive is usually undertaken by young workers; foraging for pollen and nectar in the field is usually a task for older workers.

at an early age. Human populations, of course, have both castes and status systems. As we have discussed in an earlier chapter, until recently India had an occupational caste system. In our own society, certain tasks have in the past traditionally fallen to persons of certain ages (and sex), as such terms as "paper boy" and "girl Friday" suggest. These stereotypes, like the castes of India, are now being discarded.

Human beings are remarkably plastic. Within the limitations imposed by physical strength or the exceptional skills needed by professional performers, individual men and women adapt to those roles that are necessary within their societies. Indeed, this flexibility is essential if all members of a community are to retain both social mobility and a reproductive function.

NOW, IT'S YOUR TURN (13-3)

Despite the failure of the Indian caste system to produce genetically differentiated groups of persons, with each group suited for the occupation assigned to its given caste, many persons are concerned that modern social mobility will result in genetically differentiated social strata—the so-called meritocracy. Discuss this concern and your own reaction to it. Helpful references can be obtained from your instructor.

Figure 13.4. Social organization in a pack of baboons. Normally, when the pack travels the dominant male and other adult males take up positions ahead, behind, and on either side of the females and youngsters. When a danger is encountered, the adult males move forward to meet the threat, while the others retreat to a safer area.

Diversity and Morality

Variation and diversity are essential features of all levels of higher life—individual, population, species, and ecological community. A complex individual is able to function only because its cells arrange themselves into various tissues and organs, each of which serves a special purpose.

Individuals also differ from one another; modern genetic techniques have shown that to be a fact. Why they should differ so greatly (beyond the mere accumulation of harmful mutant genes) can only be surmised. Some genetic variation may enable an individual to function normally under both of two contrasting conditions, such as (to oversimplify matters) hot and cold, wet and dry, and gluttony or starvation. If this suggestion is true, the "ideal" individual is one that, of necessity, produces offspring that differ from one another.

On a larger scale, ecological communities must be composed of many different species. It is impossible to prove by logical argument that a community of living things—trees, bushes, grass, insects, worms, birds, mice, cats, and raccoons, for example—would function better if it were to consist of only a single species. On the contrary, it is easy to prove the opposite: no matter what species is chosen as the "ideal" community-making species, other types of organisms are needed to recycle dead and dying individuals. Secondary predators are also needed to destroy primary ones, and a source of energy (green plants) is needed. By the time the list is finished, a complex ecological community has been described. No single species (including the human species and even species of green plants) can perform all the functions necessary to maintain life on Earth.

Therefore, the social roles played by individuals *must* differ if human beings are to function as a society. However, this diversity of roles brings us once more to a conflict involving the morals and ethical values of the community. The ancient Greeks, who stressed the pursuit of constant ideals and played down as accidents the imperfections of living beings, sought a code of ethics and a moral system that would also be constant. "This," they wanted to say, "is the way in which human beings should behave."

Darwinian evolutionary theory, on the other hand, considers variation to be a fact of life. Darwinian theory recognizes the existence of (and possible conflict between) different segments of human populations, but the Darwinian view undermines the basis for *absolute* or *intuitional ethics*. Any group of persons, Garrett Hardin has written, that sees itself as a distinct group and, furthermore, is seen as such by others may be called a "tribe." This group may be a religious sect, a political group, a race, or an occupational group. The essential characteristic of a tribe is that its members follow a double standard of mo-

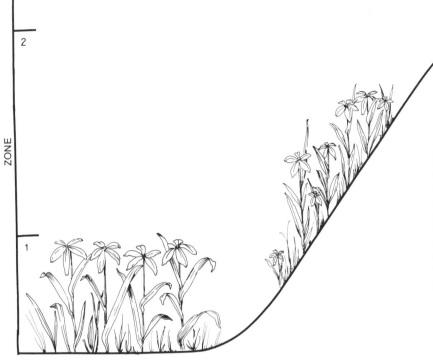

Figure 13.5. Genetic diversity that is maintained because some genotypes cope more successfully than others with particular environments. In the moist, shaded area at the bottom of the slope (zone 1), *AA* and *Aa* individuals (not identifiable by appearance) survive about equally well; in the sunny, dry area at the top of the slope (zone 3), *Aa* and *aa* individuals do about equally well. On the slope itself (zone 2), conditions are both intermediate and variable from season to season and year to year; here *Aa* plants survive better on the average than do either *AA* or *aa* individuals. Under the conditions described, a genetically variable population will be maintained in zone 2, with each genotype—*AA*, *Aa*, or *aa*—being best suited to cope with one of the many problems posed by the fluctuating environment.

rality—one kind of behavior for relations among members of the same tribe, another (and less pleasant) kind for dealing with outsiders. Groups in conflict cannot adhere to an absolute code of ethics. Only the common needs, common fears, and shared hopes of individual persons and family groups give rise to those values that are held in common by large segments of society.

What are the forces that hold a society of conflicting tribes together? Greed, for one. Individuals of all animal societies, not just of human society alone, gain more as members of a society than they lose. Among lower social animals, success and failure in group survival are the credits and debits in the ledger of life; the permanent record is maintained in the group's DNA. Stragglers, altruistic martyrs, and antisocial individuals are destroyed by predators; the cohesive members of the group survive and reproduce. Among human beings, the tally of costs and benefits is kept individually. Each of us pauses now and then to count his blessings, most of which come to us because we belong to a functioning society.

External enemies are another stabilizing force. Youth gangs may forge strong intra-gang bonds as they fight one another. In racial confrontations, alliances are made between formerly antagonistic gangs on the basis of "us" versus "them," often white versus black. On a larger scale, reports concerning the growing military strength of China and Russia cement American society, as Chinese society is cemented by reports on America and Russia, or Russian society by reports on China and America. A threatened invasion by Martians would probably quickly bond the peoples of all nations on Earth.

Neither greed nor external threats are commendable reasons for uniting diverse groups into a stable society. Far better is an understanding that interactions between differing groups, like those between the different species of a biological community, lead to cross-feeding, mutualism (in addition to conflict), and an enrichment of life beyond the level that any one group could attain alone. Unfortunately, such positive stabilizing forces (which *do* exist, because a society *must* consist of dissimilar parts) are largely concealed by intolerance, exploitation, and the fear of "others." If complex human societies are to survive, the negative forces must be overcome. Efficient communication networks between sometimes hostile groups and heroic educational efforts are needed to unify the diverse segments of modern human societies.

NOW, IT'S YOUR TURN (13-4)

One should underestimate neither the importance of what has been said in the preceding paragraph nor the number of persons who would violently disagree with what has been said. Most religions insist on absolutes when discussing morals or ethics. Situational ethics, in the minds of many, leads inevitably to nihilism—nothing is sacred. Although Fundamentalist religions—including the Christian, Jewish, and Moslem faiths—attack Darwinism most violently, some are also opposed to Einstein's theory of relativity because it also represents a threat to absolute values. The problems posed by these conflicts may be insoluble; nevertheless, attempt to sort out the issues involved in your journal. One point, not mentioned in the text, concerns the number of tribes within which each individual finds himself in a complex society: the under-thirty age group versus the over-thirty, student versus faculty, town versus gown, employee versus employer, consumer versus producer, and many more.

Is Society a Supra-organism?

Frequently one reads that communities and societies are supra-organisms. The nation is said to correspond to an individual's body.

The president and the executive branch of government are said to correspond to the brain, because they make and carry out decisions. The army, navy, and state and local police serve the same function as fists, teeth, and (especially in opposing attacks from within) the white blood cells. Merchants, farmers, truckers, railroad men, and sanitation workers have roles that correspond to those of the digestive, circulatory, and excretory systems, because they keep the community in good health. [Interestingly, the four major castes of ancient India were supposed to have arisen from the head (priests), arms (soldiers), belly (farmers), and feet (laborers) of the god Brahma.]

This type of comparison between society and the individual organism sounds reasonable only because complex, self-regulating systems face similar problems. Similar analogies can be made between the living body and an automated factory. Unfortunately, some writers are carried away by the similarities and read more into the comparisons than is really there. Evolution has proceeded from precellular material to single-celled organisms, and to multicellular plants and animals. Therefore, to some individuals, supra-organisms represent the next logical step in a natural sequence of events. Implicit in this belief are the notions that nature is "good," that evolution reveals a natural design, and therefore that evolution is "good." Such beliefs cannot be proven, and contrary notions are equally tenable.

Persons can, of course, make whatever comparisons they wish, but let's explore the implications of the supra-organism analogy of society. We should carefully examine the logical consequences of accepting the living body as a proper model for human society.

In the living body, abnormal cells and badly made protein molecules are destroyed. As protein molecules are manufactured, they are inspected by tester enzymes that are able to recognize structural errors. Molecules with flaws are not recalled for repairs like defective automobiles; on the contrary, they are destroyed and their amino acids are recycled. Indeed, natural aging is accompanied by increasing numbers of errors in protein structure; perhaps the inspection process itself develops errors as it ages.

Individual cells are inspected for errors by still other cells. In sickle-cell disease, the red blood cells, because of their odd shape, are destroyed by certain white blood cells. Persons with the sickle-cell trait are resistant to malaria because the malarial parasites within their red blood cells cause them to "sickle." Each sickled cell is destroyed. Because it reveals its presence within the cell, the malarial parasite brings on its own destruction.

Furthermore, in the normal body organ transplants are accepted only if the host's antibody system is destroyed. Otherwise, the transplant is destroyed by the host. The destruction of the antibody system, however, greatly increases the host's risk of cancer; cancer cells

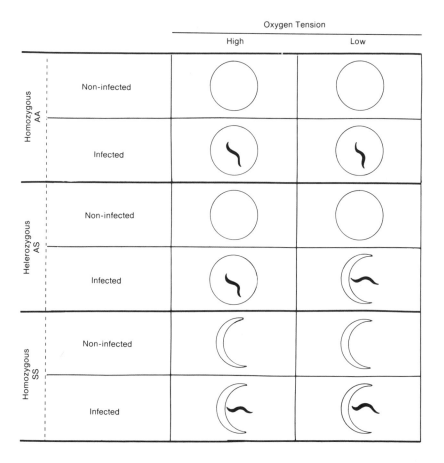

Figure 13.6. The shape of red blood cells that are either not infected or infected with malarial parasites when under high oxygen tension (as in lung and arteries) and low oxygen tension (as in tissues and spleen). The diagrams illustrate the different conditions for persons having only normal hemoglobin (*AA*), both normal and sickling hemoglobin (*AS*), or only sickling hemoglobin (*SS*). In *AA* persons, malarial parasites reproduce within red blood cells. At intervals, the parasites burst out of infected cells and each young parasite quickly enters a new cell, where it proceeds to multiply. In *AS* persons, the reproducing parasites lower the oxygen tension within the cell enough that the infected cell sickles when in surroundings where the external oxygen tension is especially low. The sickled red blood cells are destroyed by macrophages, and the malarial parasite is prematurely exposed to the individual's immune system. The parasite brings on its own destruction by revealing its presence in *AS* persons. In *SS* persons, sickled cells are always being destroyed (hence the disease name sickle-cell *anemia*); malarial parasites cannot multiply in these persons.

that normally would be destroyed by the body's own defense system are now free to multiply.

Any society that operated as ruthlessly against its citizens as the human body does against its own cells would be considered highly repressive. For example, in insect colonies, such as a hive of bees, if a foraging worker returns and behaves in an odd manner—say, because she has been affected by DDT—she is destroyed. Furthermore, if the bees that sting the offender to death are also affected by the DDT that clung to the offender's body, they too are destroyed. (In a related supra-organism analogy, Alexis Carrel, one of the nation's leading physiologists during the early 1900s, apparently viewed human populations as analogous to insect colonies. He argued that human society should not attempt to separate the criminals and the insane into the responsible and the irresponsible, because men are unable to judge men. Rather, petty criminals should be conditioned with the whip or other more scientific punishment, treated in hospitals, and released when cured. He concluded that those who have murdered, have committed armed robbery, have kidnapped, or have misled the public in important matters should be humanely and economically disposed of by euthanasia. A similar treatment, he continued, could be applied to the criminally insane.)

Human beings are aware. Consequently, societies cannot be modeled after organisms. Organisms are composed of cells that live but do not think. Society, on the contrary, is built of individuals, each of whom has a mind, and the mind is the property of the individual, not of his cells. If a society can survive only by transforming its citizens into units as mindless as the cells of the human body (or as the workers and soldiers of an ant colony), many persons would prefer to change its evolutionary course. The natural course of societal evolution can be changed because individuals understand the construction of society—what ties hold it together, and how to weaken these ties. The individual, unlike a cell, has the ability—difficult as the task may be—to alter the society within which he or she lives. Although individuals exist because of the advantages society has bestowed on them, they still retain the power to provide the guidance that determines the nature of the society. The individual human spirit need not be surrendered to society at large.

NOW, IT'S YOUR TURN (13-5)

A tumor can be looked upon as a collection of cells that have escaped from the severe regulation of cellular growth that is normally exercised by the body. Using this definition as a starting point, discuss the supra-organism concept of society from the individual's viewpoint, from society's viewpoint looking inward on itself, and from the viewpoint of societies in conflict.

The Role of Communication in Society

Although a community or a society may not be a supra-organism, some problems are common to both societies and organisms, and to that extent, the two are analogous. The main problem common to both organisms and societies is the unification of a collection of separate pieces so that they work together. When the whole unit functions properly, it represents more and accomplishes more than its individual pieces would if taken separately.

Like the parts of a body, the parts of a community communicate with one another. In the body, the left hand *must* know what the right hand is doing if they are to act together. As the body grows, organs and organ systems must keep in touch in order to maintain proper relative sizes for various parts of the body. The nervous system and hormones (chemical messengers) that circulate in the blood are the means by which communication within the body is accomplished. Circulating antibodies may be important during early development, when the growth of certain organs must be halted at the proper times. Whatever the technique, the purpose is the same: to coordinate and to synchronize the diverse functions of widely separated parts of the body.

Communities also need communication systems. Persons doing specialized tasks must remain in touch with their neighbors so that food, water, and other necessities of life can be obtained. Persons in government maintain communication with their constituencies. The defense system must be internally coordinated and, in addition, must be coordinated with the civilian branches of government. A modern society could not operate without an avalanche of reports, suggestions, decisions, and orders.

A communication network exists to convey information: facts, and decisions based on facts. A man who slaps a mosquito reacts to a perceived fact; his body's communication system is functioning properly. One who slaps an imaginary insect has a malfunctioning system.

To work properly, communication must deal with reality; it cannot serve primarily for deceit. In the body, many drugs can make reality seem remote, and under their influence the individual's sensory system transmits falsehoods. Lies thrown into the communication network of a nation also have bizarre effects. During the 1940s, Trofim Lysenko, a powerful political figure under Stalin, was in charge of agricultural development within the Soviet Union. He suggested that certain changes in farming techniques would improve the yield of wheat and other grains. Certain farms were chosen to test Lysenko's

Figure 13.7. Culture depends upon the accumulation of knowledge over generations, not merely over lifetimes. Where writing is difficult or is unknown, storytelling (oral history) is the means by which accumulated knowledge is spread both within and between generations—that is, between inhabitants of different communities or between persons of different ages (from old to young). The epic poems the *Iliad* and the *Odyssey*, originally oral recitations of their Ancient Greek author, Homer (700 BC), attest to the accuracy with which history can be transmitted without the use of writing. *(Painting by Frederic Remington entitled "The Indian's Tale"; compare with Figure 1.16.)*

Figure 13.8. Propaganda as a means of communication. The World War II posters shown here illustrate the views of Germans held by Americans and by the Germans themselves. The purpose of each poster was to unite a people—those of the United States or those of Germany. The contrasting views illustrate "tribal ethics"—tribes must have a double standard of morality: one for tribal members and the other for outsiders, who are often threatening. *(Photos courtesy of the George C. Marshall Foundation and the U.S. Military Academy Library)*

ideas; the yields of fields treated by his and standard techniques were to be compared.

Because they were afraid of Lysenko's political power and because no one could separate the grain once it was harvested, the managers of the test farms consistently reported that Lysenko's techniques improved yields by at least ten percent. After these encouraging reports, the techniques were introduced into more and more farms. Lysenko's procedures were always reported to be best. Yields, according to report after report, increased by ten percent or so if his recommendations were followed. The total effect, however, was disastrous: the greater the number of farms using Lysenko's procedures, the smaller the total harvest became. The whole calamity was caused by falsehoods, concocted by fearful persons, on which nationwide decisions were based.

In the United States we have recently seen a number of comparable perversions of communication: Congress has attempted in the past to enact legislation based on "facts" that proved later to be falsifications. It seems that military reports from Vietnam were often falsified. During the same war, intelligence reports were tailored to fit the expectations of high-ranking officials.

A society is not a supra-organism. If it is to function even as a society, however, it must (as the body does) maintain the honesty and objectivity of its internal messages. The intelligence, awareness, and social commitment of individual citizens are essential to the maintenance of an honest and realistic communication network.

NOW, IT'S YOUR TURN (13-6)

1. The following passage has been taken from E. O. Wilson's *Sociobiology:* "Deception and hypocrisy are neither absolute evils that virtuous men suppress to a minimum level nor residual animal traits waiting to be erased by further social evolution. They are very human devices for conducting the complex daily business of social life. . . . Complete honesty on all sides is not the answer. The old primate frankness would destroy the delicate fabric of social life that has built up in human populations beyond the limits of the immediate clan." The claim made in the section just concluded is that a communication system whose function is to collect and assess information for use in making decisions and initiating actions must receive factual (i.e., honest) information. Is there any means by which these two views regarding deception can be reconciled?

2. Students who are mathematically inclined may enjoy calculating the consequences of the following suggestion: Assume that the number of errors, both in sending and receiving messages, increases faster than the number of messages themselves; for example, the proportions of both kinds of errors may increase as a function of the *square* of the number of messages. What bearing would that relationship between errors and messages have on the maximum possible complexity of a modern society?

14

COMMUNITY PROBLEMS: COMMUNITY SOLUTIONS

At a symposium sponsored by the National Academy of Sciences, the medical director of one of the nation's large corporations posed this question: Can we protect the chronically ill from the ravages of air pollution, regardless of cost? He then suggested that certain victims of heart disease be provided with clean air shelters during pollution alerts, or that they be paid to move permanently to clean-air areas of the nation.

All the elements of the scene just described—the National Academy of Sciences, the industrial giant, the distinguished medical director, heart disease, and pollution alerts—are modern, but the underlying theme is as old as time. Thousands of Hindu pilgrims have toiled each year for centuries to pull the idol of their god Juggernaut from its normal resting place to a holy spot one mile distant. The huge structure, thirty-five feet square at its base and nearly fifty feet high, is rolled across the sandy ground on enormous wooden wheels. Occasionally a pilgrim is jostled or otherwise falls into the path of the slowly moving Juggernaut. As hundreds of his fellow worshipers strain to maintain the idol's slow but relentless progress, the hapless pilgrim is crushed to death beneath its tremendous weight. Indeed, the word "juggernaut" has entered the English language with the meaning "irresistible force."

Those who share the medic's views on clean air shelters for the ailing see modern society as a juggernaut that must be kept in motion. The wheels of progress, they maintain, will take their toll; a few hapless victims are to be expected. Today it is those with heart disease, since their lives depend upon clean air. Tomorrow, it may be those who earn their livings as fishermen, since their livelihoods depend upon clean water. Then it will be those who live at high altitudes, where the air is thinner and cancer-causing ultraviolet radiation is intense; they will be threatened by manmade changes in the composition of the earth's atmosphere. Always those who suffer first belong to one or another of society's many minorities. Turning to the Hindu religion once more, we find that once upon a time human sacrifices were made to the goddess Kali. It was explained to the people that she had an appetite for males. Furthermore, as the Brahmans were careful to point out, she would eat only men of lower castes. Brahmans, of course, were safely at the apex of India's social hierarchy.

The community problems that this chapter raises largely, but not exclusively, arise through environmental pollution. In this sense, "community" refers to society at large, to people in general. For example, a modern society creates all sorts of waste products that must then be disposed of somehow. Our task is to see this and similar problems in proper perspective. The phrase "people pollute" places us much too close to the problem. In our search for truly effective solutions to these enormous problems, we need a community-wide view.

Can we afford to keep the air clean? If no attempt were made to keep pollutants from the atmosphere, pure-air refuges, we are told, could be constructed for those who need them. There could be shelters for the very young, the very old, and the chronically ill.

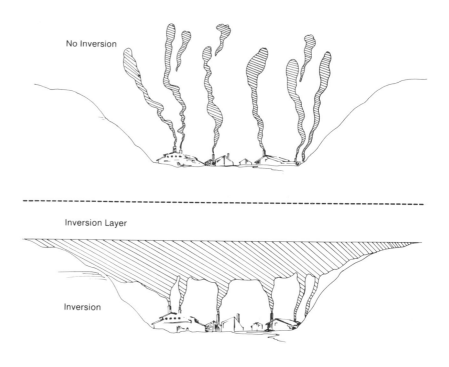

Figure 14.1. The intensification of air pollution by an atmospheric (thermal) inversion. The text discusses the possibility that air pollution should be tolerated and that clean-air shelters should be provided for the ill and the elderly. The above diagrams reveal that the level of pollution is not constant in any one place: sources of pollution that are a mere local nuisance one day may be life-threatening to entire cities on another. Atmospheric inversions (that is, layers of warm air resting upon lower layers of cooler air) act as lids that keep locally created air pollution trapped, so that the level of pollution steadily increases. In October 1948, Donora, PA, a mill town south of Pittsburgh, came within a hair's breadth of complete disaster. Twenty people died, half of the 12,000 inhabitants were ill, and hundreds of farm animals and pets died because of an atmospheric inversion. Inversions form more often in certain regions of the country than in others; Los Angeles is especially vulnerable, and the city frequently experiences smog alerts. Even in the absence of an atmospheric inversion, local pollutants do not vanish; they merely join those of other localities and circulate as general pollution.

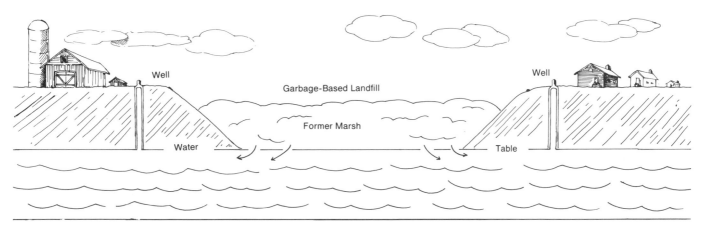

Figure 14.2. A diagram illustrating why the large-scale use of marshes as dumping areas for municipal and industrial wastes destroys the usefulness of local water sources. A marsh is merely a spot in the local terrain that is low enough to reveal the top of the local water table. Contaminants from the landfill enter the ground water directly. Most water used in the United States is drawn from the ground (ground water or aquifers). About twenty percent of this water is now said to be dangerously polluted; in addition, in many parts of the country water is being withdrawn from the ground faster than it can be replaced.

Those who recommend refuges are persuaded by imperfect calculations. To cite one example, rivers are used for the disposal of human wastes and unwanted industrial chemicals of all sorts. Throughout the world, rivers have become little more than open sewers. No matter how filthy they become, economists calculate that it is cheaper to cleanse water when and where it is needed (by a city for drinking or by an electrical power station for cooling) than it is to keep the entire river clean.

Lake Erie was called a dead lake two decades ago. Today, although considerably improved, it remains a heavily polluted lake. Most of its commercially valuable fish have disappeared. Even after the improvement, mercury pollution contaminates the fish that remain. Lakeside beaches were once closed because they constituted health hazards. During the period of the worst contamination, many persons who remembered how the lake had been in the past urged that its former usefulness be restored. Their views eventually prevailed, despite economic studies urging otherwise. The latter suggested that the lake had no commercial value other than to provide buoyancy for ships plying the St. Lawrence Seaway. Therefore, it would be far cheaper to build a swimming pool for each shoreline household than to attempt to clean up an entire lake.

The flaws of the economic analysis are obvious. Lake Erie had little economic value because its edible fish had been exterminated, either by overfishing or by toxic pollutants. At one time, however, the annual harvest of lake fish was worth millions of (uninflated) dollars. Furthermore, when the beaches were clean, the lake served the recreational needs of persons living throughout the Eastern United States and Canada—not just of those who lived on the lake shore. In making their studies, analysts often restrict their attention to the present. As a result, their analyses seldom recommend that past damage be

Figure 14.3. A photograph illustrating the futility (and absurdity) of trying to solve national and international problems at the local community level. The men in the photograph are spraying perfume on the beach at Nice, France, in order to disguise the odor of a disinfectant that was applied earlier. Raw sewage and other human wastes originating from the major cities of Italy and France, the many resort hotels bordering on the Mediterranean Sea, and freighters and passenger liners pollute every beach in southern France and Spain. The problem cannot be solved by local authorities; it requires international agreements. *(From Wide World Photos)*

undone. The lesson, then, is to fight to keep those things that we enjoy, because when they are lost, they are usually lost forever. Once people begin creeping into pure-air refuges for economic reasons, they should not expect to creep out again.

Human beings are remarkably adaptable. At any one moment some persons are sleeping on the parched, dusty, and filth-clogged streets of Calcutta and other Indian cities, others are living in smoke-filled igloos in the Arctic, and still others roam naked in the steaming jungles of tropical Brazil. Perhaps human beings adapt too easily. They can, it has been said, adapt to life on the city dump. On city dumps all over the world (and searching through street-corner trash cans in any American city) are those wretched souls whose lives confirm this claim.

Because humans know both what is and what might be, we have both the *responsibility* for choosing a course that will determine how future generations live, and the torment and the *accountability* that goes with this responsibility. Is existence in a pure-air refuge what life is about? Wouldn't life within plastic bubbles under an ocean of poisonous gases correspond to life within an enormous city dump? Still, that is only half of the problem. As biologists, we see another aspect: green plants must be preserved if future human life is to continue. Consequently, the pure-air refuge must be earth's atmosphere itself; there is no smaller, safer retreat, because the earth's plant life must also be inside.

Our discussion of pollution problems must, therefore, embrace two points of view: First, because we have not yet been driven to clean-air refuges, we argue that to be driven to them in the future would bode ill for us and our descendants. Second, we argue that atmospheric conditions that would require human beings to retreat to refuges would destroy those other natural systems that keep mankind alive. We can easily believe, as one author insists, that urban life can be pleasant. (This author implies, erroneously, that those who wish to preserve the earth's biosphere wish to preserve a rural society as well.) What we do not accept is the idea of refuges within which human beings cower while all else is destroyed. The survival of the human species is too tightly linked to the fate of that "all else" and to the ecological cycles that the "all else" generates.

If a company spokesman suggests that clean air shelters be provided for persons with heart disease, the suggestion is made so that the "acceptable" level of air pollution might be permitted to rise. Does this suggestion reveal anything about sources of pollution? It does, indeed! Contrary to the popular slogan, people do *not* pollute. People taken one at a time, that is. Individual persons litter. Individuals create nuisances. The empty beer can carelessly thrown onto a front lawn is unsightly. One family's smoky grill dirties the windows of another's house. Even so, carelessly discarded beer cans and smoky grills do not cause experts to recommend clean air shelters. Pollution at a level that calls for refuges is caused by organized society—either by the industries that sustain an industrial society or by a lack of adequate civic services. Pollution is a *community* problem (in some instances, the word "community" refers to one-half or more of the human species). The poisoning of America's rivers and lakes through the dumping of thousands of pounds of "waste" mercury cannot be controlled by beautification schemes; it must be stopped by laws prohibiting the discharge of mercury into the nation's streams. Similarly, the smogs caused by automobile exhaust gases will not be eliminated by the use of smaller cars; they will be eliminated only by making the private car unnecessary for performing routine daily tasks—going to and from work, taking children to and from school, and shopping. The elimination of smog, it seems, will require that cities be redesigned, and that once-common (and efficient) public transportation networks be put to use again.

NOW, IT'S YOUR TURN (14-1)

An early reviewer of this textbook objected to our claim that people do *not* pollute. To a large extent, the objection hinges on definitions. Take some example—from the beer can label that asks you not to litter to the automobile that will eventually be a useless hulk—and trace its fate within the constraints of society. Remember that matter cannot simply vanish. If, in your example, the object was abandoned rather than recycled, why was it necessary to abandon it?

Pollution: An Overview

Human beings destroy other forms of life, both plant and animal, in two ways. First, they move into and occupy space that once belonged to other species. Vast forests were leveled to create America's farmlands. Thousands of square miles of asphalt have been spread in creating the nation's parking lots and drive-in shopping centers. Highways and the homes that line them criss-cross what were once the foraging grounds of many animal species. As a result, these species have vanished from many areas in which they were once plentiful.

In addition, however, human beings create vast quantities of waste products. These pollutants spread far beyond human habitation and make life impossible for many plants and animals that never actually encounter a human being.

The wastes of human society may be classified with respect to their source (industry, agriculture, municipal, or individual) and their fate

Figure 14.4. An inventory of common pollutants, classified according to source and means of disposal. Not included in the figure are radioactive wastes, which, because of the enormous problems they pose, are still being "held" rather than "disposed of." The special problems involved in identifying a "safe" disposal method for radioactive wastes from nuclear power plants and defense facilities are: (1) the length of time the material remains dangerous (thousands of years for many common isotopes), (2) the extremely low level of radiation that is considered dangerous, and (3) the propensity of radioactive materials to change the physical properties of the surroundings within which they are stored, either by the heat they generate or by the radiation they emit.

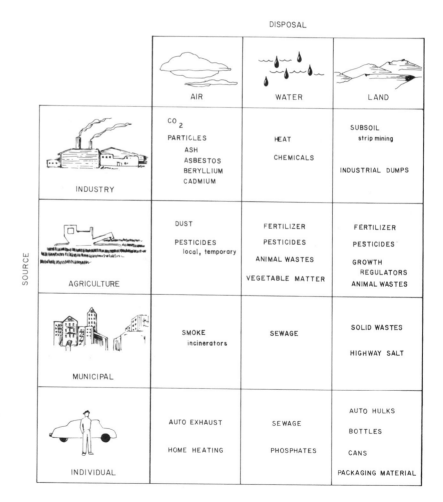

Figure 14.5. An individual's solution to a community problem. The sketch shows an apartment house's sewage drain pipe, which normally would connect with the city's sewer. Like other cities experiencing enormous increases in the numbers of its inhabitants (especially in developing countries), this city's public facilities are strained to the limit and its sewers are clogged. Because none of the sinks and toilet bowls in the apartment house drains properly when the city sewer is clogged, an apartment dweller has smashed the standpipe with a large hammer. Upon reflection, you will realize that this is the only solution for an *individual* who has been forced to solve this *community-level* problem.

(whether they pollute the air, water, or land). This two-way classification system has been used in constructing Figure 14.4. Most of the items in the figure are clear. A few, however, deserve comment. Strip mining, for example, is listed as a form of industrial pollution. During strip mining, the agriculturally valuable topsoil ordinarily is destroyed, and claylike subsoil of virtually no value is left in its place. Consequently, subsoil has been listed as the polluting substance.

During the industrial processing of farm produce, tremendous amounts of wastes are formed. If this material is dumped into streams, it can cause serious pollution. In the figure, this sort of pollution is listed under agriculture, rather than under industry.

The separation of *municipal* and *individual* sources of pollution is often arbitrary. Some cities do not provide public garbage collection. In these cities, homeowners are expected to burn their trash in backyard or apartment-house incinerators. Because the resulting pollution is caused by a lack of adequate municipal service, it has been identified in the figure as municipal. On the other hand, automobile hulks have been identified as individual pollution, although one can argue that society should also provide for their proper disposal.

Except in those houses that have their own septic tanks, the disposal of bodily wastes is identified as a municipal problem, because the individual has no biological alternative. On the other hand, the use of phosphate detergents in automatic dishwashers is not necessary, because alternative cleansers are available. Thus, one can legitimately claim that phosphate pollution comes from the individual and not from the municipality, although phosphates are discharged into rivers and lakes through municipal sewers.

Pollution: Quantities

The sheer bulk of manmade pollutants is staggering. In ages past, handfuls of human beings wandered on large, empty continents. If they cut down a forest, burned a prairie, or dug a canal between two rivers, the world was not greatly affected. The total effect was scarcely more than that created by a herd of elephants, which, in grubbing for food, may convert a young forest into an open, shrub-studded grassland.

Not only have the total amounts of manmade material increased, but also the toxicity of many of them is unbelievable. For example, many chemicals are intentionally designed to disrupt essential steps in an organism's metabolism: insecticides, pesticides, herbicides, and antibiotics are examples. Many of these biologically active compounds are effective at concentrations as low as one part per billion. It has been calculated that the pesticides that are manufactured within the United States each year could contaminate the nation's entire annual rainfall to exactly this level: one part per billion. They are not dispersed this uniformly during their use, of course.

Many of the organic chemicals used by modern industry are able to react with genetic material. Examples are the organic dyes, their chemical relatives, and the epoxies that are used in the manufacture of plastics. In university laboratories, scientists work in designated safe areas when using these chemicals in the study of the genetic control of biochemical reactions and metabolic and developmental processes in bacteria, vinegar flies, or mammalian cell cultures. If mere drops of one of the chemicals are spilled, they are carefully wiped up. If accidental contact with such chemicals occurs, skin and hands are washed thoroughly. Waste solutions are disposed of according to carefully planned safety rules. Meanwhile, tank cars of the same materials are shuttling back and forth across the nation from producer to consumer. Much of this freight travels on railroad tracks that are in disrepair and threaten to (and all too often do) collapse under the weight of passing trains. The evacuation of residents from their trackside homes has become a common occurrence in recent years.

The actual amounts of certain pollutants can be easily calculated. They stagger the imagination. Numbers are not everything, however; the eventual fate of the pollutant must be known in order to understand what the numbers mean. These points can be illustrated using three examples: the carbon monoxide produced by private automobiles, insecticides, and the now-outlawed aerosol propellants.

Carbon Monoxide

The federal emission standards for passenger automobiles permit, at most, the release of 23 grams of carbon monoxide (CO) per mile. In the United States passenger cars travel one *trillion* miles each year; therefore, as much as 23 trillion grams or 50 billion pounds of CO could be released. Because only half a gram of CO is required to inactivate all of the hemoglobin in an adult person's body, it is clear that enough carbon monoxide is released by automobile traffic in the United States to kill every person in the world—many times over. It does not kill them, however; nor is there strong evidence that CO is accumulating in the earth's atmosphere. Something that has not yet been identified, but which may be a marine organism, is constantly removing CO from the air.

Because it does not accumulate worldwide, carbon monoxide is a local problem. It affects those who are forced to spend much of their lives breathing air containing it, such as city policemen who work in tunnels or at busy intersections, garage repairmen whose employers provide inadequate ventilation, commuters who spend long hours in parking garages and heavy traffic, and those whose homes and apartments are located near heavily traveled avenues or commuter highways.

Insecticides

Synthetic insecticides, herbicides, and pesticides are complex organic compounds. The chemists who make them emphasize two properties: effectiveness and, in the case of many insecticides, persistence. Very small amounts are lethal to many organisms besides the intended targets. Once applied to a field, the poison tends to remain there unchanged, thus providing long-term protection.

The original promoters of chemical pesticides did not foresee many of the consequences of their prolonged use. First, the target pests have generally developed, through natural selection, considerable resistance to these poisons. Second, because many of these chemicals are stable, they tend to accumulate in the soils where they have been applied.

The third unforeseen consequence has proven to be a common ecological phenomenon, *biological magnification*. This effect has been illustrated for two insecticides, aldrin and dieldren. Figure 14.7 shows the passage of these insecticides through the ecological food web. The higher an organism stands within the food web (and, consequently, the more remote it seems to be from the original application of the insecticide), the higher the concentration of insecticide in its body. For example, at one time Americans had an average 5–16 ppm of DDT

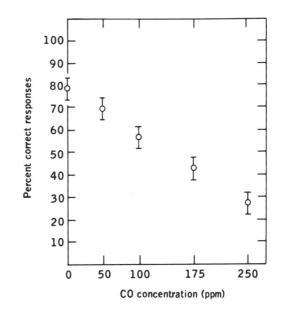

Figure 14.6. Impairment of judgment caused by a two-hour exposure to carbon monoxide. The required task was to judge which of two successive tone intervals was the longer. The ability to discriminate declined steadily with increasing concentrations of CO. In fact, the data imply something beyond an impairment of judgment. With a loss of judgment, one might expect fifty percent of the answers to be correct by chance alone. That the proportion of correct answers declines to twenty-five or thirty percent implies a degree of perverseness on the part of those being tested under the "high" CO level.

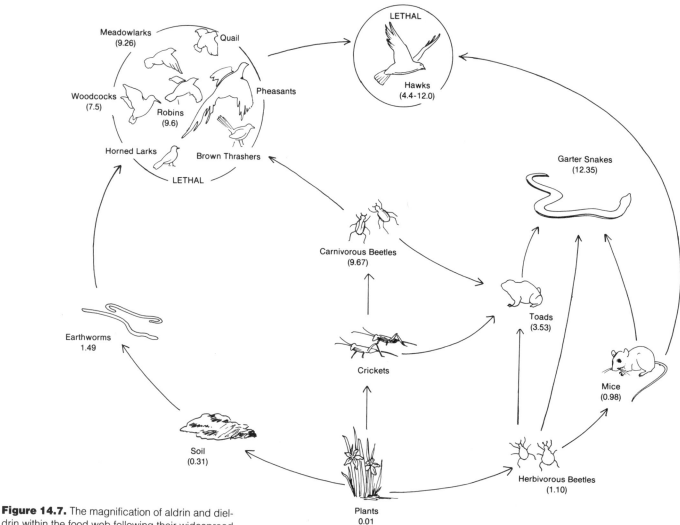

Meadowlarks
(9.26)
Quail

Woodcocks
(7.5)
Robins
(9.6)
Pheasants

Horned Larks
Brown Thrashers

LETHAL

LETHAL
Hawks
(4.4-12.0)

Garter Snakes
(12.35)

Carnivorous Beetles
(9.67)

Earthworms
1.49

Crickets

Toads
(3.53)

Mice
(0.98)

Soil
(0.31)

Herbivorous Beetles
(1.10)

Plants
0.01

Figure 14.7. The magnification of aldrin and dieldrin within the food web following their widespread use as insecticides in the 1950s until their use was banned in 1974. Notice the tendency for the concentrations of the insecticides to increase with the steps by which the organisms are removed from the plants and the soil on which they live (and die). The magnification results from the mass of food any organism must eat merely to maintain itself, as we learned for human beings in Chapter 2. Human beings, had they been included in this figure, would have been at the top, alongside the hawks.

in their body fat (and mother's milk), thus equaling (as they should because of their position in the food web) many birds of prey.

Increasing concentrations of pollutants at higher and higher levels in the food web are commonly observed. During the 1950s, radioactive substances that were produced by above-ground nuclear tests were found to increase in concentration from soil to grass to animal products. To convince the public that nuclear tests were not dangerous, physicists and engineers emphasized the small amounts of radioactive chemicals that would be found on any square foot of the earth's surface, or in any cubic meter of the earth's atmosphere. They did not know (or chose to ignore) that these chemicals would become concentrated in plants (remember that an alfalfa plant, in order to make one

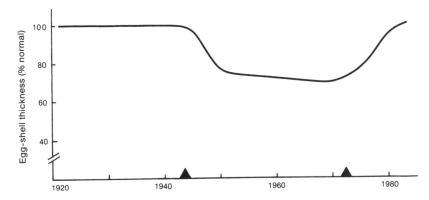

Figure 14.8. A representation of the effect DDT has on the shell thickness of eggs of falcons and similar birds of prey (a similar representation could be made for many other bird species). Before the introduction of DDT as a widely used insecticide, the eggs of these birds had a normal (average) thickness (there was variation around this average, of course). With the introduction of DDT (*left arrow*), the average thickness of the shells decreased markedly; the proportion of eggs that were crushed while being incubated became so great that hatching chicks could not replace their parents. As a result, populations of falcons, ospreys, pelicans, and other birds declined. Since the banning of DDT, shell thickness has become more nearly normal, and the bird populations are recovering.

Figure 14.9. *a:* Three young ospreys in their nest. *b:* An osprey egg that has been crushed during incubation. Thin shells, caused by DDT-induced abnormal calcium metabolism in the birds' oviducts, once caused a serious loss of hatchlings. The harm done by pesticides to birds and other carnivorous animals did not result from direct contact with these poisons; rather, the chemicals became concentrated at higher and higher levels as they were absorbed and retained by organisms higher and higher within the food web. (*Photos courtesy of J. W. Via, Virginia Tech*)

pound of dry matter, must absorb through its roots and lose through its leaves 1000 pounds of water). Nor did they know that herbivores, such as mice, woodchucks, and reindeer, would concentrate these radioactive substances still more. Nor that man and certain other predators would be most severely affected of all.

The exceptions to this general rule of ever-greater concentration occur when one of the links within the food web is killed by toxic pollutants. Cats that eat lizards that have fed on DDT-contaminated insects very often die of DDT poisoning. Cats are very sensitive to DDT. Ospreys were nearly exterminated by the DDT that was used in controlling mosquitoes. It accumulated in the birds and caused nesting females to lay thin-shelled eggs that failed to hatch. The accumulation of mercury, an element used in many herbicides, in tuna and other fish has resulted in many cases of mercury poisoning in man.

Aerosol Propellants

Each of us is likely to have a spray can of some sort at home, because nearly three billion of them are manufactured each year. That equals fifteen cans for every man, woman, and child in the United States. The contents of these cans were at one time expelled by more than 800 million pounds of "inert" propellant fluids, commonly known as fluorocarbons.

Table 14.1. An inventory of aerosol cans manufactured in the United States during a recent year

Type of can	Number produced
For personal use	
Deodorants, perfumes, etc.	1,500,000,000
Air fresheners, oven cleaners	700,000,000
Household spray paints	300,000,000
Food products	100,000,000
Pet sprays	20,000,000
Automotive	75,000,000
Industrial use	100,000,000
Miscellaneous	25,000,000
Total[a]	2,820,000,000

[a] The total number exceeds the number of acres in the continental United States.

Fluorocarbons are inert under most circumstances. That is why they can be mixed with everything from paint and hairset mixtures to perfumes and whipped cream. In the atmosphere, these substances are not inert, however. The ultraviolet rays from the sun split atoms of chlorine from these fluorocarbon molecules:

$$CFCl_3 \rightarrow CFCl_2 + Cl$$
$$CFCl_2 \rightarrow CFCl + Cl$$

This reaction takes place high in the atmosphere, in a region occupied by a layer of ozone. Ozone, a poisonous gas if inhaled, serves a vital function in the atmosphere: it absorbs dangerous cancer-producing ultraviolet rays from sunlight while allowing visible light, "tanning" ultraviolet, and heat rays to reach the surface of the earth.

The chlorine atoms that are split from fluorocarbons by ultraviolet light act as catalysts in the transformation of ozone to oxygen:

$$Cl + O_3 \rightarrow ClO + O_2$$
$$ClO + O \rightarrow Cl + O_2$$

Because the chlorine atom emerges from this reaction unchanged, the net reaction can be written:

$$O_3 + O \rightarrow 2O_2$$

Or, a molecule of ozone plus an atom of oxygen produces two molecules of oxygen; the chlorine atom is a catalyst that speeds up this reaction.

Chlorine atoms are not used up in these reactions. They are like ministers performing weddings. Although tied up momentarily, they are quickly freed to perform their catalytic function again—and again and again. That is why the release of 800 million pounds or more of fluorocarbons into the atmosphere, according to some scientists, *could* be dangerous. The pollutant, in this case, proves to be a long-lived catalyst, each molecule of which is able to mediate the destruction of thousands of molecules of ozone. A loss of one-half of the atmosphere's ozone molecules would, according to one calculation, cause a tenfold increase in the amount of dangerous, cancer-producing ultraviolet radiation reaching the earth's surface. As we suggested in the opening paragraphs of this chapter, this increase could be especially dangerous to persons, such as those living at high altitudes, who are already exposed to dangerously large amounts of ultraviolet rays.

Pollution: Biological Effects

In the next paragraphs only a few of many industrial pollutants are discussed; the entire list is too long to be completely considered. The pollutants we have chosen to discuss are *lead*, *asbestos*, and *mercury*.

Lead

In recent years, one-fourth of all the lead used in the United States—three hundred thousand tons (600,000,000 lbs) annually—has been used as a gasoline additive. When gasoline is burned in automobile and truck engines, the lead is expelled with other exhaust fumes. Scientists have shown that the lead in the upper layers of the oceans and of the glacial ice in Greenland has come from gasoline. A great many of the plants and insects found along interstate highways are (naturally selected) lead-resistant varieties; nonresistant organisms have been killed, generation after generation.

Lead from automobile exhausts (together with lead from the house paints still found in old apartments) is responsible for most cases of lead poisoning among urban children. The lead is in the air they breathe as well as the dust and dirt that they accidentally eat. Lead poisoning has a variety of symptoms, including weakness, apathy, brain damage, and night blindness. Some historians say that the fall of the Roman Empire was caused by the use of lead waterpipes and lead-lined cooking vessels. The sweet-tasting lead salts hid the unpleasant taste of copper; furthermore, the symptoms of lead poisoning also took longer to set in (and were more difficult to identify) than those of copper poisoning. In Colonial America, Benjamin Franklin warned colonists against collecting and using cistern water for cooking if rain gutters and downspouts were made of lead.

Asbestos

Throughout the world, three million tons of asbestos are mined annually. The United States uses nearly half of this amount. Asbestos is a mineral that occurs in three physical forms: solid and rocklike; thin, waferlike sheets; and single, threadlike fibers. It can be used with other fibers to make fire-resistant cloth. It can be formed into hard, fire-resistant wallboard. It can be used for insulation. Spackling compound, the do-it-yourself homeowner's plaster patcher, contains a fair amount of asbestos because presumably its fibers strengthen the patches when they harden.

The two forms of asbestos dust, both the microscopic needlelike spindles and still smaller, submicroscopic particles, cause lung and abdominal cancer and lung disease. The dust is so prevalent that most persons carry asbestos fibers in their lungs. Where does it come from? Sources include worn-out automobile brake linings, which shed asbestos particles into the air when they wear down. Demolition of urban buildings also causes clouds of asbestos dust to be released from old insulation or fireproofing. Not surprisingly, people who live near asbestos mills have higher than expected cancer rates.

Figure 14.10. A sketch based on a famous photograph by W. Eugene Smith illustrating the personal sacrifice and tragedy of industrial pollution. Tomoko Uemura, nineteen years old at the time the photograph was taken, was one of many Japanese children who were poisoned by eating fish that were contaminated by mercury. The children who did not die suffered severe mental retardation and physical malformations. The drawing, which shows a mother bathing her helpless child, emphasizes the Pietà-like quality of the scene; for the mother, this scene has occurred daily for almost all of her married life. *(Illustration courtesy of D. B. Wallace)*

Table 14.2. The quantities of heavy metals that are carried by the Rhine River as it enters the Atlantic Ocean at Rotterdam, The Netherlands

Metal	Tons/year	Metal	Tons/year
Chromium	1,000	Zinc	20,000
Manganese	6,000	Cadmium	200
Iron	80,000	Mercury	100
Nickel	2,000	Lead	2,000
Copper	2,000	Arsenic	1,000

Mercury

Many years ago, a Scotland Yard Inspector realized that he was not as incisive mentally as he had once been and that he was becoming physically less coordinated. His job involved fingerprinting persons who had been apprehended by his colleagues. His trouble was finally traced to the mercury in the fingerprinting ink; years of inhaling mercury vapor had resulted in mercury poisoning. A suburban American homeowner who was excessively proud of his lawn had a similar experience, but in his case the mercury came from the large amount of crabgrass poison he had stored in his basement.

Much more serious disorders, fatal in many instances, have afflicted those who have eaten mercury-contaminated food. On many occasions Japanese villagers have died by the dozens from eating fish taken from mercury-contaminated waters. Recently, four children in a New Mexico family suffered permanent brain damage from eating contaminated pork—their family pigs had been fed mercury-treated grain.

Sources of mercury pollution are shown in Figure 14.11. Two factors make mercury pollution so serious. First, the metal is commonly used in industry, and despite its high price, until recently many firms have found it cheaper to discard used mercury than to clean and reuse it. The nation's inventory of mercury (total produced, imported, exported, and in storage) is short by about twenty-five million pounds. Presumably this is the amount of mercury that has been thrown away with other industrial wastes. Besides the large amount that has been lost, however, mercury is a dangerous pollutant because it is extremely toxic. A steady diet of meat containing as little mercury as 0.5 parts per million could be a dangerous diet for a human being. A fish, however, can concentrate mercury as much as 5000 times; that is, as mercury moves through the food web it becomes more and more concentrated, just as many insecticides do. Consequently, a concentration of mercury in water as low as 5 parts per billion can lead to dangerous levels of mercury in fish—levels that are especially dangerous for those persons who eat fish daily. Now you can see why the contamination of the nation's rivers and streams by one million pounds of mercury each year might be dangerous; the amount of lake water and seawater that this amount could contaminate to a dangerous level is tremendous: 1,000,000 lbs of mercury divided by 5 lbs of mercury per 1 billion lbs of water equals 2×10^{14} lbs of dangerously polluted water. This much water corresponds to an arc of coastal water 1000 miles long, 10 miles wide, and 100 feet deep. Calculations of the above sort could be carried out for other chemical pollutants, but even without the calculations the basic point has been made.

In summary, Table 14.2 lists the amounts of the various metals that

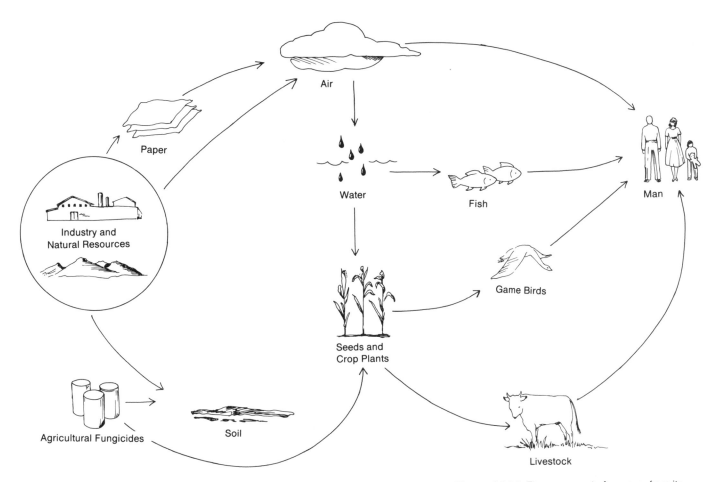

Figure 14.11. The movement of mercury from its industrial (and natural) sources through the environment. One important shortcut not illustrated is the consumption by human beings or by domesticated animals (pigs or chickens) of seeds that are coated with mercury for protection against fungal diseases. Despite the warning labels that are placed on bags of treated seeds, the temptation to use them for either human or animal food is too great for many poor families. Death or severe nervous and mental damage (as illustrated in the previous figure) is the usual outcome.

are discharged as industrial pollutants into the Rhine River, one of the major rivers of Europe. The same metals contaminate the Delaware, Hudson, Ohio, Missouri, and Mississippi Rivers in the United States. No industrial nation has escaped this type of pollution.

Heat: A Special Problem

Chemical pollutants and particulate matter like soot and asbestos fibers are measured in pounds, kilograms, or tons. Waste *heat* is energy, and it is measured in calories. Because energy can be neither created nor destroyed, every activity of a modern, energy-dependent society leads eventually to the production of heat. This heat must then be removed from the neighborhood in which it is created. Ultimately, of course, it is lost from earth by radiation into outer space. The total heat lost from the earth by radiation is the sum of that absorbed from sunlight, that released by human technology, and that arising from radioactivity of certain elements or from geological processes occurring in the earth's interior.

Modern societies run on energy. Agriculture, as we have learned, is an energy-consuming (and therefore heat-producing) activity that is powered by oil and electricity. Mining, construction, manufacturing, and home heating are other activities requiring huge quantities of energy. In Great Britain, one-third of the energy obtained from coal is used in mining more coal. The "natural" form of all but a few metals is the oxidized form that is found in native ores. Energy is required to separate metal atoms from those of the attached oxygen; still more energy is required to keep metal structures (such as bridges and buildings) from recombining with oxygen once more—that is, from rusting.

Let's explore for a moment the magnitude of a modern energy budget. A water-cooled electrical power plant requires 2 cubic feet of water each second for every million watts of installed capacity; this is cooling water for condensing steam after it has passed through the turbines. If every power station in the United States were water-cooled, the amount of water needed daily for this purpose alone would be 200 times greater than the total daily run-off the nation's rainfall. Stated differently, every drop of water falling on the continental United States would, on the average, pass through 200 power stations on its way to the sea. And, each time it passed through one, it would come out 15°–20°C hotter than when it entered. These numbers apply only to electrical power plants; the total energy budget of the United States (which includes, for instance, all automobile engines, blast furnaces in industry, and oil and gas furnaces in private homes) is ten times as great as the electrical budget alone.

Rivers, even in the United States, are not filled with boiling water, but they do get hot. In some cases, unbearably so. Even the Columbia River, which accounts for nearly one-third of the run-off from the United States, is heated several degrees by the nuclear power plants near Hanford, Washington.

Persons favoring the proliferation of nuclear power plants have calculated that the heat generated by industrialized societies will have only negligible effects on the Earth's atmosphere. Such calculations "prove," for example, that if each of twenty billion persons were supplied with twice the energy now used by American citizens, the average temperature of the earth would increase by only .25°C. Global calculations of this sort should not be trusted because they neglect too many factors. One, for example, involves the distribution of heat over the earth's surface. The calculation cited above assumes that the mixing of hot and cold air, and of hot and cold water, is thorough, which it is not. As a result, the air temperature over many American cities even now is 1°–2°C higher, on the average, than that of nearby areas.

	RETURNABLE BOTTLE	THROWAWAY BOTTLE	REMELTING THROWAWAY
MINING	990	5,195	3,635
TRANSPORT	125	650	360
MANUFACTURING	9,675	42,560	42,560
TRANSPORT	360	1,895	1,895
BOTTLING	6,100	6,100	6,100
TRANSPORT & RETURN	1,880	1,235	1,595
RETAILING	0	0	0
CONSUMER	0	0	0
TRASH COLLECTION & SORTING			6,450
COLLECTION & DISPOSAL	90	470	?
TOTALS (BTUs)	19,220	58,105	62,595

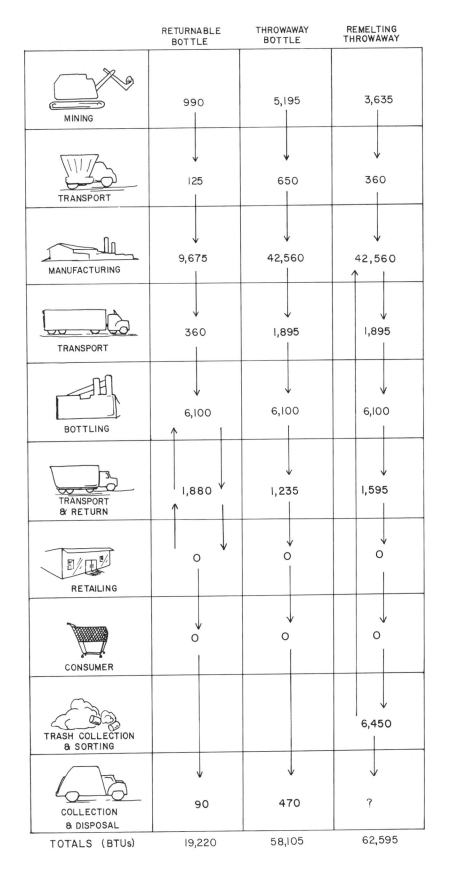

Figure 14.12. Estimates of the amount of energy required to supply each soft-drink bottle of three sorts: returnable bottles, throwaways, and recycled throwaways. Energy is given in BTUs; to convert the figures to calories, multiply by 250.

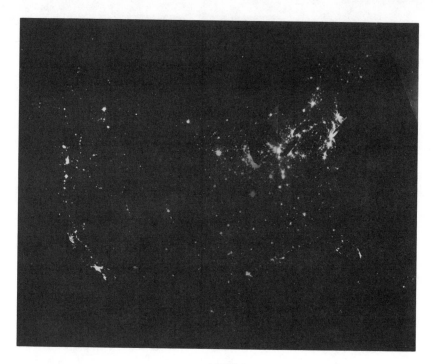

1978

Figure 14.13. Urban America as detected by satellite at night (*top*) and as defined by geographers (*bottom*). The satellite has detected the street lights and apartment and office lights of the major cities and their surrounding suburbs. The contours of the nation are clearly visible, including the outlines of three Great Lakes. This satellite "photograph" and similar ones generated by heat-sensing devices emphasize the enormous quantities of energy that are required to run a modern society. *(Produced from USAF Defense Meteorological Satellite Program film archived for NOAA/NESDIS at the University of Colorado, CIRES/National Snow and Ice Data Center)*

Similar islands or bubbles of heat would, if the use of energy were to increase greatly, spread over neighboring cities, states, or even nations. Local off-shore areas of hot ocean water would, of course, destroy much valuable marine life. As is discussed later, the cooperation of many (sometimes hostile) political units is needed to solve the problems created by the thermal pollution of the air, the rivers, and the oceans.

NOW, IT'S YOUR TURN (14-2)

Find an example of an environmental pollutant not discussed here and carry out your own research: Where is it used? In what quantities? Where does it go once used? What biological effect does it have? For example: What is the fate of vulcanized rubber?

Seeking Solutions

The pollution of the environment (air, water, and land) with poisonous and other harmful chemicals is one of the truly monumental biologically related problems facing human beings. [The other great problem (and, as is explained in the following chapter, one whose solution must be found first) concerns racial equality, the acceptance of variation as a biological fact rather than as a basis for passing judgments and justifying prejudices.] In the previous sections we have learned something of the sheer bulk of the problem: thousands of tons of one chemical and millions of pounds of another. We have learned of the cancer, lung disease, and brain damage that even the tiniest traces of these chemicals can produce. Although we mentioned it only casually, many of these contaminants (hydrocarbons, asbestos, and mercury, for example) are suspected of causing gene mutations, thus adding to the load of harmful mutations described in a previous chapter.

Now that the problem has been identified, how can we solve it? That is an extremely difficult question. Dirtying the environment in which we live saves money; cleaning it up is expensive. Clever persons know how to dirty the environments of others while not dirtying their own. For example, consider four riverside cities. Each withdraws water for household use from the upstream part of the river and discharges waste water downstream. In effect, each dumps its sewage on the next city downstream. If all four cities entered into an agreement under which the water intake pipe of each would be downstream from its own sewage discharge pipe, the river would certainly be much cleaner. Although the water taxes in the city farthest upstream would be somewhat higher because of this agreement, those of the cities farther downstream would be substantially lower. But, how can persons in downstream cities persuade those upstream to spend the extra money? At the moment, persons living farthest upstream have money to spend on libraries and local parks because they are so clever at making others clean up their sewage.

Place every community intake pipe downstream from its corresponding outlet pipe! This excellent solution to water pollution can be

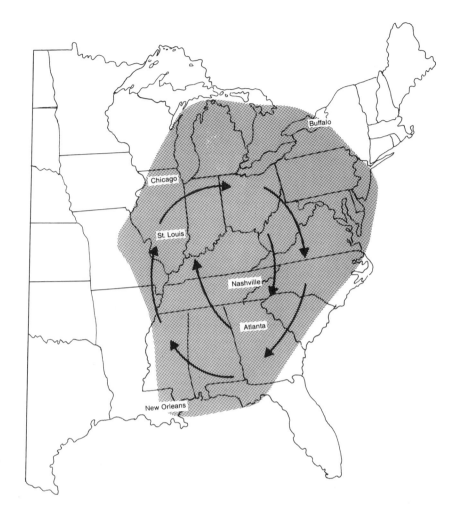

Figure 14.14. A typical distribution pattern of air-borne pollutants (smog) over the eastern half of the United States. This pattern arises when a high pressure system stalls inland before moving on toward Bermuda. Note that neither the cities nor the states involved can undertake effective local action in solving the area's pollution problem. Local regulations, for example, would merely cause the responsible industries to leave the city or the state.

enforced only under state or federal laws. The City Council of New Orleans cannot pass laws that are effective in St. Louis. Because New Orleans and St. Louis are in different states, only a federal law can help the citizens in New Orleans. At least these two cities are in the same nation. Consider the difficulty encountered by the inhabitants of Rotterdam (The Netherlands) in trying to persuade German, French, and Swiss cities and industries not to pollute the Rhine River, the only source of Rotterdam's drinking water.

Just as rivers flow past cities, states, and even nations whose citizens are jealous of their personal rights (and watchful of their money), the earth's atmosphere also moves across state and national boundaries. Air pollution caused by the discharge of millions of pounds of poisonous and corrosive chemicals from industrial plants is not a problem that local communities can solve. On the contrary, it is a problem requiring the attention of the world community. One particularly severe bout of pollution in the United States involved twenty-two states. Poor visibility in northern New York State is no longer caused by poor pollution control in Rochester and Syracuse. Pollution anywhere in the eastern United States is caused by the combined atmospheric wastes of Chicago (Illinois), New York (New York), Birmingham (Alabama), Pittsburgh (Pennsylvania), and other large eastern cities. Rochester and Syracuse contribute a relatively small share to the aerial brew that, especially during the summer, often rotates slowly for days over the entire eastern half of the United States.

Obviously, local anti-pollution measures are inadequate for cleaning up the air we breathe. Any city that undertook stringent control measures on its own would only force its local industries to move elsewhere. The polluted air would remain. As is the case in water pollution, the solutions to air pollution must come from Congress, the only law-making body that can pass laws that affect all citizens equally.

A problem of several decades' duration has only recently attracted public attention: *acid rain.* The problem has been long-standing in Europe. There are maps that show how acid rain grew as a serious international problem through the decade 1956–1966; therefore, the problem has traceable origins more than thirty years old. Evidence of the corrosive nature of acid rain can be seen on hundreds of statuettes and other limestone (or sandstone) carvings on Europe's most beautiful cathedrals, many of which date from the thirteenth and fourteenth centuries. More recently, Canada and the United States have had confrontations over the damage done in Canada by acid rain whose origins lie in the United States. Canada's forests and the wildlife in many of her lakes are endangered by acid rain. The northeastern states of the United States have now joined Canada in urging immediate action in controlling this pollutant, as opposed to the (time-consuming) studies still being recommended by federal officials.

What is acid rain? Its source is easily identified: Coal contains sulfur, which forms sulfur dioxide, an invisible gas, when the coal is burned. The solid matter of coal smoke does not pose a serious problem if it is collected and recycled through the firebox, where much of it will eventually burn. Sulfur dioxide, however, can be removed from the gases of combustion only by expensive chemical processes. Consequently, industries that burn a great deal of coal find it less expensive to build especially tall smoke stacks that propel the sulfur dioxide gas high above the ground, where it dissipates in the upper atmosphere. What you can't see, they believe, does not hurt you.

However, sulfur dioxide does not disappear. It combines with atmospheric water and forms sulfuric acid, an extremely corrosive acid. This acid is carried down to earth in rain (hence, acid rain). The effect

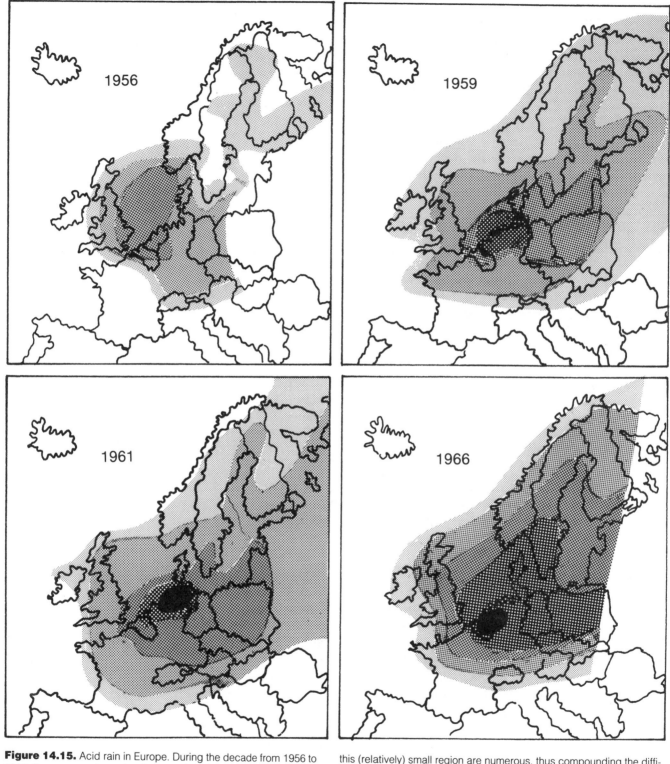

Figure 14.15. Acid rain in Europe. During the decade from 1956 to 1966, Europe experienced an increase in acid rain comparable to what occurred in the United States from 1955 to 1976; the greater effect in Europe probably hinges on the greater dependence Europeans place on coal as a source of energy. The nations involved in this (relatively) small region are numerous, thus compounding the difficulty in taking steps that might preserve the forests of Norway, Sweden, Germany, and France. The atmosphere circulates freely across national boundaries: to control what is discharged into the atmosphere requires international agreements, not local decrees.

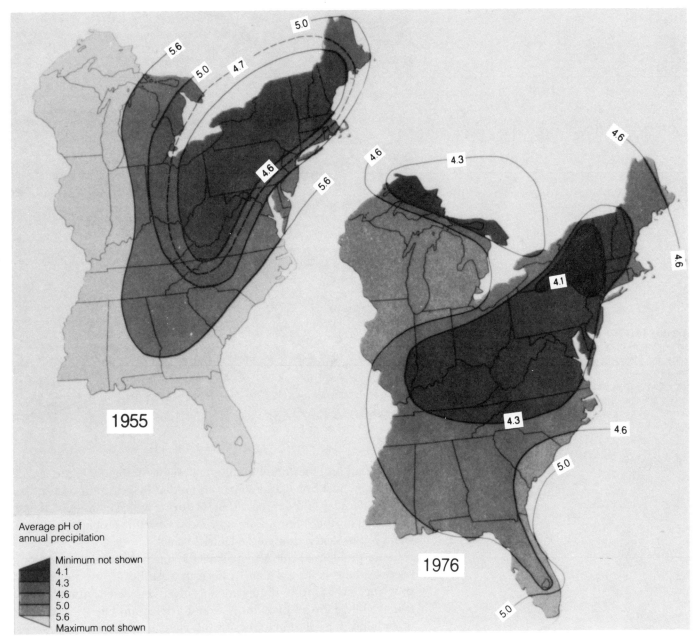

Average pH of annual precipitation

Minimum not shown
4.1
4.3
4.6
5.0
5.6
Maximum not shown

Figure 14.16. The increase in the acidity of rain in the eastern half of the United States over a 21-year interval. The decimal numbers indicate the average pH of the year's precipitation; particular rains could have been considerably more acidic than the average. Because pH is defined as the negative *logarithm* of the concentration of hydrogen ions (the smaller the pH, the higher the concentration), a change of one unit (5 to 4, for example) represents a tenfold increase in these (acidic) ions. The higher acidities of acid rain recorded in these figures are comparable to those of tomato juice, orange juice, or even club soda and other carbonated drinks.

Figure 14.17. *Top:* a lake trout taken from a lake when the pH of its water was 5.4 (1979). *Bottom:* another trout from the same lake, taken in 1982 when the pH was 5.1. The emaciated fish reflects the destruction of food sources that occurs as lakes become acidified. *(Photos courtesy of Ken Mills, Freshwater Institute, Winnipeg, Manitoba)*

of acid rain on limestone building material is dramatic. Figure 14.18 shows how just one statue has been eroded by acid rain during the last few decades of the three centuries during which it has existed. All nations in northern Europe are now faced with this problem, as are all states in the eastern half of the United States.

Can a problem of this size be solved? This is not an idle question, because there is an excellent chance that neither it nor any similar problem can be solved. If they are not solved, however, future human beings will be forced to adopt new and unpleasant lifestyles. On the other hand, the solution to these multi-state, multi-national problems will also require a new lifestyle. The question, then, is whether we will settle for new lifestyles of our own choosing, or wait until unpleasant alternatives are forced on us by circumstance. *If this textbook were to have a single purpose, it would be to make clear these two simple alternatives.* If it were to have a second purpose, it would be to educate persons so that they choose a lifestyle that is compatible with the living world of which they are but one species.

The solution to each of the pollution problems we have discussed in this chapter lies in people's willingness to sacrifice some of their lux-

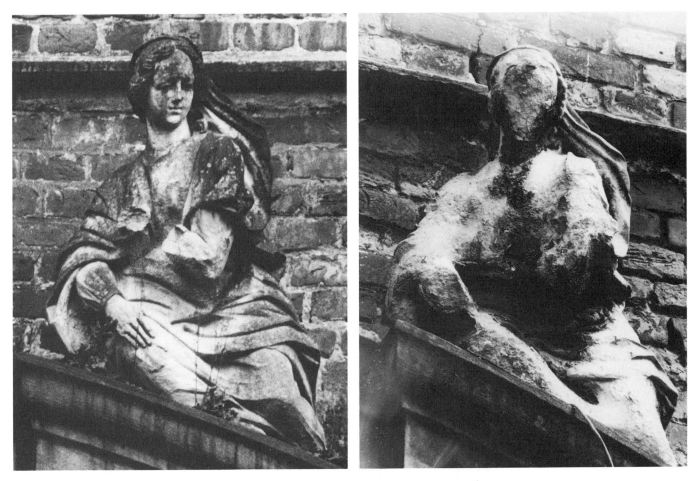

Figure 14.18. A stone figure on a castle in West Germany. The castle was built in 1702. The photograph on the left shows the appearance of the figure in 1908, two centuries later. The photograph on the right shows the same figure in 1968. The bulk of the damage (the figure was nearly completely destroyed by 1968) occurred during the two preceding decades. *(Photo courtesy of Schmidt-Thomsen and the Westfälisches Amt für Denkmalpflege)*

uries. It is tempting to say "of their happiness," but that need not be true. Despite claims to the contrary, it seems unlikely that happiness—that is, pleasurable inner contentment—has changed a great deal during human evolution. (The obvious exceptions would involve the actual oppression of one people by another or one person by another.) Sacrificing luxury does not necessarily require a corresponding sacrifice of happiness.

A personal willingness to sacrifice, however, is not enough; recall the destruction of the commons that was discussed in Chapter 1. Personal willingness is *necessary*, but not *sufficient*. The change in lifestyle must be agreed upon by society as a whole, by the community, and it must then be enforced. Otherwise, those who sacrifice are fools. If one woman gives up her car and rides the train to work, her neighbor should not now argue that he may own two cars. Sacrifices are made to better the common cause, not to encourage neighbors to take advantage of one another.

During the past ten or twenty years, solutions to environmental problems have been sought at the national level. Congress has passed pure air and water acts. The Environmental Protection Agency was established as a federal agency. Despite this progress, there are continuing (and increasingly successful) efforts by some to transfer environmental regulatory authority once more to local communities. It should be clear from the examples cited in this chapter, however, that the local community—the city, the state, and very often the nation itself—is powerless to prevent large-scale industrial pollution. Indeed, in the case of multi-national ecological problems, such as the pollution of the great rivers of every continent, of the inland seas and gulfs, and even of the earth's oceans and atmosphere, the tragedy is that international bodies like the United Nations are unable to enforce regulations or to make them binding on all nations. The worldwide community seems unable to act in response to what is a planetary threat. Unfortunately, even within the United States, if authority continues to be transferred from federal to local levels, the people may find themselves unable to respond to even a national threat.

NOW, IT'S YOUR TURN (14-3)

A plastics firm won a contract to manufacture and sell plastic containers in a foreign country. Before it could build its factory and start its sales, however, it was required to provide that country with an account of the intended fate of the discarded "empties." Do you believe the added stipulation was a fair one? If you were an official of the plastics firm, what would you have proposed?

V
EPILOGUE

In Part I of this text we stated that, in order to cope with the problems of modern society, the nonmajor biology student needs a different set of facts than does the would-be professional. Not only a different set of facts, but also a new emphasis. Simultaneously, we stressed that this "new" biology would not be an *easier* biology, just a *different* one.

The intervening thirteen chapters (Parts II, III, and IV) have presented numerous biological facts, within three contexts—self, community, and society—that emphasize their importance to human beings. Furthermore, the facts, where possible, were chosen from human physiology, human genetics, or general human biology. Occasionally, we paused to say, "This happens to other animals, too."

If, instead of writing a textbook, we were producing a romance, we might at this point feel obliged to concoct a happy ending. However, modern, realistic novels often end tragically, as do events in real life. Consequently, *Biology for Living* does not end on an optimistic note. Nor does it conclude with a series of modern-day commandments beginning "Thou shalt not . . ."

Instead, the text concludes with an attempt to predict the future. You may disagree with any or all of the predictions, but you should be able to explain in your journal and to your classmates why your conclusions differ from ours. College students are now voting members of our nation's society. A necessary first step toward coping with the complex problems of modern society is to engage in a rational, knowledgeable discussion with the persons most involved—you, your friends, and eventually your children. The second step is to do (and to encourage or persuade others to do) whatever is necessary to avert potential disasters. However, the steps that are taken must be taken in a manner that preserves the fabric of society. Thus, the people of the United States and even of the world now need, more than ever before, knowledge, foresight, sacrifice, patience, and compassion.

Earlier sections of this book deal with man's biological nature and his place within the communities of plants and animals that form the natural world. Part IV also deals with man's social and political nature, aspects of human culture that have been made possible by human biology. Unfortunately, human aspirations stemming from political or economic desires often collide with the physical limitations imposed by the biological needs of human beings. In Part V we gaze into our crystal ball to see, however dimly, what the future holds for the human species: What sorts of problems face us? What sorts of solutions are possible for problems that involve conflicts between biology and society? Thus, in this concluding chapter we ask, as Simon Peter did nearly 2000 years ago, "Quo vadis?"

15

QUO VADIS?

Looking Ahead

In looking to the future, we are not dispassionate observers staring at Earth like curious entomologists watching a flood-endangered ant-hill. On the contrary, we look to the world's future as persons greatly concerned for ourselves and our friends, our students, and our own and their children. If we predict an unpleasant future, we do so in order to avoid it—to change the course of events. Society must antici-pate, and thus avoid, the more serious future problems. Foresight and appropriate, intelligent behavior are, we hope, byproducts of education.

The predictions made in the following pages are expressed pessi-mistically. Why should this be? Because *there is only one Earth*, because we believe in and enjoy personal freedom, and because a better life for those who are still unborn depends upon sacrifices made by persons today. To counter the cynical question, "What have future generations done for me?" we can only point to past generations and what they have done for us. Unfortunately, there are many living persons whose ancestors were enslaved, exploited, or otherwise ill-treated. These persons who were, in effect, robbed by history, may not be grateful for the past; on that account they may feel small concern for everyone's future. Yet, if age-old conflicts continue to fester, the future looks dim indeed. The problem that must be solved before all others—the one whose solution must precede lasting solutions for those others—is racial harmony: the acceptance of human variation as variation, not as a basis for moral or value judgments, and the willingness to concede that other people's rights are equal to our own.

If persons are unwilling to make voluntary sacrifices, sacrifices will be forced upon them by the world in which they live. In the latter case, personal freedom may be lost. Some persons fear even now that the problems of a complex society can be solved only by the imposition of dictatorship. Knowing that such fears exist is depressing. The authors' pessimism, then, can be ascribed to an unwillingness to renounce personal freedom, added to a nagging suspicion that lasting solutions to societal problems cannot be arrived at by free citizens acting in traditionally selfish ways.

The Art of Fortunetelling

Many persons attempt to predict the future. Some are optimists, others are pessimists. A few defy classification: One avowed "optimist" has calmly predicted that the coming population/resources crisis should be no worse than the Black Plague—an epidemic that killed one person of every four in many European villages. In making their forecasts, many persons are critical of everything about them; others, like the three monkeys, see, hear, and speak no evil. In general, though, most predictions are foolish. Why? Well, let's look at them.

Many predictions are Utopian. They describe a world that someone believes *ought* to be. Invariably, the Utopian fails to allow for human variation. Furthermore, he usually creates a fine niche for himself in his brave new world. Almost always, all forms of life other than our own are ignored; hence, the glass, chrome, and stainless steel landscapes of science fiction.

In 1935, H. J. Muller foretold the development of transistorized radios, tape recorders, videotapes, portable TV sets, and miniature calculators. His understanding of developing technologies was uncanny. However, he was completely mistaken about the uses to which the gadgets would be put. He spoke of literature, concerts, and legislative debates. He saw universities, theaters, and libraries as the sources of beautiful and interesting programs. In reality, portable radios are frequently banned from recreational areas as nuisances. Those that are equipped with headphones threaten the hearing of an entire generation of young persons. Television has often been called a wasteland. Videotapes are best known for the instant replay of key moments in sporting events: football, tennis, and boxing. Tape recorders are notorious for their use in eavesdropping, even within the Oval Office. Muller's Utopian vision assumed that everyone enjoys the intellectual world that represented his *personal* Utopia.

Some persons make "predictions" in order to shape future events for their own financial gain. When experts speak, we must ask what masters they serve. The Futurama display at the 1939 World's Fair was intended to promote automobile sales and highway construction. The Futurama was meant to be a self-fulfilling prediction; consequently, it

Figure 15.1. Highways of the future, as foretold at the New York World's Fair in 1939. Automobiles shown are not in numbers sufficient to make the proposed highway system economically feasible. Not foreseen, it seems, were (1) the commercialization that would occur at intersections like the one shown here, and (2) the distribution pattern of private homes that would ensue, along with the concomitant abandonment of the railroads, even for hauling freight, their preferred source of income. *(Photos courtesy of the General Motors Company)*

Figure 15.2. A multi-lane highway (the Pasadena Freeway in Los Angeles) during rush hour. Traffic like this and the delays caused by congestion at exit ramps were not foreseen by the designers of the early models of what is now the Interstate Highway system. *(Photo courtesy of the U.S. Department of Housing and Urban Development)*

was overly attractive. The twenty-lane model highways of the display were almost empty of toy automobiles. Exhaust-free, make-believe cars traveled on spotless highways that wove in and out among model skyscrapers and high-rise apartment houses. Present-day realities— battered and rusty guardrails, grease and grime, potholes, abandoned auto parts, salt-damaged steel pillars, animal carcasses, decaying slum areas, exhaust fumes, and smog—were not shown in the Futurama display. If these realities had been foreseen, they would still have been omitted from the exhibit, since its purpose was to sell cars, trucks, and buses and to promote gigantic highway construction programs.

Population

Eventually, the number of human beings on earth will stabilize. This is prediction number one. It must be correct, because there is no alternative. "Stabilization" may come in one of three forms: (1) a stable, constant population, (2) a number that fluctuates between ex-

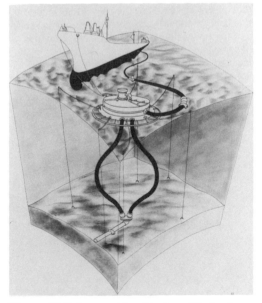

Figure 15.3. A seaport for oil tankers. Missing from the sketch are realistic depictions of the waves that might toss the ship around, the oil leaks that could ooze from the numerous connections, or the marine life that should be flourishing in the crystal clear water. *(Illustration courtesy of the Dow Chemical Company)*

Clouded Crystal Ball

Modern agriculture has been a topic of discussion at various times in this book. We have described how it demands vast areas of uninterrupted farmland, how it presents a tremendous food resource for the exponential growth of insect pests and plant pathogens, and how it is an energy-consuming, rather than an energy-producing, enterprise.

The modern agriculture that we have described is not that foreseen by the editors of *Life* magazine in 1939. In the June 5th issue of *Life* magazine of that year, we read: THE FARMERS OF THE FUTURE WILL HAVE SMALLER FARMS, BETTER CROPS. The trend of the 1930s, according to the article, had been for an ever greater reliance on tractors and trucks on America's farms, but the trend was to be reversed by "agrobiology." Farmers of the future were going to turn to smaller farms and concentrate on quality produce.

The future, of course, is always before us, and the *Life* magazine prediction may yet prove to be right. Why has it been so completely wrong? First, in the United States the land is still there to be farmed. Israel, in contrast, has a limited land area, and it has turned to trickle irrigation (where water is dribbled directly onto each plant), hydroponics (where nutrient chemical solutions replace soil as the growth medium), and the extensive use of greenhouses for growing agricultural produce. Israel, that is, does fit the agrobiology prediction rather well.

A second misjudgment in the *Life* prediction concerns labor and labor costs. During the 1930s, a "hired man" was paid 50¢ per day, plus room and board. Such wages made labor-intensive farming conceivable. In Western countries, however, labor has become a major expense in the production of manufactured and agricultural goods, and in rendering hospital, custodial, or nursing care. The high cost of labor has, of course, driven agricultural practices in precisely the opposite direction from that predicted in 1939 by *Life* magazine.

treme values, or (3) an approximately constant, worldwide average number that conceals wide local fluctuations. Whatever the details, the number of persons that can be fed, housed, clothed, and maintained in fair health is limited. The earth as a whole, like each of its smaller parts, has a limited carrying capacity.

The persistent growth of human populations during the past two or more centuries will cease when birth and death rates are equal. That is a mathematical certainty. If, through indifference, this equalization is left to disease, famine, and war, the total number of persons on earth might be ten times larger than today's four billion. These persons would be forced to live simply, eking out a marginal existence, because the earth's resources cannot provide an elaborate, energy-consuming technology for so many people. Recent events in several countries suggest that political "stability" under crowded, impoverished conditions might be possible only under dictatorial governments.

If, on the other hand, people consciously lower birth rates until they equal the present (low) death rates, the total number of persons on earth need not be large. The precise number could, moreover, be whatever the future human society preferred. If, through a series of rewards and acceptable penalties, birth rates were placed under rational control, populations could be allowed to dwindle, to remain constant, or to grow at will. It is important to notice that a population can dwindle without increasing its death rate (say, by allowing persons to starve, as many persons seem to believe). Current death rates are low, but populations would dwindle if birth rates were even lower.

Provided that persons in the future were satisfied with a level of technology more modest and less wasteful than that prevailing in the present-day United States, we predict that the total number of persons on earth will eventually shrink to about two billion. Unfortunately, the path that leads to this number is an unbelievably difficult one to follow, and so this may be a *hope* rather than a *prediction*. The desires of various groups within any one society, as we have seen, are in conflict. Conflict, not harmony, governs human behavior. Political power within democratic nations is determined by numbers of votes. The rival aims of different nations lead to wars, the outcomes of which are largely determined by sizes of armies. To imagine that the peoples of today's world would agree to a population policy stabilizing (or decreasing) numbers of persons is to imagine a near-miracle. Only the horrors of the alternative possibilities make the vision credible.

Technology: Energy

Energy is the key to modern technology. Pro-energy articles appearing in newspapers and magazines suggest that unlimited energy would provide a good life for all persons. Some persons claim that un-

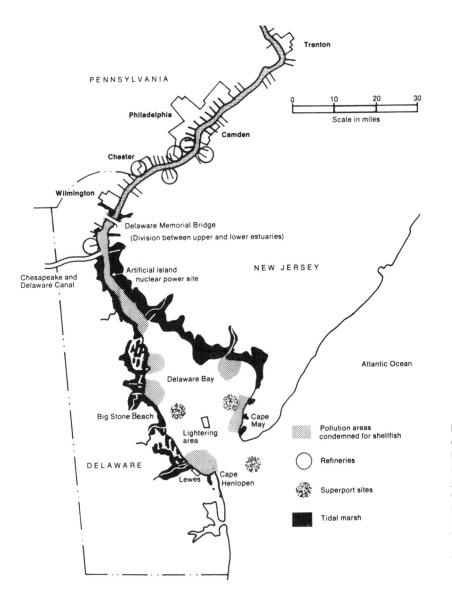

PENNSYLVANIA

Trenton

Philadelphia

Camden

Chester

Wilmington

Delaware Memorial Bridge
(Division between upper and lower estuaries)

Chesapeake and
Delaware Canal

Artificial island
nuclear power site

NEW JERSEY

0 10 20 30
Scale in miles

Atlantic Ocean

Delaware Bay

Big Stone Beach

Cape
May

Lightering
area

DELAWARE

Lewes

Cape
Henlopen

Pollution areas
condemned for shellfish

Refineries

Superport sites

Tidal marsh

Figure 15.4. A map showing the oil refineries located along the Delaware River and the areas of pollution that have led to a ban on shellfishing near the river. The pollution has contaminated much of the adjoining tidal marshlands. In advance of the refineries' construction, the usual assurances were given that such unwanted effects would not occur, but the map illustrates what actually has happened in the Delaware Bay. Similar maps could be drawn for Chesapeake Bay, the areas around Galveston and Houston, Texas, and many West Coast harbors.

precedented abundance awaits everyone. On the contrary, the claim that unlimited energy would bring unlimited abundance either to a single nation or to all mankind is false.

Power-plant owners are especially proud of the slave-equivalents that they provide for each family. The modern household, they say, has the equivalent of several slaves helping with the cleaning, the cooking, and the laundry. Convenient? Yes. Notice, however, that these slaves are not owned by the household. They are owned by the electrical companies and are *rented* to those who need them. The rental fee increases with the depletion of fuel supplies. At one time, rates increased with demand; now, because of fixed overhead ex-

Figure 15.5. Hunger as a social problem. In an earlier chapter, hunger was presented in physiological terms, as an individual ordeal. Today, hunger has become a commonplace threat to entire populations. As these words are being written, a newscaster is reporting that the derailment of a single locomotive may have endangered one million lives: relief food being carried by the train cannot get to the starving persons who desperately need it. Mass hunger is a societal problem because, unless transportation is adequate and a reliable (not charitable) source of food is guaranteed, the number of persons in an area should not habitually exceed the level that local resources can support.

penses, rates increase even when energy-conscious customers *lower* their demands.

Energy is power. This does not mean only that energy can be expressed in such terms as slave-power or horsepower. The person who controls energy possesses *political* power. The nation that has abundant energy is a powerful nation; it (the *nation*, acting as a coherent social unit) overwhelms politically, economically, or militarily smaller, more poorly endowed nations. In such a world it is foolhardy to seek an unlimited supply of energy for all. Throughout the world—Europe, the Middle East, Africa, Asia, and the Americas—individuals and small groups of persons have assassinated, terrorized, and disrupted recognized governments. To speak of unlimited energy for each citizen, when energy is power, is to speak of anarchy.

Mankind's possession of unlimited energy would surely lead to the destruction of most other species of plants and animals on earth, and to the destruction of the thin layer of topsoil that supports them, and us, as well. As we have learned in earlier chapters, the community of living organisms and the earth's fertile soil form the magic carpet upon which human beings live. Destroy the carpet, and mankind is destroyed. Unlimited energy would lead to the unlimited exploitation of both land and sea. It would portend a desolated planet.

Abundant energy would be provided for all peoples of the world if they were less numerous, even if the per capita demand were large. However, the problem of population growth has not been solved; consequently, energy production (and the search for new energy sources) proceeds at a wild pace in an effort to stave off—on a day-by-day, year-by-year basis—hunger and disease. The outcome of this heroic effort is preordained. Hunger and disease (or war) will eventually stabilize the world's population, but, if they do become the stabilizing forces, the total population will be too large to permit plentiful energy *per person*. Thus, without voluntary control of population size, we foresee a future world in which impoverished persons will scratch in destroyed soil for food.

Technology: Food

At the present time, the world contains about one acre (one-half hectare) of arable land for each living person. This is equivalent to a football field. Should the human population continue to grow at its present rate, the world's population will double within thirty-five years. At that moment, the arable land per person will have been reduced to one-half acre, an amount scarcely adequate for providing a year's supply of food for a single person. Each of us will have been pushed back to the 50-yard line.

According to its advocates, modern agriculture saves the world's hungry from starvation. Never, they say, have so many been fed by so

'GET HIM TO TELL YOU THE ONE ABOUT THE BIG OIL AND WHEAT DEAL WITH RUSSIA!'

Figure 15.6. A famous cartoonist's view of a trade agreement under which American wheat was to be exchanged for Soviet oil. The irony of the exchange lies in the energy demands of modern agriculture. Unless the calculations are made with extreme care, more oil could be consumed raising the exported wheat than is received from the Soviet Union. (OLIPHANT. *Copyright 1976, Washington Star; reprinted with permission of Universal Press Syndicate; all rights reserved*)

few. Nor have they! Surplus grain from the United States and Canada has been shipped to nation after nation as crops elsewhere in the world have failed. The yields of North America's farmlands, especially its grain fields, have in past years saved millions from starvation.

The high yields of modern agriculture have been attained at a price, however. Modern agriculture is an energy-*consuming* enterprise. Ordinarily, we think of farming as an occupation that *yields* energy— more energy than is required by those who labor in the fields or barns. If agriculture did not yield surplus energy, it would never have replaced hunting as a human way of life. Today, however, agricultural practices no longer yield excess energy. The American diet, with its ample portions of meat and dairy products, consumes nearly 5 calories of energy in the form of electricity or fossil fuels for every one it provides for the body. If meat were to be omitted from the diet, so that grain became the main source of energy, the input and output of energy would be more nearly equal. Dietary practices, however, are not the problem: modern agricultural practices are based on a too liberal use of oil, gas, and coal.

Suppose that modern agricultural techniques were adopted by all persons on earth, each of whom wished to eat as well as Americans do (too well, many nutritionists would claim). What would be the energy required annually to feed four billion persons? The total is arrived at by the following calculations:

> 2500 kcal/person/day
> > × 4,000,000,000 persons
> > × 1000 calories/kcal
> > × 5 calories required/calorie yielded
> > × 365 days/year, or
> > > 2×10^{19} calories/year (approximately)

Figure 15.7. A fruit and vegetable market in Paris. In most European cities, even the large ones, fresh food arrives daily from nearby farms and orchards. Although this food may not retain its high quality long, its inability to do so lies in part in those characteristics that are responsible for its excellent flavor. The agriculture that produces this high-quality food consumes relatively little energy, but it does require human labor and care bordering on devotion. *(Photo courtesy of the Dow Chemical Company)*

This energy is equivalent to 2×10^{13} kilowatt-hours: ten times the electrical energy produced in the entire United States in 1972, and an amount equal to the *entire* energy budget of the United States for that year.

The above calculation allows us to predict that a "lower" standard of eating than that of modern America will eventually prevail throughout the world. The term "lower" is misleading, however. Modern agriculture has emphasized the mechanization of farms and has stressed the appearance, uniformity, and resistance to spoilage of farm produce. Therefore, the choice of fruits and vegetables available to the modern shopper in much of the United States is severely limited. Frozen food is more economical for home use than fresh because less labor goes into its processing, packaging, and distribution for sale. Fresh produce is either not available at supermarkets, or it can be found only in single varieties, shipped in from elsewhere. Many persons throughout the world would consider this a needlessly *low* standard of eating.

Our calculation of the energy needed for extending modern agriculture to all peoples prompts us to predict that in the future more agricultural energy will again be supplied by farm laborers and animals. Although the use of human labor is a characteristic of a "low" standard of living, the quality of the food produced is generally high. Even now, more and more people are recognizing the benefits that accompany organic farming. Farmers' markets have become commonplace in many of the nation's smaller cities. The chief benefits of these developments (other than the satisfaction that goes with the effort of gardening) is that once more flavorful varieties of vegetables and fruits are chosen for planting, rather than varieties that have been bred specifically for mass production.

Agricultural technology eventually will take full advantage of recyclable materials. The manufacture of chemical fertilizers is the largest energy consumer in modern agriculture: ammonia, a widely used fertilizer, is manufactured by burning premium natural gas. One-half of the burned gas is wasted as heat; the other half changes from CH_4 to NH_4. With an increased use of manure as a fertilizer, cattle fattening will become a local, rather than a centralized, business. Land-based municipal sewage disposal systems will be developed in order to conserve the nutrients and essential trace elements in human waste.

No mention has been made yet of either farming the sea or making deserts bloom through the use of desalinated seawater. This silence stems from pessimism. Except for a narrow band of shallow water along the continental shorelines, the ocean is a relatively unproductive place. One-half of the marine harvest is taken from less than one-tenth of one percent of the ocean's surface. The bulk of all commercially important marine fish breed in salt marshes and tidal flats.

These are precisely the areas that are continually threatened by real estate developers, offshore oil drillers, airport engineers, and promoters of nuclear parks (combinations of power stations and desalinizing plants). Those who would farm the oceans would merely lengthen the list of activities that prevent the reproduction of the sea animals that we now harvest. The productivity of the sea would be increased more by restoring coastal breeding areas and lessening the dumping of toxic pollutants into the sea than by corralling and fattening whales in offshore pens, as some persons have suggested.

The proposed use of desalinated seawater for the widespread irrigation of farmlands does not seem profitable, even in terms of energy alone. At best, desalination can be used in extremely restricted localities. This limitation can be explained in two phrases: (1) one thousand gallons of water must pass through an alfalfa plant in order that this plant manufacture one pound of dry matter; (2) one thousand gallons of seawater contain three hundred pounds of salt. Thus, for each pound of plant material (alfalfa) that is harvested as food through the use of desalinated seawater, three hundred pounds of salt must be discarded. The economically minded (and biologically ignorant) engineer would merely dump the salt offshore, thus destroying even more of the ocean's breeding grounds (in effect, trading animal protein for plant—alfalfa—protein). Also interesting are the energy relationships between desalinizing water on the one hand and the caloric content of the plant matter on the other. The distillation of one thousand gallons of seawater requires more than two *billion* calories. The input:output ratio of energy in this case exceeds 1000:1. Although counterflow procedures would provide a saving, it seems highly unlikely that many deserts will be encouraged to bloom for long, even at a 100:1 ex-

Table 15.1. Fish production in various regions of the ocean

Region	Area (square kilometers)	Annual fish production (metric tons)	Pounds/km^2
Open ocean	326,000,000	1,600,000	10
Coastal zone	36,000,000	120,000,000	7,500
Coastal up-welling areas	360,000	120,000,000	750,000

Figure 15.8. An artist's conception of an agro-industrial complex built around a large nuclear power plant. By being built on a desolate region of the seashore, the power plant could generate energy to provide fresh water, fertilizer, chemicals, metals, and other valuable materials. The authors of the article for which this drawing was prepared suggested that 4000 such industrial complexes could be built on the shorelines of the world. *(Photo courtesy of The Institute for Energy Analysis; reprinted with permission of The American Scientist)*

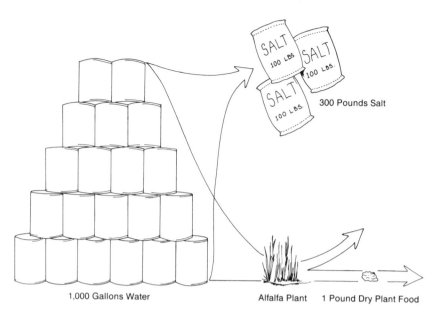

Figure 15.9. An example of the logistics involved in the agricultural use of desalinated seawater. One thousand gallons (shown as twenty 50-gallon drums) of water must pass through an alfalfa plant for each pound of dry plant food it provides as fodder for animals. One thousand gallons of seawater contain three hundred pounds of salt; this salt must be returned to the sea. What effect might this salt have on the marine life living near where it is dumped? Make your calculations not on the basis of three hundred pounds of salt, but on the farmer's expectation of raising 100 *acres* of alfalfa, which might reasonably yield four tons of alfalfa *per acre*.

change. Deserts that have been made to bloom today bloom not only because of water but also because of the *oil* that keeps the giant stills running.

Technology: Cities

The forces that determine the residence patterns of human beings are complex. In the United States, a highway leading into rural farmland pulls from each town and city a long, threadlike "miracle" strip containing drive-in snack bars, roadside inns, used-car lots, convenience stores, mobile homes, and ranch-style houses. The private car, combined with an urge to own an outside grill and a private swimming pool, has caused a mass exodus of city dwellers into the Levittowns, Forest Manors, and other housing developments of the suburbs. These dispersive forces are currently weakening, though, and we predict that in the near future, high-density population centers will be reestablished.

Once energy sources are depleted, modern energy-consuming agricultural practices will not be equal to sustaining us. Because the United States' population continues to grow, we must find new, more efficient patterns of distribution. These patterns will, we believe, place each community near its own operating farms. Large metropolitan centers and their sprawling suburbs consume more farmland than the nation can afford to lose. Furthermore, they require giant transportation systems for the dual purposes of bringing food to city dwellers and carrying away their garbage. Energy resources will eventually be too scarce and too costly to sustain a nation of urban and suburban dwellers permanently.

Obviously, no person or group of persons (especially college professors) can dictate how other people should live. Because many alternative patterns exist, each of which appears to be about as good as the others, prediction is difficult. In the past, planners who were responsible for urban renewal have uniformly failed to improve city life. First, they lacked modesty. For example, no landscape architect has ever laid out sidewalks on a college campus in a pattern that has met the needs of the local students. Spider webs of footpaths form between buildings that professional designers have neglected to connect. Now, if the pedestrian patterns set by students going to and from scheduled classes that meet in known classroom buildings cannot be predicted, how can a livable city be designed on a drawing board?

Second, urban planning has never been viewed as an *experiment*. If it had been, we might by now have learned something. Instead, a uniformly dismal pattern has been imposed upon the center of one city after another throughout the United States: ramps leading from overhead highways spiral down over deserted areas of sumac, ragweed, and discarded auto parts. These ramps lead eventually to free-standing office and apartment buildings, each of which is surrounded by an enormous parking area. Walking is discouraged because it is unpleasant at best, and dangerous at worst. If cities are supposed to be centers of urban life, then modern planners are guilty of urbanicide.

If future planners were to test several possible urban styles in many communities on a long-term experimental basis, we predict that the modern equivalent of the ancient European city (see Figure 15.10) would prove to be by far the most popular. Examination of this seemingly exotic prediction reveals its many benefits and advantages:

• Densely clustered apartment houses six to eight stories high can accommodate about 100,000 persons in a circle one mile in diameter.

• The farms surrounding each city of this sort would be close to their market; long-distance hauling of agricultural produce would be unnecessary.

• The private automobile would not be needed for routine daily functions, such as going to or from work, taking children to or from school, or shopping.

• High-density communities encourage the development and efficient operation of inter-urban transportation systems.

• Wiring, plumbing, and other building material, as well as fuel for heating, are all conserved by high-density apartment living.

Americans are traditionally a rural people. Most Americans prefer a suburban house and lot to a city apartment. However, this preference could be satisfied by the European-type city of 100,000 persons, since it places every individual within one-half mile or less of genuine farmland and country life. This claim cannot be made for most sprawling

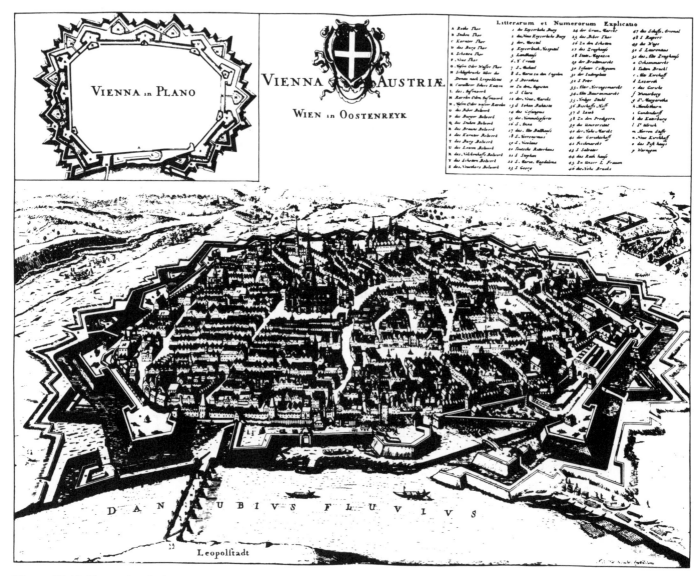

Figure 15.10. Vienna, Austria, in 1677. The text suggests that, with the proper use of modern technology, the design of early European cities might eventually prove to be a favorite model for urban renewal if it were conducted on a long-term experimental basis. The text discusses several reasons for this prediction. Two more reasons can be added here: (1) The interior of the city is not congested; its streets may be comparatively narrow, but green parks abound within the center of Vienna. (2) Because the streets are not laid out at right angles, many varied routes are equally convenient in going from one place to another; this is important when people commute to work or school on foot. *(Illustration courtesy of J. W. Reps, Cornell University)*

metropolitan areas, including Detroit, Los Angeles, Denver, and New York City.

Will living patterns within the United States evolve toward the culturally efficient, energy-saving form that we have described? Perhaps, but not easily. This city design distresses those who advocate unrestricted ownership and use of private land, real estate speculation, highway construction, and the private automobile. These people are powerful. They include bankers, who until recently determined the fate of large sections of modern cities by "redlining": that is, by marking off areas encompassing entire city blocks within which all home mortgages were stopped by an agreement between bankers and despite the wishes of the homeowners. As a result of redlining, neighborhood communities decay, houses are sold for demolition, and profitable office buildings are erected in their place. In addition, discouraging the use of automobiles invites the wrath of oilmen, automakers, auto dealers, auto service personnel, and concrete and asphalt manufacturers. Five million union members (who, with their families, represent twenty to twenty-five million Americans) depend on the Interstate Highway system and motor vehicles for their livelihoods.

Attitudes do change, however, although slowly. As recently as 1940, laws prohibiting the removal of topsoil on Long Island (this island would be a barren sandbar if its thin layer of excellent topsoil were removed) were regarded by persons living farther inland as "communistic." Today, even politically conservative persons often oppose the large-scale destruction of range- and farmland by strip miners. Topsoil is a national resource.

Odysseus, returning home following a voyage of many years, killed his wife's suitors who had moved into his house. As further retribution, he also hanged a dozen or more servant women. No one objected, because in ancient Greece servants were personal property, to be treated or disposed of as their owners saw fit. Slavery was acceptable in many societies for centuries. Changed attitudes now extend to the treatment of pets and livestock. Wondrously, a book arguing that rocks, trees, and scenic views should have legal rights of their own has recently been written. Human beings do seem to be developing a

Figure 15.11. A modernistic proposal for high-density living in a congested world. The tetrahedronal city (shown moored in San Francisco Bay) is designed to float at sea, thus making the ocean's surfaces available for human habitation. Before a million or more persons put to sea in a steel and concrete box, they should ponder the source of their food: Is its source certain? Is its arrival guaranteed? *(Background photo courtesy of the Redwood Empire Association)*

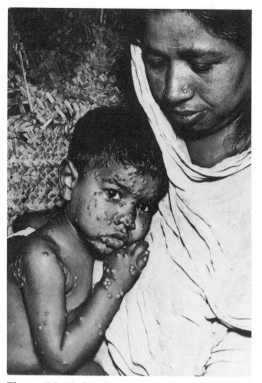

Figure 15.12. Smallpox is one human disease that seems to have been eliminated. The last known victim recovered in 1975. The elimination of smallpox was made possible largely because the virus had no nonhuman reservoir from which it might return to reinfect human beings. Flu viruses pose a much more difficult problem for public health officials because the periodic epidemics seem to have their origins in viruses that attack domesticated animals. *(Photo courtesy of the World Health Organization; photographer, L. Matlowsky)*

more friendly, more responsible attitude toward their surroundings and toward other life forms. The being who seems most resistant to change in the industrialized Western world is that legally fabricated man, that artificial person known as the corporation. In time, however, even the morals of corporate bodies (and of the courts that act as the consciences of corporations) will change.

Disease

If human beings fail to bring their numbers under rational control, death rates will eventually rise to counterbalance birth rates. Death rates may be increased by disease, starvation, war, or any combination of these. The nature of future wars is not revealed in our crystal ball, although some experts have said that World War IV will be fought with clubs and slingshots. The course of starvation will never change—the same wasted bodies, bloated stomachs, and vacant stares. As always, the children and the aged will die first. But what about disease? What sorts of diseases can we expect to face in the future?

Under *severely* crowded conditions, epidemics (especially those spread by contaminated water supplies) are always possible. A breakdown in the manufacture and delivery of antibiotics or vaccines could lead to a resurgence of today's "conquered" diseases. In the United States, gonorrhea, syphilis, and streptococcal infections are common, and bacteria like the *Staphylococci* and *Pneumococcus* still cause pimples, boils, sinusitis, and ear and other infections. Toxic shock, an often fatal syndrome striking women using certain types of tampons, is caused by a *Staphylococcus* species. For persons in close contact with one another, and weakened by hunger, bacterial epidemics are as always a constant threat.

However, during the past several decades, developed nations have experienced a steady decline in deaths due to diseases caused by infections and an increase in those caused by functional disorders, such as stroke, heart disease, and cancer. Even before the discovery of modern "miracle" drugs, infectious diseases were on the decline. Their conquest came from a knowledge of their ecology. Water sources were protected against bacterial contamination, human wastes were disposed of more carefully than before, and the roles of insects and other carriers of disease were better understood. In short, sanitation had improved. Because every person must die eventually, each medical conquest has, of course, yielded new importance to previously incidental causes of death, especially the physical dysfunctions that occur in advanced age. The conquest of these functional disorders will probably bring still other causes of death to the fore. The degeneration of the central nervous system and a loss of all coordinated mental and physical functions in very old persons would be our prediction.

Although bacterial diseases are declining in Western societies, viral diseases seem to be increasing. In part, an improved knowledge of viruses is responsible. An ailment that, twenty or thirty years ago, would have been blamed on "something I ate," is now likely to be known to be caused by viruses. The 24-hour intestinal flus that cause vomiting and diarrhea are of that sort. On the other hand, viruses are becoming more important bcause sanitary techniques suitable for controlling bacteria are often ineffective in controlling viruses. Filtration and other water purification techniques remove bacteria, but not viruses. Nor do the procedures used in many doctors' offices for sterilizing thermometers, hypodermic needles, and other medical apparatus kill viruses.

There is some evidence that viruses may be evolving in response to new opportunities that human beings offer them. Serum hepatitis got its name because it was once a hospital-transmitted disease; one caught it only through medical injections. The intravenous injection of illicit drugs and the use of primitive sterilization procedures (or none at all) by addicts enabled serum hepatitis to move from the hospital into the darkened stairwell and the street. Subsequently, medical journals reported serum hepatitis cases among wives and husbands of drug users, wives and husbands who themselves had never taken drugs. Serum hepatitus had evolved into a venereal disease that could be transmitted (it seemed) by sexual intercourse. More recently, an article has appeared raising the possibility that, like mononucleosis, serum hepatitis has become a "kissing" disease that is transmitted orally. Some believe that it can now be transmitted by mosquitoes as well. In this case the mosquito acts like a hypodermic needle. The changing modes of transmission lead from infrequent events (hospital treatments and mainlining) to more frequent ones (sex and kissing). Eventually, the virus may be spread by the most frequent social act of all: the friendly handshake. Faint evidence suggests that the AIDS virus discussed earlier in this text is also evolving new transmission routes.

Manmade Diseases

Is there any possibility that the human species may be threatened in the future by a disease organism created by man himself? Since 1975, many conferences have been held at which scientists have discussed just that risk. A prominent worry stems from the use of the common colon bacterium (*Escherichia coli*) as the experimental organism into which fragments of DNA from other species are introduced for "artificial" replication.

There is little doubt that *E. coli* (normally an inhabitant of the large intestine) is adapting to a changing human environment. Especially in women, bladder infections are becoming more common, and *E. coli*

What Is Happening to Our Health Care System?
Arnold S. Relman

With all its limitations, however, I believe the old system was better than what appears to be replacing it now. . . . Further discussion of this point should be deferred until I explain how the old system began to destroy itself.

The seeds were all there: A rapidly expanding technological base; a growing and increasingly specialized corps of medical professionals trained to practice high-technology medicine and reimbursed on a piecework basis; an open-ended insurance system based on payment of customary charges; and more than two decades of essentially unregulated proliferation of facilities.

In any event health care has now become the largest sector of our national economy. If it were the automobile industry or the computer industry or any other domestic market, such rapid growth would be hailed as an industrial triumph. The jobs and general prosperity generated by this expansion would be a source of great rejoicing. Why, then, is the growth of the health care sector so widely regarded as a national financial disaster? The primary answer to that question, I think, is that those who are paying for most of the costs (i.e., the federal government and large businesses) say they can't afford it any longer. The recipients of health care do not pay most of the bill, but to the extent that they do share in the cost, they also think that costs are too high.

Inevitably, however, we will come to recognize that we cannot live peacefully without an acceptable health care system that embraces all our citizens. This will mean substantially more taxes for medical care and that in turn will probably mean some sort of National Health Insurance system as a means of controlling costs, quality, and access. I don't believe that a National Health Service (i.e., federal ownership of the system) is in the cards, but I can't see how we will ultimately avoid a tax-based comprehensive insurance system. What is political anathema now is likely to become political necessity.

Arnold S. Relman is Editor of the *New England Journal of Medicine*. © 1986 *Bulletin*, The American Academy of Arts and Sciences; reprinted with permission.

A Manmade Disease of Dogs?

For a century or more, cats have been known to suffer from a severe viral disease. The causative agent, a parvovirus, can apparently infect any member of the cat family from tigers to house cats. Effective vaccines have been developed, but the virus remains a major cause of illness among felines.

In 1947, outbreaks of viral infections struck mink farms in Canada, the United States, and in the rest of the world. The responsible virus was shown to be serologically indistinguishable from the virus that affects cats. The nature of the change that enabled the virus to spread from cats to mink is not known; perhaps the virus first encountered farm mink in 1947, and then spread from ranch to ranch.

In 1978, reports of a "killer virus" affecting dogs first appeared. The resultant disease had two major manifestations, both of which were generally lethal: a fulminating diarrhea that removed much of the gut lining, and a generalized destruction of heart muscle. Death could occur as quickly as one hour after the first onset of symptoms.

Where did the dog virus come from? Molecular tests have revealed that the virus differs only slightly from those affecting cats and mink. The spread of this new virus worldwide was spectacular: within two years, dogs of all nations proved to carry antibodies against the new virus. Previously, dogs had been immune to cat and mink virus, and lacked antibodies against them.

Two possible sources of the deadly (to dogs) virus have been suggested. First, an accidental contamination within a research laboratory of dog tissue-culture cells by cat virus, followed by the adaptation of the viruses to dog cells and their subsequent distribution in veterinary medicines. Second, a deliberate attempt to attenuate cat viruses by passage through cells of other mammals (including those of dogs) so that the live, attenuated viruses could be used to induce antibody production in vaccinated cats.

One author has written: "The scenario of an epidemic disease with the features of feline panleukopenia, mink enteritis, or canine parvovirus disease in valuable animal livestock or even in man is highly worrying." This author recommends that attenuated viral vaccines no longer be used, because of the possible dissemination of mutant viruses of unknown host ranges.

is the most frequent cause of these infections. *E. coli* has also been implicated in kidney infections. Strains of *E. coli* that are resistant to antibiotics are frequently resistant to heavy metals as well. This latter resistance (which may be favored by the presence of lead and mercury in the environment) permits these strains to live in the kidney, where concentrations of heavy metals are rather high. The use of certain antibiotics has been known to allow *E. coli* to infect the central nervous system. These infections seem to follow the elimination of other bacteria and the exploitation of uninhabited parts of the patient's body by the more resistant *E. coli*.

Two types of experiments that might have been done (but that were not actually carried out) can illustrate the possibility of creating a manmade disease. *Staphylococcus* is a bacterium that is responsible for pimples and boils; it also causes serious throat and sinus infections. Some strains of this genus carry a gene that causes the synthesis of an enzyme that enables these bacteria to cling to teeth, where they cause tooth decay. Without this enzyme, staphylococci do not affect teeth. Still other strains carry a gene that causes the synthesis of a second enzyme; these bacteria lodge in the valves of the heart and cause endocarditis (inflammation of the interior of the heart). Without this second enzyme, bacteria cannot cling to the heart tissue; instead, they are swept away by the bloodstream.

The inheritance of these genes could be easily studied by hybridizing the staphylococcal DNA with that of *E. coli*. Suppose, however, that the newly created strains of *E. coli* were to escape from laboratory culture dishes. They would now possess the ability to cause either tooth decay or heart disease. The danger of creating such new and potentially harmful disease germs led in each instance to a decision not to carry out the experiment.

Accidentally created manmade diseases may eventually strike individual persons, such as laboratory workers, their friends, or their immediate families. We consider it unlikely, however, that a manmade genetic change would generate a highly fatal, worldwide epidemic. On the other hand, the less serious the "new" disease, the more widespread it might become.

A newly created, highly virulent disease should not spread, because the responsible bacterium would kill its host. The bacterium would also lack the ability to cause a behavior—sneezing or diarrhea, for example—in an infected person that would spread the infection. The possible exception, an *E. coli* strain capable of causing tooth decay, could have become widespread because oral contact through kissing is a common aspect of human behavior.

Some of the problems that we see affecting our future—overpopulation, overcrowding, exhaustion of fossil fuels, limited food resources, and disease—are major problems, which will determine our destiny as a civilization. It is appropriate to end this book in the last course you will probably ever take in general biology with the same question we ask students as they begin their careers, "Where are you going?" If you are not reasonably sure of the answer, then you are like the plankton of the seas, driven, moved about, herded by the physical forces of winds, tides, and currents. But human beings have another choice. They can take responsibility for their future and the future of following generations. Just as it takes personal energy, careful study, planning, creativity, and long hours of hard work to prepare for your own career, so it is that the same processes must be used to ensure the success of our civilization and of those that will follow us.

Unlike our ancestors, we have access to detailed historical data. We now know that the resources necessary for life on this planet are finite; we know that such resources can be exhausted through over-population, misuse, and pollution. We have the medical skills and technology to track the life cycle of disease organisms, and we have the technological ability to re-use nonrenewable resources in ways that do not contaminate or pollute the environment that sustains us. The accumulation of our knowledge can give us the confidence to take charge of the destiny of our civilization just as we have done in forging our careers as individuals.

To do so requires an informed, involved, determined citizenry. It takes a community of individuals who are willing to think in creative ways, study issues in depth, and ponder the consequences of alternatives, and who are willing to *vote* their convictions. Because most of the major decisions affecting the outcome of our civilization will have a biological basis that must be understood by an informed citizenry, the science of biology will be extremely important in guiding our course for the future.

REFERENCE MATERIAL

SUGGESTED READINGS

General

Many of the readily available scientific journals publish special editions dealing with topics that are pertinent to those emphasized in this textbook. Without attempting to be complete, the following can be cited as important examples:

Readings from *Scientific American*
  "Thirty-nine Steps to Biology" (1968)
  "Science, Conflict, and Society" (1969)
  "Facets of Genetics" (1970)
  "Man and the Ecosphere" (1971)
  "Genetics" (1981)
Special issues of *Scientific American*
  "Technology and Economic Development" (Sept. 1963)
  "The Ocean" (Sept. 1969)
  "The Biosphere" (Sept. 1970)
  "Energy and Power" (Sept. 1971)
  "Life and Death and Medicine" (Sept. 1973)
  "The Human Population" (Sept. 1974)
  "Food and Agriculture" (Sept. 1976)
  "The Molecules of Life" (Oct. 1985)
Special issues of *Science*
  "Energy" (19 Apr. 1974)
  "Biotechnology" (11 Feb. 1983)
  "Biological Frontiers" (18 Nov. 1983)
  "Neuroscience" (21 Sept. 1984)
Special issues of *Bioscience*
  "Behavioral Endocrinology" (Oct. 1983)
  "Developmental Neurobiology" (May 1984)
  "Global Ecology" (Sept. 1984)
Special Issues of the *American Biology Teacher*
  "Acid Rain" (Apr.–May 1983)
  "Human Ecology" (Sept. 1984)
  "Genetic Engineering" (Oct. 1984)
  "History, Philosophy, and Sociology of Biology" (Apr. 1985)

An important adjunct to this textbook and the course is the *World Almanac and Book of Facts* (New York: Newspaper Enterprise Association, Inc., published annually).

Following are suggested readings for individual chapters. Few or none of these references are limited to matters that are of concern only to the chapter under which they are listed. At times a few words are added explaining why a book has been suggested as further reading.

Chapter 1

Bronowski, Jacob. *Magic, Science, and Civilization*. New York: Columbia University Press, 1978. An articulate case for the interdependence of science and ethics.

Bush, Vannevar. *Science Is Not Enough*. New York: William Morrow and Co., 1967.

Carson, Rachel. *Silent Spring*. Boston: Houghton Mifflin Co., 1962.

David, R. E. "Legends and Myths: A Basis for Scientific Research." *Perspectives in Biology and Medicine* 26 (1983): 198–203.

Grobman, Arnold B., ed. *Social Implications of Biological Education*. Washington, D.C.: National Association of Biology Teachers, 1970.

Hearnshaw, Leslie S. *Cyril Burt: Psychologist*. Ithaca: Cornell University Press, 1979. An extensive investigation of the origin and course of one of the greatest scientific frauds: Burt's "studies" of the IQ of twins raised together and apart.

Knight, J. A. "Exploring the Compromise of Ethical Principles in Science." *Perspectives in Biology and Medicine* 27 (1984): 432–42.

Ramsey, Paul. *Fabricated Man: The Ethics of Genetic Control*. New Haven: Yale University Press, 1970.

Simpson, George G. *The Meaning of Evolution*. New Haven: Yale University Press, 1952. The last section of the book presents one thoughtful biologist's views concerning the meaning of evolution with respect to human values.

Wang, H. "The Formal and the Intuitive in the Biological Sciences." *Perspectives in Biology and Medicine* 27 (1984): 525–42.

Watson, James D. *The Double Helix*. New York: New American Library, 1968.

Chapters 2 and 3

Students using *Biology for Living* should have standard high-school and college biology textbooks available for handy reference; possible titles are listed here:

Abramson, D. I., and P. B. Dobrin, eds. *Blood Vessels and Lymphatics in Organ Systems*. Orlando, Fla.: Academic Press, 1984.

Biological Science: An Ecological Approach. Chicago: Rand McNally, 1978.

Biological Science: An Inquiry into Life. New York: Harcourt, Brace, Jovanovich, 1973.

Biological Science: Molecules to Man. Boston: Houghton Mifflin Co., 1973. The above three books are the Biological Sciences Curriculum Study's Green, Yellow, and Blue Versions.

Currey, John. *The Mechanical Adaptations of Bones*. Princeton: Princeton University Press, 1984.

Davis, J. *Endomorphins: New Waves in Brain Chemistry*. New York: Doubleday and Co., 1984.

Gallup, G. G., and S. D. Suarez. "Homosexuality as a By-Product of Selection for Optimal Heterosexual Strategies." *Perspectives in Biology and Medicine* 26 (1983): 315–22.

Gilbert, S. F. *Developmental Biology*. Sunderland, Mass.: Sinauer Associates, 1985.

Keeton, William T. *Biological Science*. New York: W. W. Norton and Co., 1980.

Kuffler, S. W., and J. G. Nicholls. *From Neuron to Brain: A Cellular Approach to the Function of the Nervous System,* 2d ed. Sunderland, Mass.: Sinauer Associates, 1984.

Raven, Peter H., and George B. Johnson. *Biology.* St. Louis: Times Mirror/Mosby College Publishing, 1986. This book and Keeton's 1980 text are standard college-level biology texts chosen from many that could have been listed.

Shapiro, J. A., ed. *Mobile Genetic Elements.* Orlando, Fla.: Academic Press, 1983.

Vander, A., J. Sherman, and D. Luciano. *Human Physiology: The Mechanisms of Body Function,* 4th ed. New York: McGraw-Hill Book Co., 1985.

Wallace, Bruce. *Essays in Social Biology.* Vol. 1. *People and Their Needs, the Environment, and Ecology.* Vol. 2. *Genetics, Evolution, Race, Radiation Biology.* Englewood Cliffs, N.J.: Prentice-Hall, 1972. The essays and literary selections that appear in these texts are related to many topics that are discussed in the first and later chapters of *Biology for Living.*

Zaner, R. M. "A Criticism of Moral Conservatism's View of *in vitro* Fertilization and Embryo Transfer." *Perspectives in Biology and Medicine* 27 (1984): 200–212.

Chapter 4

Borgstrom, Georg. *The Hungry Planet.* New York: Collier Books, 1965.

Cairns-Smith, A. G. *Genetic Takeover and the Mineral Origins of Life.* Cambridge: Cambridge University Press, 1982.

Edwards, Gerry, and David Walker. C_3, C_4: *Mechanisms, and Cellular and Environmental Regulation, of Photosynthesis.* Berkeley and Los Angeles: University of California Press, 1983.

Little, Colin. *The Colonisation of Land: Origins and Adaptations of Terrestrial Animals.* Cambridge: Cambridge University Press, 1984.

Postgate, J. R. *The Fundamentals of Nitrogen Fixation.* Cambridge: Cambridge University Press, 1982.

Stanhill, G., ed. *Energy and Agriculture.* Berlin: Springer-Verlag, 1984.

Chapter 5

Charig, Alan. *A New Look at Dinosaurs.* New York: Facts on File, 1983.

Ehrlich, Paul R., and Anne H. Ehrlich. *Population, Resources, Environment.* San Francisco: W. H. Freeman and Co., 1972.

Ehrlich, Paul R., Anne H. Ehrlich, and J. P. Holden. *Human Ecology: Problems and Solutions.* San Francisco: W. H. Freeman and Co., 1973.

Gates, D. M. *Energy and Ecology.* Sunderland, Mass.: Sinauer Associates, 1985.

Henderson, L. J. *The Fitness of the Environment.* New York: Macmillan Co., 1913.

Sears, Paul B. *Deserts on the March,* 4th rev. ed. Norman: University of Oklahoma Press, 1980.

Chapter 6

Cockburn, F., and R. Gitzelmann, eds. *Inborn Errors of Metabolism in Humans.* New York: Alan R. Liss, 1982.

Gonick, Larry, and Mark Wheelis. *The Cartoon Guide to Genetics.* New York: Barnes and Noble/Harper and Row, 1983.

Hildemann, W. H. *Essentials of Immunology.* New York: Elsevier Science Publishing Co., 1983.

Lawrence, C. W., ed. *Induced Mutagenesis: Molecular Mechanisms and Their Implications for Environmental Protection.* New York: Plenum Press, 1983.

Tu, A. T. *Handbook of Natural Toxins*, Vol. 2. *Insect Poisons, Allergens, and Other Invertebrate Venoms*. New York: Marcel Dekker, 1984.

Chapter 7

Begon, M., J. L. Harper, and C. R. Townsend. *Ecology: Individuals, Populations, and Communities*. Sunderland, Mass.: Sinauer Associates, 1986.

Wallace, Bruce. *Population Genetics*. Pamphlet #12. Boulder, Colo.: Biological Sciences Curriculum Study, 1964. See Chapter 2.

Chapter 8

Briggs, David, and S. M. Walters. *Plant Variation and Evolution*, 2d ed. Cambridge: Cambridge University Press, 1984.

Dobzhansky, Theodosius. *Evolution, Genes, and Man*. New York: John Wiley and Sons, 1955.

Drlica, Karl. *Understanding DNA and Gene Cloning: A Guide for the Curious*. New York: John Wiley and Sons, 1984.

Ginzburg, L. R. *Theory of Natural Selection and Population Growth*. Series in Evolutionary Biology. Menlo Park, Calif.: Benjamin/Cummings Publishing Co., 1983.

McDonald, J. F. "The Molecular Basis of Adaptation: A Critical Review of Relevant Ideas and Observations." *Annual Review of Ecology and Semantics* 14 (1983): 77–102.

Sober, E. *The Nature of Selection: Evolutionary Theory in Philosophical Focus*. Cambridge, Mass.: MIT Press, 1984.

Wallace, Bruce, and Adrian M. Srb. *Adaptation*. Englewood Cliffs, N.J.: Prentice-Hall, 1964.

Chapter 9

Bates, Marston, and P. S. Humphrey, eds. *The Darwin Reader*. New York: Charles Scribner's Sons, 1956.

Futuyma, Douglas. *Science on Trial: The Case for Evolution*. New York: Pantheon Books, 1983.

Grene, Marjorie, ed. *Dimensions of Darwinism: Themes and Counter Themes in Twentieth-Century Evolutionary Theory*. Cambridge: Cambridge University Press, 1983.

Keller, Evelyn F. *A Feeling for the Organism: The Life and Work of Barbara McClintock*. San Francisco: W. H. Freeman and Co., 1983.

Lowenstein, J. M., and A. L. Zihlman. "Human Evolution and Molecular Biology." *Perspectives in Biology and Medicine* 27 (1984): 611–22.

Morehead, Alan. *Darwin and the Beagle*. New York: Harper and Row, 1969.

Stebbins, G. Ledyard. *Darwin to DNA: Molecules to Humanity*. San Francisco: W. H. Freeman and Co., 1982.

Stebbins, G. Ledyard. *Processes of Organic Evolution*. Englewood Cliffs, N.J.: Prentice-Hall, 1966.

Volpe, E. Peter. *Understanding Evolution*. Dubuque, Ia.: William C. Brown, 1984.

Wallace, Bruce. *Chromosomes, Giant Molecules, and Evolution*. New York: W. W. Norton and Co., 1966.

Chapter 10

Bates, Marston. *Man in Nature*. Englewood Cliffs, N.J.: Prentice-Hall, 1964.

Bentley, Barbara, and Thomas Elias, eds. *The Biology of Nectaries*. New York: Columbia University Press, 1983.

Georghiou, G. P., and Tetsuo Saito, eds. *Pest Resistance to Pesticides*. New York: Plenum Publishing Corp., 1983.

Hardin, Garrett J. *Naked Emperors: Essays of a Taboo-Stalker*. Los Altos, Calif.: William Kaufmann, 1982.

Leopold, Aldo. *A Sand County Almanac*. New York: Sierra Club/Ballantine Books, 1966.

Myers, Norman. *A Wealth of Wild Species: Storehouse for Human Welfare*. Boulder, Colo.: Westview Press, 1983.

Quich, H. F. *Population Ecology*. Indianapolis: Pegasus, 1974.

Chapter 11

Dobzhansky, Theodosius, *Mankind Evolving: The Evolution of the Human Species*. New Haven: Yale University Press, 1964.

Johanson, D., and M. Edey. *Lucy: The Beginnings of Humankind*. New York: Simon and Schuster, 1981.

King, James C. *The Biology of Race*, rev. ed. Berkeley and Los Angeles: University of California Press, 1982.

Leakey, R. E. *The Making of Mankind*. New York: E. P. Dutton, 1981.

Lewontin, Richard C. *The Genetic Basis of Evolutionary Change*. New York: Columbia University Press, 1974. Several tables from this book have been cited in Chapter 11.

Lewontin, Richard C. *Human Diversity*. San Francisco: Scientific American Library/W. H. Freeman and Co., 1982.

Richardson, K., D. Spears, and M. Richards, eds. *Race and Intelligence*. Baltimore: Penguin Books, 1972.

Smith, F. H., and F. Spencer, eds. *The Origin of Modern Humans: A World Survey of Fossil Evidence*. New York: Alan R. Liss, 1984.

Chapter 12

Gould, Stephen J. *The Mismeasure of Man*. New York: W. W. Norton and Co., 1981.

Molnar, S. *Races, Types, and Ethnic Groups*, 2d ed. Englewood Cliffs, N.J.: Prentice-Hall, 1983.

Pilbeam, D. "The Descent of Hominoids and Hominids." *Scientific American* (1984): 84–96.

Sonneborn, T. M., ed. *The Control of Human Heredity and Evolution*. New York: Macmillan Co., 1965. One of the earliest conferences dealing with what is now known as "bioethics."

Wallace, Bruce. *Genetic Load*. Englewood Cliffs, N.J.: Prentice-Hall, 1970.

Chapter 13

Dobzhansky, Theodosius, and Ernest Boesiger. Ed. Bruce Wallace. *Human Culture: A Moment in Evolution*. New York: Columbia University Press, 1983.

Sahlins, Marshall. *The Use and Abuse of Biology*. Ann Arbor: University of Michigan Press, 1977.

Todd, N. J., and J. Todd. *Bioshelter, Ocean Arks, City Farming: Ecology as the Basis of Design*. San Francisco: Sierra Club Books, 1984.

Wallace, Bruce. *Essays in Social Biology*. Vol. 3. *Disease, Sex, Communication, Behavior*. Englewood Cliffs, N.J.: Prentice-Hall, 1972.

Wilson, E. O. *Sociobiology*. Cambridge: Belknap Press/Harvard University Press, 1975.

Chapter 14

Epstein, S. S., L. O. Brown, and C. Pope. *Hazardous Waste in America*. San Francisco: Sierra Club Books, 1982.

Evenari, Michael, L. Shanan, and N. Tadmor. *The Negev: The Challenge of a Desert*, 2d ed. Cambridge, Mass.: Harvard University Press, 1982.

Finley, Jeanne, and Aileen Smith. *Minamata.* Tucson: University of Arizona, Center for Creative Photography, 1981. Distributed by the University of Arizona Press.

Turk, J., A. Turk, and K. Arms. *Environmental Science,* 3d ed. Philadelphia: W. B. Saunders Co., 1984.

Chapter 15

Dubos, René. *The Dreams of Reason.* New York: Columbia University Press, 1961.

Dubos, René. *A God Within.* New York: Charles Scribner's Sons, 1972.

Norman, M. J. T., C. J. Pearson, and P. G. E. Searle. *The Ecology of Tropical Food Crops.* Cambridge: Cambridge University Press, 1984.

ReVelle, P., and C. ReVelle. *The Environment: Issues and Choices for Society,* 2d ed. Boston: Willard Grant Press, 1984.

Shanks, B. *This Land Is Your Land: The Struggle to Save America's Public Lands.* San Francisco: Sierra Club Books, 1984. Distributed by Random House, New York.

Soulé, M. E., ed. *Conservation Biology: The Science of Scarcity and Diversity.* Sunderland, Mass.: Sinauer Associates, 1986.

Southwick, C. H., ed. *Global Ecology.* Sunderland, Mass.: Sinauer Associates, 1985.

INDEX

INDEX